STUDENT'S SOLUTIONS MANUAL

DAVID ATWOOD

Rochester Community and Technical College

INTERMEDIATE ALGEBRA WITH APPLICATIONS AND VISUALIZATION

FOURTH EDITION

Gary Rockswold

Minnesota State University, Mankato

Terry Krieger

Rochester Community and Technical College

PEARSON

Boston Columbus Indianapolis New York San Francisco Upper Saddle River
Amsterdam Cape Town Dubai London Madrid Milan Munich Paris Montreal Toronto
Delhi Mexico City Sao Paulo Sydney Hong Kong Seoul Singapore Taipei Tokyo

Copyright © 2013, 2009, 2005 Pearson Education, Inc.
Publishing as Pearson, 75 Arlington Street, Boston, MA 02116.

ISBN-13: 978-0-321-74784-6
ISBN-10: 0-321-74784-4

2 3 4 5 6 EBM 16 15 14

www.pearsonhighered.com

Contents

Chapter 1   1

Chapter 2   31

Chapter 3   67

Chapter 4   111

Chapter 5   163

Chapter 6   201

Chapter 7   255

Chapter 8   293

Chapter 9   343

Chapter 10   379

Chapter 11   409

Chapter 1: Real Numbers and Algebra

Section 1.1: Describing Data with Sets of Numbers

1. The natural numbers are the counting numbers and do not include zero. $N = \{1, 2, 3, \ldots\}$

   The whole numbers are the counting numbers along with zero. $W = \{0, 1, 2, 3, \ldots\}$

3. The rational numbers are numbers which can be written in the form $\dfrac{p}{q}$ where p and $q \neq 0$ are

   integers. An example is $\dfrac{3}{4}$. *Answers may vary.*

5. An example of the commutative property for addition is $3 + 2 = 2 + 3$. *Answers may vary.*

7. The number 1 is called the multiplicative identity since $1 \cdot a = a$ for any number a.

9. An example of the associative property for multiplication is $2 \cdot (3 \cdot 4) = (2 \cdot 3) \cdot 4$.

   Answers may vary.

11. $\dfrac{6 + 9 + 11 + 22}{4} = \dfrac{48}{4} = 12$

13. The property illustrated is the commutative property for addition, because the result of adding the amounts in a different order is the same.

15. The number 33,000 is a natural number, integer, rational number and a real number.

17. The number $\dfrac{89}{3687}$ is a rational number and a real number.

19. The number 7.5 is a rational number and a real number.

21. The number $90\sqrt{2}$ is a real number.

23. Natural numbers: 6   Whole numbers: 6   Integers: –5, 6

    Rational Numbers: $-5, 6, \dfrac{1}{7}, 0.2$ Irrational Numbers: $\sqrt{7}$

25. Natural numbers: $\dfrac{3}{1}$ Whole numbers: $\dfrac{3}{1}$ Integers: $\dfrac{3}{1}$

    Rational Numbers: $\dfrac{3}{1}, -\dfrac{5}{8}, 0.\overline{45}$ Irrational Numbers: $\sqrt{5}, \pi$

27. Natural numbers: $\sqrt{9} = 3$ Whole numbers: $\sqrt{9} = 3$ Integers: $-2, \sqrt{9} = 3$

    Rational numbers: $-2, \dfrac{1}{2}, \sqrt{9} = 3, 0.\overline{26}$ Irrational numbers: none

29. Rational numbers. A shoe size could be a fraction such as $8\dfrac{1}{2}$.

31. Natural numbers. Speed limits must be positive and are not fractions.

33. Integers. Winter temperatures in Montana could be either positive or negative but are not given as fractions.

35. Identity

37. Commutative

39. Distributive

41. Associative

43. Commutative

45. Associative

47. Distributive

49. $a+4$

51. $\dfrac{1}{3}a$

53. $x+1$

55. xy

57. $(4+5)+b=9+b$

59. $(5\cdot10)x=50x$

61. $x+(y+z)$

63. $x\cdot(3\cdot4)=x\cdot12$

65. $4x+4y$

67. $5x-35$

69. $-x-1$

71. $a(x-y)$

73. $3(4+x)$

75. $(3+7)x=10x$

77. $(8-2)t=6t$

79. $\left(1-\dfrac{3}{4}\right)x=\dfrac{1}{4}x$

81. $3-1+2z=2+2z$

83. $(3+4+1)z=8z$

85. $(10-1-4)t=5t$

87. $7(11-3)=7\cdot8=56$ or $7(11-3)=7\cdot11-7\cdot3=77-21=56$

89. $13(16+23)=13\cdot39=507$ or $13(16+23)=13\cdot16+13\cdot23=208+299=507$

91. $5(19-7)=5\cdot12=60$ or $5(19-7)=5\cdot19-5\cdot7=95-35=60$

93. $(3+4+5+8)\div4=5$; natural number, integer, and rational number

95. $(45+33+52) \div 3 = 43.\overline{3}$; rational number

97. $(121.5+45.7+99.3+45.9) \div 4 = 78.1$; rational number

99. $8+3+2+7 = (8+2)+(3+7) = 10+10 = 20$

101. $50 \cdot 198 = 50(200-2)+(10,000-100) = 9900$

103. $\dfrac{2}{9} \cdot 8 \cdot 9 \cdot \dfrac{3}{8} = \left(\dfrac{2}{9} \cdot 9\right) \cdot \left(8 \cdot \dfrac{3}{8}\right) = 2 \cdot 3 = 6$

105. $4 \cdot 16 - 4 \cdot 6 = 4(16-6) = 4 \cdot 10 = 40$

107. $\dfrac{4}{5}\left(\dfrac{5}{2}+5+\dfrac{15}{4}\right) = \left(\dfrac{4}{5} \cdot \dfrac{5}{2}\right)+\left(\dfrac{4}{5} \cdot 5\right)+\left(\dfrac{4}{5} \cdot \dfrac{15}{4}\right) = \dfrac{4}{2}+4+\dfrac{15}{5} = 2+4+3 = 9$

109. The property used is the identity property of 1, because $\dfrac{a}{a} = 1$ and $1 \cdot \dfrac{b}{c} = \dfrac{b}{c}$. *Answers may vary.*

111. (a)   From the table, enrollment in 2016 might be 20.3 million.

    (b)   The average of the 4 numbers is about 20.1 million. *Answers may vary.*

    (c) $(19.5+19.9+20.3+20.6) \div 4 = 20.075$ million; Yes, the answers nearly agree.

113. No; $8 \cdot 4 \text{ GB} = 4 \cdot 8 \text{ GB} = 32 \text{ GB}$. The commutative property for multiplication applies.

115. (a)   Valid; $2r\pi = 2\pi r$

    (b)   Valid; $2(\pi r) = 2\pi r$

    (c)   Not valid; $2\pi + 2r \neq 2\pi r$

Section 1.2 Operations on Real Numbers

1. right

3. positive

5. negative

7. $-a$

9. positive

11. 8

13.

15.

17.

19. $|-6.1| = 6.1$

21. $\left|-\dfrac{2}{3}\right| = \dfrac{2}{3}$

23. $|\pi-1| = \pi-1$

25. $-|-8| = -8$

27. $|-8| - |2| = 8 - 2 = 6$

29. $|8 - 11| = |-3| = 3$

31. $|x| = x$ when $x > 0$

33. $|x - y| = x - y$ when $x > 0$ and $y < 0$

35. Positive. Area can never be negative.

37. Both. Temperature can be both above and below zero.

39. Positive. A car's mileage can never be negative.

41. The additive inverse of 56 is $-(56) = -56$.

43. The additive inverse of -6.9 is $-(-6.9) = 6.9$.

45. The additive inverse of $-\pi + 2$ is $-(-\pi + 2) = \pi - 2$.

47. The additive inverse of $a - b$ is $-(a - b) = -a + b$.

49. The additive inverse of $-(x - y)$ is $-(-(x - y)) = x - y$.

51. The additive inverse of $z - (2 - z)$ is $-(z - (2 - z)) = -(2z - 2) = -2z + 2$.

53. $\dfrac{1}{3}$

55. $-\dfrac{3}{2}$

57. $\dfrac{1}{\pi}$

59. $\dfrac{1}{a + 3}$

61. $-\dfrac{b}{2a}$

63. $x - 7$

65. $4 + (-6) = -2$

67. $-7.4 + (-9.2) = -16.6$

69. $-8 - 22 = -8 + (-22) = -30$

71. $-1.1 - (-3.4) = -1.1 + (3.4) = 2.3$

73. $-\dfrac{3}{4} - \left(-\dfrac{1}{4}\right) = -\dfrac{3}{4} + \dfrac{1}{4} = -\dfrac{2}{4} = -\dfrac{1}{2}$

75. $-9 + 1 + (-2) + 5 = -8 + (-2) + 5 = -10 + 5 = -5$

77. $-6 \cdot -12 = 72$

79. $-\dfrac{1}{2} \cdot \dfrac{5}{7} = -\dfrac{5}{14}$

81. $-\dfrac{1}{2} \div -\dfrac{3}{4} = -\dfrac{1}{2} \cdot -\dfrac{4}{3} = \dfrac{4}{6} = \dfrac{2}{3}$

83. $\dfrac{3}{4} \div (-2) = \dfrac{3}{4} \cdot -\dfrac{1}{2} = -\dfrac{3}{8}$

85. $-\dfrac{8}{2} = -4$

87. $-25 \div -5 = 5$

89. $\dfrac{1}{0}$ is undefined because division by zero is not possible.

91. $-4 \cdot 7 \cdot (-5) = -28 \cdot (-5) = 140$

93. $6 \cdot \dfrac{2}{3} \cdot 3 \cdot \left(-\dfrac{1}{6}\right) = \left[6 \cdot \left(-\dfrac{1}{6}\right)\right] \cdot \left(\dfrac{2}{3} \cdot 3\right) = -1 \cdot 2 = -2$

95. $4x - 9x = (4 - 9)x = -5x$

97. $z - 4z + 2z = (1 - 4 + 2)z = -z$

99. Entering $-23.1 + 45.7 - (-34.6)$ into a graphing calculator gives 57.2.

101. Entering $(1/2) + (2/3) - (5/7)$ ▶Frac into a graphing calculator gives $\dfrac{19}{42}$.

103. Entering $(-3/4)(4/5)/(5/3)$ ▶Frac into a graphing calculator gives $-\dfrac{9}{25}$.

105. Entering $(1/2)((4/11) + (2/5))$ ▶Frac into a graphing calculator gives $\dfrac{21}{55}$.

107. $\left[9 + 5 + 15 + (-9)\right] \div 4 = 5$

109. $\left[(-2) + 12 + 7 + (-17)\right] \div 4 = 0$

111. $\left[101 + 99 + (-42) + 82\right] \div 4 = 60$

113. $\left[(a+b) + (a+b) + (a+b)\right] \div 6 = \dfrac{a+b}{2}$

115. $-2 + 8 + 3 + 2 - 11 = (8 + 2 + 3) + \left[(-2) + (-11)\right] = 13 - 13 = 0$

117. $103 - 44 + 97 - 56 = (103 + 97) + \left[(-44) + (-56)\right] = 200 - 100 = 100$

119. $\dfrac{1}{5} \cdot \dfrac{2}{3} \cdot \dfrac{1}{7} \cdot \dfrac{1}{9} \cdot 5 \cdot \dfrac{3}{2} \cdot 7 \cdot 9 = \left(\dfrac{1}{5} \cdot 5\right) \cdot \left(\dfrac{2}{3} \cdot \dfrac{3}{2}\right) \cdot \left(\dfrac{1}{7} \cdot 7\right) \cdot \left(\dfrac{1}{9} \cdot 9\right) = 1 \cdot 1 \cdot 1 \cdot 1 = 1$

121. $\left(\dfrac{1}{2}-\dfrac{1}{3}\right)+\left(\dfrac{1}{3}-\dfrac{1}{4}\right)+\left(\dfrac{1}{4}-\dfrac{1}{5}\right)+\left(\dfrac{1}{5}-\dfrac{1}{6}\right)=\dfrac{1}{2}+\left(\dfrac{1}{3}-\dfrac{1}{3}\right)+\left(\dfrac{1}{4}-\dfrac{1}{4}\right)+\left(\dfrac{1}{5}-\dfrac{1}{5}\right)-\dfrac{1}{6}=\dfrac{1}{2}-\dfrac{1}{6}=\dfrac{1}{3}$

123. $a\cdot b\cdot c\cdot\dfrac{1}{a}\cdot\dfrac{1}{b}\cdot\dfrac{1}{c}=\left(a\cdot\dfrac{1}{a}\right)\cdot\left(b\cdot\dfrac{1}{b}\right)\cdot\left(c\cdot\dfrac{1}{c}\right)=1\cdot1\cdot1=1$

125. The box contains 1200 cubic inches because if $200=\text{Length}\times\text{Width}\times\text{Height}$ then

$2\cdot3\cdot200=2\cdot\text{Length}\times3\cdot\text{Width}\times\text{Height}$ and $2\cdot3\cdot200=1200$.

127. Since $\text{Volume}=\text{Length}\times\text{Width}\times\text{Height}$, the larger pool has a volume $4\cdot2=8$ times greater than

the smaller pool. It will take 8 times longer to fill it, or $8\cdot2\text{ days}=16\text{ days}$.

129. We may estimate the cost to be \$2600 since $\$200+\$100\cdot24=\$2600$. *Answers may vary.*

The actual cost is $\$202+\$98.99\cdot24=\$2577.76$.

131. (a)   No, the data is changing by different numbers each year. Yes, the number of references is
    getting larger as we progress in time.

    (b) $(795+2179+5591+9903+13,000)\div5=6293.6$

    (c)   Answers may vary.

133. (a)   Data are increasing.

    (b)   *Answers may vary.*

Checking Basic Concepts Sections 1.1 and 1.2

1. (a)   The number –9 is an integer, a rational number and a real number.

   (b)   The number $\dfrac{8}{4}$ is a natural number, integer, rational number and a real number.

   (c)   The number $\sqrt{5}$ is a real number.

   (d)   The number 0.5 is a rational number and a real number.

2. (a)   Commutative property; $a+b=b+a$

   (b)   Associative property; $a\cdot(b\cdot c)=(a\cdot b)\cdot c$

   (c)   Distributive property; $a(b+c)=ab+ac$

   (d)   Distributive property; $a(b-c)=ab-ac$ with $a=-1$

3. (a) $-3+4+(-5)=1+(-5)=-4$

   (b) $5.1\cdot(-4)\cdot2=-20.4\cdot2=-40.8$

   (c) $-\dfrac{2}{3}\cdot\left(\dfrac{1}{4}\div\dfrac{2}{5}\right)=-\dfrac{2}{3}\cdot\left(\dfrac{1}{4}\cdot\dfrac{5}{2}\right)=-\dfrac{2}{3}\cdot\dfrac{5}{8}=-\dfrac{10}{24}=-\dfrac{5}{12}$

4. Since there are 3 times as many acres and the water is 2 times deeper, we multiply the total gallons
   by 6. The larger lake has about 12 billion gallons.

Section 1.3 Integer Exponents

1. The base is 8 and the exponent is 3.

3. 7^3

5. No. $-4^2 = -(4 \cdot 4) = -16$ and $(-4)^2 = (-4) \cdot (-4) = 16$

7. $\dfrac{1}{7^n}$

9. 5^{m-n}

11. 2^{mk}

13. x^n

15. $\left(\dfrac{z}{y}\right)^n$ or $\dfrac{z^n}{y^n}$

17. $8 = 2 \times 2 \times 2 = 2^3$

19. $256 = 4 \times 4 \times 4 \times 4 = 4^4$

21. $1 = 6^0$

23. (a)	$4^3 = 4 \cdot 4 \cdot 4 = 64$

   (b)	$3^4 = 3 \cdot 3 \cdot 3 \cdot 3 = 81$

25. (a)	$\dfrac{1}{4^{-2}} = 4^2 = 4 \cdot 4 = 16$

   (b)	$\dfrac{1}{3^{-3}} = 3^3 = 3 \cdot 3 \cdot 3 = 27$

27. (a)	$\left(\dfrac{2}{3}\right)^3 = \dfrac{2}{3} \cdot \dfrac{2}{3} \cdot \dfrac{2}{3} = \dfrac{8}{27}$

   (b)	$\left(-\dfrac{2}{3}\right)^{-3} = \left(-\dfrac{3}{2}\right)^3 = \left(-\dfrac{3}{2}\right) \cdot \left(-\dfrac{3}{2}\right) \cdot \left(-\dfrac{3}{2}\right) = -\dfrac{27}{8}$

29. (a)	$\dfrac{3^{-2}}{2^{-4}} = \dfrac{2^4}{3^2} = \dfrac{2 \cdot 2 \cdot 2 \cdot 2}{3 \cdot 3} = \dfrac{16}{9}$

   (b)	$\dfrac{4^{-3}}{5^{-2}} = \dfrac{5^2}{4^3} = \dfrac{5 \cdot 5}{4 \cdot 4 \cdot 4} = \dfrac{25}{64}$

31. (a)	$\dfrac{1}{2x^{-3}} = \dfrac{x^3}{2}$

   (b)	$\dfrac{1}{(ab)^{-1}} = ab$

33. (a)	$3^5 \cdot 3^{-3} = 3^{5+(-3)} = 3^2 = 9$

(b) $x^2 x^5 = x^{2+5} = x^7$

35. (a) $\left(-3x^{-2}\right)\left(5x^5\right) = (-3)(5)x^{-2+5} = -15x^3$

(b) $(ab)\left(a^2 b^{-3}\right) = a^{1+2}b^{1+(-3)} = a^3 b^{-2} = \dfrac{a^3}{b^2}$

37. (a) $5^{-2} \cdot 5^3 \cdot 2^{-4} \cdot 2^3 = 5^{-2+3} \cdot 2^{-4+3} = 5^1 \cdot 2^{-1} = 5 \cdot \dfrac{1}{2} = \dfrac{5}{2}$

(b) $2a^3 \cdot b^2 \cdot 4a^{-4} \cdot b^{-5} = 2 \cdot 4 \cdot a^{3+(-4)} \cdot b^{2+(-5)} = 8a^{-1}b^{-3} = \dfrac{8}{ab^3}$

39. (a) $\dfrac{4^3}{4^2} = 4^{3-2} = 4^1 = 4$

(b) $\dfrac{10^{-3}}{10^{-5}} = 10^{-3-(-5)} = 10^2 = 100$

41. (a) $\dfrac{b^{-3}}{b^2} = b^{-3-2} = b^{-5} = \dfrac{1}{b^5}$

(b) $\dfrac{24x^3}{6x} = \dfrac{24}{6}x^{3-1} = 4x^2$

43. (a) $\dfrac{12a^2 b^3}{18a^4 b^2} = \dfrac{12}{18}a^{2-4} \cdot b^{3-2} = \dfrac{2}{3}a^{-2}b^1 = \dfrac{2b}{3a^2}$

(b) $\dfrac{21x^{-3}y^4}{7x^4 y^{-2}} = \dfrac{21}{7}x^{-3-4} \cdot y^{4-(-2)} = 3x^{-7}y^6 = \dfrac{3y^6}{x^7}$

45. (a) $\left(3^2\right)^4 = 3^{2 \cdot 4} = 3^8 = 6561$

(b) $\left(x^3\right)^{-2} = x^{3 \cdot (-2)} = x^{-6} = \dfrac{1}{x^6}$

47. (a) $\left(4y^2\right)^3 = 4^3 \cdot y^{2 \cdot 3} = 64y^6$

(b) $\left(-2xy^3\right)^{-4} = \dfrac{1}{\left(-2xy^3\right)^4} = \dfrac{1}{(-2)^4 \cdot x^4 \cdot y^{3 \cdot 4}} = \dfrac{1}{16x^4 y^{12}}$

49. (a) $\left(\dfrac{4}{x}\right)^3 = \dfrac{4^3}{x^3} = \dfrac{64}{x^3}$

(b) $\left(\dfrac{2x}{z^4}\right)^{-5} = \left(\dfrac{z^4}{2x}\right)^5 = \dfrac{\left(z^4\right)^5}{(2x)^5} = \dfrac{z^{4 \cdot 5}}{2^5 \cdot x^5} = \dfrac{z^{20}}{32x^5}$

51. $\dfrac{12m^2 n^{-5}}{8mn^{-2}} = \dfrac{12}{8}m^{2-1}n^{-5-(-2)} = \dfrac{3}{2}mn^{-3} = \dfrac{3m}{2n^3}$

53. $\left(2x^3y^{-2}\right)^{-2} = 2^{-2}\left(x^3\right)^{-2}\left(y^{-2}\right)^{-2} = \dfrac{1}{2^2}\, x^{3\cdot(-2)}y^{(-2)\cdot(-2)} = \dfrac{1}{4}x^{-6}y^4 = \dfrac{y^4}{4x^6}$

55. $\dfrac{\left(b^2\right)^3}{\left(b^{-1}\right)^2} = \dfrac{b^{2\cdot3}}{b^{-1\cdot2}} = \dfrac{b^6}{b^{-2}} = b^{6-(-2)} = b^8$

57. $\dfrac{\left(-3ab^2\right)^3}{\left(a^2b\right)^2} = \dfrac{(-3)^3\,a^3\left(b^2\right)^3}{\left(a^2\right)^2b^2} = \dfrac{-27a^3b^{2\cdot3}}{a^{2\cdot2}b^2} = \dfrac{-27a^3b^6}{a^4b^2} = -27a^{3-4}b^{6-2} = -27a^{-1}b^4 = -\dfrac{27b^4}{a}$

59. $\dfrac{\left(-m^2n^{-1}\right)^{-2}}{(mn)^{-1}} = \dfrac{(mn)^1}{\left(-m^2n^{-1}\right)^2} = \dfrac{mn}{\left(-m^2\right)^2\left(n^{-1}\right)^2} = \dfrac{mn}{m^{2\cdot2}n^{-1\cdot2}} = \dfrac{mn}{m^4n^{-2}} = m^{1-4}n^{1-(-2)} = m^{-3}n^3 = \dfrac{n^3}{m^3}$

61. $\left(\dfrac{2a^3}{6b}\right)^4 = \left(\dfrac{a^3}{3b}\right)^4 = \dfrac{\left(a^3\right)^4}{3^4b^4} = \dfrac{a^{3\cdot4}}{81b^4} = \dfrac{a^{12}}{81b^4}$

63. $\left(\dfrac{t^{-3}}{t^{-4}}\right)^2 = \left(t^{-3-(-4)}\right)^2 = \left(t^1\right)^2 = t^2$

65. $\dfrac{8x^{-3}y^2}{4x^{-2}y^{-4}} = \dfrac{8}{4}x^{-3-(-2)}y^{2-(-4)} = 2x^{-1}y^2 = \dfrac{2y^2}{x}$

67. $\left(\dfrac{2t}{-r^2}\right)^{-3} = \left(\dfrac{-r^2}{2t}\right)^3 = \dfrac{\left(-r^2\right)^3}{2^3t^3} = \dfrac{-r^{2\cdot3}}{8t^3} = -\dfrac{r^6}{8t^3}$

69. $\dfrac{\left(r^2t^2\right)^{-2}}{\left(r^3t\right)^{-1}} = \dfrac{r^3t}{\left(r^2t^2\right)^2} = \dfrac{r^3t}{\left(r^2\right)^2\left(t^2\right)^2} = \dfrac{r^3t}{r^{2\cdot2}t^{2\cdot2}} = \dfrac{r^3t}{r^4t^4} = r^{3-4}t^{1-4} = r^{-1}t^{-3} = \dfrac{1}{rt^3}$

71. $\dfrac{4x^{-2}y^3}{\left(2x^{-1}y\right)^2} = \dfrac{4x^{-2}y^3}{2^2\left(x^{-1}\right)^2y^2} = \dfrac{4x^{-2}y^3}{4x^{-1\cdot2}y^2} = \dfrac{4x^{-2}y^3}{4x^{-2}y^2} = \left(\dfrac{4x^{-2}}{4x^{-2}}\right)y^{3-2} = 1y = y$

73. $\left(\dfrac{-15r^2t}{3r^{-3}t^4}\right)^3 = \left(\dfrac{-15}{3}r^{2-(-3)}t^{1-4}\right)^3 = \left(-5r^5t^{-3}\right)^3 = (-5)^3\left(r^5\right)^3\left(t^{-3}\right)^3 = -125r^{5\cdot3}t^{-3\cdot3} =$

$-125r^{15}t^{-9} = -\dfrac{125r^{15}}{t^9}$

75. $4 + 5\cdot6 = 4 + 30 = 34$

77. $2\left(4+(-8)\right) = 2\cdot(-4) = -8$

79. $5\cdot2^3 = 5\cdot8 = 40$

81. $\dfrac{-2^4-3^2}{4} + \dfrac{1+2}{4} = \dfrac{-16-9}{4} + \dfrac{3}{4} = \dfrac{-25}{4} + \dfrac{3}{4} = \dfrac{-22}{4} = -\dfrac{11}{2}$

83. $\dfrac{1-2\cdot4^2}{5^{-1}} = \dfrac{1-2\cdot16}{\frac{1}{5}} = \dfrac{1-32}{\frac{1}{5}} = \dfrac{-31}{\frac{1}{5}} = \dfrac{-31}{1}\cdot\dfrac{5}{1} = -155$

85. $7^2 - 3(4+2\cdot5) = 49 - 3(4+10) = 49 - 3(14) = 49 - 42 = 7$

87. $4 + 6 - 3\cdot5 \div 3 = 4 + 6 - 15 \div 3 = 4 + 6 - 5 = 10 - 5 = 5$

89. $\dfrac{-3^2+3}{3} = \dfrac{-9+3}{3} = \dfrac{-6}{3} = -2$

91. $-4^2 + \dfrac{15+2}{8-7} = -16 + \dfrac{17}{1} = 1$

93. $\left|7 - 2^2\cdot3\right| = \left|7 - 4\cdot3\right| = \left|7 - 12\right| = \left|-5\right| = 5$

95. $\sqrt{4^2 + 3^2} = \sqrt{16+9} = \sqrt{25} = 5$

97. Move the decimal point 6 places to the left. 5.9×10^6

99. Move the decimal point 7 places to the left. 5×10^7

101. Move the decimal point 2 places to the right. 3.2×10^{-2}

103. Move the decimal point 6 places to the right. 1.0×10^{-6}

105. 500,000

107. 9,300,000

109. –0.006

111. 0.00005876

113. $\left(2\times10^4\right)\left(3\times10^2\right) = (2\cdot3)\times10^{4+2} = 6\times10^6 = 6,000,000$

115. $\left(4\times10^{-4}\right)\left(2\times10^{-2}\right) = (4\cdot2)\times10^{-4+(-2)} = 8\times10^{-6} = 0.000008$

117. $\dfrac{6.2\times10^3}{3.1\times10^{-2}} = \dfrac{6.2}{3.1}\times10^{3-(-2)} = 2\times10^5 = 200,000$

119. (a)   By trial and error $k = 10$

  (b) $\dfrac{1024}{52} \approx 20$ years

121. $V = (2a)^3 = 2^3 a^3 = 8a^3$

123. $A = (3ab)^2 = 3^2 a^2 b^2 = 9a^2 b^2$

125. (a) $12 \div 6 = 2 \Rightarrow \left(\dfrac{1}{2}\right)^2 = \dfrac{1}{4}$

  (b) $30 \div 6 = 5 \Rightarrow \left(\dfrac{1}{2}\right)^5 = \dfrac{1}{32}$

127. $\left(4.72 \times 10^{12}\right) \div \left(2.96 \times 10^{8}\right) \approx \$15,946$ per person

129. $256 \times 2^{20} = 268,435,456$ bytes

131. See Figure 131

Country	2000	2025
China	1.2689×10^{9}	1.48×10^{9}
Germany	8.22×10^{7}	8.09×10^{7}
India	1.0041×10^{9}	1.3302×10^{9}
Mexico	9.99×10^{7}	1.302×10^{8}
U.S.	2.821×10^{8}	3.325×10^{8}

Figure 131

Section 1.4 Variables, Equations, and Formulas

1. variable

3. equals

5. x and y

7. $3x = 15$ *Answers may vary.*

9. $y = 2x = 2(3) = 6$

11. 3

13. b

15. $b^2 = a$

17. False

19. $-\sqrt{49} = -7$

21. $\pm\sqrt{5^2 + 12^2} = \pm\sqrt{25 + 144} = \pm\sqrt{169} = \pm13$

23. $\sqrt{10 - 2 \cdot 3} = \sqrt{10 - 6} = \sqrt{4} = 2$

25. $\sqrt{-16}$ is not a real number since there is no real number multiplied by itself that will equal -16.

27. $\sqrt[3]{27} = 3$

29. $\sqrt[3]{1 - 3^2} = \sqrt[3]{1 - 9} = \sqrt[3]{-8} = -2$

31. $\sqrt[3]{2 \cdot 8 - 4^2} = \sqrt[3]{16 - 16} = \sqrt[3]{0} = 0$

33. $y = 5280x$

35. $A = s^2$

37. $y = 3600x$

39. $A = \dfrac{1}{2}bh$

41. $r = \frac{1}{2}d$ and $A = \pi r^2$, so $A = \pi\left(\frac{1}{2}d\right)^2 = \pi\left(\frac{1}{4}d^2\right) = \frac{1}{4}\pi d^2$.

43. $y = 5(6) = 30$

45. $y = -3.1 + 5 = 1.9$

47. $d = (-3)^2 + 1 = 9 + 1 = 10$

49. $z = \sqrt{2(18)} = \sqrt{36} = 6$

51. $y = -\frac{1}{2}\sqrt[3]{\frac{1}{8}} = -\frac{1}{2} \cdot \frac{1}{2} = -\frac{1}{4}$

53. $N = 3\left(\frac{1}{3}\right)^3 - 1 = 3\left(\frac{1}{27}\right) - 1 = \frac{1}{9} - 1 = \frac{1}{9} - \frac{9}{9} = -\frac{8}{9}$

55. $P = |5 - 4.7| = |0.3| = 0.3$

57. $A = \frac{1}{2}(3)(6) = \frac{1}{2}(18) = 9$

59. $V = \pi\left(\frac{1}{2}\right)^2(5) = \pi\left(\frac{1}{4}\right)(5) = \frac{5}{4}\pi$

61. (ii)

63. (iii)

65. $a = -3$

67. $a = 2$

69. See Figure 69.

x	0	2	4	6	8
y	-0.5	4.5	9.5	14.5	19.5

Figure 69

x	-3	-1	1	3
y	15	5	5	15

Figure 71

71. See Figure 71.

73. See Figure 73.

x	-1	0	1	8
y	-3	-2	-1	0

Figure 73

x	a	$a-1$	a^2-1
y	$\sqrt{a+1}$	$\sqrt{a}$	a

Figure 75

75. See Figure 75.

77.

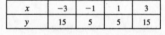

79.

81. The formula for the volume of a pyramid is $V = \frac{1}{3}b^2h$. When $b = 756$ ft and $h = 481$ ft the volume

    is $V = \frac{1}{3} \cdot 756^2 \cdot 481 \approx 91,636,272$ ft^3.

83. Since the time is being multiplied by 60 for $t = 1,2,3,4$ we have $y = 60t$

    $t = 6 \Rightarrow y = 60(6) = 360$ mi

85. (a) $\dfrac{1000 \text{ songs}}{4 \text{ GB}} = 250$ songs per GB and $\dfrac{15,000 \text{ songs}}{60 \text{ GB}} = 250$ songs per GB, so $S = 250x$.

    (b) $S = 250(30) = 7500$ songs

87. (a)   See Figure 87.

    (b) $d = \dfrac{40^2}{12} = \dfrac{1600}{12} = 133.\overline{3}$ ft

    (c)   It quadruples.

    (d) $\dfrac{60^2}{9} - \dfrac{60^2}{12} = 400 - 300 = 100$ ft

Speed (mph)	10	20	30	40	50	60	70
Braking Distance (ft)	8.3	33.3	75	133.3	208.3	300	408.3

Figure 87

89. (a)   Multiply the number of pages viewed by the number of months. $P = 662x$

    (b)   Two years is equivalent to 24 months and we have $P = 662(24) = 15,888$ pages.

91. (a) $W = 1.1(0.75)^3 \approx 0.464$ kg

    (b) $W = 1.1(1.5)^3 = 3.7125$ kg

    (c)   It increases 8 times.

93. (a) $C = 336x$

    (b) $C = 336 \cdot 30 = 10,080$ calories

95. (a) $S = 0.11\sqrt[3]{0.5^2} \approx 0.069$ m^2

    (b)   No. For example, when $W = 1$, $S = 0.11$ but when $W = 2$, $S \approx 0.175$. The value of S does not

           double.

97. If the shorter animal has a shoulder height of h, then its stepping frequency is $F_1 = \dfrac{0.87}{\sqrt{h}}$.

    For the taller animal with shoulder height $4h$, $F_2 = \dfrac{0.87}{\sqrt{4h}} = \dfrac{0.87}{2\sqrt{h}} = \dfrac{1}{2}F_1$.

    The stepping frequency of the taller animal is half the stepping frequency of the shorter animal.

99. (a) $N = \dfrac{885}{\sqrt{25}} = 177$ bpm

(b) $N = \dfrac{885}{\sqrt{1600}} \approx 22 \text{ bpm}$

101. (a) $A = \dfrac{\sqrt{3}}{4}(2)^2 = \dfrac{\sqrt{3}}{4} \cdot 4 = \sqrt{3} \approx 1.73 \text{ ft}^2$

(b) $A = \dfrac{\sqrt{3}}{4}(4)^2 = \dfrac{\sqrt{3}}{4} \cdot 16 = 4\sqrt{3} \approx 6.93 \text{ m}^2$

103. (a) $C = 2\pi(14) = 28\pi \approx 88 \text{ inches}$

(b) $C = 2\pi(1.3) = 2.6\pi \approx 8.2 \text{ miles}$

105. If $s^2 = 81$, then $s = 9$ inches.

107. If $s^3 = 27$, then $s = 3$ meters.

Checking Basic Concepts Sections 1.3 and 1.4

1. (a) $2^4 = 2 \cdot 2 \cdot 2 \cdot 2 = 16$

(b) $3^{-2} \cdot 2^0 = \dfrac{1}{3^2} \cdot 1 = \dfrac{1}{9}$

(c) $\dfrac{2^4}{2^2 \cdot 2^{-3}} = \dfrac{2^4}{2^{2+(-3)}} = \dfrac{2^4}{2^{-1}} = 2^{4-(-1)} = 2^5 = 32$

(d) $x^3 x^{-4} x^2 = x^{3+(-4)+2} = x^1 = x$

(e) $\left(\dfrac{2x^3}{y^{-4}}\right)^2 = \dfrac{2^2 \cdot x^{3 \cdot 2}}{y^{-4 \cdot 2}} = \dfrac{4x^6}{y^{-8}} = 4x^6 y^8$

2. (a) $4 + 5 \cdot (-2) = 4 + (-10) = -6$

(b) $\dfrac{1+3}{-4+3} = \dfrac{4}{-1} = -4$

(c) $2^3 - 5(2 - 3 \cdot 4) = 8 - 5(2 - 12) = 8 - 5(-10) = 8 + 50 = 58$

3. (a) 1.03×10^5

(b) 5.23×10^{-4}

(c) 6.7×10^0

4. (a) 5,430,000

(b) 0.0098

5. See Figure 5. Each person requires 900 ft^3 of ventilation per hour.

People	10	20	30	40
Ventilation (ft³/hr)	9000	18,000	27,000	36,000

Figure 5

Section 1.5 Introduction to Graphing

1. A relation is a set of ordered pairs (x, y).

3. See Figure 3.

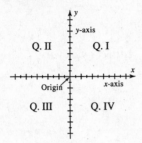

Figure 3

5. No

7. 3

9. $D = \{1, 3, 5\}$; $R = \{-4, 2, 6\}$

11. $D = \{-2, -1, 0, 1, 2\}$; $R = \{0, 1, 2, 3\}$

13. $D = \{41, 87, 96\}$; $R = \{24, 53, 67, 88\}$

15. $S = \{(1, 3), (3, 7), (5, 11), (7, 15), (9, 19)\}$; $D = \{1, 3, 5, 7, 9\}$; $R = \{3, 7, 11, 15, 19\}$

17. $S = \{(2006, 139), (2007, 217), (2008, 325), (2009, 474), (2010, 560)\}$;

    $D = \{2006, 2007, 2008, 2009, 2010\}$; $R = \{139, 217, 325, 474, 560\}$

19. See Figure 19. (1, 2) is in QI, (–3, 0) is on the x-axis, (0, –2) is on the y-axis, (–1, 3) is in QII.

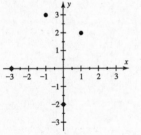

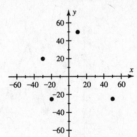

Figure 19   Figure 21

21. See Figure 21. (10, 50) is in QI, (–30, 20) is in QII, (50, –25) is in QIV, (–20, –25) is in QIII.

23. Substituting 0, 1, 2, 3 for x gives the ordered pairs $(0, 0)$, $(1, 4)$, $(2, 8)$, $(3, 12)$.

    Answers may vary.

25. Substituting 0, 1, 2, 3 for t gives the ordered pairs $(0, 4)$, $(1, 3)$, $(2, 0)$, $(3, -5)$.

    Answers may vary.

27. Substituting 0, 1, 2, 3 for r gives the ordered pairs $(0, 1)$, $\left(1, \frac{1}{2}\right)$, $\left(2, \frac{1}{5}\right)$, $\left(3, \frac{1}{10}\right)$.

    Answers may vary.

29. $S = \{(-3, 2), (-2, 1), (2, -3), (3, 3)\}$; $D = \{-3, -2, 2, 3\}$; $R = \{-3, 1, 2, 3\}$

31. $S = \{(-4, 4), (-3, 2), (-2, 0), (0, -3), (2, 4), (4, 4)\}$;

     $D = \{-4, -3, -2, 0, 2, 4\}$; $R = \{-3, 0, 2, 4\}$

33. $S = \{(Facebook, 662), (MySpace, 262), (Twitter, 67), (Digg, 18)\}$

     $D = \{Facebook, MySpace, Twitter, Digg\}$; $R = \{662, 262, 67, 18\}$. *Answers may vary.*

35. $a = b$

37. (a) $D = \{-3, -2, 0, 1\}$; $R = \{-4, -3, 0, 2, 4\}$

     (b) x-min: -3, x-max: 1; y-min: -4, y-max: 4

     (c)& (d) See Figure 37.

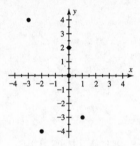

Figure 37   Figure 39

39. (a) $D = \{-30, 10, 20, 30\}$; $R = \{-50, 20, 40, 50\}$

     (b) x-min: -30, x-max: 30; y-min: -50, y-max: 50

     (c)& (d) See Figure 39.

41. See Figure 41.

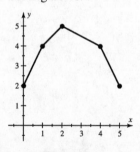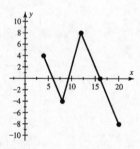

Figure 41   Figure 43

43. See Figure 43.

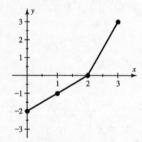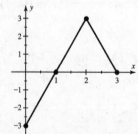

Figure 45   Figure 47

45. See Figure 45.

47. See Figure 47.

49. $(-2, -6)$, $(-1, -3)$, $(0, 0)$, $(1, 3)$, $(2, 6)$. See Figure 49.

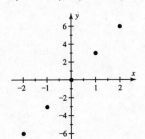
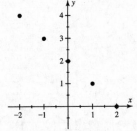

Figure 49   Figure 51

51. $(-2, 4)$, $(-1, 3)$, $(0, 2)$, $(1, 1)$, $(2, 0)$. See Figure 51.

53. $(-2, 1)$, $\left(-1, \frac{1}{2}\right)$, $(0, 0)$, $\left(1, -\frac{1}{2}\right)$, $(2, -1)$. See Figure 53.

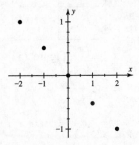
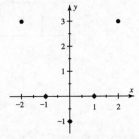

Figure 53   Figure 55

55. $(-2, 3)$, $(-1, 0)$, $(0, -1)$, $(1, 0)$, $(2, 3)$. See Figure 55.

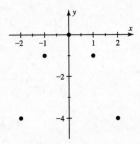
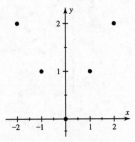

Figure 57   Figure 59

57. $(-2, -4)$, $(-1, -1)$, $(0, 0)$, $(1, -1)$, $(2, -4)$. See Figure 57.

59. $(-2, 2)$, $(-1, 1)$, $(0, 0)$, $(1, 1)$, $(2, 2)$. See Figure 59.

61. See Figure 61. There are 10 tick marks on the positive x-axis and 10 tick marks on the positive y-axis.

$[-10, 10, 1]$ by $[-10, 10, 1]$ $[0, 100, 10]$ by $[-50, 50, 10]$

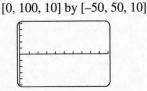

Figure 61   Figure 63

63. See Figure 63. There are 10 tick marks on the positive x-axis and 5 tick marks on the positive y-axis.

65. See Figure 65. There are 15 tick marks on the positive *x*-axis and 4 tick marks on the positive *y*-axis.

[1980, 1995, 1] by [12,000, 16,000, 1000]

Figure 65

67. $S = \{(-2, 1), (-1, 0), (1, -1), (2, 1)\}$

69. $S = \{(-4, -2), (-2, -2), (0, 2), (2, 1), (4, -1)\}$

71. See Figure 71.

[–6, 6, 1] by [–6, 6, 1]   [–30, 30, 5] by [–50, 50, 5]   [–200, 200, 50] by [–250, 250, 50]

                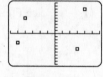   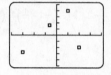

Figure 71   Figure 73   Figure 75

73. See Figure 73.

75. See Figure 75.

77. (a) See Figure 77.

(b)   The number of text messages sent has increased.

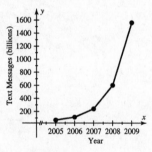   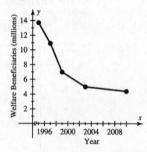

Figure 77   Figure 79

79. (a)   See Figure 79.

(b)   The number of welfare beneficiaries decreased.

81. (a)   See Figure 81.

(b)   The Asian-American population was increasing steadily.

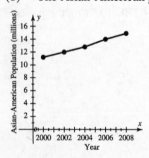

Figure 81

Checking Basic Concepts Section 1.5

1. $D = \{-5,\ 1,\ 2\}$; $R = \{-1,\ 3,\ 4\}$

2. See Figure 2. (1, 4) is in QI, (0, –3) is on the y-axis, (2, –2) is in QIV, (–2, 3) is in QII.

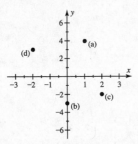

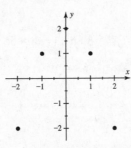

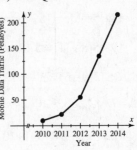

Figure 2   Figure 3   Figure 4

3. $(-2,\ -2),\ (-1,\ 1),\ (0,\ 2),\ (1,\ 1),\ (2,\ -2)$. See Figure 3.

4. See Figure 4. The mobile data traffic has steadily increased over time.

5. There are 10^{15} bytes in a petabyte. This number is read as 1 quadrillion.

Chapter 1 Review

1. Natural numbers: 9   Whole numbers: 9   Integers: –2, 9

    Rational Numbers: $-2, 9, \dfrac{2}{5}, 2.68$ Irrational Numbers: $\sqrt{11},\ \pi$

2. Natural numbers: $\dfrac{6}{2}$ Whole numbers: $\dfrac{6}{2}, \dfrac{0}{4}$ Integers: $\dfrac{6}{2}, \dfrac{0}{4}$

    Rational Numbers: $\dfrac{6}{2}, -\dfrac{2}{7}, 0.\overline{3}, \dfrac{0}{4}$ Irrational Numbers: $\sqrt{6}$

3. False

4. True

5. 9

6. ± 3

7. –4

8. –3

9. Identity

10. Commutative property

11. Associative property

12. Distributive property

13. Distributive property

14. Distributive property

15. $1 \cdot (a + 0) = 1 \cdot a = a$

16. $x \cdot \dfrac{1}{4} = \dfrac{1}{4}x$

17. $8(10x) = (8 \cdot 10)x = 80x$

18. $5z - 3z = (5 - 3)z = 2z$

19. $5(8 + 11) = 5 \cdot 19 = 95$ or $5(8 + 11) = 5 \cdot 8 + 5 \cdot 11 = 40 + 55 = 95$

20. $3(9 - 5) = 3 \cdot 4 = 12$ or $3(9 - 5) = 3 \cdot 9 - 3 \cdot 5 = 27 - 15 = 12$

21. $(6 + 9 + 3 + 11 + 5 + 20) \div 6 = 9$

22. $(3.2 + 6.8 + 6.1 + 10.8 + 1.7) \div 5 = 5.72$

23. $12 + 23 + (-2) + 7 = (12 + (-2)) + (23 + 7) = 10 + 30 = 40$

24. $\dfrac{3}{2} \cdot \dfrac{2}{5} \cdot \dfrac{5}{7} \cdot \dfrac{7}{3} = \left(\dfrac{3}{2} \cdot \dfrac{2}{5}\right)\left(\dfrac{5}{7} \cdot \dfrac{7}{3}\right) = \dfrac{3}{5} \cdot \dfrac{5}{3} = \dfrac{3 \cdot 5}{5 \cdot 3} = \dfrac{15}{15} = 1$

25. $5 \cdot 23 + 5 \cdot 7 = 5(23 + 7) = 5 \cdot 30 = 150$

26. $45 - 34 + 55 - 66 = (45 + 55) + (-34 - 66) = 100 - 100 = 0$

27.

28. $|-7.2 + 4| = |-3.2| = 3.2$

29. $\dfrac{2}{3}$

30. $\dfrac{5}{4}$

31. The opposite of $-2x + 3$ is $-(-2x + 3) = 2x - 3$.

32. The reciprocal of $-\dfrac{1}{a+b}$ is $-\dfrac{a+b}{1} = -(a+b) = -a - b$.

33. $-5 + (-7) + 8 = -12 + 8 = -4$

34. $-9 + 11 = 2$

35. $-12 - (-8) = -12 + 8 = -4$

36. $\dfrac{1}{2} + (-2) + \dfrac{3}{4} = \dfrac{2}{4} + \left(-\dfrac{8}{4}\right) + \dfrac{3}{4} = -\dfrac{6}{4} + \dfrac{3}{4} = -\dfrac{3}{4}$

37. $\dfrac{2}{3} \div (-4) - \dfrac{1}{3} = \dfrac{2}{3} \cdot \left(-\dfrac{1}{4}\right) - \dfrac{1}{3} = -\dfrac{2}{12} - \dfrac{4}{12} = -\dfrac{6}{12} = -\dfrac{1}{2}$

38. $-\dfrac{7}{11} + \dfrac{\frac{1}{5}}{\frac{2}{9}} = -\dfrac{7}{11} + \dfrac{1}{5} \cdot \dfrac{9}{2} = -\dfrac{7}{11} + \dfrac{9}{10} = -\dfrac{70}{110} + \dfrac{99}{110} = \dfrac{29}{110}$

39. $-7 \cdot 4 \cdot -\dfrac{1}{7} = -7 \cdot -\dfrac{1}{7} \cdot 4 = 1 \cdot 4 = 4$

40. $\dfrac{3}{4} \cdot \dfrac{4}{7} + 1 = \dfrac{3}{7} + 1 = \dfrac{3}{7} + \dfrac{7}{7} = \dfrac{10}{7}$

41. $4x + 5x = (4+5)x = 9x$

42. $2z - 3z + 8z = (2-3+8)z = 7z$

43. Entering $1/5 - 3/7 + (2/13)/(4/5)$ ▶Frac into a graphing calculator gives $-\dfrac{33}{910}$.

44. $-4 + 9 + 4 + 11 - 6 + 16 = 30$; the average is $30 \div 6 = 5$.

45. Base: 4; exponent: -2

46. They are not equal since $3^\pi \approx 31.54$ but $\pi^3 \approx 31.01$.

47. $-2^4 = -(2 \cdot 2 \cdot 2 \cdot 2) = -16$

48. $(-2)^4 = (-2) \cdot (-2) \cdot (-2) \cdot (-2) = 16$

49. $9^0 = 1$

50. $\left(\dfrac{2}{3}\right)^{-3} = \left(\dfrac{3}{2}\right)^3 = \dfrac{3^3}{2^3} = \dfrac{3 \cdot 3 \cdot 3}{2 \cdot 2 \cdot 2} = \dfrac{27}{8}$

51. $4^{-3} = \dfrac{1}{4^3} = \dfrac{1}{4 \cdot 4 \cdot 4} = \dfrac{1}{64}$

52. $\dfrac{1}{5^{-2}} = 5^2 = 5 \cdot 5 = 25$

53. $\dfrac{5^{-3}}{3^{-2}} = \dfrac{3^2}{5^3} = \dfrac{3 \cdot 3}{5 \cdot 5 \cdot 5} = \dfrac{9}{125}$

54. $\dfrac{1}{2 \cdot 4^{-2}} = \dfrac{4^2}{2} = \dfrac{4 \cdot 4}{2} = \dfrac{16}{2} = 8$

55. $4^3 \cdot 4^{-5} = 4^{3+(-5)} = 4^{-2} = \dfrac{1}{4^2} = \dfrac{1}{16}$

56. $10^4 \cdot 10^{-2} = 10^{4+(-2)} = 10^2 = 100$

57. $x^7 \cdot x^{-2} = x^{7+(-2)} = x^5$

58. $\dfrac{3^4}{3^{-7}} = 3^{4-(-7)} = 3^{11} = 177{,}147$

59. $\dfrac{5a^{-4}}{10a^2} = \left(\dfrac{5}{10}\right) \cdot a^{-4-2} = \left(\dfrac{1}{2}\right) \cdot a^{-6} = \left(\dfrac{1}{2}\right) \cdot \left(\dfrac{1}{a^6}\right) = \dfrac{1}{2a^6}$

60. $\dfrac{15a^4 b^3}{3a^2 b^6} = \left(\dfrac{15}{3}\right) \cdot a^{4-2} \cdot b^{3-6} = 5a^2 b^{-3} = \dfrac{5a^2}{b^3}$

61. $\left(2^2\right)^4 = 2^{2 \cdot 4} = 2^8 = 256$

62. $\left(x^{-3}\right)^5 = x^{-3 \cdot 5} = x^{-15} = \dfrac{1}{x^{15}}$

63. $\left(4x^{-2}y^3\right)^2 = 4^2 \cdot x^{-2 \cdot 2} \cdot y^{3 \cdot 2} = 16x^{-4}y^6 = \dfrac{16y^6}{x^4}$

64. $(4a)^5 = 4^5 \cdot a^5 = 1024a^5$

65. $\left(\dfrac{5x^3}{3z^4}\right)^3 = \dfrac{5^3 \cdot x^{3 \cdot 3}}{3^3 \cdot z^{4 \cdot 3}} = \dfrac{125x^9}{27z^{12}}$

66. $\left(\dfrac{-3x^4y^3}{z}\right)^{-2} = \left(\dfrac{z}{-3x^4y^3}\right)^2 = \dfrac{z^2}{(-3)^2 \cdot x^{4 \cdot 2} \cdot y^{3 \cdot 2}} = \dfrac{z^2}{9x^8y^6}$

67. $\left(\dfrac{3a^{-4}}{4b^{-7}}\right)^2 = \dfrac{3^2\left(a^{-4}\right)^2}{4^2\left(b^{-7}\right)^2} = \dfrac{9a^{-4 \cdot 2}}{16b^{-7 \cdot 2}} = \dfrac{9a^{-8}}{16b^{-14}} = \dfrac{9b^{14}}{16a^8}$

68. $\left(\dfrac{3m^2n^{-4}}{9m^3n}\right)^{-1} = \dfrac{9m^3n}{3m^2n^{-4}} = \dfrac{9}{3}m^{3-2}n^{1-(-4)} = 3mn^5$

69. $\left(\dfrac{rt}{2r^3t^{-1}}\right)^{-3} = \left(\dfrac{2r^3t^{-1}}{rt}\right)^3 = \dfrac{2^3\left(r^3\right)^3\left(t^{-1}\right)^3}{r^3t^3} = \dfrac{8r^{3 \cdot 3}t^{-1 \cdot 3}}{r^3t^3} = \dfrac{8r^9t^{-3}}{r^3t^3} = 8r^{9-3}t^{-3-3} = 8r^6t^{-6} = \dfrac{8r^6}{t^6}$

70. $\left(\dfrac{3r^2}{4t^{-3}}\right)^2 = \dfrac{3^2\left(r^2\right)^2}{4^2\left(t^{-3}\right)^2} = \dfrac{9r^{2 \cdot 2}}{16t^{-3 \cdot 2}} = \dfrac{9r^4}{16t^{-6}} = \dfrac{9r^4t^6}{16}$

71. $2 + 3 \cdot 9 = 2 + 27 = 29$

72. $4 - 1 - 6 = 3 - 6 = -3$

73. $5 \cdot 2^3 = 5 \cdot 8 = 40$

74. $\dfrac{2+4}{2} + \dfrac{3-1}{3} = \dfrac{6}{2} + \dfrac{2}{3} = 3 + \dfrac{2}{3} = \dfrac{9}{3} + \dfrac{2}{3} = \dfrac{11}{3}$

75. $20 \div 4 \div 2 = 5 \div 2 = \dfrac{5}{2}$

76. $\dfrac{3^3 - 2^4}{4 - 3} = \dfrac{27 - 16}{1} = \dfrac{11}{1} = 11$

77. 1.86×10^5

78. 3.4×10^{-4}

79. $45,000$

80. 0.00923

81. Because there are 12 inches in one foot, the formula is $y = 12x$.

82. Because the area formula for one circle is $A = \pi r^2$, the formula for the area of six circles is

$A = 6\pi r^2$.

83. $y = 12(3) = 36$

84. $d = \sqrt{67 - 3} = \sqrt{64} = 8$

85. $N = \left(\dfrac{3}{2}\right)^2 - \dfrac{3}{4} = \dfrac{9}{4} - \dfrac{3}{4} = \dfrac{6}{4} = \dfrac{3}{2}$

86. $P = (-2)^3 - 2 = -8 - 2 = -10$

87. $A = \dfrac{1}{2}(4)(5) = 2(5) = 10$

88. $V = (3)^2(3) = 9(3) = 27$

89. (iii)

90. $a = \dfrac{3}{2}$

91. See Figure 91.

x	-2	-1	0	1	2
y	-7	0	1	2	9

Figure 91

92. See Figure 92.

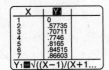

Figure 92

93. $D = \{-1, 2, 3\}$; $R = \{-6, 1, 3, 7\}$

94. $S = \{(-8, 4), (-4, -4), (4, 0), (8, 4)\}$; $D = \{-8, -4, 4, 8\}$; $R = \{-4, 0, 4\}$

95. $(-2, 6), (-1, 3), (0, 0), (1, -3), (2, -6)$. See Figure 95.

96. $(-2, -2), (-1, -1.5), (0, -1), (1, -0.5), (2, 0)$. See Figure 96.

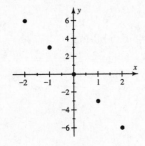

Figure 95

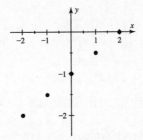

Figure 96

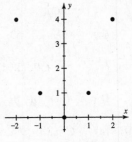

Figure 97

97. $(-2, 4), (-1, 1), (0, 0), (1, 1), (2, 4)$. See Figure 97.

98. $(-2, 1)$, $(-1, 2.5)$, $(0, 5)$, $(1, 2.5)$, $(2, 1)$. See Figure 98.

99. See Figure 99. $(-2, 2)$ is in QII, $(-1, -3)$ is in QIII, $(0, 1)$ is on the y-axis, $(2, -1)$ is in QIV.

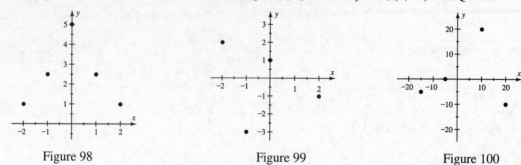

Figure 98   Figure 99   Figure 100

100. See Figure 100. $(-15, -5)$ is in QIII, $(-5, 0)$ is on the x-axis, $(10, 20)$ is in QI, $(20, -10)$ is in QIV.

101. See Figure 101. There are 9 tick marks on the positive x-axis and 2 tick marks on the positive y-axis.

$[-9, 9, 1]$ by $[-6, 6, 3]$ $[-20, 20, 5]$ by $[-12, 12, 4]$

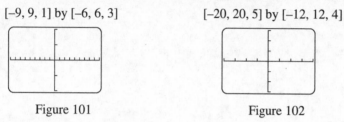

Figure 101   Figure 102

102. See Figure 102. There are 4 tick marks on the positive x-axis and 3 tick marks on the positive y-axis.

103. See Figure 103. $D = \{-1, 0, 2, 3, 4\}$; $R = \{-1, 0, 2, 3, 4\}$.

104. See Figure 104. $D = \{-20, -10, 45, 50\}$; $R = \{-30, -25, 10, 20\}$.

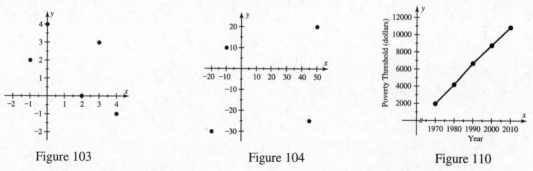

Figure 103   Figure 104   Figure 110

105. (a) The approximate average is $(30 + 50 + 50 + 50) \div 4 = (180 \div 4) = 45$, which is less than \$50 billon.

   (b) $(29 + 47 + 53 + 53.5) \div 4 = (182.5 \div 4) = 45.625$. The result is \$45.625 billion.

106. (a) $A = 2500(1.01)^1 = \$2525$

   (b) $A = 800(1.01)^2 = 800(1.01)(1.01) = \$816.08$

107. (a) $C = 2\pi \cdot (9.3 \times 10^7) \approx 5.84 \times 10^8$ miles

   (b) $\dfrac{5.84 \times 10^8 \text{ mi}}{1 \text{ year}} \cdot \dfrac{1 \text{ year}}{365 \text{ days}} \cdot \dfrac{1 \text{ day}}{24 \text{ hr}} \approx 66,700$ mph

108. $d = 40t$

109. For a 16-pound cat the pulse rate is $N = \dfrac{885}{\sqrt{16}} = \dfrac{885}{4} \approx 221$ bpm.

For a 144-pound person the pulse rate is $N = \dfrac{885}{\sqrt{144}} = \dfrac{885}{12} \approx 74$ bpm.

110. See Figure 110. The poverty threshold increased during this time period.

111. $A = (4ab)^2 = 4^2 a^2 b^2 = 16a^2 b^2$

112. $V = (5z)^3 = 5^3 z^3 = 125z^3$

Chapter 1 Test

1. Natural numbers: $\sqrt{9}$ Whole numbers: $\sqrt{9}$ Integers: -5, $\sqrt{9}$

   Rational Numbers: -5, $\dfrac{2}{3}$, $\sqrt{9}$, -1.83 Irrational Numbers: $-\dfrac{1}{\sqrt{5}}$, π

2. (a)   Identity

   (b)   Associative property

   (c)   Commutative property

   (d)   Distributive property

3. (a) $0 + (a \cdot 1) \cdot 1 = 0 + a \cdot 1 = 0 + a = a$

   (b) $2x - 5x + 4x = (2 - 5 + 4)x = 1x = x$

   (c) $2(x \cdot 3) + 2x = 2(3 \cdot x) + 2x = (2 \cdot 3)x + 2x = 6x + 2x = (6 + 2)x = 8x$

   (d) $(a + b)a - ab = a \cdot a + ab - ab = a^2 + 0 = a^2$

4. $(34 + 15 + 96 + 11 + 0) \div 5 = 156 \div 5 = 31.2$

5.

6. (a) $\left| \dfrac{1}{2} + \dfrac{2}{3} - \dfrac{8}{3} \right| = \left| \dfrac{3}{6} + \dfrac{4}{6} - \dfrac{16}{6} \right| = \left| -\dfrac{9}{6} \right| = \left| -\dfrac{3}{2} \right| = \dfrac{3}{2}$

   (b) $12 + (-3) + 18 + (-7) = 9 + 18 + (-7) = 27 + (-7) = 20$

   (c) $\dfrac{1}{2} \cdot \dfrac{2}{3} \cdot \dfrac{3}{4} \cdot \dfrac{3}{2} \cdot 2 = \dfrac{3}{4}$

   (d) $\dfrac{1}{2} + \dfrac{2}{3} - \dfrac{3}{2} + \dfrac{1}{3} = \dfrac{3}{6} + \dfrac{4}{6} - \dfrac{9}{6} + \dfrac{2}{6} = 0$

7. (a) $\sqrt{36} = 6$

   (b) $\sqrt[3]{-8} = -2$

8. (a) $-\dfrac{4}{5}$

(b) $\dfrac{b+1}{2}$

9. (a) $-\dfrac{1}{2}+\dfrac{2}{3}\div 3 = -\dfrac{1}{2}+\dfrac{2}{3}\cdot\dfrac{1}{3} = -\dfrac{1}{2}+\dfrac{2}{9} = -\dfrac{9}{18}+\dfrac{4}{18} = -\dfrac{5}{18}$

(b) $-4+\dfrac{\frac{2}{3}}{-\frac{1}{4}} = -4+\dfrac{2}{3}\cdot\left(-\dfrac{4}{1}\right) = -4-\dfrac{8}{3} = -\dfrac{12}{3}-\dfrac{8}{3} = -\dfrac{20}{3}$

(c) $5-2\cdot 5^2\div 5 = 5-2\cdot 25\div 5 = 5-50\div 5 = 5-10 = -5$

(d) $(6-4\cdot 5)\div(-7) = (6-20)\div(-7) = (-14)\div(-7) = 2$

10. (a) $5^{-2} = \dfrac{1}{5^2} = \dfrac{1}{25}$

(b) $\pi^0 = 1$

(c) $\left(-\dfrac{2}{5}\right)^4 = \left(-\dfrac{2}{5}\right)\cdot\left(-\dfrac{2}{5}\right)\cdot\left(-\dfrac{2}{5}\right)\cdot\left(-\dfrac{2}{5}\right) = \dfrac{16}{625}$

(d) $\dfrac{6^{-2}}{2^{-4}} = \dfrac{2^4}{6^2} = \dfrac{2\cdot 2\cdot 2\cdot 2}{6\cdot 6} = \dfrac{16}{36} = \dfrac{4}{9}$

(e) $\dfrac{1}{5^{-3}} = 5^3 = 5\cdot 5\cdot 5 = 125$

(f) $2^2\cdot 2^{-4} = 2^{[2+(-4)]} = 2^{-2} = \dfrac{1}{2^2} = \dfrac{1}{4}$

11. (a) $x^6\cdot x^{-4}\cdot y^3 = x^{6+(-4)}\cdot y^3 = x^2 y^3$

(b) $\dfrac{16x^{-2}y^8}{6xy^{-7}} = \left(\dfrac{16}{6}\right)\cdot x^{-2-1}\cdot y^{8-(-7)} = \dfrac{8}{3}x^{-3}y^{15} = \dfrac{8y^{15}}{3x^3}$

(c) $\left(2yz^{-2}\right)^3 = 2^3\cdot y^3\cdot z^{-2\cdot 3} = 8y^3 z^{-6} = \dfrac{8y^3}{z^6}$

(d) $\left(\dfrac{15x^4}{10xy^{-2}}\right)^{-2} = \left(\dfrac{10xy^{-2}}{15x^4}\right)^2 = \dfrac{10^2\cdot x^2\cdot y^{-2\cdot 2}}{15^2\cdot x^{4\cdot 2}} = \dfrac{100}{225}\cdot x^{2-8}\cdot y^{-4} = \dfrac{4}{9}x^{-6}y^{-4} = \dfrac{4}{9x^6 y^4}$

12. 0.00052

13. 3.4×10^6

14. $H = \dfrac{x}{60}$

15. $y = \sqrt{3+1}+3 = \sqrt{4}+3 = 2+3 = 5;\quad y = \sqrt{-1+1}+(-1) = \sqrt{0}+(-1) = 0+(-1) = -1$

16. $D = \{-3,\ -1,\ 2\},\ R = \{-4,\ 2,\ 3\}$

17. $(-2, 3)$: QII; $(-1, -2)$: QIII; $(0, 2)$: y-axis; $(1, -1)$: QIV; $(2, 1)$: QI See Figure 17.

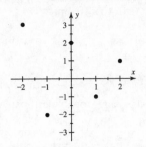

       Figure 17   Figure 19

18. $S = \{(-30, 20), (-20, 20), (-10, 10), (10, 30), (20, 10), (30, -20)\}$

    $D = \{-30, -20, -10, 10, 20, 30\}$; $R = \{-20, 10, 20, 30\}$

19. See Figure 19. The equation $y = 1.25x$ models these data.

20. See Figure 20.

    $[-20, 20, 5]$ by $[-5, 40, 5]$

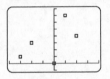

       Figure 20

21. $C = \dfrac{5}{9}(5 - 32) = \dfrac{5}{9}(-27) = -15°C$

22. $\left(2.41 \times 10^8\right) \cdot 2.14 = 5.1574 \times 10^8$ gal

23. $r = \sqrt{\dfrac{25}{\pi}} \approx 2.82$ ft

24. $M = 61t$, because $244 \div 4 = 366 \div 6 = 610 \div 10 = 61$.

Chapter 1 Extended and Discovery Exercises

1. (a)   The surface area of the earth is $A = 4\pi \cdot 3960^2 \approx 1.97 \times 10^8$ mi^2.

  (b)   The surface area of the oceans is $A = \left(1.97 \times 10^8 \text{ mi}^2\right) \cdot 0.71 \approx 1.40 \times 10^8$ mi^2.

  (c)   If the Arctic ice cap were to melt, the rise in sea level would be $\dfrac{6.8 \times 10^5 \text{ mi}^3}{1.40 \times 10^8 \text{ mi}^2} \approx 0.00486$ mi.

        That is $0.00486 \text{ mi} \cdot \dfrac{5280 \text{ ft}}{1 \text{ mi}} \approx 25.7$ ft.

  (d)   If the Arctic ice cap melted, cities such as Boston, New Orleans and San Diego would be

        flooded.

(e)   If the Antarctic ice cap were to melt, the rise in sea level would be $\dfrac{6.3\times10^6 \text{ mi}^3}{1.40\times10^8 \text{ mi}^2} = 0.045$ mi.

            That is 0.045 mi $\cdot \dfrac{5280 \text{ ft}}{1 \text{ mi}} \approx 238$ ft.

2. (a)   See Figure 2. The injury rate has decreased over this period of time.

   (b)   The rate of change decreases by about 0.3 for each 1-year period, so a reasonable estimate might be about 4.5.

   (c)   The rate of change decreases by about 0.3 for each 1-year period, so a reasonable estimate might be about 3.0.

   (d)   It is not valid to try to estimate when the injury rate might reach 0 because the apparent trend cannot continue. There will always be injuries of this type.

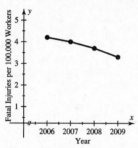

Figure 2

3. $y = \dfrac{0.455W}{(0.0254H)^2} \approx \dfrac{705W}{H^2}$

4. Noting that 5 feet 8 inches = 68 inches, Anna Kournikova's BMI is $y = \dfrac{0.455(123)}{(0.0254(68))^2} \approx 18.8$

5. Noting that 6 feet 1 inch = 73 inches, Venus Williams' BMI is $y = \dfrac{0.455(160)}{(0.0254(73))^2} \approx 21.2.$

6. Noting that 7 feet 1 inch = 85 inches, Shaquille O'Neal's BMI is $y = \dfrac{0.455(300)}{(0.0254(85))^2} \approx 29.3.$

7. When x increases by 2, y decreases by 3, so $a = -\dfrac{3}{2}.$

   By using substitution, $5 = -\dfrac{3}{2}\cdot(-3)+b \Rightarrow 5 = \dfrac{9}{2}+b \Rightarrow b = \dfrac{1}{2}.$

8. By using substitution we find that $a+b=1$ and $2a+b=-1$. By trial-and-error, $a = -2$ and $b = 3$.

9. By using substitution we find that $a+b=1$ and $4a+2b=3$. By trial-and-error, $a = \dfrac{1}{2}$ and $b = \dfrac{1}{2}.$

10. Substituting $x=0$ and $y=3$ gives $b=3$. Then by substituting $x=1$ and $y=6$ we find that $a=2$.

Chapter 2: Linear Functions and Models

2.1: Functions and Their Representations

1. function

3. symbolic

5. domain

7. one

9. True

11. (a, b)

13. 1

15. Yes, there is only one output for each input.

17. No, one exam can have many students who pass.

19. Yes, there is only one sales tax for each purchase.

21. $f(-1) = 4(-1) - 2 = -6$; $f(0) = 4(0) - 2 = -2$

23. $f(0) = \sqrt{0} = 0$; $f\left(\dfrac{9}{4}\right) = \sqrt{\dfrac{9}{4}} = \dfrac{3}{2}$

25. $f(-5) = (-5)^2 = 25$; $f\left(\dfrac{3}{2}\right) = \left(\dfrac{3}{2}\right)^2 = \dfrac{9}{4}$

27. $f(-8) = 3$; $f\left(\dfrac{7}{3}\right) = 3$

29. $f(-2) = 5 - (-2)^3 = 5 - (-8) = 13$; $f(3) = 5 - 3^3 = 5 - 27 = -22$

31. $f(-5) = \dfrac{2}{-5+1} = \dfrac{2}{-4} = -\dfrac{1}{2}$; $f(4) = \dfrac{2}{4+1} = \dfrac{2}{5}$

33. (a) Because there are 36 inches in 1 yard, the formula is $I(x) = 36x$.

   (b) $I(10) = 36(10) = 360$. There are 360 inches in 10 yards.

35. (a) Because there are 5280 feet in 1 mile, the formula is $M(x) = \dfrac{x}{5280}$.

   (b) $M(10) = \dfrac{10}{5280} \approx 0.0019$. Ten feet is equivalent to about 0.0019 mile.

37. (a) Because there are 43,560 square feet in 1 acre, the formula is $A(x) = 43,560x$.

   (b) $A(10) = 43,560(10) = 435,600$. There are 435,600 square feet in 10 acres.

39. $f = \{(1,3), (2,-4), (3,0)\}$; $D = \{1,2,3\}$, $R = \{-4,0,3\}$

41. $f = \{(a,b), (c,d), (e,a), (d,b)\}$; $D = \{a,c,d,e\}$, $R = \{a,b,d\}$

43. See Figure 43.

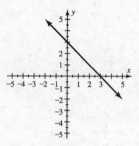

Figure 43

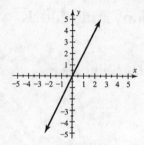

Figure 45

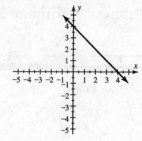

Figure 47

45. See Figure 45.

47. See Figure 47.

49. See Figure 49.

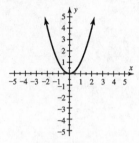

Figure 49

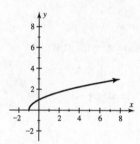

Figure 51

51. See Figure 51.

53. $f(0) = 3; f(2) = -1$

55. $f(-2) = 0; f(1) = 2$

57. $f(1) = -4; f(2) = -3$

59. $f(0) = 5.5; f(2) = 3.7$

61. $f(1990) = 26.9$ mpg ; In 1990 average fuel efficiency was 26.9 mpg.

63. Symbolic: $y = x + 5$. Numerical: See the table in Figure 63a.

Graphical: See the graph in Figure 63b.

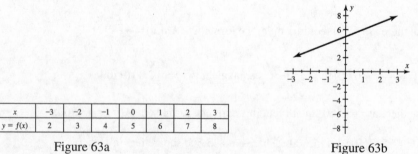

x	−3	−2	−1	0	1	2	3
$y = f(x)$	2	3	4	5	6	7	8

Figure 63a   Figure 63b

65. Symbolic: $y = 5x - 2$. Numerical: See the table in Figure 65a.

Graphical: See the graph in Figure 65b.

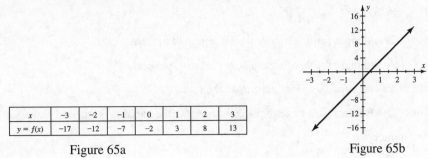

x	−3	−2	−1	0	1	2	3
y = f(x)	−17	−12	−7	−2	3	8	13

Figure 65a   Figure 65b

67. Numerical   Graphical

[−10, 10, 1] by [−10, 10 1]

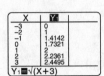   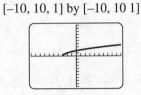

69. Numerical   Graphical

[−10, 10, 1] by [−10, 10, 1]

   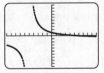

71. Subtract $\frac{1}{2}$ from the input x to obtain the output y.

73. Divide the input x by 3 to obtain the output y.

75. Subtract 1 from the input x and then take the square root to obtain the output y.

77. Symbolic: $f(x) = 0.50x$. Graphical: See the graph in Figure 77a. Numerical: See the table in Figure 77b.

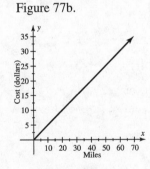

Miles	10	20	30	40	50	60	70
Cost	$5	$10	$15	$20	$25	$30	$35

Figure 77a   Figure 77b

79. $S(2011) = 225(2011) - 450,650 = 452,475 - 450,650 = 1825$ In 2011 there were 1825 billion or 1.825 trillion World Wide Web searches.

81. $D: -2 \le x \le 2; \ R: 0 \le y \le 2$

83. $D: -2 \le x \le 4; \ R: -2 \le y \le 2$

85. D: all real numbers; $R: \ y \ge -1$

87. $D: -3 \le x \le 3; \ R: -3 \le y \le 2$

89. $D = \{1, 2, 3, 4\}; R = \{5, 6, 7\}$

91. Any real number is a valid input for this function. $D:$ all real numbers.

93. Any real number is a valid input for this function. $D:$ all real numbers.

95. The denominator of this function cannot equal zero. $D: x \neq 5$.

97. The denominator of this function will never equal zero because the variable is squared and added to one. $D:$ all real numbers.

99. The radicand must be greater than or equal to zero. $D: x \geq 1$.

101. Any real number is a valid input for this function. $D:$ all real numbers.

103. The denominator of this function cannot equal zero. $D: x \neq 0$.

105. (a) $W(2008) = 1726$; In 2008 there were 1726 whales sighted in Maui.

   (b) $D = \{2005, 2006, 2007, 2008, 2009\}$; $R = \{649, 1265, 959, 1726, 1010\}$

   (c) The number of whale sightings increased every other year.

107. $D = \{1, 2, 3, ..., 20\}$; $R = \{200, 400, 600, ..., 4000\}$

109. No. The value 1 in the domain corresponds to more than one value in the range.

111. Yes. Each value in the domain corresponds to exactly one value in the range.

113. (a) May is month number 5. The corresponding value for P is 0.2.

   (b) Yes. Each month has exactly one average precipitation.

   (c) Months 2, 3, 7 and 11.

115. Yes. The graph passes the vertical line test. $D:$ all real numbers; $R:$ all real numbers

117. No. The graph does not pass the vertical line test.

119. Yes. The graph passes the vertical line test. $D: -4 \leq x \leq 4$; $R: 0 \leq y \leq 4$

121. Yes. The graph passes the vertical line test. $D:$ all real numbers; $R: y = 3$

123. No. The graph does not pass the vertical line test when $x = -2$.

125. Yes. The graph passes the vertical line test. $D = \{-6, -4, 2, 4\}$; $R = \{-4, 2\}$

127. Yes. Each value in the domain corresponds to exactly one value in the range.

129. No. The value 5 in the domain corresponds to more than one value in the range.

131. Walks away from home, then turns around and walks back a little slower.

133. See Figure 133.

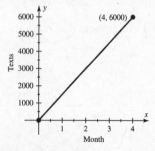

Figure 133

Section 2.2 Linear Functions

1. $mx+b$

3. line

5. 7

7. True

9. Carpet costs $2 per square foot. Ten square feet of carpet costs $20.

11. Yes; $m = \dfrac{1}{2}$, $b = -6$

13. No, the variable is raised to the power of 2

15. Yes; $m = 0$, $b = -9$

17. Yes; $m = -9$, $b = 0$

19. Yes. The graph is a straight line.

21. No. The graph is not a straight line.

23. Yes. For each unit increase in x, the values of $f(x)$ increase by 3 units, so $m = \dfrac{3}{1} = 3$.

   Because $f(x) = -6$ when $x = 0$, the y-intercept is $b = -6$. The function can be written

   $f(x) = 3x - 6$.

25. Yes. For each 2-unit increase in x, the values of $f(x)$ decrease by 3 units, so $a = -\dfrac{3}{2}$.

   Because $f(x) = 3$ when $x = 0$, the y-intercept is $b = 3$. The function can be written

   $f(x) = -\dfrac{3}{2}x + 3$.

27. No. For each unit increase in x, the values of $f(x)$ do not increase by a constant amount.

29. Yes. For each unit increase in x, the values of $f(x)$ increase by 2 units, so $m = \dfrac{2}{1} = 2$.

   Because $f(x) = 0$ when $x = 0$, the y-intercept is $b = 0$ The function can be written $f(x) = 2x$.

31. Yes. For each unit increase in x, the values of $f(x)$ increase by 0 units, so $m = 0$.

   Because $f(x) = -4$ when $x = 0$, the y-intercept is $b = -4$ The function can be written $f(x) = -4$.

33. $f(-4) = 4(-4) = -16$; $f(5) = 4(5) = 20$

35. $f\left(-\dfrac{2}{3}\right) = 5 - \left(-\dfrac{2}{3}\right) = \dfrac{15}{3} + \dfrac{2}{3} = \dfrac{17}{3}$; $f(3) = 5 - 3 = 2$

37. $f\left(-\dfrac{3}{4}\right) = -22$; $f(13) = -22$

39. $f(-1) = -2$; $f(0) = 0$

41. $f(-2) = -1$; $f(4) = -4$

43. $f(-3) = 1$; $f(1) = 1$

45. $f(x) = 6x$; $f(3) = 6(3) = 18$

47. $f(x) = \dfrac{x}{6} - \dfrac{1}{2}$; $f(3) = \dfrac{(3)}{6} - \dfrac{1}{2} = \dfrac{1}{2} - \dfrac{1}{2} = 0$

49. The graph should have a positive slope and pass through $(0, 0)$. d

51. The graph should have a positive slope and pass through $(0, -2)$. b

53. See Figure 53.

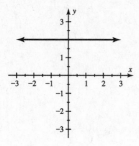

Figure 53

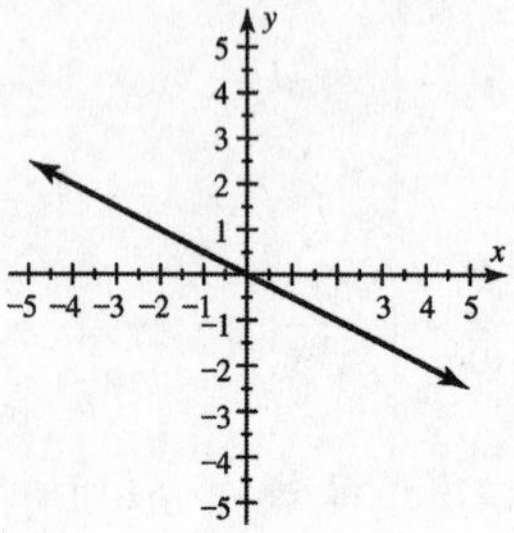

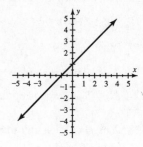

Figure 55

Figure 57

55. See Figure 55.

57. See Figure 57.

59. See Figure 59.

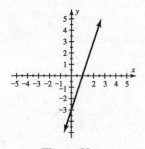

Figure 59

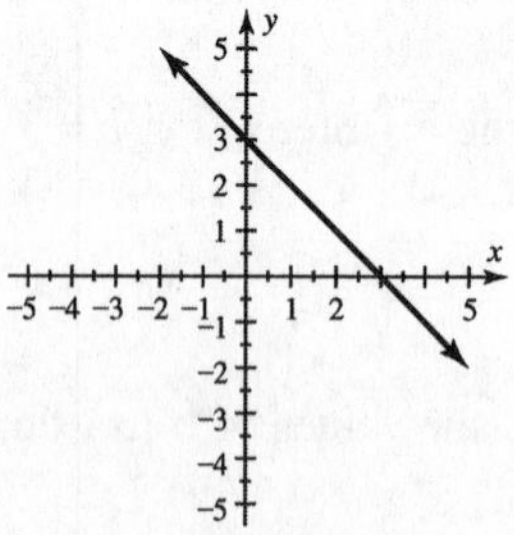

Figure 61

61. See Figure 61.

63. Since each pound is divided into 16 ounces: $f(x) = \dfrac{1}{16}x$

65. Since the car travels 65 miles each hour: $f(t) = 65t$

67. Since every day has 24 hours: $f(x) = 24$

69. *a*

71. (a)   *f* multiplies the input *x* by –2 and then adds 1 to obtain the output *y*.

    (b)   See Figure 71b.

    (c)   See Figure 71c.

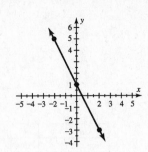

x	−2	0	2
y = f(x)	5	1	−3

Figure 71b   Figure 71c

73. (a) f multiplies the input x by $\frac{1}{2}$ and then subtracts 1 to obtain the output y.

    (b)   See Figure 73b.

    (c)   See Figure 73c.

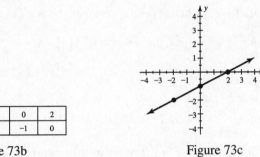

x	−2	0	2
y = f(x)	−2	−1	0

Figure 73b   Figure 73c

75.   (a)   (b)

                              [−6, 6, 1] by [−4, 4, 1]

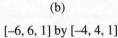
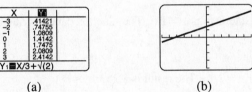

77.   (a)   (b)

                              [−6, 6, 1] by [−4, 4, 1]

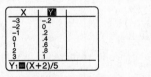
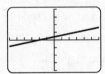

79. The graph should increase since the cost of tuition has been rising but it should not start at zero. b

81. The graph should be a horizontal line since this distance has not changed over the past 10 years. c

83. (a)   We divide the number of miles traveled by the miles per gallon. $G(x) = \dfrac{x}{E}$

    (b)   We will multiply the function that was created in part (a) by 3. $C(x) = \dfrac{3x}{E}$

85. (a)   Symbolic: $f(x) = 70$ Graphical: A graph of the function is shown in Figure 85a.

    (b)   A table of the function is shown in Figure 85b.

    (c)   The function f is a constant function.

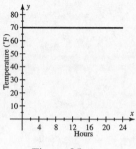

Figure 85a

Hours	0	4	8	12	16	20	24
Temp. (°F)	70	70	70	70	70	70	70

Figure 85b

87. For each 2-hour increase in t, the distance increases by 120 miles, so $m = \dfrac{120}{2} = 60$. Because

$D = 50$ when $t = 0$, the y-intercept is $b = 50$. The function can be written $D(t) = 60t + 50$.

89. (a) Because the average kid under 18 sent 93 texts per day, the function is $K(x) = 93x$.

    (b) Because the average adult sent 1 text per day, the function is $A(x) = x$.

    (c) $K(365) = 93(365) = 33,945$, On average, someone under 18 sends 33,945 texts in 1 year,

        while someone over 65 sends 365 texts.

91. (a) According to the table the sales in 2000 were 1.6 million.

    (b) For each unit increase in x, the values of $f(x)$ increase by 0.1 units, so $m = \dfrac{0.1}{1} = 0.1$.

    (c) For each unit increase in x, the values of $f(x)$ increase by 0.1 units, so $m = \dfrac{0.1}{1} = 0.1$.

        Because $f(x) = 1.6$ when $x = 0$, the y-intercept is $b = 1.6$ The function can be written

        $f(x) = 0.1x + 1.6$

    (d) Let $x = 6 \Rightarrow f(6) = 0.1(6) + 1.6 = 0.6 + 1.6 = 2.2$. The result is 2.2 million.

93. (a) Let $x = 4 \Rightarrow S(4) = 110(4) + 123 = 440 + 123 = 563$ million

    (b) In 2006 there were about 123 million people who used Skype.

    (c) The number of Skype users increased, on average, by 110 million per year.

95. (a) Because each 1°C increase in temperature results in a 0.5 cubic centimeter increase in volume,

        $m = 0.5$. Because the volume is 137 cubic centimeters when the temperature is 0°C, the y-

        intercept is $b = 137$. The formula is $V(T) = 0.5T + 137$.

    (b) $V(50) = 0.5(50) + 137 = 25 + 137 = 162$ cubic centimeters.

97. $f(x) = 40x$ In 30 days each additional pound of muscle will burn $40(30) = 1200$ calories. Then

the amount of muscle necessary to lose 1 pound of fat is $3500 \div 1200 \approx 2.92$ lb.

99. (a) According to the table 55% of those with a cell phone also subscribed to a data package.

    (b) The percentage changed by 4% per year.

(c)   For each year increase the percentage increases by 4, so $m = \dfrac{4}{1} = 4.$ Because in 2007 (the

starting year) the percentage is 55 the y-intercept is $b = 55$. The function can be written

$P(x) = 4x + 55$.

(d)   Let $x = 3 \Rightarrow P(3) = 4(3) + 55 = 12 + 55 = 67$ The result is 67% in 2010.

Checking Basic Concepts Sections 2.1 & 2.2

1. Symbolic: $f(x) = x^2 - 1$ Graphical: The graph is shown in Figure 1.

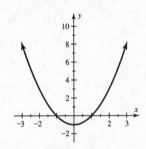
Figure 1

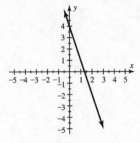
Figure 4

2. (a) $D : -3 \le x \le 3;\ R : -4 \le y \le 4$

  (b) $f(0) = 0;\ f(2) = 4$

  (c)   No. The graph is not a straight line.

3. (a)   Yes. The function is of the form $f(x) = mx + b.$

  (b)   No. The function can not be written in the form $f(x) = mx + b.$

  (c)   Yes. The function could be written $f(x) = 0x - 7$ which is of the form $f(x) = mx + b.$

  (d)   Yes. The function could be written $f(x) = 3x + 9$ which is of the form $f(x) = mx + b.$

4. The graph is shown in Figure 4. $f(-2) = 4 - 3(-2) = 4 + 6 = 10.$

5. For each 2-unit increase in x, the values of $f(x)$ increase by 1 unit, so $m = \dfrac{1}{2}.$ Because

$f(x) = -1$ when $x = 0$, the y-intercept is $b = -1$. The function can be written $f(x) = \dfrac{1}{2}x - 1.$

6. (a) $f(20) = 0.225(20) + 27.7 = 4.5 + 27.7 = 32.2;$ In 1990 the median age was about 32 years.

  (b)   The number 0.225 means that the median age increased by 0.225 year each year. The number 27.7 means that the initial median age in 1970 was 27.7 years.

Section 2.3 The Slope of a Line

1. y; x

3. rises

5. vertical

7. In 1999, \$9 billion was spent on pet health care.

9. In 2006, 123 million people used Skype.

11. 2. The y-value increases 2 units for every 1 unit increase in the x-value.

13. $-\dfrac{2}{3}$. The y-value decreases 2 units for every 3 units of increase in the x-value.

15. 0. This is a horizontal line.

17. $m = \dfrac{4-2}{2-1} = \dfrac{2}{1} = 2$

19. $m = \dfrac{3-1}{-1-2} = \dfrac{2}{-3} = -\dfrac{2}{3}$

21. $m = \dfrac{6-6}{4-(-3)} = \dfrac{0}{7} = 0$

23. $m = \dfrac{\frac{13}{17}-\left(-\frac{2}{7}\right)}{\frac{1}{2}-\frac{1}{2}} = \dfrac{\frac{125}{119}}{0} \Rightarrow$ Undefined

25. $m = \dfrac{\frac{1}{3}-\left(-\frac{4}{3}\right)}{\frac{1}{6}-\frac{1}{3}} = \dfrac{\frac{1}{3}+\frac{4}{3}}{\frac{1}{6}-\frac{2}{6}} = \dfrac{\frac{5}{3}}{-\frac{1}{6}} = \dfrac{5}{3}\cdot\left(-\dfrac{6}{1}\right) = -\dfrac{30}{3} = -10$

27. $m = \dfrac{16-10}{1999-1989} = \dfrac{6}{10} = \dfrac{3}{5}$

29. $m = \dfrac{4.3-3.6}{-1.2-2.1} = \dfrac{0.7}{-3.3} = -0.\overline{21}$

31. $m = \dfrac{4b-b}{3a-2a} = \dfrac{3b}{a}$

33. $m = \dfrac{b-0}{a-(a+b)} = \dfrac{b}{-b} = -1$

35. (a) $m = \dfrac{-1-2}{3-(-1)} = \dfrac{-3}{4}$

   (b) See Figure 35b.

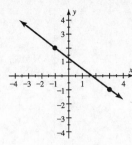

Figure 35b

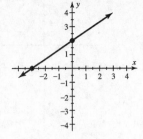

Figure 37b

(c)   The graph falls 3 units for every 4-unit increase in *x*.

37. (a) $m = \dfrac{0-2}{-3-0} = \dfrac{2}{3}$

(b)   See Figure 37b.

(c)   The graph rises 2 units for every 3-unit increase in *x*.

39. See Figure 39.

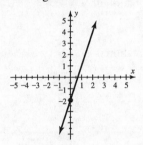

Figure 39

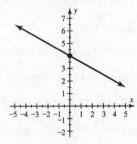

Figure 41

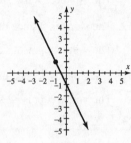

Figure 43

41. See Figure 41.

43. See Figure 43.

45. See Figure 45.

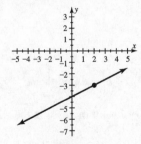

Figure 45

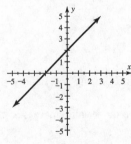

Figure 47

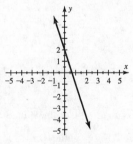

Figure 49

47. (a) $m = 1;\ b = 2$

(b)   See Figure 47.

49. (a) $m = -3;\ b = 2$

(b)   See Figure 49.

51. (a) $m = \dfrac{1}{3};\ b = 0$

(b)   See Figure 51.

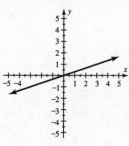

Figure 51

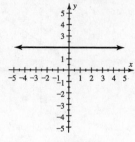

Figure 53

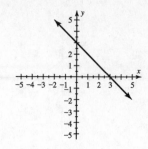

Figure 55

53. (a) $m = 0;\ b = 2$

(b)   See Figure 53.

55. (a) $m = -1;\ b = 3$

(b)   See Figure 55.

57. The slope is -1 since the y-value decreases 1 unit for every 1–unit increase in the x-value.

The y-intercept is 4 since the graph crosses the y-axis at $(0, 4)$. The equation is $y = -x + 4$.

59. The slope is 0 since this is a horizontal line.

The y-intercept is 2 since the graph crosses the y-axis at $(0, 2)$. The equation is $y = 2$.

61. The slope is 1 since the y-value increases 1 unit for every 1–unit increase in the x-value.

The y-intercept is -2 since the graph passes through $(0, -2)$. The equation is $y = x - 2$.

63. $y = 3x - 5$

65. The slope is $m = \dfrac{-\frac{3}{2} - 0}{0 - 1} = \dfrac{-\frac{3}{2}}{-1} = \dfrac{3}{2}$. The y-intercept is $-\dfrac{3}{2}$ since the graph crosses the y-axis at

$\left(0, -\dfrac{3}{2}\right)$. The equation is $y = \dfrac{3}{2}x - \dfrac{3}{2}$.

67. $m = \dfrac{0 - (-2)}{1 - 0} = \dfrac{2}{1} = 2;\ b = -2$

69. $m = \dfrac{-10 - (-3)}{2 - 1} = \dfrac{-7}{1} = -7$; By substituting $x = 1$ and $y = -3$ into $y = -7x + b$, we find that $b = 4$.

71. (a)   The missing value is 3 since the y-value increases 2 units for every 1–unit increase in the x

value $(m = 2)$.

(b)   The y-intercept is -1 since the graph passes through $(0, -1)$. The equation is $f(x) = 2x - 1$.

73. (a)   The missing number is 5 since the y-value increases 9 units for every 6 units of increase in the

x-value $\left(m = \dfrac{3}{2}\right)$.

(b)   The y-intercept is 5 since the graph passes through $(0, 5)$. The equation is $f(x) = \dfrac{3}{2}x + 5$.

75. The graph should be increasing since a person's pay increases with time and it should start at $(0, 0)$. c

77. The graph should be increasing since the world's population is increasing but it should not start at $(0, 0)$. b

79. (a) $m_1 = 50$, $m_2 = 0$, $m_3 = 150$ $\left(m_1 = \dfrac{300 - 100}{4 - 0} = \dfrac{200}{4} = 50, \, m_3 = \dfrac{600 - 300}{8 - 6} = \dfrac{300}{2} = 150 \right)$

   (b) m_1: the pump is adding water at the rate of 50 gallons per hour. m_2: the pump is turned off

   m_3: the pump is adding water at the rate of 150 gallons per hour.

   (c) Initially the pool contained 100 gallons of water. The pump added 200 gallons of water over the first 4 hours. Then the pump was turned off for 2 hours. Finally, the pump added 300 gallons of water over the last 2 hours.

81. (a) $m_1 = 100$, $m_2 = 25$, $m_3 = -100$

   $\left(m_1 = \dfrac{300 - 100}{2 - 0} = \dfrac{200}{2} = 100, \, m_2 = \dfrac{400 - 300}{6 - 2} = \dfrac{100}{4} = 25, \, m_3 = \dfrac{200 - 400}{8 - 6} = \dfrac{-200}{2} = -100 \right)$

   (b) m_1: the pump is adding water at the rate of 100 gallons per hour.

   m_2: the pump is adding water at the rate of 25 gallons per hour.

   m_3: the pump is removing water at the rate of 100 gallons per hour.

   (c) Initially the pool contained 100 gallons of water. The pump added 200 gallons of water over the first 2 hours. Then the pump added 100 gallons of water over the next 4 hours. Finally, the pump removed 200 gallons of water over the last 2 hours.

83. (a) $m_1 = 50$, $m_2 = 0$, $m_3 = -20$, $m_4 = 0$ $\left(m_1 = \dfrac{50 - 0}{1 - 0} = \dfrac{50}{1} = 50, \, m_3 = \dfrac{10 - 50}{4 - 2} = \dfrac{-40}{2} = -20 \right)$

   (b) m_1: the car is moving away from home at a rate of 50 miles per hour.

   m_2: the car is not moving.

   m_3: the car is moving toward home at a rate of 20 miles per hour.

   m_4: the car is not moving.

   (c) The car started at home and moved away from home at 50 mph for 1 hour to a location 50 miles from home. The car was then parked for 1 hour. Next the car moved toward home at 20 mph for 2 hours to a location 10 miles from home. Finally, the car was parked for 1 hour.

85. See Figure 85.

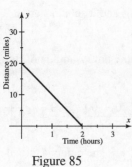

Figure 85

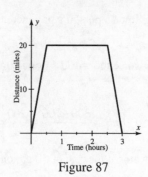

Figure 87

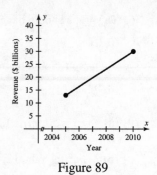

Figure 89

87. See Figure 87.

89. (a) $R(2005) = \dfrac{10}{3}(2005) - 6670 \approx 13.3$ billion, $R(2008) = \dfrac{10}{3}(2008) - 6670 \approx 23.3$ billion

   (b) See Figure 89.

   (c) $m = \dfrac{10}{3}$

   (d) Revenue increased by $\$3\dfrac{1}{3}$ billion per year.

91. (a) $t = 9 \Rightarrow M(9) = 18.8(9) + 299 = 468.2$ The result is approximately 468,000

   (b) The slope is 18.8 and the y-intercept is 299.

   (c) On average, the number of finishers increased by 18,800 per year.

   (d) In 2000 there were 299,000 finishers.

93. At the beginning of the 5-hour period, the pool contained $\dfrac{1}{4} \cdot 20,000 = 5000$ gallons of water. At the

   end of the 5-hour period, the pool contained $\dfrac{5}{8} \cdot 20,000 = 12,500$ gallons of water. That is, the

   amount of water in the pool increased by $12,500 - 5000 = 7500$ gallons in 5 hours. The rate is

   $\dfrac{7500}{5} = 1500$ gallons per hour.

95. Since the graph passes through the points (0, 0) and (1, 10), the slope is $m = \dfrac{10 - 0}{1 - 0} = \dfrac{10}{1} = 10.$

   The carpet costs $10 per square yard.

97. (a) $m = \dfrac{12 - 4}{2008 - 1984} = \dfrac{8}{24} = \dfrac{1}{3}$

   (b) On average, spending on pet health care increased by $\$\dfrac{1}{3}$ billion per year from 1984 to 2008.

99. (a) The value of m is equal to the slope of the line passing through the points (0, 35,000)

      and (17, 60,000). $m = \dfrac{60,000 - 35,000}{17 - 0} = \dfrac{25,000}{17} \approx 1470.6$

      The value of b is the initial value in 1990, $b = 35,000.$

(b) The year 2000 corresponds to $x = 10$.

$$f(10) = 1470.6(10) + 35,000 = 14,706 + 35,000 = \$49,706$$

Section 2.4 Equations of Lines and Linear Models

1. One

3. $y = mx + b$

5. $x = 5$

7. $y = b$

9. $m_1 = m_2$

11. The line must have slope $-\dfrac{3}{4}$. One example is $y = -\dfrac{3}{4}x - 3$. *Answers may vary.*

13. The line must be vertical and have an equation of the form $x = h$. For example, $x = 1$.
    Answers may vary.

15. From the given point, count 3 units down and 4 units to the right to return to the line. Thus

    $m = \dfrac{-3}{4} = -\dfrac{3}{4}$. Using the given point in the point-slope form gives

    $y = -\dfrac{3}{4}(x - (-3)) + 2$ or $y = -\dfrac{3}{4}x - \dfrac{1}{4}$.

17. From the given point, count 1 unit down and 3 units to the left to return to the line. Thus

    $m = \dfrac{-1}{-3} = \dfrac{1}{3}$. Using the given point in the point-slope form gives $y = \dfrac{1}{3}(x - 1) + 3$ or $y = \dfrac{1}{3}x + \dfrac{8}{3}$.

19. Yes. By substituting for x in the equation we see that when $x = -3$, $y = 4$.

21. No. By substituting for x in the equation we see that when $x = -4$, $y = 2 \neq 3$.

23. Since $m > 0$, and $b \neq 0$ the graph must be increasing. d

25. Since $m < 0$ and $b = 0$ the graph must be decreasing and it must pass through the origin. a

27. Since $m > 0$ and $b = 0$ the graph must be increasing and it must pass through the origin. f

29. Here $m = -2$, $x_1 = 2$ and $y_1 = -3$. The equation is $y = -2(x - 2) - 3$.

31. Here $m = 1.3$, $x_1 = 1990$ and $y_1 = 25$. The equation is $y = 1.3(x - 1990) + 25$.

33. Here $m = \dfrac{-1 - 3}{-5 - 1} = \dfrac{-4}{-6} = \dfrac{2}{3}$. Then either $x_1 = 1$ and $y_1 = 3$ or $x_1 = -5$ and $y_1 = -1$.

    The equation can be written as either $y = \dfrac{2}{3}(x - 1) + 3$ or $y = \dfrac{2}{3}(x + 5) - 1$.

35. Here $m = \dfrac{45 - 5}{2000 - 1980} = \dfrac{40}{20} = 2$. Then either $x_1 = 1980$ and $y_1 = 5$ or $x_1 = 2000$ and $y_1 = 45$.

    The equation can be written as either $y = 2(x - 1980) + 5$ or $y = 2(x - 2000) + 45$.

37. Here $m = \dfrac{4-0}{0-6} = \dfrac{4}{-6} = -\dfrac{2}{3}$. Then either $x_1 = 6$ and $y_1 = 0$ or $x_1 = 0$ and $y_1 = 4$.

    The equation can be written as either $y = -\dfrac{2}{3}(x-6)+0$ or $y = -\dfrac{2}{3}(x-0)+4$.

39. See Figure 39. $y = -\dfrac{1}{2}(x-0)+2$; Substitute the point $(-2,3) \Rightarrow$

    $3 = -\dfrac{1}{2}(-2-0)+2 \Rightarrow 3 = -\dfrac{1}{2}(-2)+2 \Rightarrow 3 = 1+2 \Rightarrow 3 = 3$ The point does lie on the line.

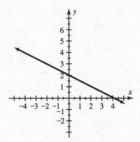

    Figure 39

41. $y = 2(x-1)-2 \Rightarrow y = 2x-2-2 \Rightarrow y = 2x-4$

43. $y = \dfrac{1}{2}(x+4)+1 \Rightarrow y = \dfrac{1}{2}x+2+1 \Rightarrow y = \dfrac{1}{2}x+3$

45. $y = 22(x-1.5)-10 \Rightarrow y = 22x-33-10 \Rightarrow y = 22x-43$

47. $m = \dfrac{-8-7}{2-(-3)} = \dfrac{-15}{5} = -3$; $y = mx+b \Rightarrow 7 = -3(-3)+b \Rightarrow 7 = 9+b \Rightarrow -2 = b \Rightarrow y = -3x-2$

49. $m = \dfrac{-5-1}{4-(-4)} = \dfrac{-6}{8} = -\dfrac{3}{4}$, $y = mx+b \Rightarrow 1 = -\dfrac{3}{4}(-4)+b \Rightarrow 1 = 3+b \Rightarrow -2 = b \Rightarrow y = -\dfrac{3}{4}x-2$

51. $y = -\dfrac{1}{3}(x-0)-5 \Rightarrow y = -\dfrac{1}{3}x+0-5 \Rightarrow y = -\dfrac{1}{3}x-5$

53. $m = \dfrac{-1-(-2)}{2-3} = \dfrac{1}{-1} = -1 \Rightarrow y = -(x-3)-2 \Rightarrow y = -x+3-2 \Rightarrow y = -x+1$

55. The y-intercept is given. The point-slope form is not needed. $m = \dfrac{-\frac{2}{3}-0}{0-2} = \dfrac{-\frac{2}{3}}{-2} = \dfrac{1}{3} \Rightarrow y = \dfrac{1}{3}x - \dfrac{2}{3}$

57. The y-intercept is given. The point-slope form is not needed. $m = \dfrac{3-0}{0-1} = \dfrac{3}{-1} = -3 \Rightarrow y = -3x+3$

59. Parallel lines have the same slope. $m = 4 \Rightarrow y = 4(x-1)+3 \Rightarrow y = 4x-4+3 \Rightarrow y = 4x-1$

61. Parallel lines have the same slope. $m = \dfrac{-2-3}{1-(-2)} = -\dfrac{5}{3}$, $y = mx+b \Rightarrow 2 = -\dfrac{5}{3}(-3)+b \Rightarrow$

    $2 = 5+b \Rightarrow -3 = b \Rightarrow y = -\dfrac{5}{3}x-3$

63. The product of the slopes of two perpendicular lines is –1.

$$m = 3 \Rightarrow y = 3(x+3)+5 \Rightarrow y = 3x+9+5 \Rightarrow y = 3x+14$$

65. The line passing through (–1, 6) and (8, –4) has slope $\dfrac{-4-6}{8-(-1)} = -\dfrac{10}{9}$. The product of the slopes of

two perpendicular lines is –1. $m = \dfrac{9}{10} \Rightarrow y = \dfrac{9}{10}\left(x+\dfrac{1}{2}\right)-2 \Rightarrow y = \dfrac{9}{10}x+\dfrac{9}{20}-\dfrac{40}{20} \Rightarrow y = \dfrac{9}{10}x-\dfrac{31}{20}$

67. The product of the slopes of two perpendicular lines is –1. $m = -\dfrac{1}{c} \Rightarrow y = -\dfrac{1}{c}x+b$

69. (a)   The slope of the perpendicular line must be the negative reciprocal of $\dfrac{1}{2}$. That is $m = -2$.

    Here $x_1 = 0$ and $y_1 = 2$. The equation is $y = -2(x-0)+2$ or $y = -2x+2$.

    (b)   See Figure 69.

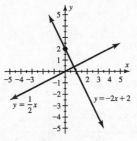   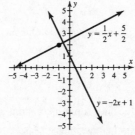   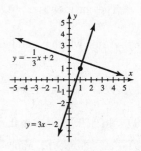

Figure 69   Figure 71   Figure 73

71. (a)   The slope of the perpendicular line must be the negative reciprocal of –2. That is $m = \dfrac{1}{2}$.

    Here $x_1 = -1$ and $y_1 = 2$. The equation is $y = \dfrac{1}{2}(x+1)+2$ or $y = \dfrac{1}{2}x+\dfrac{5}{2}$.

    (b)   See Figure 71.

73. (a)   The slope of the perpendicular line must be the negative reciprocal of $-\dfrac{1}{3}$. That is $m = 3$.

    Here $x_1 = 1$ and $y_1 = 1$. The equation is $y = 3(x-1)+1$ or $y = 3x-2$.

    (b)   See Figure 73.

75. (a)   The slope of y_1 is $m_1 = \dfrac{2-0}{1-0} = \dfrac{2}{1} = 2$ and the y-intercept is $b = 0$. The equation is $y_1 = 2x$.

    The slope of y_2 is $m_2 = \dfrac{1-0}{-2-0} = \dfrac{1}{-2} = -\dfrac{1}{2}$ and the y-intercept is $b = 0$. The equation is

    $y_2 = -\dfrac{1}{2}x$.

    (b)   The two lines are perpendicular since $m_1 m_2 = 2\left(-\dfrac{1}{2}\right) = -1$.

77. (a)   The slope of y_1 is $m_1 = \dfrac{2-(-1)}{-2-0} = \dfrac{3}{-2} = -\dfrac{3}{2}$ and the y-intercept is $b = -1$. The equation is

$y_1 = -\dfrac{3}{2}x - 1$. The slope of y_2 is $m_2 = \dfrac{1-(-1)}{3-0} = \dfrac{2}{3}$ and the y-intercept is $b = -1$. The

equation is $y_2 = \dfrac{2}{3}x - 1$.

   (b)   The two lines are perpendicular since $m_1 m_2 = -\dfrac{3}{2}\left(\dfrac{2}{3}\right) = -1$.

79. The equation of a vertical line is $x = h$. The equation is $x = -1$.

81. The equation of a horizontal line is $y = k$. The equation is $y = -\dfrac{5}{6}$.

83. A line which is perpendicular to a horizontal line is vertical and has equation $x = h$.

    The equation is $x = 4$.

85. A line which is parallel to a vertical line is vertical and has equation $x = h$. The equation is $x = -\dfrac{2}{3}$.

87. Yes. $m = \dfrac{0-(-4)}{2-1} = \dfrac{4}{1} = 4 \;\Rightarrow\; y = 4(x-2) + 0 \;\Rightarrow\; y = 4x - 8$

89. No. The change in the y-values does not remain constant as the x-values increase.

91. (a)   Since the graph is decreasing, water is leaving the tank. There are 50 gallons in the tank after 3

         minutes.

   (b)   The x-intercept is 8: After 8 minutes the tank is empty.

         The y-intercept is 80: Initially the tank contains 80 gallons of water.

   (c) $m = \dfrac{80-0}{0-8} = \dfrac{80}{-8} = -10 \;\Rightarrow\; y = -10x + 80$

         The water is leaving the tank at a rate of 10 gallons per minute.

   (d) $D: 0 \le x \le 8;\; R: 0 \le y \le 80$

93. (a)   Two acres of land have 100 people and 4 acres of land have 200 people, on average.

   (b)   A zero-acre parcel has no people.

   (c)   The slope is $m = \dfrac{200-100}{4-2} = \dfrac{100}{2} = 50$, and the y-intercept is $b = 0$. The equation is $y = 50x$.

   (d)   The land had 50 people per acre, on average.

   (e) $P(x) = 50x$

95. (a)

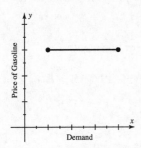

(b)   The graph is a horizontal line.

97. (a)   The slope is $m = \dfrac{38,000 - 28,000}{2011 - 2006} = \dfrac{10,000}{5} = 2000$. Using the point (2006, 28,000) the

equation is $y = 2000(x - 2006) + 28,000$ or $y = 2000x - 3,984,000$.

(b)   The cost is increasing by \$2000 per year, on average.

(c)   In 2015 the cost is estimated to be $y = 2000(2015) - 3,984,000 = \$46,000$.

99. First find a linear function that can be used to model these data. Then use the function to find the

value in 2008. The slope is $m = \dfrac{60 - 57}{2010 - 2005} = \dfrac{3}{5} = 0.6$. Using the point (2005, 57) the equation is

$f(x) = 0.6(x - 2005) + 57$ or $f(x) = 0.6x - 1146$. The estimated chicken consumption in 2008

is $f(2008) = 0.6(2008) - 1146 = 58.8$ pounds.

101. First find a linear function that can be used to model these data. Then use the function to find the

value in 1980. The slope is $m = \dfrac{78 - 50}{2010 - 1910} = \dfrac{28}{100} = 0.28$. Using the point (1910, 50) the equation

is $f(x) = 0.28(x - 1910) + 50$ or $f(x) = 0.28x - 484.8$. The estimated life expectancy of a baby

born in 1980 is $f(1980) = 0.28(1980) - 484.8 = 69.6$ years.

103. Using the points (1, 1) and (4, 13) the slope is $m = \dfrac{13 - 1}{4 - 1} = \dfrac{12}{3} = 4$.

Then using the point (1, 1) the equation is $y = 4(x - 1) + 1$ or $y = 4x - 3$.

105. Using the points (1, 8) and (5, –6) the slope is $m = \dfrac{-6 - 8}{5 - 1} = \dfrac{-14}{4} = -3.5$.

Then using the point (1, 8) the equation is $y = -3.5(x - 1) + 8$ or $y = -3.5x + 11.5$.

107. Using the points (1970, 1.4) and (2010, 5.8) the slope is $m = \dfrac{5.8 - 1.4}{2010 - 1970} = \dfrac{4.4}{40} = 0.11$.

Then using the point (1970, 1.4) the equation is $y = 0.11(x - 1970) + 1.4$ or $y = 0.11x - 215.3$.

109. Using the points (1950, 1.7) and (2010, 7.1) the slope is $m = \dfrac{7.1 - 1.7}{2010 - 1950} = \dfrac{5.4}{60} = 0.09$.

Then using the point (1950, 1.7) the equation is $y = 0.09(x - 1950) + 1.7$ or $y = 0.09x - 173.8$.

111. (a) The cost per mile is $a = 0.30$. The fixed cost is $b = 189.20$.

(b) The y-intercept represents the fixed cost of owning the car for one month.

113. (a) $m = \dfrac{7 - 5}{5.4 - 4.9} = \dfrac{2}{0.5} = 4 \Rightarrow R(c) = 4(c - 4.9) + 5 \Rightarrow R(c) = 4c - 19.6 + 5 \Rightarrow R(c) = 4c - 14.6$

(b) $R(6.16) = 4(6.16) - 14.6 = 10.04$; the ring size is about 10.

115. (a) See Figure 115.

(b) Using the points (1980, 227) and (2010, 308) the slope is $m = \dfrac{308 - 227}{2010 - 1980} = \dfrac{81}{30} = 2.7$.

Then using the point (1980, 227), $x_1 = 1980$ and $y_1 = 227$. *Answers may vary.*

(c) The population in 2020 is estimated to be $f(2020) = 2.7(2020 - 1980) + 227 = 335$ million.

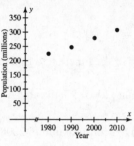

Figure 115

117. (a) Using the points (2008, 12) and (1984, 4) the slope is $m = \dfrac{12 - 4}{2008 - 1984} = \dfrac{8}{24} = \dfrac{1}{3}$.

Then using the point (1984, 4) $\Rightarrow f(x) = \dfrac{1}{3}(x - 1984) + 4$.

(b) $f(2014) = \dfrac{1}{3}(2014 - 1984) + 4 = 14$, The result is \$14 billion.

Checking Basic Concepts Sections 2.3 and 2.4

1. (a) The equation is in point-slope form, $m = -3$. One point on this line is (5, 7). *Answers may vary.*

(b) This is a horizontal line, $m = 0$. One point on this line is (1, 10). *Answers may vary.*

(c) This is a vertical line, m is undefined. One point on this line is (–5, 0). *Answers may vary.*

(d) The equation is in slope-intercept form, $m = 5$. One point on this line is (0, 3). *Answers may vary.*

2. (a) $m = \dfrac{2 - (-4)}{5 - 2} = \dfrac{6}{3} = 2$

(b) For the point-slope form of the equation either $y = 2(x - 2) - 4$ or $y = 2(x - 5) + 2$.

The slope-intercept form is $y = 2(x - 2) - 4 \Rightarrow y = 2x - 4 - 4 \Rightarrow y = 2x - 8$.

(c)   x-intercept: $0 = 2x - 8 \Rightarrow 8 = 2x \Rightarrow x = 4$; y-intercept: $y = 2(0) - 8 \Rightarrow y = -8$

3. A vertical line has an equation of the form $x = h$. The equation is $x = -2$.

   A horizontal line has an equation of the form $y = k$. The equation is $y = 5$.

4. The given line has slope $m = -\dfrac{1}{2}$; thus a parallel line has slope $m = -\dfrac{1}{2}$ and a perpendicular line

   has slope $m = 2$. Perpendicular: $y = 2(x - 2) + (-4) \Rightarrow y = 2x - 4 - 4 \Rightarrow y = 2x - 8$

   Parallel: $y = -\dfrac{1}{2}(x - 2) + (-4) \Rightarrow y = -\dfrac{1}{2}x + 1 - 4 \Rightarrow y = -\dfrac{1}{2}x - 3$

5. (a)   Since the line is decreasing, the distance from home is decreasing. The car is moving toward home.

   (b) $m = -50$ since the y-value decreases 50 units for every unit increase in the x-value. The car is moving toward home at 50 mph.

   (c)   x-intercept: 5; after 5 hours the car is home. y-intercept: 250; The car is initially 250 miles from home.

   (d)   The slope is $a = -50$ and the y-intercept is $b = 250$.

   (e) $D: 0 \le x \le 5; R: 0 \le y \le 250$

Chapter 2 Review

1. $f(-2) = 3(-2) - 1 = -7;\ f\left(\dfrac{1}{3}\right) = 3\left(\dfrac{1}{3}\right) - 1 = 0$

2. $f(-3) = 5 - 3(-3)^2 = 5 - 27 = -22;\ f(1) = 5 - 3(1)^2 = 5 - 3 = 2$

3. $f(0) = \sqrt{0} - 2 = -2;\ f(9) = \sqrt{9} - 2 = 3 - 2 = 1$

4. $f(-5) = 5;\ f\left(\dfrac{7}{5}\right) = 5$

5. (a)   Since there are 2 pints in a quart, $P(q) = 2q$.

   (b) $P(5) = 2(5) = 10$. There are 10 pints in 5 quarts.

6. (a)   Three less than four times a number is written $f(x) = 4x - 3$.

   (b) $f(5) = 4(5) - 3 = 17$. Three less than four times five is 17.

7. $(3, -2)$

8. $f(4) = -6$; the answers are 4 and −6.

9. See Figure 9.

10. See Figure 10.

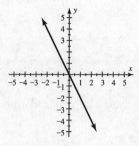

Figure 9

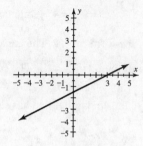

Figure 10

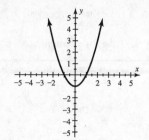

Figure 11

11. See Figure 11.

12. See Figure 12.

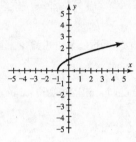

Figure 12

13. $f(0) = 1$; $f(-3) = 4$

14. $f(-2) = 1$; $f(1) = -2$

15. $f(-1) = 7$; $f(3) = -1$

16. Numerical: The table is shown in Figure 16a.   Symbolic: $f(x) = 3x - 2$

    Graphical: The graph is shown in Figure 16b.

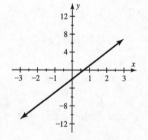

x	-3	-2	-1	0	1	2	3
$y = f(x)$	-11	-8	-5	-2	1	4	7

Figure 16a   Figure 16b

17. D: all real numbers; R: $y \le 4$

18. D: $-4 \le x \le 4$; R: $-4 \le y \le 0$

19. Yes. The graph passes the vertical line test.

20. No. The graph does not pass the vertical line test.

21. $D = \{-3, -1, 2, 4\}$; $R = \{-1, 3, 4\}$; Yes, S is a function since each input has exactly one output.

22. $D = \{-1, 0, 1, 2\}$; $R = \{-2, 2, 3, 4, 5\}$; No, S is not a function because the input -1 has more than one

    output

23. Any real number is a valid input for this function. D: all real numbers.

24. The radicand must be greater than or equal to zero. $D: x \geq 0$.

25. The denominator of this function cannot equal zero. $D: x \neq 0$.

26. Any real number is a valid input for this function because the variable is squared.

    D: all real numbers.

27. The radicand must be greater than or equal to zero. $D: x \leq 5$

28. The denominator of this function cannot equal zero. $D: x \neq -2$

29. Any real number is a valid input for this function. D: all real numbers

30. Any real number is a valid input for this function. D: all real numbers

31. No. The graph is not a straight line.

32. Yes. The graph is a straight line.

33. This function is linear because it is in the form $f(x) = mx + b$ with $m = -4$ and $b = 5$.

34. This function is linear because it can be written in the form $f(x) = mx + b$ with $m = -1$ and $b = 7$.

35. This function is not linear because it contains a square root.

36. This function is linear because it can be written in the form $f(x) = mx + b$ with $m = 0$ and $b = 6$.

37. Yes. For each 2-unit increase in x, the values of $f(x)$ increase by 3 units, so $m = \dfrac{3}{2}$.

    Because $f(x) = -3$ when $x = 0$, the y-intercept is $b = -3$.

    The function can be written $f(x) = \dfrac{3}{2}x - 3$.

38. No. For each unit increase in x, the values of $f(x)$ do not increase by a constant amount.

39. $f(-4) = \dfrac{1}{2}(-4) + 3 = -2 + 3 = 1$

40. $f(-2) = -3$ and $f(1) = 0$

41. See Figure 41.

42. See Figure 42.

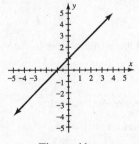

Figure 41

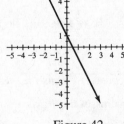

Figure 42

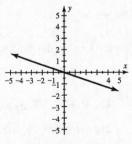

Figure 43

43. See Figure 43.

44. See Figure 44.

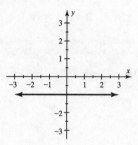 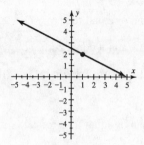 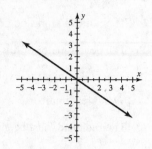

Figure 44   Figure 55   Figure 56

45. There are 24 hours in each day, so $H(x) = 24x$. $H(2) = 24(2) = 48$; there are 48 hours in 2 days.

46. (a)   See Figure 46a.   (b) See Figure 46b. Domain: $x \geq -2$

$[-10, 10, 1]$ by $[-10, 10, 1]$

 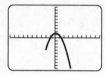

Figure 46a   Figure 46b

47. -2. The y-value decreases 2 units for every 1–unit increase in the x-value.

48. 0. This is a horizontal line.

49. $\dfrac{1}{3}$. The y-value increases 1 unit for every 3 units of increase in the x-value.

50. Undefined. This is a vertical line.

51. $m = \dfrac{8-2}{3-(-1)} = \dfrac{6}{4} = \dfrac{3}{2}$

52. $m = \dfrac{-\frac{1}{2} - \frac{5}{2}}{1-(-3)} = \dfrac{-\frac{6}{2}}{4} = \dfrac{-3}{4} = -\dfrac{3}{4}$

53. $m = \dfrac{-4-(-4)}{5-3} = \dfrac{0}{2} = 0$

54. $m = \dfrac{8-6}{-2-(-2)} = \dfrac{2}{0} \Rightarrow$ undefined

55. See Figure 55.

56. By writing the equation in the form $y = -\dfrac{2}{3}x + 0$, the slope is $-\dfrac{2}{3}$ and the y-intercept is 0. See

Figure 56.

57. The slope is -3 since the y-value decreases 3 units for every 1–unit increase in the x-value.

The y-intercept is 1 since the graph crosses the y-axis at $(0, 1)$. The equation is $y = -3x + 1$.

58. The slope is $m = \dfrac{2-0}{0-(-1)} = \dfrac{2}{1} = 2$. The y-intercept is 2. The equation is $y = 2x + 2$.

59. The slope is –1 since the y-value decreases 1 unit for every 1–unit increase in the x-value. The y-intercept is 1 since the graph crosses the y-axis at (0, 1).

60. (a) $m_1 = \dfrac{1000 - 500}{2 - 0} = \dfrac{500}{2} = 250$; Similarly, $m_2 = 500$, $m_3 = 0$, $m_4 = -750$

(b) m_1 : Water is being added to the pool at a rate of 250 gallons per hour. m_2 : Water is being added to the pool at a rate of 500 gallons per hour. m_3 : No water is being added or removed. m_4 : Water is being removed from the pool at a rate of 750 gallons per hour.

(c) Initially the pool contains 500 gallons of water. For the first two hours water is added to the pool at a rate of 250 gallons per hour until there are 1000 gallons in the pool. Then water is added at a rate of 500 gallons per hour for 1 hour until there are 1500 gallons in the pool. For the next hour, no water is added or removed. Finally, water is removed from the pool at a rate of 750 gallons per hour for 1 hour until it contains 750 gallons.

61. See Figure 61.

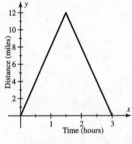

Figure 61   Figure 62

62. (a) The equation is in the form $f(x) = mx + b$. The slope is $\dfrac{1}{2}$ and the y-intercept is -2.

(b) See Figure 62.

63. Yes, the point (2, 1) lies on the graph because when $x = 2$, $y = \dfrac{3}{2}(2) - 2 = 3 - 2 = 1$.

64. The slope is $m = \dfrac{4 - 6}{0 - (-1)} = \dfrac{-2}{1} = -2$. From the table, the x-intercept is 2 and the y-intercept is 4.

65. $y = -3(x - 2) + 1 \Rightarrow y = -3x + 6 + 1 \Rightarrow y = -3x + 7$

66. $y = -3(x + 2) + 3$; $y = -3x - 6 + 3 \Rightarrow y = -3x - 3$

67. Here $m = \dfrac{-3 - 0}{0 - 2} = \dfrac{-3}{-2} = \dfrac{3}{2}$. Use $x_1 = 2$ and $y_1 = 0$ in the point-slope form to get the slope-intercept form. The slope-intercept equation is $y = \dfrac{3}{2}(x - 2) + 0 \Rightarrow y = \dfrac{3}{2}x - 3$.

68. Here $m = \dfrac{-2 - 4}{2 - (-1)} = \dfrac{-6}{3} = -2$. Use $x_1 = -1$ and $y_1 = 4$ in the point-slope form to get the slope intercept form. The slope-intercept equation is $y = -2(x + 1) + 4 \Rightarrow y = -2x + 2$.

69. Parallel lines have the same slope, thus $m = 4$. Use $x_1 = -\dfrac{3}{5}$ and $y_1 = \dfrac{1}{5}$ in the point-slope form to get the slope-intercept form. The slope-intercept equation is

$$y = 4\left(x + \frac{3}{5}\right) + \frac{1}{5} \;\Rightarrow\; y = 4x + \frac{12}{5} + \frac{1}{5} \;\Rightarrow\; y = 4x + \frac{13}{5}.$$

70. The product of the slopes of two perpendicular lines is -1 thus $m = -2$. Then $x_1 = -1$ and $y_1 = 1$. The slope-intercept equation is $y = -2(x+1) + 1 \;\Rightarrow\; y = -2x - 2 + 1 \;\Rightarrow\; y = -2x - 1.$

71. The slope of the first line is $m = \dfrac{3-1}{-2-(2)} = \dfrac{2}{-4} = -\dfrac{1}{2}$. The product of the slopes of two perpendicular lines is -1 thus $m = 2$. Then $x_1 = 3$ and $y_1 = 1$. The slope-intercept equation is

$$y = 2(x-3) + 1 \;\Rightarrow\; y = 2x - 6 + 1 \;\Rightarrow\; y = 2x - 5.$$

72. The slope of the first line is $m = \dfrac{-4-0}{-3-(3)} = \dfrac{-4}{-6} = \dfrac{2}{3}$. The slopes of parallel lines are equal thus

$m = \dfrac{2}{3}$. Then $x_1 = 3$ and $y_1 = 3$ and the slope-intercept equation is

$$y = \frac{2}{3}(x-3) + 3 \;\Rightarrow\; y = \frac{2}{3}x - 2 + 3 \;\Rightarrow\; y = \frac{2}{3}x + 1.$$

73. By inspection we see that the slope is $m = -1$. The y-intercept is 2. The equation is $y = -x + 2$.

74. By inspection we see that the slope is $m = 2$. The y-intercept is -3. The equation is $y = 2x - 3$.

75. No. By substituting for x in the equation we see that when $x = 2$, $y = 1 \neq -1$.

76. Yes. By substituting for x in the equation we see that when $x = 4$, $y = -\dfrac{5}{2}$.

77. The equation of a vertical line is $x = h$. The equation is $x = -4$.

78. The equation of a horizontal line is $y = k$. The equation is $y = -\dfrac{7}{13}$.

79. The line $x = -3$ is vertical, so a line perpendicular to it is horizontal. The horizontal line through $(-2, 1)$ is $y = 1$.

80. The line $y = 5$ is horizontal, so a line parallel to it is also horizontal. The horizontal line through $(4, -8)$ is $y = -8$.

81. No. The change in the y-values does not remain constant as the x-values increase.

82. Yes. $m = \dfrac{5-1}{0-(-2)} = \dfrac{4}{2} = 2 \;\Rightarrow\; y = 2(x-0) + 5 \;\Rightarrow\; y = 2x + 5$

83. (a) $m_1 = 1.3, m_2 = 2.35, m_3 = 2.4, m_4 = 2.7$ $\left(\text{i.e. } m_1 = \dfrac{132 - 106}{1940 - 1920} = \dfrac{26}{20} = 1.3\right)$

    (b) m_1: from 1920 to 1940 the population increased on average by 1.3 million per year.

m_2 : from 1940 to 1960 the population increased on average by 2.35 million per year.

m_3 : from 1960 to 1980 the population increased on average by 2.4 million per year.

m_4 : from 1980 to 2000 the population increased on average by 2.7 million per year.

84. For the first 2 minutes the inlet pipe is open. For the next 3 minutes both pipes are open. For the next 2 minutes only the outlet pipe is open. Finally, for the last 3 minutes both pipes are closed.

85. (a) $f(1910) = -0.0492(1910) + 119.1 \approx 25.1$ years

    (b)   Graph $Y_1 = -0.0492X + 119.1$ in [1885, 1965, 10] by [22, 26, 1]. See Figure 85.

          The median age at first marriage for males has decreased over this time period.

    (c)   The slope is –0.0492. The median age decreased by about 0.0492 year per year.

[1885, 1965, 10] by [22, 26, 1]

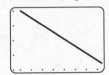

Figure 85

86. (a) $f(2006) = 2.2$ million

    (b)   The number of marriages each year did not change over this time period.

87. (a) $f(x) = 8x$

    (b)   The slope of the graph of f is 8.

    (c)   The total fat changes at a rate of 8 grams per cup of milk.

88. (a)   See Figure 88.

    (b)   Using the points (1950, 24.1) and (2010, 13.5) the slope is $m = \dfrac{13.5 - 24.1}{2010 - 1950} = \dfrac{-10.6}{60} \approx 0.176$.

          Then using the point (1950, 24.1) the equation is

          $y = -0.176(x - 1950) + 24.1$ or $y = -0.176x + 367.3$.

    (c) $f(2000) = -0.176(2000 - 1950) + 24.1 = -0.176(50) + 24.1 = 15.3$

          The birth rate is about 15 per 1000 people. *Answers may vary.*

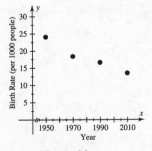

Figure 88

89. (a)   From the table $f(1995) = 113$. In 1995 there were 113 unhealthy days.

    (b) $D = \{1995, 1999, 2000, 2003, 2007\}$; $R = \{56, 87, 88, 100, 113\}$

(c)   The number of unhealthy days decreased and then increased over this time period.

90. (a)   See Figure 90. There is a linear relationship.

(b)   Using the points (0, 32) and (100, 212) the slope is $m = \dfrac{212 - 32}{100 - 0} = \dfrac{180}{100} = \dfrac{9}{5}$.

The y-intercept is 32. The function is given by $f(x) = \dfrac{9}{5}x + 32$. The slope of $\dfrac{9}{5}$ means that a

1°C change equals a $\dfrac{9}{5}$°F change.

(c) $f(20) = \dfrac{9}{5}(20) + 32 = 36 + 32 = 68°$ F.

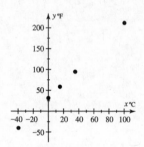
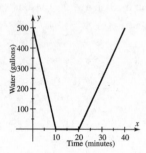

Figure 90   Figure 97

91. The graph should be increasing and should initially be a positive amount. b

92. The graph should be a horizontal line above the x-axis since this distance is a positive constant. e

93. The graph should be increasing and should start at zero. a

94. The graph should be decreasing since this film format is nearly extinct. d

95. The graph should be a horizontal line below the x-axis since this temperature is a negative constant. f

96. The graph should start below the x-axis and then increase to a value above the x-axis. c

97. See Figure 97.

98. (a)   Because the rate of change is 56,000 cases per year, $m = 56,000$. Here

$x_1 = 2006$ and $y_1 = 1,100,000$ The equation is $f(x) = 56,000(x - 2006) + 1,100,000$.

(b) $f(2015) = 56,000(2015 - 2006) + 1,100,000 = 1,604,000$ cases. This is the total (cumulative)

number of U.S. AIDS cases reported in 2015.

Chapter 2 Test

1. $f(4) = 3(4)^2 - \sqrt{4} = 3(16) - 2 = 48 - 2 = 46; \; (4, 46)$

2. $C(x) = 4x; \; C(5) = 4(5) = 20; \;$ 5 pounds of candy costs \$20.

3. (a)   See Figure 3a.

(b)   See Figure 3b.

(c)   See Figure 3c.

(d)   See Figure 3d.

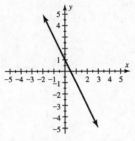

Figure 3a

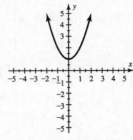

Figure 3b

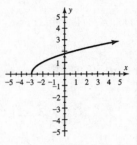

Figure 3c

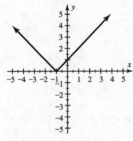

Figure 3d

4. $f(-3)=0;\ f(0)=-3;\ D:\{x|-3\le x\le 3\}$ and $R:\{x|-3\le y\le 0\}$

5. Symbolic: $f(x)=x^2-5$ Numerical: The table is shown in Figure 5a.

Graphical: The graph is shown in Figure 5b.

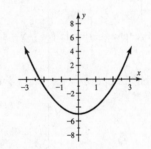

x	-3	-2	-1	0	1	2	3
$y=f(x)$	4	-1	-4	-5	-4	-1	4

Figure 5a   Figure 5b

6. No. It does not pass the vertical line test.

7. (a) $D=\{-2,-1,0,5\},\ R=\{3,5,7\}$

    (b)   Any real number is a valid input for this function. D: all real numbers

    (c)   The radicand must be greater than or equal to zero. $D: x \ge -4$

    (d)   Any real number is a valid input for this function. D: all real numbers

    (e)   The denominator of this function cannot equal zero. $D: x \ne 5$

8. It is; $f(x)=-8x+6$

9. $m=\dfrac{-2-7}{6-(-3)}=\dfrac{-9}{9}=-1$

10. $-\dfrac{1}{2}$. The y-value decreases 1 unit for every 2 units of increase in the x-value.

11. (a)   The line passes through $(-2, 0)$ and $(0, -4)$. $m = \dfrac{-4-0}{0-(-2)} = \dfrac{-4}{2} = -2 \Rightarrow y = -2x - 4$

    (b)   The line passing through $(-2, 5)$ and $(-1, 3)$ has slope $\dfrac{3-5}{-1-(-2)} = \dfrac{-2}{1} = -2$. The product of

    the slopes of perpendicular lines is -1.

    $$m = \dfrac{1}{2} \Rightarrow y = \dfrac{1}{2}(x+5)+2 \Rightarrow y = \dfrac{1}{2}x + \dfrac{5}{2} + 2 \Rightarrow y = \dfrac{1}{2}x + \dfrac{9}{2}$$

    (c) $m = \dfrac{\frac{3}{2}-(-2)}{-5-1} = \dfrac{\frac{3}{2}+\frac{4}{2}}{-6} = \dfrac{\frac{7}{2}}{-6} = \dfrac{7}{2} \cdot \left(-\dfrac{1}{6}\right) = -\dfrac{7}{12}$

    $$\Rightarrow y = -\dfrac{7}{12}(x-1)-2 \Rightarrow y = -\dfrac{7}{12}x - \dfrac{17}{12}$$

12. The slope is $m = \dfrac{0-4}{2-0} = \dfrac{-4}{2} = -2$. From the table, the x-intercept is 2 and the y-intercept is 4.

13. A line parallel to the given line has slope $m = -3$.

    The slope-intercept equation is $y = -3\left(x - \dfrac{1}{3}\right) + 2 \Rightarrow y = -3x + 1 + 2 \Rightarrow y = -3x + 3$.

14. By inspection the line has slope $m = \dfrac{2}{3}$. The y-intercept is -2.

    The slope-intercept equation is $y = \dfrac{2}{3}x - 2$. A line perpendicular to this line that passes through the

    origin would have slope $m = -\dfrac{3}{2}$ and y-intercept 0. The equation of the perpendicular line

    is $y = -\dfrac{3}{2}x$.

15. A vertical line has equation $x = h$. The equation is $x = \dfrac{2}{3}$. A horizontal line has equation

    $y = k$. The equation is $y = -\dfrac{1}{7}$.

16. (a) $m_1 = \dfrac{11-7}{1980-1970} = \dfrac{4}{10} = 0.4$; Similarly, $m_2 = 0$, $m_3 = -0.4$

    (b) m_1: From 1970 to 1980 the number of welfare beneficiaries increased by 0.4 million per year

          on average.

          m_1: From 1980 to 1990 there was no change in the number of welfare beneficiaries

          m_1: From 1990 to 2000 the number of welfare beneficiaries decreased by 0.4 million per year

          on average.

17. See Figure 17.

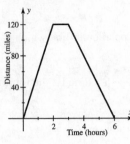

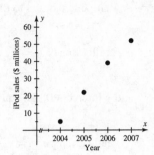

Figure 17   Figure 18

18. (a)   See Figure 18.

(b)   Using the points (2004, 5) and (2007, 52): $m = \dfrac{52-5}{2007-2004} = \dfrac{47}{3} \approx 15.67,$

$x_1 = 2004,\ y_1 = 5$ *Answers may vary.*

(c) $f(2008) = 15.67(2008-2004)+5 = 15.67(4)+5 = 62.68+5 \approx 67.68$ In 2008, IPod sales
were about 67.7 million.

Chapter 2: Extended and Discovery Exercises

1. (a)   The graph for tank A is linear. See Figure 1a. The graph for tank B is nonlinear.
See Figure 1b.

(b)   Tank B flows faster at first so it is the first to be half empty.

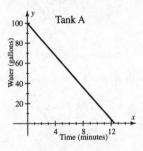

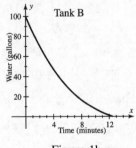

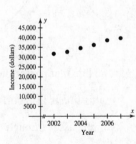

Figure 1a   Figure 1b   Figure 2

2. (a)   See Figure 2.

(b)   Since the data appears to be nearly linear with an annual increase of about $1800 per year we
may find a point-slope form of a linear equation using the point (2003, 32,271) and slope
$m = 1800$ The equation is $f(x) = 1800(x-2003)+32,271.$ *Answers may vary.*

(c) $f(2000) = 1800(2000-2003)+32,371 = \$26,971$

3. The line that passes through (0, 0) and (5, 3) has slope $m = \dfrac{3-0}{5-0} = \dfrac{3}{5}$ and y-intercept 0.

The equation for this line is $y = \dfrac{3}{5}x$. The line parallel to this line has the same slope and

y-intercept 5. The equation for this parallel line is $y = \dfrac{3}{5}x + 5$. The perpendicular line which passes

through $(0, 0)$ has slope $m = -\dfrac{5}{3}$ and y-intercept 0. Its equation is $y = -\dfrac{5}{3}x$. Finally, the fourth side

of the rectangle passes through $(5, 3)$ and has slope $m = -\dfrac{5}{3}$. Its equation is

$$y = -\frac{5}{3}(x-5) + 3 \;\Rightarrow\; y = -\frac{5}{3}x + \frac{34}{3}.$$

4. The line that passes through $(0, 0)$ and $(2, 2)$ has slope $m = \dfrac{2-0}{2-0} = \dfrac{2}{2} = 1$ and y-intercept 0.

The equation for this line is $y = x$. The line parallel to this line has the same slope and passes

through $(1, 3)$. The equation for this parallel line is $y = 1(x-1) + 3 \;\Rightarrow\; y = x + 2$. The perpendicular

line which passes through $(0, 0)$ has slope $m = -1$ and y-intercept 0. Its equation is $y = -x$. Finally,

the fourth side of the rectangle passes through $(1, 3)$ and has slope $m = -1$. Its equation is

$y = -1(x-1) + 3 \;\Rightarrow\; y = -x + 4$. See Figure 4.

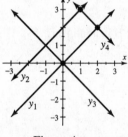   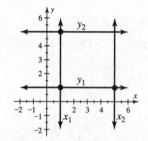   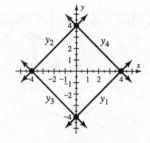

Figure 4   Figure 5   Figure 6

5. The line that passes through $(1, 1)$ and $(5, 1)$ is horizontal and has equation $y = 1$. The line parallel

to this line is also horizontal and passes through $(5, 5)$. Its equation is $y = 5$. The perpendicular line

which passes through the point $(1, 1)$ is vertical and has equation $x = 1$. Finally, the other vertical

line passes through $(5, 1)$ and has equation $x = 5$. See Figure 5.

6. The line that passes through $(4, 0)$ and $(0, 4)$ has slope $m = \dfrac{4-0}{0-4} = \dfrac{4}{-4} = -1$ and y-intercept 4.

The equation for this line is $y = -x + 4$. The line parallel to this line has the same slope and y-

intercept -4. The equation for this parallel line is $y = -x - 4$. The perpendicular line which passes

through $(0, 4)$ has slope $m = 1$ and y-intercept 4. Its equation is $y = x + 4$. Finally, the fourth side of

the rectangle has slope $m = 1$ and y-intercept -4. Its equation is $y = x - 4$. See Figure 6.

7. (a)   See Figure 7.

(b) $m_1 = -0.24, m_2 = -0.98, m_3 = -0.97, m_4 = -0.95, m_5 = -0.93,$

$m_6 = -0.85, m_7 = -0.76, m_8 = -0.63$ ie. we may calculate the slope of the first line segment as

follows: $m_1 = \dfrac{69.9 - 72.3}{10 - 0} = \dfrac{-2.4}{10} = -0.24$ m_1 : from age 0 to age 10 the remaining life

expectancy decreases at the rate of 0.24 years per year The remaining slopes may be

interpreted in a similar way.

(c)   A 20-year-old woman is expected to live 60.1 more years making her total life expectancy 80.1

years. A 70-year-old woman is expected to live 15.5 more years making her total life

expectancy 85.5 years. The 70-year-old woman has already lived through much of the risk that

a 20-year-old must still face.

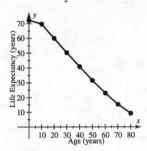

Figure 7

8. (a)   When the fish hatches it weighs about 7 mg. It weighs about 105 mg at 6 weeks of age and

about 158 mg at 12 weeks of age. *Answers may vary.*

(b)   From hatching to 6 weeks: $\dfrac{105 - 7}{6 - 0} = \dfrac{98}{6} \approx 16.3$ mg/week.

From 6 weeks to 12 weeks: $\dfrac{158 - 105}{12 - 6} = \dfrac{53}{6} \approx 8.8$ mg/week.

(c)   On average the fish gains about 16.3 mg per week during the first 6 weeks of its life and about

8.8 mg per week during the second 6 weeks of its life.

(d)   The fish gains weight the fastest during the first 6 weeks of its life.

9. (a)   The carbon dioxide levels are increasing.

(b)   The carbon dioxide levels oscillate each year.

(c)   Seasonal changes in plant growth affect the carbon dioxide levels on a small scale. This can be

attributed to the use of carbon dioxide by many plants during the process of photosynthesis.

10. (a)   The graph shows a general increasing trend which is similar to the graph for Hawaii but the

yearly oscillations are larger.

(b)   The climate in Alaska is more extreme so the seasonal changes in plant growth are more

dramatic. Both graphs show an overall increasing trend due to the use of fossil fuels and other

factors.

Chapters 1-2 Cumulative Review Exercises

1. Natural number: $\sqrt[3]{8}$ Whole number: $\sqrt[3]{8}$ Integer: $-3, \sqrt[3]{8}$

   Rational number: $-3, \frac{3}{4}, -5.8, \sqrt[3]{8}$ Irrational number: $\sqrt{7}$

2. Natural number: $\frac{50}{10}$ Whole number: $0, \frac{50}{10}$ Integer: $0, -5, -\sqrt{9}, \frac{50}{10}$

   Rational number: $0, -5, \frac{1}{2}, -\sqrt{9}, \frac{50}{10}$ Irrational number: $\sqrt{8}$

3. Distributive property; $6x + 8x = (6+8)x = 14x$

4. Commutative property for addition

5. $-5^2 + 3 + 4 \cdot 5 = -25 + 3 + 4 \cdot 5 = -25 + 3 + 20 = -2$

6. $(x+1)x + (x+1)5 = (x+1)(x+5)$

7. $-(a-b) = -a - (-b) = -a + b$

8. $\dfrac{c}{a+b}$

9. $\dfrac{1}{3} \div \left(\dfrac{3}{4} \cdot \dfrac{7}{8} \right) + \dfrac{5}{6} = \dfrac{1}{3} \div \dfrac{21}{32} + \dfrac{5}{6} = \dfrac{1}{3} \cdot \dfrac{32}{21} + \dfrac{5}{6} = \dfrac{32}{63} + \dfrac{5}{6} = \dfrac{64}{126} + \dfrac{105}{126} = \dfrac{169}{126}$

10. (a) $\left(\dfrac{2^{-3}}{3^{-2}} \right)^2 = \left(\dfrac{3^2}{2^3} \right)^2 = \left(\dfrac{9}{8} \right)^2 = \dfrac{81}{64}$

    (b) $\dfrac{\left(3x^2 y^{-3}\right)^4}{x^3 \left(y^4\right)^{-2}} = \dfrac{3^4 x^8 y^{-12}}{x^3 y^{-8}} = 81 x^{8-3} y^{-12-(-8)} = 81 x^5 y^{-4} = \dfrac{81 x^5}{y^4}$

    (c) $\left(\dfrac{ab^{-2}}{a^{-3} b^4} \right)^{-2} = \left(\dfrac{a^{-3} b^4}{ab^{-2}} \right)^2 = \dfrac{\left(a^{-3} b^4\right)^2}{\left(ab^{-2}\right)^2} = \dfrac{a^{-6} b^8}{a^2 b^{-4}} = a^{-6-2} b^{8-(-4)} = a^{-8} b^{12} = \dfrac{b^{12}}{a^8}$

11. 9.54×10^3

12. $f(4) = 2(4)^2 + \sqrt{4} = 2(16) + 2 = 32 + 2 = 34$

13. $B = \dfrac{1}{2}(4)(3)^2 = \dfrac{1}{2}(4)(9) = 18$

14. (a) $D = \{-3, 0, 2\}$

    (b)   The denominator of the function cannot equal zero. $D: x \ne -6$

    (c)   The radicand of the function must be greater than or equal to zero. $D: x \ge -4$

15. See Figure 15.

16. See Figure 16.

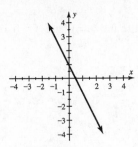

Figure 15

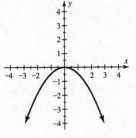

Figure 16

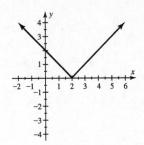

Figure 17

17. See Figure 17.

18. See Figure 18.

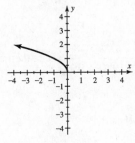

Figure 18

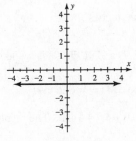

Figure 19

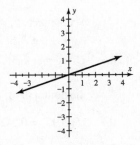

Figure 20

19. See Figure 19.

20. See Figure 20.

21. (a) Yes, the graph passes the vertical line test so it represents a function.

　 (b) D: all real numbers; R: all real numbers

　 (c) From the graph, $f(-1)=1.5$ and $f(0)=1$.

　 (d) x-intercept: $(2, 0)$; y-intercept: $(0, 1)$

　 (e) For each 2-unit increase in x, the values of $f(x)$ decrease by 1 unit, so $m=-\dfrac{1}{2}$.

　 (f) $f(x)=-\dfrac{1}{2}x+1$

22. $m=\dfrac{-4-5}{2-(-4)}=\dfrac{-9}{6}=-\dfrac{3}{2}\ \Rightarrow\ y=-\dfrac{3}{2}(x+4)+5\ \Rightarrow\ y=-\dfrac{3}{2}x-6+5\ \Rightarrow\ y=-\dfrac{3}{2}x-1$

23. The slope of the line passing through $(1,-3)$ and $(2,0)$ is $\dfrac{0-(-3)}{2-1}=\dfrac{3}{1}=3$. The slope of the

perpendicular line must be the negative reciprocal of 3, thus

$$m=-\dfrac{1}{3}\ \Rightarrow\ y=-\dfrac{1}{3}(x+1)+2\ \Rightarrow\ y=-\dfrac{1}{3}x-\dfrac{1}{3}+\dfrac{6}{3}\ \Rightarrow\ y=-\dfrac{1}{3}x+\dfrac{5}{3}.$$

24. A line perpendicular to the x-axis must be vertical. The vertical line passing through $(-2, 4)$ is

$x=-2.$

25. $m = \dfrac{-3-0}{2-1} = \dfrac{-3}{1} = -3$; the x-intercept is given in the table: 1; By substituting $x = 1$ and $y = 0$ into

$y = -3x + b$, we find that $b = 3$.

26. For each 2-unit increase in x, the y- values decrease by 3 units, so $m = -\dfrac{3}{2}$. The y-intercept is $\dfrac{3}{2}$, so

$y = -\dfrac{3}{2}x + \dfrac{3}{2}$.

27. Commutative property for addition; $120 + 80 = 200 = 80 + 120$

28. Since Volume $= \text{Length} \times \text{Width} \times \text{Height}$, $2(\text{Length}) \times \text{Width} \times 2(\text{Height}) = 4(\text{Volume})$ so

$4(400 \text{ cubic inches}) = 1600 \text{ cubic inches}$.

29. (a)   Using the points (2010, 12.7) and (1970, 4.7) we have $m = \dfrac{12.7 - 4.7}{2010 - 1970} = \dfrac{8}{40} = 0.2$, then using

the point (1970, 4.7) we have $f(x) = 0.2(x - 1970) + 4.7$.

    (b) $f(2017) = 0.2(2017 - 1970) + 4.7 = 14.1$, The result is about 14.1%.

30. The person earns $9 per hour.

31. See Figure 31.

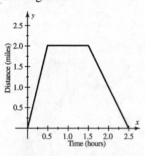

Figure 31

32. (a)   The function is in the form $f(x) = mx + b$ and is a linear function.

    (b) $f(2005) = 0.243(2005) - 450.8 \approx 36.4$, In 2005 the median age was 36.4

    (c)   The slope is 0.243.

    (d)   Each year the median age is increasing by 0.243 year on average.

33. (a) $f(x) = 10x$

    (b)   See Figure 33b.

    (c)   10

    (d)   Total fat increases at a rate of 10 g of fat per slice of pizza.

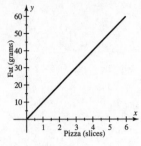

Figure 33b   Figure 34b

34. (a) $f(6) = \dfrac{4}{25}(6) = \dfrac{24}{25} = 0.96$ pint

   (b) See Figure 34b.

   (c) $\dfrac{4}{25} = 0.16$

   (d) 0.16 pint of oil should be added per gallon of gasoline.

35. (a) $f(1960) = 3.4(1960 - 1960) + 87.8 = 3.4(0) + 87.8 = 87.8.$ In 1960, there were 87.8 million

   tons of waste.

   (b) See Figure 35b.

Figure 35b

   (c) 3.4; solid waste increased by 3.4 million tons per year.

36. (a) $m_1 = \dfrac{200 - 300}{2 - 0} = \dfrac{-100}{2} = -50$, $m_2 = 0$, $m_3 = \dfrac{0 - 200}{7 - 3} = \dfrac{-200}{4} = -50$

   (b) m_1: The driver is traveling toward home at a rate of 50 miles per hour.

   m_2: The car is not moving.

   m_3: The driver is traveling toward home at a rate of 50 miles per hour.

   (c) The car started 300 miles from home and moved toward home at 50 mph for 2 hours to a

   location 200 miles from home. The car was then parked for 1 hour. Finally, the car moved

   toward home at 50 mph for 4 hours, at which point it was home.

Chapter 3 Linear Equations and Inequalities

Section 3.1 Linear Equations

1. $ax+b=0$, where $a \neq 0$

3. Yes, since $4(1)-1=3$ and $3(1)=3$.

5. 4

7. An equals sign $(=)$

9. The solution is the x-coordinate of the intersection point. That is, the solution is 3.

11. $a+c=b+c$

13. True

15. False

17. No, since $5-6=-1 \neq -2$.

19. Yes, since $3\left[2\left(-\dfrac{2}{3}\right)+3\right]=6\left(-\dfrac{2}{3}\right)+9=-4+9=5$ and $\dfrac{13}{3}-\left(-\dfrac{2}{3}\right)=\dfrac{15}{3}=5$.

21. Yes, since $-\left(2(-2)-3\right)+2(-2)=3$ and $1-(-2)=3$.

23. $x-4=10 \Rightarrow x=14$; Check: $14-4=10 \Rightarrow 10=10$, so $x=14$

25. $5+x=3 \Rightarrow x=-2$; Check: $5+(-2)=3 \Rightarrow 3=3$, so $x=-2$

27. $\dfrac{1}{2}-x=2 \Rightarrow \dfrac{1}{2}-x=\dfrac{4}{2} \Rightarrow -x=\dfrac{3}{2} \Rightarrow x=-\dfrac{3}{2}$; Check: $\dfrac{1}{2}-\left(-\dfrac{3}{2}\right)=2 \Rightarrow \dfrac{4}{2}=2 \Rightarrow 2=2$, so $x=-\dfrac{3}{2}$

29. $-6x=-18 \Rightarrow x=3$; Check: $-6(3)=-18 \Rightarrow -18=-18$, so $x=3$

31. $-\dfrac{1}{3}x=4 \Rightarrow x=-12$; Check: $-\dfrac{1}{3}(-12)=4 \Rightarrow 4=4$, so $x=-12$

33. $2x+3=13 \Rightarrow 2x=10 \Rightarrow x=5$; Check: $2(5)+3=13 \Rightarrow 10+3=13 \Rightarrow 13=13$, so $x=5$

35. $-3x+7=-8 \Rightarrow -3x=-15 \Rightarrow x=5$; Check: $-3(5)+7=-8 \Rightarrow -15+7=-8 \Rightarrow -8=-8$, so $x=5$

37. $2x=8-\dfrac{1}{2}x \Rightarrow 2x+\dfrac{1}{2}x=8 \Rightarrow \dfrac{5}{2}x=8 \Rightarrow x=8\left(\dfrac{2}{5}\right) \Rightarrow x=\dfrac{16}{5}$;

   Check: $2\left(\dfrac{16}{5}\right)=8-\dfrac{1}{2}\left(\dfrac{16}{5}\right) \Rightarrow \dfrac{32}{5}=8-\dfrac{8}{5} \Rightarrow \dfrac{32}{5}=\dfrac{40}{5}-\dfrac{8}{5} \Rightarrow \dfrac{32}{5}=\dfrac{32}{5}$, so $x=\dfrac{16}{5}$

39. $3x-1=11(1-x) \Rightarrow 3x-1=11-11x \Rightarrow 14x=12 \Rightarrow x=\dfrac{12}{14} \Rightarrow x=\dfrac{6}{7}$;

   Check: $3\left(\dfrac{6}{7}\right)-1=11\left(1-\dfrac{6}{7}\right) \Rightarrow \dfrac{18}{7}-\dfrac{7}{7}=11\left(\dfrac{7}{7}-\dfrac{6}{7}\right) \Rightarrow \dfrac{11}{7}=11\left(\dfrac{1}{7}\right) \Rightarrow \dfrac{11}{7}=\dfrac{11}{7}$, so $x=\dfrac{6}{7}$

41. $x + 4 = 2 - \dfrac{1}{3}x \Rightarrow x + \dfrac{1}{3}x = -2 \Rightarrow \dfrac{4}{3}x = -2 \Rightarrow x = -2\left(\dfrac{3}{4}\right) \Rightarrow x = -\dfrac{3}{2}$;

Check: $-\dfrac{3}{2} + 4 = 2 - \dfrac{1}{3}\left(-\dfrac{3}{2}\right) \Rightarrow -\dfrac{3}{2} + \dfrac{8}{2} = \dfrac{4}{2} + \dfrac{1}{2} \Rightarrow \dfrac{5}{2} = \dfrac{5}{2}$, so $x = -\dfrac{3}{2}$

43. $2(x-1) = 5 - 2x \Rightarrow 2x - 2 = 5 - 2x \Rightarrow 4x = 7 \Rightarrow x = \dfrac{7}{4}$;

Check: $2\left(\dfrac{7}{4} - 1\right) = 5 - 2\left(\dfrac{7}{4}\right) \Rightarrow 2\left(\dfrac{7}{4} - \dfrac{4}{4}\right) = \dfrac{20}{4} - \dfrac{14}{4} \Rightarrow 2\left(\dfrac{3}{4}\right) = \dfrac{6}{4} \Rightarrow \dfrac{6}{4} = \dfrac{6}{4}$, so $x = \dfrac{7}{4}$

45. $\dfrac{2x+1}{3} = \dfrac{2x-1}{2} \Rightarrow 2(2x+1) = 3(2x-1) \Rightarrow 4x + 2 = 6x - 3 \Rightarrow 5 = 2x \Rightarrow x = \dfrac{5}{2}$;

Check: $\dfrac{2\left(\frac{5}{2}\right)+1}{3} = \dfrac{2\left(\frac{5}{2}\right)-1}{2} \Rightarrow \dfrac{5+1}{3} = \dfrac{5-1}{2} \Rightarrow \dfrac{6}{3} = \dfrac{4}{2} \Rightarrow 2 = 2$, so $x = \dfrac{5}{2}$

47. $4.2x - 6.2 = 1 - 1.1x \Rightarrow 5.3x = 7.2 \Rightarrow x = \dfrac{7.2}{5.3} \Rightarrow x = \dfrac{72}{53} \approx 1.36$;

Check: $4.2\left(\dfrac{72}{53}\right) - 6.2 = 1 - 1.1\left(\dfrac{72}{53}\right) \Rightarrow \dfrac{302.4}{53} - \dfrac{328.6}{53} = \dfrac{53}{53} - \dfrac{79.2}{53} \Rightarrow -\dfrac{26.2}{53} = -\dfrac{26.2}{53}$, so $x = \dfrac{72}{53}$

49. $\dfrac{1}{2}x - \dfrac{3}{2} = 4 \Rightarrow \dfrac{1}{2}x = 4 + \dfrac{3}{2} \Rightarrow \dfrac{1}{2}x = \dfrac{11}{2} \Rightarrow x = \dfrac{11}{2}\left(\dfrac{2}{1}\right) \Rightarrow x = 11$;

Check: $\dfrac{1}{2}(11) - \dfrac{3}{2} = 4 \Rightarrow \dfrac{11}{2} - \dfrac{3}{2} = 4 \Rightarrow \dfrac{8}{2} = 4 \Rightarrow 4 = 4$, so $x = 11$

51. $4(x - 1980) + 6 = 18 \Rightarrow 4x - 7920 + 6 = 18 \Rightarrow 4x = 7932 \Rightarrow x = \dfrac{7932}{4} \Rightarrow x = 1983$;

Check: $4(1983 - 1980) + 6 = 18 \Rightarrow 4(3) + 6 = 18 \Rightarrow 12 + 6 = 18 \Rightarrow 18 = 18$, so $x = 1983$

53. $2(y - 3) + 5(1 - 2y) = 4y + 1 \Rightarrow 2y - 6 + 5 - 10y = 4y + 1 \Rightarrow -12y = 2 \Rightarrow y = -\dfrac{1}{6}$;

Check: $2\left(-\dfrac{1}{6} - 3\right) + 5\left(1 - 2\left(-\dfrac{1}{6}\right)\right) = 4\left(-\dfrac{1}{6}\right) + 1 \Rightarrow 2\left(-\dfrac{1}{6} - \dfrac{18}{6}\right) + 5\left(\dfrac{6}{6} + \dfrac{2}{6}\right) = -\dfrac{4}{6} + \dfrac{6}{6}$

$\Rightarrow 2\left(-\dfrac{19}{6}\right) + 5\left(\dfrac{8}{6}\right) = \dfrac{2}{6} \Rightarrow -\dfrac{38}{6} + \dfrac{40}{6} = \dfrac{2}{6} \Rightarrow \dfrac{2}{6} = \dfrac{2}{6}$, so $y = -\dfrac{1}{6}$

55. $-(5 - (z + 1)) = 9 - (5 - (2z - 3)) \Rightarrow -(5 - z - 1) = 9 - (5 - 2z + 3) \Rightarrow$

$-5 + z + 1 = 9 - 5 + 2z - 3 \Rightarrow z - 4 = 2z + 1 \Rightarrow -z = 5 \Rightarrow z = -5$;

Check: $-(5 - (-5 + 1)) = 9 - (5 - (2(-5) - 3)) \Rightarrow -(5 - (-4)) = 9 - (5 - (-10 - 3))$

$\Rightarrow -9 = 9 - (5 - (-13)) \Rightarrow -9 = 9 - 18 \Rightarrow -9 = -9$, so $z = -5$

57. Clear fractions by multiplying each side of the equation by 6.

$$\frac{2}{3}(t-3)+\frac{1}{2}t=5 \Rightarrow 4(t-3)+3t=30 \Rightarrow 4t-12+3t=30 \Rightarrow 7t=42 \Rightarrow t=6 \,;$$

Check: $\frac{2}{3}(6-3)+\frac{1}{2}(6)=5 \Rightarrow \frac{2}{3}(3)+3=5 \Rightarrow 2+3=5 \Rightarrow 5=5,$ so $t=6$

59. Clear fractions by multiplying each side of the equation by 12.

$$\frac{3k}{4}-\frac{2k}{3}=\frac{1}{6} \Rightarrow 9k-8k=2 \Rightarrow k=2 \,;$$

Check: $\frac{3(2)}{4}-\frac{2(2)}{3}=\frac{1}{6} \Rightarrow \frac{6}{4}-\frac{4}{3}=\frac{1}{6} \Rightarrow \frac{18}{12}-\frac{16}{12}=\frac{2}{12} \Rightarrow \frac{2}{12}=\frac{2}{12},$ so $k=2$

61. $0.2(n-2)+0.4n=0.05 \Rightarrow 0.2n-0.4+0.4n=0.05 \Rightarrow 0.6n=0.45 \Rightarrow n=\dfrac{0.45}{0.6} \Rightarrow n=0.75 \,;$

Check: $0.2(0.75-2)+0.4(0.75)=0.05 \Rightarrow 0.2(-1.25)+0.3=0.05$

$\Rightarrow -0.25+0.3=0.05 \Rightarrow 0.05=0.05,$ so $n=0.75$

63. $0.7y-0.8(y-1)=2 \Rightarrow 0.7y-0.8y+0.8=2 \Rightarrow -0.1y=1.2 \Rightarrow y=\dfrac{1.2}{-0.1} \Rightarrow y=-12 \,;$ Check:

$0.7(-12)-0.8(-12-1)=2 \Rightarrow -8.4-0.8(-13)=2 \Rightarrow -8.4+10.4=2 \Rightarrow 2=2,$ so $y=-12$

65. $ax+b=0 \Rightarrow ax=-b \Rightarrow x=-\dfrac{b}{a}, a\neq 0$

67. The completed table is shown in Figure 67. From the table we see that $-4x+8=0$ when $x=2$.

x	1	2	3	4	5
$-4x+8$	4	0	-4	-8	-12

x	-2	-1	0	1	2
$4-2x$	8	6	4	2	0
$x+7$	5	6	7	8	9

　　Figure 67　　　　　　　　　　　　　　　　　Figure 69

69. The completed table is shown in Figure 69. From the table we see that $4-2x=x+7$ when $x=-1$.

71. From a table of values (not shown) we find that $x-3=7$ when $x=10$.

73. From a table of values (not shown) we find that $2y-\dfrac{1}{2}=\dfrac{3}{2}$ when $y=1$.

75. From a table of values (not shown) we find that $3(z-1)+1.5=2z$ when $z=1.5$.

77. The graphs intersect when $x=-2$.

79. The graphs intersect when $x=-1$.

81. Graph $Y_1=5-2X$ and $Y_2=7$ in $[-10, 10, 1]$ by $[-10, 10, 1]$. See Figure 81.

　　The solution is $x=-1$.

[–10, 10, 1] by [–10, 10, 1]　　　　[–10, 10, 1] by [–10, 10, 1]

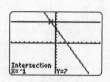　　　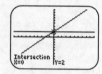

Figure 81　　　　　　　　Figure 83

83.　Graph $Y_1 = 2X - (X - 2)$ and $Y_2 = 2$ in [–10, 10, 1] by [–10, 10, 1]. See Figure 83.

The solution is $x = 0$.

85.　Graph $Y_1 = 3(X + 2) + 1$ and $Y_2 = X + 1$ in [–10, 10, 1] by [–10, 10, 1]. See Figure 85.

The solution is $x = -3$.

[–10, 10, 1] by [–10, 10, 1]　　　　[1980, 2020, 10] by [0, 150, 10]

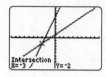　　　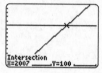

Figure 85　　　　　　　　Figure 87

87.　Graph $Y_1 = 5(X - 1990) + 15$ and $Y_2 = 100$ in [1980, 2020, 10] by [0, 150, 10]. See Figure 87.

The solution is $x = 2007$.

89.　Graph $Y_1 = \sqrt{2} + 4(X - \pi)$ and $Y_2 = 1/7 X$ in [–2, 8, 1] by [–5, 5, 1]. See Figure 89.

The solution is $x \approx 2.89$.

[–2, 8, 1] by [–5, 5, 1]

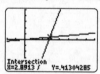

Figure 89

91.　(a)　Table $Y_1 = 2X - 1$ and $Y_2 = 13$ with TblStart = 1 and ΔTbl = 1. See Figure 91a. $x = 7$

　　(b)　Graph $Y_1 = 2X - 1$ and $Y_2 = 13$ in [0, 10, 1] by [0, 20, 2]. See Figure 91b. $x = 7$

　　(c)　$2x - 1 = 13 \Rightarrow 2x = 14 \Rightarrow x = \dfrac{14}{2} \Rightarrow x = 7$

[0, 10, 1] by [0, 20, 2]

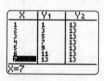　　　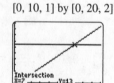

Figure 91a　　　　　　　Figure 91b

93.　(a)　Table $Y_1 = 3X - 5 - (X + 1)$ and $Y_2 = 0$ with TblStart = 0 and ΔTbl = 1. See Figure 93a. $x = 3$

　　(b)　Graph $Y_1 = 3X - 5 - (X + 1)$ and $Y_2 = 0$ in [–10, 10, 1] by [–10, 10, 1]. See Figure 93b. $x = 3$

(c) $3x - 5 - (x + 1) = 0 \Rightarrow 3x - 5 - x - 1 = 0 \Rightarrow 2x = 6 \Rightarrow x = \dfrac{6}{2} \Rightarrow x = 3$

$[-10, 10, 1]$ by $[-10, 10, 1]$

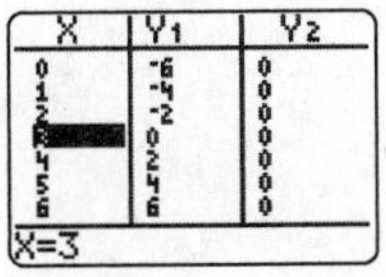

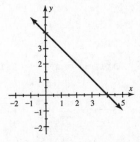

Figure 93a

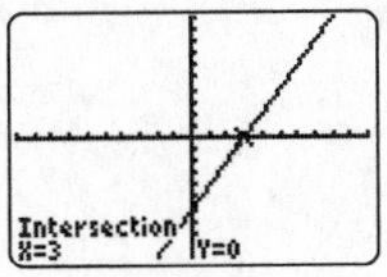

Figure 93b

95. $2x + 3 = 2x \Rightarrow 3 = 0$, false; contradiction

97. $2x + 3x = 5x \Rightarrow (2 + 3)x = 5x \Rightarrow 5x = 5x$; identity

99. $2(x - 1) = 2x - 2 \Rightarrow 2x - 2 = 2x - 2 \Rightarrow -2 = -2$; identity

101. $5 - 7(4 - 3x) = x - 4(2 - 4x) \Rightarrow 5 - 28 + 21x = x - 8 + 16x$

$\Rightarrow -23 + 21x = 17x - 8 \Rightarrow -23 + 4x = -8 \Rightarrow 4x = 15 \quad x = \dfrac{15}{4}$; conditional

103. Graph of f: $m = \dfrac{-1 - 9}{1 - (-4)} = \dfrac{-10}{5} = -2;\quad y - 9 = -2(x - (-4)) \Rightarrow y - 9 = -2x - 8 \Rightarrow y = -2x + 1$

Graph of g: $m = \dfrac{13 - 1}{0 - (-3)} = \dfrac{12}{3} = 4;\quad y - 13 = 4(x - 0) \Rightarrow y - 13 = 4x \Rightarrow y = 4x + 13$

$-2x + 1 = 4x + 13 \Rightarrow -6x + 1 = 13 \Rightarrow -6x = 12 \Rightarrow x = -2;$

$y = -2(-2) + 1 \Rightarrow y = 4 + 1 \Rightarrow y = 5; (-2, 5)$

105. (a) To find the x-intercept, let $y = 0$ and solve for x: $x + 0 = 4 \Rightarrow x = 4$; to find the y-intercept, let $x = 0$ and solve for y: $0 + y = 4 \Rightarrow y = 4$

(b) See Figure 105b.

(c) $x + y = 4 \Rightarrow y = -x + 4$

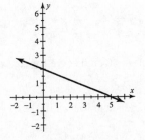

Figure 105b

Figure 107b

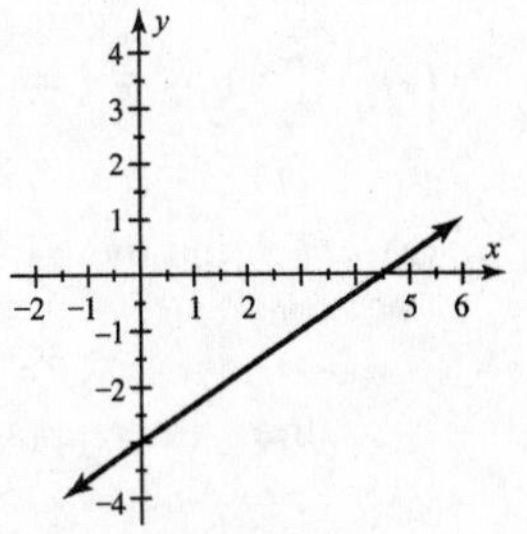

Figure 109b

107. (a) To find the x-intercept, let $y = 0$ and solve for x: $2x + 5(0) = 10 \Rightarrow 2x = 10 \Rightarrow x = 5$; to find the y-intercept, let $x = 0$ and solve for y: $2(0) + 5y = 10 \Rightarrow 5y = 10 \Rightarrow y = 2$

(b) See Figure 107b.

(c) $2x + 5y = 10 \Rightarrow 5y = -2x + 10 \Rightarrow y = -\dfrac{2}{5}x + 2$

109. (a) To find the x-intercept, let $y = 0$ and solve for x: $2x - 3(0) = 9 \Rightarrow 2x = 9 \Rightarrow x = \dfrac{9}{2}$; to find the

     y-intercept, let $x = 0$ and solve for y: $2(0) - 3y = 9 \Rightarrow -3y = 9 \Rightarrow y = -3$

   (b) See Figure 109b.

   (c) $2x - 3y = 9 \Rightarrow -3y = -2x + 9 \Rightarrow y = \dfrac{2}{3}x - 3$

111. (a) To find the x-intercept, let $y = 0$ and solve for x: $-\dfrac{2}{3}x - 0 = 2 \Rightarrow -\dfrac{2}{3}x = 2 \Rightarrow x = -3$; to find

     the y-intercept, let $x = 0$ and solve for y: $-\dfrac{2}{3}(0) - y = 2 \Rightarrow -y = 2 \Rightarrow y = -2$

   (b) See Figure 111b.

   (c) $-\dfrac{2}{3}x - y = 2 \Rightarrow -y = \dfrac{2}{3}x + 2 \Rightarrow y = -\dfrac{2}{3}x - 2$

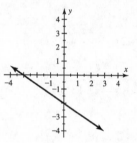   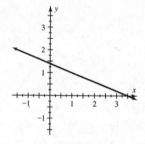   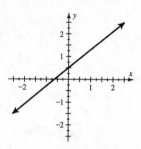

   Figure 111b   Figure 113b   Figure 115b

113. (a) To find the x-intercept, let $y = 0$ and solve for x: $2x + 5(0) = 7 \Rightarrow 2x = 7 \Rightarrow x = \dfrac{7}{2}$ to find

     the y-intercept, let $x = 0$ and solve for y: $2(0) + 5y = 7 \Rightarrow 5y = 7 \Rightarrow y = \dfrac{7}{5}$

   (b) See Figure 113b.

   (c) $2x + 5y = 7 \Rightarrow 5y = -2x + 7 \Rightarrow y = -\dfrac{2}{5}x + \dfrac{7}{5}$

115. (a) To find the x-intercept, let $y = 0$ and solve for x: $5x - 6(0) = -3 \Rightarrow 5x = -3 \Rightarrow x = -\dfrac{3}{5}$ to find

     the y-intercept, let $x = 0$ and solve for y: $5(0) - 6y = -3 \Rightarrow -6y = -3 \Rightarrow y = \dfrac{1}{2}$

   (b) See Figure 115b.

   (c) $5x - 6y = -3 \Rightarrow -6y = -5x - 3 \Rightarrow y = \dfrac{5}{6}x + \dfrac{1}{2}$

117. (a) $m = \dfrac{157 - 42}{2008 - 1960} = \dfrac{115}{48} \approx 2.4$; $S(x) = 2.4(x - 1960) + 42$

   (b) $S(2000) = 2.4(2000 - 1960) + 42 = 2.4(40) + 42 = 96 + 42 = 138$ million tons

(c) $157 + 45 = 2.4(x - 1960) + 42 \Rightarrow 160 = 2.4(x - 1960) \Rightarrow$

$$\frac{160}{2.4} = x - 1960 \Rightarrow \frac{160}{2.4} + 1960 = x \Rightarrow x \approx 2027$$

119. (a) $0.0338x - 66.5 = 1.3 \Rightarrow 0.0338x = 67.8 \Rightarrow x = \dfrac{67.8}{0.0338} \Rightarrow x \approx 2006$

(b) Table $Y_1 = 0.0338X - 66.5$ and $Y_2 = 1.3$. See Figure 119b. $x \approx 2006$

(c) Graph Table $Y_1 = 0.0338X - 66.5$ and $Y_2 = 1.3$. See Figure 119c. $x \approx 2006$

[2000, 2010, 1] by [0, 3, 1]   [15, 40, 1] by [0, 2, 1]

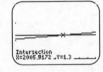   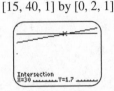

Figure 119b   Figure 119c   Figure 121c

121. (a) $m = \dfrac{1.9 - 1.1}{2010 - 1970} = \dfrac{0.8}{40} = 0.02$ Since x is the number of years since 1970 the y-intercept is

$1.1 \Rightarrow D(x) = 0.02x + 1.1$

(b) In 1970 ($x = 0$) 1.1 million degrees were awarded.

(c) $1.7 = 0.02x + 1.1 \Rightarrow 0.6 = 0.02x \Rightarrow 30 = x$ Since x represents the number of years since 1970

the result is 2000.

(d) See Figure 121c.

123. (a) $N(3) = -3.285(3) + 47.5 = 37.645$ The result is $37.645 billion.

(b) Online ad revenue is increasing while newspaper ad revenue is decreasing.

(c) See Figure 123.

(d) $-3.285x + 47.5 = 2.244x + 13 \Rightarrow 34.5 = 5.529x \Rightarrow \dfrac{34.5}{5.529} = x \Rightarrow 6.24 \approx x$ Since x

represents the number of years since 2005 the result is 2011. Yes, the answers agree.

[0, 10, 1] by [20,000, 50,000, 1]

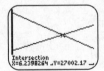

Figure 123

125. (a) $S(2) = 9.76(2) + 10 = 29.52$ In 2010 about 29.5% of mobile phone owners had smart phones.

(b) In 2008 ($x = 0$) 10% of mobile phone owners had smartphones.

(c) $9.76x + 10 = -9.76x + 90 \Rightarrow 19.52x = 80 \Rightarrow x = \dfrac{80}{19.52} \Rightarrow x \approx 4.098$ Since x represents the

number of years since 2008 the result is 2012.

127. $51.6(x-1985)+9.1=-31.9(x-1985)+167.7 \Rightarrow 51.6x-102,416.9=-31.9x+63,489.2 \Rightarrow$

$83.5x=165,906.1 \Rightarrow x \approx 1986.89.$ Sales were about equal in late 1986 or around 1987.

129. Using 1990 and 2005 data, about 1986 (answers will vary)

Section 3.2 Introduction to Problem Solving

1. We divide each side by 2 to obtain $y=\dfrac{x}{2}$.

3. The perimeter is $P=2W+2L.$

5. $43.1\%=0.431$

7. $4x+3y=12 \Rightarrow 3y=-4x+12 \Rightarrow y=\dfrac{-4x+12}{3} \Rightarrow y=-\dfrac{4}{3}x+4$

9. $5(2x-3y)=2x \Rightarrow 10x-15y=2x \Rightarrow 8x=15y \Rightarrow x=\dfrac{15}{8}y$

11. $S=6ab \Rightarrow \dfrac{S}{6a}=\dfrac{6ab}{6a} \Rightarrow b=\dfrac{S}{6a}$

13. $\dfrac{r+t}{2}=7 \Rightarrow r+t=14 \Rightarrow t=14-r$

15. $-3x+y=8 \Rightarrow y=3x+8 \Rightarrow f(x)=3x+8$

17. $4x=2\pi y \Rightarrow \dfrac{4x}{2\pi}=\dfrac{2\pi y}{2\pi} \Rightarrow \dfrac{2x}{\pi}=y \Rightarrow f(x)=\dfrac{2x}{\pi}$

19. $\dfrac{3y}{8}=x \Rightarrow 3y=8x \Rightarrow y=\dfrac{8x}{3} \Rightarrow f(x)=\dfrac{8x}{3}$

21. $3(2x+3y)=-x \Rightarrow 2x+3y=-\dfrac{x}{3} \Rightarrow 3y=-\dfrac{x}{3}-\dfrac{6x}{3} \Rightarrow 3y=-\dfrac{7x}{3} \Rightarrow y=-\dfrac{7x}{9} \Rightarrow f(x)=-\dfrac{7x}{9}$

23. $\dfrac{2x-3y}{5}=4 \Rightarrow 2x-3y=20 \Rightarrow -3y=-2x+20 \Rightarrow y=\dfrac{2x-20}{3} \Rightarrow f(x)=\dfrac{2x-20}{3}$

25. (a) $x+2=12$

   (b) $x+2=12 \Rightarrow x=10$

27. (a) $\dfrac{x}{5}=x+1$

   (b) $\dfrac{x}{5}=x+1 \Rightarrow x=5x+5 \Rightarrow -4x=5 \Rightarrow x=-\dfrac{5}{4}$

29. (a) $\dfrac{x+5}{2}=7$

   (b) $\dfrac{x+5}{2}=7 \Rightarrow x+5=14 \Rightarrow x=9$

31. (a) $\dfrac{x}{2} = 17$

    (b) $\dfrac{x}{2} = 17 \Rightarrow x = 34$

33. (a) $x + (x+1) + (x+2) = 30$

    (b) $x + (x+1) + (x+2) = 30 \Rightarrow 3x + 3 = 30 \Rightarrow 3x = 27 \Rightarrow x = \dfrac{27}{3} \Rightarrow x = 9$

35. Move the decimal point two places to the left. $55\% = 0.55$

37. Move the decimal point two places to the left. $0.04\% = 0.0004$

39. Move the decimal point two places to the left. $341\% = 3.41$

41. Move the decimal point two places to the left. $-1.67\% = -0.0167$

43. Move the decimal point two places to the left. $\dfrac{1}{4}\% = 0.25\% = 0.0025$

45. (a) Solve $P = 2L + 2W$ for L. $P = 2L + 2W \Rightarrow \dfrac{P}{2} = L + W \Rightarrow L = \dfrac{P}{2} - W$

    (b) $L = \dfrac{86}{2} - 19 = 43 - 19 = 24$ feet

47. $A = 2\pi r h \Rightarrow \dfrac{A}{2\pi r} = \dfrac{2\pi r h}{2\pi r} \Rightarrow h = \dfrac{A}{2\pi r}$

49. $F = \dfrac{9}{5}C + 32 \Rightarrow F - 32 = \dfrac{9}{5}C \Rightarrow \dfrac{5}{9}(F - 32) = C \Rightarrow C = \dfrac{5}{9}(F - 32)$

51. $A = \dfrac{1}{2}bh \Rightarrow 2A = bh \Rightarrow \dfrac{2A}{b} = \dfrac{bh}{b} \Rightarrow h = \dfrac{2A}{b}$

53. Let x represent the smallest integer. Then $x+1$ and $x+2$ represent the next consecutive integers.

    $x + (x+1) + (x+2) = 135 \Rightarrow 3x + 3 = 135 \Rightarrow 3x = 132 \Rightarrow x = 44$. The integers are 44, 45, and 46.

55. Let x represent the smallest odd integer. Then $x+2$ and $x+4$ represent the next consecutive odd

    integers. $2[x + (x+2) + (x+4)] = 150 \Rightarrow 3x + 6 = 75 \Rightarrow 3x = 69 \Rightarrow x = 23$.

    The integers are 23, 25, and 27.

57. Two sides of length x plus two sides of length $x+8$ equals 48.

    $2x + 2(x+8) = 48 \Rightarrow 2x + 2x + 16 = 48 \Rightarrow 4x = 32 \Rightarrow x = \dfrac{32}{4} \Rightarrow x = 8$ feet

59. Let x represent the length of a side of the square (and the diameter of each circle). Then

    $2\pi x + 4x = 41 \Rightarrow (2\pi + 4)x = 41 \Rightarrow x = \dfrac{41}{2\pi + 4} \Rightarrow x \approx 4$. The wire should be cut at about $4(4) = 16$

    inches for the square and $\pi(4) \approx 12.5$ inches for each circle.

61. Use the formula $d = rt$. Let x represent the speed of one car and $x+6$ the speed of the other

    car. $355 = x(2.5) + (x+6)(2.5) \Rightarrow 355 = 2.5x + 2.5x + 15 \Rightarrow 340 = 5x \Rightarrow x = 68$. One car is traveling at 68 miles per hour, and the other car is traveling at 74 miles per hour.

63. Mixing 3 liters of 30% acid with x liters of 80% acid results in $x+3$ liters of 60% acid.

$$0.30(3)+0.80x = 0.60(x+3) \Rightarrow 0.9+0.8x = 0.6x+1.8 \Rightarrow 0.2x = 0.9 \Rightarrow x = 4.5$$

The chemist used 4.5 liters of 80% sulfuric acid.

65. Using the formula $d = rt$, the car travels $74x$ miles at 74 mph and $63(3.5-x)$ at 63 mph. The total

must be 248. $74x+63(3.5-x) = 248 \Rightarrow 74x+220.5-63x = 248 \Rightarrow 11x = 27.5 \Rightarrow x = 2.5$ hour. The

car traveled for 2.5 hour at 74 mph and $3.5-2.5 = 1$ hour at 63 mph.

67. Mixing 6 liters of 50% acid with x liters of 20% acid results in $x+6$ liters of 38% acid.

$$0.50(6)+0.20x = 0.38(x+6) \Rightarrow 3+0.2x = 0.38x+2.28 \Rightarrow 0.72 = 0.18x \Rightarrow x = 4$$

The chemist used 4 liters of 20% acid.

69. Let x represent the loan amount at 6% interest. Then the remaining amount at 4% interest is

$5000-x$. $0.06x+0.04(5000-x) = 244 \Rightarrow 0.06x+200-0.04x = 244 \Rightarrow 0.02x = 44 \Rightarrow x = 2200$

The loans were for $2200 at 6% and $2800 at 4%.

71. Let x represent the measure of the smallest angle. Then the largest angle has measure $2x$ and the third

angle has measure $2x-10$. $x+2x+2x-10 = 180 \Rightarrow 5x = 190 \Rightarrow x = 38$. The angle measures are

$38°$, $66°$, and $76°$.

73. (a) $900(40) = 36,000 \text{ ft}^3/\text{hr}$

    (b) $900x = 60,000 \Rightarrow x = \dfrac{60,000}{900} \Rightarrow x \approx 66.7$ There should be no more than 66 people in the

     room.

75. Let x represent the cost of a gallon of gasoline. $1050 = \dfrac{15,000}{40}x \Rightarrow x = \$2.80$

77. Let x represent the area of Nevada. $0.09x = 9900 \Rightarrow x = 110,000$ square miles

79. Let x represent the amount spent by Sprint. $x+4.92x = 7.1 \Rightarrow 5.92x = 7.1 \Rightarrow x \approx 1.2$

The result is $x \approx \$1.2$ billion

81. Let x be the number of people surveyed. $0.32x = 480 \Rightarrow x = \dfrac{480}{0.32} \Rightarrow x = 1500$ people

83. Let x be the average GPA at private colleges and universities. Then $1-0.091 = 0.909$ of the GPA for

private colleges and universities will be equal to the GPA of public colleges and universities.

$$0.909x = 3 \Rightarrow x = \dfrac{3}{0.909} \Rightarrow x \approx 3.3$$

Checking Basic Concepts   Sections 3.1 and 3.2

   1. A linear equation has one solution.

   2. Multiply each side of the equation by 6 to clear fractions.

$$\frac{1}{2}z - (1-2z) = \frac{2}{3}z \Rightarrow 3z - 6(1-2z) = 4z \Rightarrow 3z - 6 + 12z = 4z \Rightarrow 11z = 6 \Rightarrow z = \frac{6}{11}$$

3. (a) $2(3x+4)+3 = -1 \Rightarrow 6x+8+3 = -1 \Rightarrow 6x = -12 \Rightarrow x = \frac{-12}{6} \Rightarrow x = -2$

   (b) Graph $Y_1 = 2(3X+4)+3$ and $Y_2 = -1$ in [–10, 10, 1] by [–10, 10, 1]. See Figure 3b. $x = -2$

   (c) Table $Y_1 = 2(3X+4)+3$ and $Y_2 = -1$ with TblStart = –5 and ΔTbl = 1.

   See Figure 3c. $x = -2$ The answers agree.

[–10, 10, 1] by [–10, 10, 1]

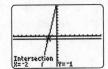

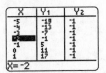

Figure 3b   Figure 3c

4. Let x represent the time the athlete spends jogging at 10 miles per hour. Then $1.2 - x$ represents the time spent jogging at 8 miles per hour.

   $$10x + 8(1.2 - x) = 10.2 \Rightarrow 10x + 9.6 - 8x = 10.2 \Rightarrow 2x = 0.6 \Rightarrow x = 0.3$$

   The athlete jogged for 0.3 hour at 10 miles per hour and 0.9 hour at 8 miles per hour.

5. To find the x-intercept, Let $y = 0$ and solve for x: $5x - 4(0) = 20 \Rightarrow 5x = 20 \Rightarrow x = 4$; to find the y intercept, let $x = 0$ and solve for y: $5(0) - 4y = 20 \Rightarrow -4y = 20 \Rightarrow y = -5$.

6. (a) $3(x-2) = 3x - 6 \Rightarrow 3x - 6 = 3x - 6 \Rightarrow -6 = -6$; identity

   (b) $2x + 3x = 5x - 2 \Rightarrow (2+3)x = 5x - 2 \Rightarrow 5x = 5x - 2 \Rightarrow 0 = -2$; contradiction

7. Let x represent the width of the room. Then $x + 5$ represents its length.

   $$2x + 2(x+5) = 82 \Rightarrow 2x + 2x + 10 = 85 \Rightarrow 4x = 72 \Rightarrow x = 18 \quad \text{The room is 18 ft. by 23ft.}$$

Section 3.3 Linear Inequalities

1. An example of a linear inequality is $3x + 2 < 5$. *Answers may vary.*

3. Yes. Subtract 5 from each side.

5. Yes. They have the same solution set.

7. An equation has an equals sign while an inequality has an inequality symbol.

9. $\{x \mid x < 6\}$

11. Yes. $2 - 5 \le 3$

13. No. $2(5) - 3 \not> 5(5) - \left[2(5) + 1\right]$

15. (a) $x - 2 = 0 \Rightarrow x = 2 \Rightarrow \{x \mid x = 2\}$

(b) $x - 2 < 0 \Rightarrow x < 2 \Rightarrow \{x \mid x < 2\}$

(c) $x - 2 > 0 \Rightarrow x > 2 \Rightarrow \{x \mid x > 2\}$

17. (a) $-2x + 6 = 0 \Rightarrow -2x = -6 \Rightarrow x = 3 \Rightarrow \{x \mid x = 3\}$

     (b) $-2x + 6 < 0 \Rightarrow -2x < -6 \Rightarrow x > 3 \Rightarrow \{x \mid x > 3\}$

     (c) $-2x + 6 > 0 \Rightarrow -2x > -6 \Rightarrow x < 3 \Rightarrow \{x \mid x < 3\}$

19. $x + 3 \le 5 \Rightarrow x \le 2 \Rightarrow \{x \mid x \le 2\}$

21. $\dfrac{1}{4}x > 9 \Rightarrow x > 36 \Rightarrow \{x \mid x > 36\}$

23. $4 - 3x \le -\dfrac{2}{3} \Rightarrow -3x \le -\dfrac{2}{3} - 4 \Rightarrow -3x \le -\dfrac{14}{3} \Rightarrow x \ge -\dfrac{1}{3}\left(-\dfrac{14}{3}\right) \Rightarrow x \ge \dfrac{14}{9} \Rightarrow \left\{x \mid x \ge \dfrac{14}{9}\right\}$

25. $7 - \dfrac{1}{2}x < x - \dfrac{3}{2} \Rightarrow 14 - x < 2x - 3 \Rightarrow -3x < -17 \Rightarrow x > \dfrac{17}{3} \Rightarrow \left\{x \mid x > \dfrac{17}{3}\right\}$

27. $\dfrac{5}{2}(2x - 3) < 6 - 2x \Rightarrow 5x - \dfrac{15}{2} < 6 - 2x \Rightarrow 7x < \dfrac{27}{2} \Rightarrow x < \dfrac{1}{7}\left(\dfrac{27}{2}\right) \Rightarrow x < \dfrac{27}{14} \Rightarrow \left\{x \mid x < \dfrac{27}{14}\right\}$

29. $\dfrac{3x - 2}{-2} \le \dfrac{x - 4}{-5} \Rightarrow -15x + 10 \le -2x + 8 \Rightarrow -13x \le -2 \Rightarrow x \ge \dfrac{2}{13} \Rightarrow \left\{x \mid x \ge \dfrac{2}{13}\right\}$

31. $3(x - 2000) + 15 < 45 \Rightarrow 3x < 6030 \Rightarrow x < \dfrac{6030}{3} \Rightarrow x < 2010 \Rightarrow \{x \mid x < 2010\}$

33. $0.4x - 0.7 < 1.3 \Rightarrow 0.4x < 2 \Rightarrow x < 5 \Rightarrow \{x \mid x < 5\}$

35. $\dfrac{4}{5}x - \dfrac{1}{5} \ge -5 \Rightarrow 4x - 1 \ge -25 \Rightarrow 4x \ge -24 \Rightarrow x \ge -6 \Rightarrow \{x \mid x \ge -6\}$

37. $-\dfrac{1}{3}(z - 3) - \dfrac{1}{4} \ge \dfrac{1}{4}(5 - z) \Rightarrow -4(z - 3) - 3 \ge 3(5 - z) \Rightarrow -4z + 12 - 3 \ge 15 - 3z \Rightarrow$

     $-4z + 9 \ge 15 - 3z \Rightarrow -z \ge 6 \Rightarrow z \le -6 \Rightarrow \{z \mid z \le -6\}$

39. $\dfrac{3}{4}(2t - 5) \le \dfrac{1}{2}(4t - 6) + 1 \Rightarrow 3(2t - 5) \le 2(4t - 6) + 4 \Rightarrow 6t - 15 \le 8t - 12 + 4 \Rightarrow$

     $6t - 15 \le 8t - 8 \Rightarrow -2t \le 7 \Rightarrow t \ge -\dfrac{7}{2} \Rightarrow \left\{t \mid t \ge -\dfrac{7}{2}\right\}$

41. $\dfrac{1}{2}(4 - (x + 3)) + 4 > -\dfrac{1}{3}(2x - (1 - x)) \Rightarrow 3(4 - x - 3) + 24 > -2(2x - 1 + x) \Rightarrow$

     $12 - 3x - 9 + 24 > -4x + 2 - 2x \Rightarrow -3x + 27 > -6x + 2 \Rightarrow 3x > -25 \Rightarrow x > -\dfrac{25}{3} \Rightarrow \left\{x \mid x > -\dfrac{25}{3}\right\}$

43. $0.05 + 0.08x < 0.01x - 0.04(3 - 4x) \Rightarrow 5 + 8x < x - 4(3 - 4x)$

$\Rightarrow 5+8x < x-12+16x \Rightarrow -9x < -17 \Rightarrow x > \frac{17}{9} \Rightarrow \left\{x \middle| x > \frac{17}{9}\right\}$

45. We cannot have a negative value under the radical. $2x-1 \geq 0 \Rightarrow x \geq \frac{1}{2}$ Domain: $\left\{x \middle| x \geq \frac{1}{2}\right\}$

47. We cannot have zero in the denominator of a fraction. $5-2x \neq 0 \Rightarrow x \neq \frac{5}{2}$ Domain: $\left\{x \middle| x \neq \frac{5}{2}\right\}$

49. We cannot have a negative value under the radical. $5(x-2)+1 \geq 0 \Rightarrow x \geq \frac{9}{5}$ Domain: $\left\{x \middle| x \geq \frac{9}{5}\right\}$

51. We cannot have a negative value under the radical nor zero in the denominator. $1+2x > 0 \Rightarrow x > -\frac{1}{2}$

Domain: $\left\{x \middle| x > -\frac{1}{2}\right\}$

53. See Figure 53. $\{x | x \geq 3\}$

x	1	2	3	4	5
$-2x+6$	4	2	0	−2	−4

Figure 53

55. Table $Y_1 = X-3$ and $Y_2 = 0$ with TblStart = 0 and ΔTbl = 1. See Figure 55. $\{x | x > 3\}$

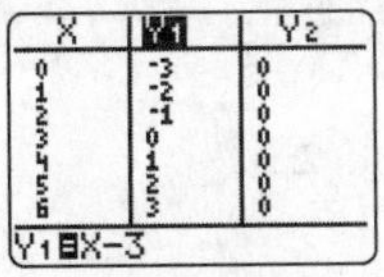

X	Y1	Y2
0	-3	0
1	-2	0
2	-1	0
3	0	0
4	1	0
5	2	0
6	3	0

Y1=X-3

Figure 55

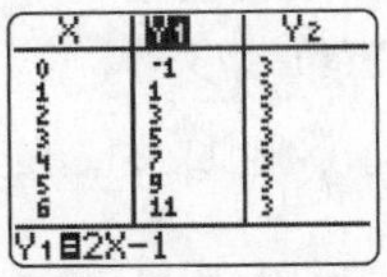

X	Y1	Y2
0	-1	3
1	1	3
2	3	3
3	5	3
4	7	3
5	9	3
6	11	3

Y1=2X-1

Figure 57

57. Table $Y_1 = 2X-1$ and $Y_2 = 3$ with TblStart = 0 and ΔTbl = 1. See Figure 57. $\{x | x \geq 2\}$

59. (a) The graph crosses the x-axis at the point $(-1, 0)$. The solution set is $\{x | x = -1\}$.

(b) The graph is below the x-axis whenever $x < -1$. The solution set is $\{x | x < -1\}$.

(c) The graph is above the x-axis whenever $x > -1$. The solution set is $\{x | x > -1\}$.

61. (a) The graph crosses the x-axis at the point $\left(\frac{1}{2}, 0\right)$. The solution set is $\left\{x \middle| x = \frac{1}{2}\right\}$.

(b) The graph is below the x-axis whenever $x > \frac{1}{2}$. The solution set is $\left\{x \middle| x > \frac{1}{2}\right\}$.

(c) The graph is above the x-axis whenever $x < \frac{1}{2}$. The solution set is $\left\{x \middle| x < \frac{1}{2}\right\}$.

63. The graph of y_1 is below or equal to the graph of y_2 whenever $x \geq 1$. The solution set is $\{x | x \geq 1\}$.

65. The graph of y_1 is above the graph of y_2 whenever $x > 2$. The solution set is $\{x | x > 2\}$.

67. (a) Car 1 is traveling faster. The slope of the line for Car 1 is greater than the slope of the line for

Car 2.

(b)   They are the same distance from St. Louis when $x = 5$ hours. The distance from St. Louis is 400 miles.

(c)   Car 2 is farther from St. Louis than Car 1 for times in the interval $0 \le x < 5$.

69. (a)   The two graphs intersect at the point $(1, 1)$ and $f(x) = g(x)$ when $x = 1$

(b)   The graph of $f(x)$ is below the graph of $g(x)$ when $x > 1$. The solution set is $\{x \mid x > 1\}$.

(c)   The graph of $f(x)$ is above or equal to the graph of $g(x)$ when $x \le 1$. The solution set is $\{x \mid x \le 1\}$.

(d)   The graph of $f(x)$ is below or equal to the graph of $g(x)$ when $x \ge 1$. The solution set is $\{x \mid x \ge 1\}$.

71. Graph $Y_1 = X - 1$ and $Y_2 = 0$ in $[-10, 10, 1]$ by $[-10, 10, 1]$. See Figure 71. $\{x \mid x < 1\}$

$[-10, 10, 1]$ by $[-10, 10, 1]$ $[-10, 10, 1]$ by $[-10, 10, 1]$

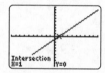   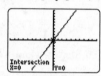

Figure 71   Figure 73

73. Graph $Y_1 = 2X$ and $Y_2 = 0$ in $[-10, 10, 1]$ by $[-10, 10, 1]$. See Figure 73. $\{x \mid x \ge 0\}$

75. Graph $Y_1 = 4 - 2X$ and $Y_2 = 8$ in $[-10, 10, 1]$ by $[-10, 10, 1]$. See Figure 75. $\{x \mid x \ge -2\}$

$[-10, 10, 1]$ by $[-10, 10, 1]$ $[-10, 10, 1]$ by $[-10, 10, 1]$

Figure 75   Figure 77

77. Graph $Y_1 = X - (2X + 4)$ and $Y_2 = 0$ in $[-10, 10, 1]$ by $[-10, 10, 1]$. See Figure 77. $\{x \mid x < -4\}$

79. Graph $Y_1 = 2(X + 2) + 5$ and $Y_2 = -X$ in $[-10, 10, 1]$ by $[-10, 10, 1]$. See Figure 79. $\{x \mid x < -3\}$

$[-10, 10, 1]$ by $[-10, 10, 1]$ $[1985, 2000, 5]$ by $[-50, 50, 10]$

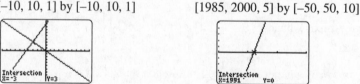

Figure 79   Figure 81

81. Graph $Y_1 = 25(X - 1995) + 100$ and $Y_2 = 0$ in $[1985, 2000, 5]$ by $[-50, 50, 10]$.

See Figure 81. $\{x \mid x \le 1991\}$

83. Graph $Y_1 = \pi\left(\sqrt{(2)}X - 1.2\right) + \sqrt{(3)}X$ and $Y_2 = (\pi + 1)/3$ in $[-10, 10, 1]$ by

$[-10, 10, 1]$.   See Figure 83. $\{x \mid x > 0.834\}$

$[-10, 10, 1]$ by $[-10, 10, 1]$.

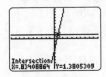

Figure 83

85. Reverse the inequality symbol: $\{x \mid x < a\}$

87. After about 1975.

89. (a)   Table $Y_1 = 5X - 2$ and $Y_2 = 8$ with TblStart $= 0$ and ΔTbl $= 1$. See Figure 89a. $\{x \mid x < 2\}$

   (b)   Graph $Y_1 = 5X - 2$ and $Y_2 = 8$ in $[-10, 10, 1]$ by $[-10, 10, 1]$. See Figure 89b. $\{x \mid x < 2\}$

   (c) $5x - 2 < 8 \Rightarrow 5x < 10 \Rightarrow x < \dfrac{10}{5} \Rightarrow x < 2 \Rightarrow \{x \mid x < 2\}$

$[-10, 10, 1]$ by $[-10, 10, 1]$

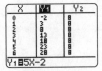   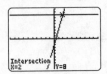

Figure 89a   Figure 89b

91. (a)   Table $Y_1 = 3X - 3$ and $Y_2 = 2X$ with TblStart $= 0$ and ΔTbl $= 1$. See Figure 91a. $\{x \mid x \le 3\}$

   (b)   Graph $Y_1 = 3X - 3$ and $Y_2 = 2X$ in $[-10, 10, 1]$ by $[-10, 10, 1]$. See Figure 91b. $\{x \mid x \le 3\}$

   (c) $2x \ge 3x - 3 \Rightarrow x \le 3 \Rightarrow \{x \mid x \le 3\}$

$[-10, 10, 1]$ by $[-10, 10, 1]$

Figure 91a   Figure 91b

93. (a) $x - 5 = 7 \Rightarrow x = 12 \Rightarrow \{x \mid x = 12\}$

   (b) $x - 5 > 7 \Rightarrow x > 12 \Rightarrow \{x \mid x > 12\}$

95. (a) $4 - 3x \ge 11 \Rightarrow -3x \ge 7 \Rightarrow x \le -\dfrac{7}{3} \Rightarrow \left\{x \mid x \le -\dfrac{7}{3}\right\}$

   (b) $4 - 3x \le 11 \Rightarrow -3x \le 7 \Rightarrow x \ge -\dfrac{7}{3} \Rightarrow \left\{x \mid x \ge -\dfrac{7}{3}\right\}$

97. (a) $3(x+7)+1 \leq -5 \Rightarrow 3x+22 \leq -5 \Rightarrow 3x \leq -27 \Rightarrow x \leq -9 \Rightarrow \{x \mid x \leq -9\}$

    (b) $3(x+7)+1 \geq -5 \Rightarrow 3x+22 \geq -5 \Rightarrow 3x \geq -27 \Rightarrow x \geq -9 \Rightarrow \{x \mid x \geq -9\}$

99. (a)   The band paid \$200 for the blank DVDs because $P=-200$ when $x=0$.

    (b)   The band charged \$5 for each recorded DVD because the slope of the line is 5.

    (c)   Since the slope of the line is 5 and the y-intercept is -200, the equation is $P=5x-200$.

    (d)   The break-even point occurs when $P=0$. From the graph, this is $x=40$ DVDs.

    (e)   The band will make a profit if it sells more than 40 DVDs or when $x>40$.

101. Two widths plus two lengths must be less than 50.

$$2x+2(x+5) < 50 \Rightarrow 2x+2x+10 < 50 \Rightarrow 4x < 40 \Rightarrow x < \frac{40}{4} \Rightarrow x < 10 \text{ ft}$$

103. (a)   The first car is traveling 70 mph and the second car is traveling 60 mph.

    (b)   They are the same distance away when $70x = 60x+35 \Rightarrow 10x = 35 \Rightarrow x = \frac{35}{10} \Rightarrow x = 3.5$ hr.

    (c)   The first car is farther away when time has elapsed past 3.5 hours.

105. $60-19x > 0 \Rightarrow 60 > 19x \Rightarrow \dfrac{60}{19} > x \Rightarrow x < 3.16$ (approximate). Below about 3.16 miles.

107. $19 \leq \dfrac{705W}{(70)^2} \Rightarrow 19 \leq \dfrac{705W}{4900} \Rightarrow 93,100 \leq 705W \Rightarrow 132 \leq W$

$$25 \geq \frac{705W}{(70)^2} \Rightarrow 25 \geq \frac{705W}{4900} \Rightarrow 122,500 \geq 705W \Rightarrow 174 \geq W$$

W is between 132 pounds and 174 pounds.

109. (a)   See Figure 109a.

    (b)   See Figure 109b.

    (c) $4124x + 45,000 \geq -5176x + 75,000 \Rightarrow 9300x \geq 30,000 \Rightarrow x \geq 3.23$ Since x represents the

          number of years after 2009, the result is from 2009 to 2012.

    [0,5,1 by 45,000, 60000, 10]

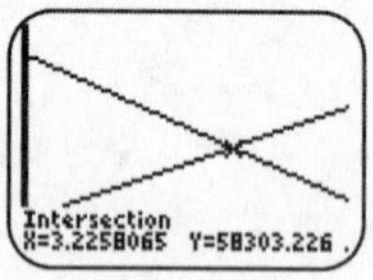

Figure 109a

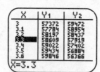

Figure 109b

Section 3.4 Compound Inequalities

1. An example of a compound inequality containing the word *and* is $x>1$ and $x \leq 7$;

   Answers may vary.

3. No, $1 \not> 3$.

5. Yes. The inequality can be written in either form.

7. Yes, $x = 2$ satisfies both inequalities. No, $x = 6$ does not satisfy $x - 1 < 5$.

9. No, $x = 0$ does not satisfy either inequality. Yes, $x = 3$ satisfies $2x \geq 3$.

11. No, $x = -3$ does not satisfy $2 - x \leq 4$. Yes, $x = 0$ satisfies both inequalities.

13. $[2, 10]$

15. $(5, 8]$

17. $(-\infty, 4)$

19. $(-2, \infty)$

21. $[-2, 5)$

23. $(-8, 8]$

25. $(3, \infty)$

27. $(-\infty, -2] \cup [4, \infty)$

29. $(-\infty, 1) \cup [5, \infty)$

31. $(-3, 5]$

33. $(-\infty, -2)$

35. $(-\infty, 4)$

37. $(-\infty, 1) \cup (2, \infty)$

39. $x \leq 3$ and $x \geq -1 \Rightarrow -1 \leq x \leq 3; \{x \mid -1 \leq x \leq 3\}$; See Figure 39.

Figure 39

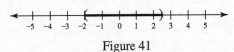
Figure 41

41. $2x < 5$ and $2x > -4 \Rightarrow x < \dfrac{5}{2}$ and $x > -2 \Rightarrow -2 < x < 2.5; \{x \mid -2 < x < 2.5\}$; See Figure 41.

43. $x + 2 > 5$ and $3 - x < 10 \Rightarrow x > 3$ and $-x < 7 \Rightarrow x > -7 \Rightarrow x > 3; \{x \mid x > 3\}$; See Figure 43.

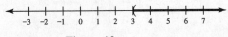

Figure 43

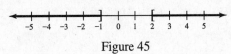
Figure 45

45. $x \leq -1$ or $x \geq 2; \{x \mid x \leq -1 \text{ or } x \geq 2\}$; See Figure 45.

47. $5 - x > 1$ or $x + 3 \geq -1 \Rightarrow -x > -4$ or $x \geq -4 \Rightarrow x < 4$ or $x \geq -4$; All real numbers; See Figure 47.

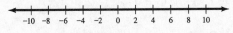
Figure 47

49. $x-3 \le 4 \Rightarrow x \le 7$ and $x+5 \ge -1 \Rightarrow x \ge -6$

The solutions must satisfy both of these inequalities. The interval is $[-6, 7]$.

51. $3t-1 > -1 \Rightarrow 3t > 0 \Rightarrow t > 0$ and $2t-\dfrac{1}{2} > 6 \Rightarrow 2t > \dfrac{13}{2} \Rightarrow t > \dfrac{13}{4}$

The solutions must satisfy both of these inequalities. The interval is $\left(\dfrac{13}{4}, \infty\right)$.

53. $x-4 \ge -3 \Rightarrow x \ge 1$ or $x-4 \le 3 \Rightarrow x \le 7$

The solutions may satisfy either one or both of these inequalities. The interval is $(-\infty, \infty)$.

55. $-x < 1 \Rightarrow x > -1$ or $5x+1 < -10 \Rightarrow 5x < -11 \Rightarrow x < -\dfrac{11}{5}$

The solutions may satisfy either one or both of these inequalities. The interval is

$\left(-\infty, -\dfrac{11}{5}\right) \cup (-1, \infty)$.

57. $1-7x < -48 \Rightarrow -7x < -49 \Rightarrow x > 7$ and $3x+1 \le -9 \Rightarrow 3x \le -10 \Rightarrow x \le -\dfrac{10}{3}$

The solutions must satisfy both of these inequalities. This is not possible. No solutions.

59. $-2 \le t+4 < 5 \Rightarrow -6 \le t < 1; [-6, 1)$

61. $-\dfrac{5}{8} \le y - \dfrac{3}{8} < 1 \Rightarrow -\dfrac{1}{4} \le y < \dfrac{11}{8}; \left[-\dfrac{1}{4}, \dfrac{11}{8}\right)$

63. $-27 \le 3x \le 9 \Rightarrow -9 \le x \le 3; [-9, 3]$

65. $\dfrac{1}{2} < -2y \le 8 \Rightarrow -\dfrac{1}{4} > y \ge -4 \Rightarrow -4 \le y < -\dfrac{1}{4}; \left[-4, -\dfrac{1}{4}\right)$

67. $-4 < 5z+1 \le 6 \Rightarrow -5 < 5z \le 5 \Rightarrow -1 < z \le 1; (-1, 1]$

69. $3 \le 4-n \le 6 \Rightarrow -1 \le -n \le 2 \Rightarrow 1 \ge n \ge -2 \Rightarrow -2 \le n \le 1; [-2, 1]$

71. $-1 < 2z-1 < 3 \Rightarrow 0 < 2z < 4 \Rightarrow 0 < z < 2; (0, 2)$

73. $-2 \le 5 - \dfrac{1}{3}m < 2 \Rightarrow -7 \le -\dfrac{1}{3}m < -3 \Rightarrow 21 \ge m > 9 \Rightarrow 9 < m \le 21; (9, 21]$

75. $100 \le 10(5x-2) \le 200 \Rightarrow 10 \le 5x-2 \le 20 \Rightarrow 12 \le 5x \le 22 \Rightarrow \dfrac{12}{5} \le x \le \dfrac{22}{5}; \left[\dfrac{12}{5}, \dfrac{22}{5}\right]$

77. $-3 < \dfrac{3z+1}{4} < 1 \Rightarrow -12 < 3z+1 < 4 \Rightarrow -13 < 3z < 3 \Rightarrow -\dfrac{13}{3} < z < 1; \left(-\dfrac{13}{3}, 1\right)$

79. $-\dfrac{5}{2} \le \dfrac{2-m}{4} \le \dfrac{1}{2} \Rightarrow -10 \le 2-m \le 2 \Rightarrow -12 \le -m \le 0 \Rightarrow 12 \ge m \ge 0 \Rightarrow 0 \le m \le 12; [0, 12]$

81. The values of $3x$ are between -3 and 6 when $-1 \le x \le 2$. The interval is $[-1, 2]$.

83. The values of $1-x$ are between -1 and 2 when $-1 < x < 2$. The interval is $(-1, 2)$.

85. The values of y_1 are between the lines $y = -2$ and $y = 2$ when $-3 \le x \le 1$. The interval is $[-3, 1]$.

87. The values of y_1 are below the line $y = -2$ or above the line $y = 2$ when $x < -2$ or $x > 0$.

    The interval is $(-\infty, -2) \cup (0, \infty)$.

89. (a) The car is moving toward Omaha since the distance is decreasing.

    (b) The car is 100 miles from Omaha when $x = 4$ hr. The car is 200 miles from Omaha when

         $x = 2$ hr.

    (c) The car is 100 to 200 miles from Omaha when the elapsed time is between 2 and 4 hours.

    (d) The car's distance from Omaha is greater than or equal to 200 miles during the first 2 hours.

91. (a) Graphs y_1 and y_2 intersect when $x = 2$.

    (b) Graphs y_2 and y_3 intersect when $x = 4$.

    (c) The graph of y_2 is between the graphs of y_1 and y_3 when $x \le 4$ and $x \ge 2$. The solution

         set is $\{x \,|\, 2 \le x \le 4\}$.

    d)   The graph of y_2 is below the graph of y_1 when $x < 2$ and $x \ge 0$. The solution set is

         $\{x \,|\, 0 \le x < 2\}$.

93. Numerical: Table $Y_1 = 2X - 4$ with TblStart $= 0$ and ΔTbl $= 1$. See Figure 93a.

    Graphical: Graph $Y_1 = -2$, $Y_2 = 2X - 4$ and $Y_3 = 4$ in [–5, 5, 1] by [–5, 5, 1]. See Figure 93b.

    The solution is $[1, 4]$.

[–5, 5, 1] by [–5, 5, 1]		[–5, 5, 1] by [–5, 5, 1]

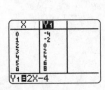 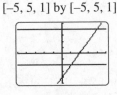 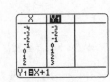 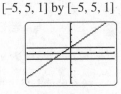

     Figure 93a   Figure 93b   Figure 95a   Figure 95b

95. Numerical: Table $Y_1 = X + 1$ with TblStart $= -4$ and ΔTbl $= 1$. See Figure 95a.

    Graphical: Graph $Y_1 = -1$, $Y_2 = X + 1$ and $Y_3 = 1$ in [–5, 5, 1] by [–5, 5, 1]. See Figure 95b.

    The solution is $(-\infty, -2) \cup (0, \infty)$.

97. Numerical: Table $Y_1 = 25(X - 2000) + 45$ with TblStart $= 2000$ and ΔTbl $= 1$. See Figure 97a.

    Graphical: Graph $Y_1 = 95$, $Y_2 = 25(X - 2000) + 45$ and $Y_3 = 295$ in [2000, 2020, 1] by [0, 325, 25].

    See Figure 97b. The solution is $[2002, 2010]$.

[2000, 2020, 1] by [0, 325, 25]

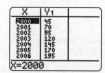

Figure 97a   Figure 97b

99. $5 - x \geq 0 \Rightarrow x \leq 5 \Rightarrow (-\infty, 5]$

101. $x \neq 1 \Rightarrow (-\infty, -1) \cup (-1, \infty)$

103. $x > 0 \Rightarrow (0, \infty)$

105. $4 \leq 5x - 1 \leq 14 \Rightarrow 5 \leq 5x \leq 15 \Rightarrow 1 \leq x \leq 3; [1, 3]$

Graphical: Graph $Y_1 = 4$, $Y_2 = 5X - 1$ and $Y_3 = 14$ in [0, 10, 1] by [0, 20, 1]. See Figure 105a.

Numerical: Table $Y_1 = 5X - 1$ with TblStart = 0 and ΔTbl = 1. See Figure 105b.

The solution is $[1, 3]$.

[0, 10, 1] by [0, 20, 1]   [−10, 10, 1] by [−10, 10, 1]

 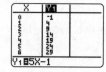   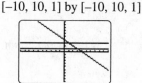 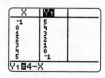

Figure 105a   Figure 105b   Figure 107a   Figure 107b

107. $4 - x \geq 1$ or $4 - x < 3 \Rightarrow -x \geq -3$ or $-x < -1 \Rightarrow x \leq 3$ or $x > 1; (-\infty, \infty)$

Graphical: Graph $Y_1 = 1$, $Y_2 = 4 - X$ and $Y_3 = 3$ in [−10, 10, 1] by [−10, 10, 1]. See Figure 107a.

Numerical: Table $Y_1 = 4 - X$ with TblStart = −1 and ΔTbl = 1. See Figure 107b.

The solution is $(-\infty, \infty)$.

109. $2x + 1 < 3$ or $2x + 1 \geq 7 \Rightarrow x < 1$ or $x \geq 3; (-\infty, 1) \cup [3, \infty)$

Graphical: Graph $Y_1 = 3$, $Y_2 = 2X + 1$ and $Y_3 = 7$ in [−10, 10, 1] by [−10, 10, 1]. See Figure 109a.

Numerical: Table $Y_1 = 2X + 1$ with TblStart = 0 and ΔTbl = 1. See Figure 109b.

The solution is $(-\infty, 1) \cup [3, \infty)$.

[−10, 10, 1] by [−10, 10, 1]

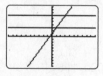 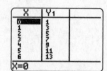

Figure 109a   Figure 109b

111. $c < x + b \leq d \Rightarrow c - b < x \leq d - b; (c - b, d - b]$

113. (a)   See Figure 113a.

(b)   See Figures 113b.

(c) $15 \leq 2.5x - 5000 \leq 20 \Rightarrow 5015 \leq 2.5x \leq 5020 \Rightarrow 2006 \leq x \leq 2008$

[2000, 2010, 1] by [10, 25, 1]   [2000, 2010, 1] by [10, 25, 1]

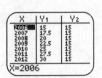   

Figure 113a   Figure 113b

115. $51.3 \leq 60 - 5.8x \leq 57.1 \Rightarrow -8.7 \leq -5.8x \leq -2.9 \Rightarrow 1.5 \geq x \geq 0.5$; The dew point ranges from 57.1°F

to 51.3°F from 0.5 to 1.5 miles high.

117. The perimeter of the rectangle is given by $2(x+3)+2(2x)$.

$$40 \leq 2(x+3)+2(2x) \leq 60 \Rightarrow 40 \leq 6x+6 \leq 60 \Rightarrow 34 \leq 6x \leq 54 \Rightarrow 5.\overline{6} \leq x \leq 9$$

119. $22.5 \leq 70 - 19x \leq 41.5 \Rightarrow -47.5 \leq -19x \leq -28.5 \Rightarrow 2.5 \geq x \geq 1.5 \Rightarrow 1.5 \leq x \leq 2.5$

The air temperature is from 22.5°F to 41.5°F at altitudes from 1.5 to 2.5 miles.

121. (a) $m = \dfrac{500-250}{2010-2000} = \dfrac{250}{10} = 25$. Since x represents the number of years after 2000 the

y- intercept is 250. The function is $M(x) = 25x + 250$.

(b) $300 \leq 25x + 250 \leq 400 \Rightarrow 50 \leq 25x \leq 150 \Rightarrow 2 \leq x \leq 6$. The result is 2002 to 2006.

Checking Basic Concepts Sections 3.3 and 3.4

1. $4 - 3x < \dfrac{1}{2}x \Rightarrow 8 - 6x < x \Rightarrow 8 < 7x \Rightarrow \dfrac{8}{7} < x \Rightarrow \left\{ x \middle| x > \dfrac{8}{7} \right\}$

2. (a)   The two lines intersect at (2, 0). The solution is $\left\{ x \middle| x = 2 \right\}$.

(b)   The line $y = \dfrac{1}{2}x - 1$ is below the line $y = 2 - x$ when $x < 2$. The solution set is $\left\{ x \middle| x < 2 \right\}$.

(c)   The line $y = \dfrac{1}{2}x - 1$ intersects and is above the line $y = 2 - x$ when $x \geq 2$. The solution set is

$\left\{ x \middle| x \geq 2 \right\}$.

3. (a)   Yes, since $2(3) - 1 = 5 \geq 3$.

(b)   No, since $(3) + 2 = 5 \nless 4$.

4. (a) $-5 \leq 2x + 1 \leq 3 \Rightarrow -6 \leq 2x \leq 2 \Rightarrow -3 \leq x \leq 1; [-3, 1]$

(b) $1 - x \leq -2 \Rightarrow -x \leq -3 \Rightarrow x \geq 3$ or $1 - x > 2 \Rightarrow -x > 1 \Rightarrow x < -1; (-\infty, -1) \cup [3, \infty)$

(c) $-2 < \dfrac{4-3x}{2} \leq 6 \Rightarrow -4 < 4 - 3x \leq 12 \Rightarrow -8 < -3x \leq 8 \Rightarrow \dfrac{8}{3} > x \geq -\dfrac{8}{3}; \left[-\dfrac{8}{3}, \dfrac{8}{3} \right)$

Section 3.5 Absolute Value Equations and Inequalities

1. An example of an absolute value equation is $|3x+2|=6.$ *Answers may vary.*

3. Yes, since $|-3|=3.$

5. Yes

7. 2 times

9. No, $|2(-3)-5|=|-11|=11\neq 1.$ Yes, $|2(3)-5|=|1|=1.$

11. No, $|7-4(-2)|=|15|=15\nleq 5.$ Yes, $|7-4(2)|=|-1|=1\leq 1.$

13. Yes, $\left|7\left(-\dfrac{4}{7}\right)+4\right|=|-4+4|=|0|=0>-1.$ Yes, $|7(2)+4|=|18|=18>-1.$

15. $x=0$ and $x=4$

17. $x=-3$ and $x=3$

19. $-3<x<3\Rightarrow(-3,3)$

21. $x<-3$ or $x>3\Rightarrow(-\infty,-3)\cup(3,\infty)$

23. $x=-7$ and $x=7$

25. An absolute value can not be negative. No solution.

27. $4x=9\Rightarrow x=\dfrac{9}{4}$ or $4x=-9\Rightarrow x=-\dfrac{9}{4}$

29. Since $|-2x|-6=2\Rightarrow|-2x|=8:$ $-2x=8\Rightarrow x=-4$ or $-2x=-8\Rightarrow x=4$

31. $2x+1=11\Rightarrow 2x=10\Rightarrow x=5$ or $2x+1=-11\Rightarrow 2x=-12\Rightarrow x=-6$

33. Since $|-2x+3|+3=4\Rightarrow|-2x+3|=1:$

$-2x+3=1\Rightarrow-2x=-2\Rightarrow x=1$ or $-2x+3=-1\Rightarrow-2x=-4\Rightarrow x=2$

35. $|3-4x|=0\Rightarrow 3-4x=0\Rightarrow 3=4x\Rightarrow\dfrac{3}{4}=x$

37. $\dfrac{1}{2}x-1=5\Rightarrow\dfrac{1}{2}x=6\Rightarrow x=12$ or $\dfrac{1}{2}x-1=-5\Rightarrow\dfrac{1}{2}x=-4\Rightarrow x=-8$

39. An absolute value can not be negative. No solution.

41. Since $\left|\dfrac{2}{3}z-1\right|-3=8\Rightarrow\left|\dfrac{2}{3}z-1\right|=11:$

$\dfrac{2}{3}z-1=11\Rightarrow\dfrac{2}{3}z=12\Rightarrow z=18$ or $\dfrac{2}{3}z-1=-11\Rightarrow\dfrac{2}{3}z=-10\Rightarrow z=-15$

43. $z-1=2z\Rightarrow-z=1\Rightarrow z=-1$ or $z-1=-2z\Rightarrow 3z=1\Rightarrow z=\dfrac{1}{3}$

45. $3t+1=2t-4\Rightarrow t=-5$ or $3t+1=-2t+4\Rightarrow 5t=3\Rightarrow t=\dfrac{3}{5}$

47. $\frac{1}{4}x = 3 + \frac{1}{4}x \Rightarrow 0 = 3$ (no solution) or $\frac{1}{4}x = -3 - \frac{1}{4}x \Rightarrow \frac{1}{2}x = -3 \Rightarrow x = -6$

49. (a) $2x = 8 \Rightarrow x = 4$ or $2x = -8 \Rightarrow x = -4$

   (b) $\{x | -4 < x < 4\}$

   (c) $\{x | x < -4 \text{ or } x > 4\}$

51. (a) $5 - 4x = 3 \Rightarrow -4x = -2 \Rightarrow x = \frac{1}{2}$ or $5 - 4x = -3 \Rightarrow -4x = -8 \Rightarrow x = 2$

   (b) $\left\{x \Big| \frac{1}{2} \leq x \leq 2\right\}$

   (c) $\left\{x \Big| x \leq \frac{1}{2} \text{ or } x \geq 2\right\}$

53. The solutions to $|x| \leq 3$ satisfy $c \leq x \leq d$ where c and d are the solutions to $|x| = 3$.

   $|x| = 3$ is equivalent to $x = -3$ and $x = 3$. The interval is $[-3, 3]$.

55. The solutions to $|k| > 4$ satisfy $k < c$ or $k > d$ where c and d are the solutions to $|k| = 4$.

   $|k| = 4$ is equivalent to $k = -4$ and $k = 4$. The interval is $(-\infty, -4) \cup (4, \infty)$.

57. The inequality $|t| \leq -3$ has no solutions because absolute value is never negative.

59. The inequality $|z| > 0$ is true for any value of z except $z = 0$. The interval is $(-\infty, 0) \cup (0, \infty)$.

61. The solutions to $|2x| > 7$ satisfy $x < c$ or $x > d$ where c and d are the solutions to $|2x| = 7$.

   $|2x| = 7$ is equivalent to $2x = -7 \Rightarrow x = -\frac{7}{2}$ or $2x = 7 \Rightarrow x = \frac{7}{2}$.

   The interval is $\left(-\infty, -\frac{7}{2}\right) \cup \left(\frac{7}{2}, \infty\right)$.

63. The solutions to $|-4x + 4| < 16$ satisfy $c < x < d$ where c and d are the solutions to $|-4x + 4| = 16$.

   $|-4x + 4| = 16$ is equivalent to $-4x + 4 = -16 \Rightarrow x = 5$ or $-4x + 4 = 16 \Rightarrow x = -3$.

   The interval is $(-3, 5)$.

65. First divide each side of $2|x + 5| \geq 8$ by 2 to obtain $|x + 5| \geq 4$.

   The solutions to $|x + 5| \geq 4$ satisfy $x \leq c$ or $x \geq d$ where c and d are the solutions to $|x + 5| = 4$.

   $|x + 5| = 4$ is equivalent to $x + 5 = -4 \Rightarrow x = -9$ or $x + 5 = 4 \Rightarrow x = -1$.

   The interval is $(-\infty, -9] \cup [-1, \infty)$.

67. First add 1 to each side of $|8 - 6x| - 1 \leq 2$ to obtain $|8 - 6x| \leq 3$.

   The solutions to $|8 - 6x| \leq 3$ satisfy $c \leq x \leq d$ where c and d are the solutions to $|8 - 6x| = 3$.

$|8-6x|=3$ is equivalent to $8-6x=-3 \Rightarrow x = \dfrac{11}{6}$ or $8-6x=3 \Rightarrow x = \dfrac{5}{6}$. The interval is $\left[\dfrac{5}{6}, \dfrac{11}{6} \right]$.

69. First subtract 5 from each side of $5 + \left| \dfrac{2-x}{3} \right| \le 9$ to obtain $\left| \dfrac{2-x}{3} \right| \le 4$.

The solutions to $\left| \dfrac{2-x}{3} \right| \le 4$ satisfy $c \le x \le d$ where c and d are the solutions to $\left| \dfrac{2-x}{3} \right| = 4$.

$\left| \dfrac{2-x}{3} \right| = 4$ is equivalent to $\dfrac{2-x}{3} = -4 \Rightarrow x = 14$ or $\dfrac{2-x}{3} = 4 \Rightarrow x = -10$.

The interval is $[-10, 14]$.

71. The inequality $|2x-1| \le -3$ has no solutions because absolute value is never negative.

73. First add 1 to each side of $|x+1| - 1 > -3$ to obtain $|x+1| > -2$.

The inequality $|x+1| > -2$ is true for all values of x because absolute value is never negative.

The interval is $(-\infty, \infty)$.

75. The inequality $|2z-4| \le -1$ has no solutions because absolute value is never negative.

77. The inequality $|3z-1| > -3$ is true for all values of z because absolute value is never negative.

The interval is $(-\infty, \infty)$.

79. The solutions to $\left| \dfrac{2-t}{3} \right| \ge 5$ satisfy $t \le c$ or $t \ge d$ where c and d are the solutions to $\left| \dfrac{2-t}{3} \right| = 5$.

$\left| \dfrac{2-t}{3} \right| = 5$ is equivalent to $\dfrac{2-t}{3} = -5 \Rightarrow t = 17$ or $\dfrac{2-t}{3} = 5 \Rightarrow x = -13$.

The interval is $(-\infty, -13] \cup [17, \infty)$.

81. The solutions to $|t-1| \le 0.1$ satisfy $c \le t \le d$ where c and d are the solutions to $|t-1| = 0.1$.

$|t-1| = 0.1$ is equivalent to $t-1 = -0.1 \Rightarrow t = 0.9$ or $t-1 = 0.1 \Rightarrow t = 1.1$. The interval is $[0.9, 1.1]$.

83. The solutions to $|b-10| > 0.5$ satisfy $b < c$ or $b > d$ where c and d are the solutions to $|b-10| = 0.5$.

$|b-10| = 0.5$ is equivalent to $b-10 = -0.5 \Rightarrow b = 9.5$ or $b-10 = 0.5 \Rightarrow b = 10.5$.

The interval is $(-\infty, 9.5) \cup (10.5, \infty)$.

85. (a) From the table, $y = 2$ when $x = -1$ or $x = 3$.

    (b) $y < 2$ when $-1 < x < 3$. The interval is $(-1, 3)$.

    (c) $y > 2$ when $x < -1$ or $x > 3$. The interval is $(-\infty, -1) \cup (3, \infty)$.

87. (a) From the graph $y_1 = 1$ when $x = -1$ or $x = 0$.

    (b) $y_1 \le 1$ when $-1 \le x \le 0$. The interval is $[-1, 0]$.

(c) $y_1 \geq 1$ when $x \leq -1$ or $x \geq 0$. The interval is $(-\infty, \ -1] \cup [0, \ \infty)$.

89. Graph $Y_1 = abs(X)$ and $Y_2 = 1$ in $[-3, 3, 1]$ by $[-3, 3, 1]$. See Figures 89a and 89b.

$(-\infty, \ -1] \cup [1, \ \infty)$

[–3, 3, 1] by [–3, 3, 1]   [–3, 3, 1] by [–3, 3, 1]

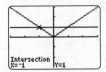   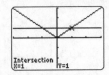

Figure 89a   Figure 89b

91. Graph $Y_1 = abs(X-1)$ and $Y_2 = 3$ in $[-5, 5, 1]$ by $[-5, 5, 1]$. See Figures 91a and 91b. $[-2, \ 4]$

[–5, 5, 1] by [–5, 5, 1]   [–5, 5, 1] by [–5, 5, 1]

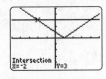   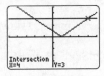

Figure 91a   Figure 91b

93. Graph $Y_1 = abs(4-2X)$ and $Y_2 = 2$ in $[0, 5, 1]$ by $[0, 5, 1]$. See Figures 93a and 93b.

$(-\infty, \ 1) \cup (3, \ \infty)$

[0, 5, 1] by [0, 5, 1]   [0, 5, 1] by [0, 5, 1]

Figure 93a   Figure 93b

95. Graph $Y_1 = abs(10-3X)$ and $Y_2 = 4$ in $[0, 6, 1]$ by $[0, 6, 1]$. See Figures 95a and 95b. $\left(2, \ 4.\overline{6}\ \right)$

[0, 6, 1] by [0, 6, 1]   [0, 6, 1] by [0, 6, 1]

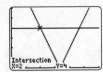   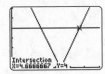

Figure 95a   Figure 95b

97. Graph $Y_1 = abs(8.1-X)$ and $Y_2 = -2$ in $[0, 15, 1]$ by $[-5, 5, 1]$. See Figure 97. $(-\infty, \ \infty)$

[0, 15, 1] by [–5, 5, 1]

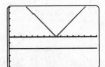

Figure 97

99. (a) $3x = 9 \Rightarrow x = 3$ or $3x = -9 \Rightarrow x = -3$; $\{x \mid -3 \leq x \leq 3\}$

     (b)   Graph $Y_1 = \text{abs}(3X)$ and $Y_2 = 9$ in $[-5, 5, 1]$ by $[-5, 15, 1]$. See Figures 99a and 99b.

     (c)   Table $Y_1 = \text{abs}(3X)$ with TblStart $= -7$ and ΔTbl $= 2$. See Figure 99c.

     The solution set is $\{x \mid -3 \leq x \leq 3\}$.

$[-5, 5, 1]$ by $[-5, 15, 1]$ $[-5, 5, 1]$ by $[-5, 15, 1]$

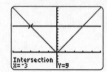   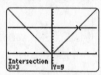   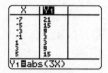

Figure 99a   Figure 99b   Figure 99c

101. (a) $2x - 5 = 1 \Rightarrow 2x = 6 \Rightarrow x = 3$ or $2x - 5 = -1 \Rightarrow 2x = 4 \Rightarrow x = 2$; $\left\{x \mid x < 2 \text{ or } x > 3\right\}$

     (b)   Graph $Y_1 = \text{abs}(2X - 5)$ and $Y_2 = 1$ in $[0, 5, 1]$ by $[-1, 2, 1]$. See Figures 101a and 101b.

     (c)   Table $Y_1 = \text{abs}(2X - 5)$ with TblStart $= 1$ and ΔTbl $= 0.5$. See Figure 101c.

     The solution set is $\left\{x \mid x < 2 \text{ or } x > 3\right\}$.

$[0, 5, 1]$ by $[-1, 2, 1]$ $[0, 5, 1]$ by $[-1, 2, 1]$

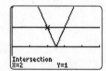   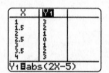

Figure 101a   Figure 101b   Figure 101c

103. $|x| \leq 4$

105. $|y| > 2$

107. $|2x + 1| \leq 0.3$

109. $|\pi x| \geq 7$

111. two

113. (a) $T - 43 = 24 \Rightarrow T = 67$ or $T - 43 = -24 \Rightarrow T = 19$; $\{T \mid 19 \leq T \leq 67\}$

     (b)   The monthly average temperatures in Marquette, Michigan vary from 19°F to 67°F.

115. (a) $T - 10 = 36 \Rightarrow T = 46$ or $T - 10 = -36 \Rightarrow T = -26$; $\{T \mid -26 \leq T \leq 46\}$

     (b)   The monthly average temperatures in Chesterfield, Canada vary from −26°F to 46°F.

117. (a) $A = (29,028 + 22,834 + 20,320 + 19,340 + 18,510 + 16,066 + 7,310) \div 7 \approx 19,058$ feet

     (b)   Africa and Europe have elevations within 1000 feet of A.

     (c)   South America, North America, Africa, Europe and Antarctica have elevations within 5000 feet of A.

     (d) $|E - A| \leq 5000$

119. $d - 2.5 = 0.002 \Rightarrow d = 2.502$ or $d - 2.5 = -0.002 \Rightarrow d = 2.498$; $\{d \mid 2.498 \le d \le 2.502\}$

The diameter can vary from 2.498 inches to 2.502 inches.

121. $|d - 3.8| \le 0.03$

123. The solutions to $\left|\dfrac{x-20}{20}\right| < 0.05$ satisfy $c < x < d$ where c and d are the solutions to $\left|\dfrac{x-20}{20}\right| = 0.05$.

$\left|\dfrac{x-20}{20}\right| = 0.05$ is equivalent to $\dfrac{x-20}{20} = -0.05 \Rightarrow x = 19$ or $\dfrac{x-20}{20} = 0.05 \Rightarrow x = 21$.

The interval is $(19, \ 21)$. The values must be between 19 and 21, exclusively.

Checking Basic Concepts Section 3.5

1. $\left|\dfrac{3}{4}x - 1\right| - 3 = 5 \Rightarrow \left|\dfrac{3}{4}x - 1\right| = 8 \Rightarrow \dfrac{3}{4}x - 1 = -8 \Rightarrow \dfrac{3}{4}x = -7 \Rightarrow$

$x = -\dfrac{28}{3}$ or $\dfrac{3}{4}x - 1 = 8 \Rightarrow \dfrac{3}{4}x = 9 \Rightarrow x = 12$

2. (a) $3x - 6 = 8 \Rightarrow 3x = 14 \Rightarrow x = \dfrac{14}{3}$ or $3x - 6 = -8 \Rightarrow 3x = -2 \Rightarrow x = -\dfrac{2}{3}$

(b) The solutions to $|3x - 6| < 8$ satisfy $c < x < d$ where c and d are the solutions to $|3x - 6| = 8$.

From part (a), the interval is $\left(-\dfrac{2}{3}, \ \dfrac{14}{3}\right)$.

(c) The solutions to $|3x - 6| > 8$ satisfy $x < c$ or $x > d$ where c and d are the solutions

to $|3x - 6| = 8$. From part (a), the interval is $\left(-\infty, \ -\dfrac{2}{3}\right) \cup \left(\dfrac{14}{3}, \ \infty\right)$.

3. The solutions to $|-2(3-x)| < 6$ satisfy $c < x < d$ where c and d are the solutions to $|-2(3-x)| = 6$.

$|-2(3-x)| = 6$ is equivalent to $-2(3-x) = -6 \Rightarrow x = 0$ or $-2(3-x) = 6 \Rightarrow x = 6$.

The interval is $(0, \ 6)$. Similarly, the solution to $|-2(3-x)| \ge 6$ is the interval $(-\infty, \ 0] \cup [6, \ \infty)$.

4. (a) From the graph $y = 2$ when $x = 1$ or $x = 3$.

(b) $y \le 2$ when $1 \le x \le 3$. The interval is $[1, \ 3]$.

(c) $y \ge 2$ when $x \le 1$ or $x \ge 3$. The interval is $(-\infty, \ 1] \cup [3, \ \infty)$.

Chapter 3 Review

1. The completed table is shown in Figure 1. From the table we see that $3x - 6 = 0$ when $x = 2$.

x	0	1	2	3	4
$3x - 6$	-6	-3	0	3	6

Figure 1

x	-1	0	1	2	3
$5 - 2x$	7	5	3	1	-1

Figure 2

2. The completed table is shown in Figure 2. From the table we see that $5 - 2x = 3$ when $x = 1$.

3. Yes it is a solution since $\dfrac{1}{2}\left(-\dfrac{2}{3}\right) - 2\left(2\left(-\dfrac{2}{3}\right) + 3\right) = -\dfrac{1}{3} - 2\left(-\dfrac{4}{3} + 3\right) = -\dfrac{1}{3} - 2\left(\dfrac{5}{3}\right) = -\dfrac{11}{3}$

and $4\left(-\dfrac{2}{3}\right) - 1 = -\dfrac{11}{3}$.

4. Since the graphs intersect at the point $(1, 3)$, the solution is $x = 1$.

5. Graph $Y_1 = 5 - 2X$ and $Y_2 = -1 + X$ in $[-10, 10, 1]$ by $[-10, 10, 1]$. See Figure 5. $x = 2$

$[-10, 10, 1]$ by $[-10, 10, 1]$

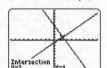

Figure 5   Figure 6

6. Table $Y_1 = 4X - 3$ and $Y_2 = 5 + 2X$ with TblStart = 0 and ΔTbl = 1. See Figure 6. $x = 4$

7. $2x - 7 = 21 \ \Rightarrow \ 2x = 28 \ \Rightarrow \ x = \dfrac{28}{2} \ \Rightarrow \ x = 14$

8. $1 - 7x = -\dfrac{5}{2} \ \Rightarrow \ -7x = -\dfrac{5}{2} - 1 \ \Rightarrow \ -7x = -\dfrac{7}{2} \ \Rightarrow \ x = -\dfrac{1}{7}\left(-\dfrac{7}{2}\right) \ \Rightarrow \ x = \dfrac{1}{2}$

9. $-2(4x - 1) = 1 - x \ \Rightarrow \ -8x + 2 = 1 - x \ \Rightarrow \ 1 = 7x \ \Rightarrow \ \dfrac{1}{7} = x \ \Rightarrow \ x = \dfrac{1}{7}$

10. $-\dfrac{3}{4}(x - 1) + 5 = 6 \ \Rightarrow \ -\dfrac{3}{4}x + \dfrac{3}{4} + 5 = 6 \ \Rightarrow \ -\dfrac{3}{4}x = \dfrac{1}{4} \ \Rightarrow \ x = -\dfrac{4}{3}\left(\dfrac{1}{4}\right) \ \Rightarrow \ x = -\dfrac{1}{3}$

11. $\dfrac{x - 4}{3} = 2 \ \Rightarrow \ x - 4 = 6 \ \Rightarrow \ x = 10$

12. $\dfrac{2x - 3}{2} = \dfrac{x + 3}{5} \ \Rightarrow \ 5(2x - 3) = 2(x + 3) \ \Rightarrow \ 10x - 15 = 2x + 6 \ \Rightarrow \ 8x = 21 \ \Rightarrow \ x = \dfrac{21}{8}$

13. $-2(z - 1960) + 32 = 8 \ \Rightarrow \ -2(z - 1960) = -24 \ \Rightarrow \ z - 1960 = 12 \ \Rightarrow \ z = 1972$

14. $5 - (3 - 2z) + 4 = 5(2z - 3) - (3z - 2) \ \Rightarrow \ 2z + 6 = 7z - 13 \ \Rightarrow \ -5z = -19 \ \Rightarrow \ z = \dfrac{19}{5}$

15. $\dfrac{2}{3}\left(\dfrac{t - 3}{2}\right) + 4 = \dfrac{1}{3}t - (1 - t) \ \Rightarrow \ \dfrac{t - 3}{3} + 4 = \dfrac{4}{3}t - 1 \ \Rightarrow \ t - 3 + 12 = 4t - 3 \ \Rightarrow$

$t + 9 = 4t - 3 \ \Rightarrow \ -3t = -12 \ \Rightarrow \ t = 4$

16. $0.3r - 0.12(r-1) = 0.4r + 2.1 \implies 0.18r + 0.12 = 0.4r + 2.1 \implies -0.22r = 1.98 \implies r = -9$

17. $-4(5-3x) = 12x - 20 \implies -20 + 12x = 12x - 20 \implies -20 = -20;$ identity

18. $4x - (x-3) = 3x - 3 \implies 4x - x + 3 = 3x - 3 \implies 3x + 3 = 3x - 3 \implies 3 = -3;$ contradiction

19. To find the x-intercept, let $y = 0$ and solve for $x: 4x - 5(0) = 20 \implies 4x = 20 \implies x = 5;$ to find the

    y-intercept, let $x = 0$ and solve for $y: 4(0) - 5y = 20 \implies -5y = 20 \implies y = -4;$

    $$4x - 5y = 20 \implies -5y = -4x + 20 \implies y = \frac{4}{5}x - 4$$

20. To find the x-intercept, let $y = 0$ and solve for $x: 2x + \frac{1}{2}(0) = -6 \implies 2x = -6 \implies x = -3;$ to find the

    y-intercept, let $x = 0$ and solve for $y: 2(0) + \frac{1}{2}y = -6 \implies \frac{1}{2}y = -6 \implies y = -12;$

    $$2x + \frac{1}{2}y = -6 \implies \frac{1}{2}y = -2x - 6 \implies y = -4x - 12$$

21. $5x - 4y = 20 \implies 5x - 20 = 4y \implies \dfrac{5x - 20}{4} = y \implies y = \dfrac{5}{4}x - 5$

22. $-\dfrac{1}{3}x + \dfrac{1}{2}y = 1 \implies \dfrac{1}{2}y = \dfrac{1}{3}x + 1 \implies y = \dfrac{2}{3}x + 2$

23. $2a + 3b = a \implies a = -3b$

24. $4m - 5n = 6m + 2n \implies -7n = 2m \implies n = -\dfrac{2}{7}m$

25. $A = \dfrac{1}{2}h(a+b) \implies 2A = h(a+b) \implies \dfrac{2A}{h} = a + b \implies \dfrac{2A}{h} - a = b \implies b = \dfrac{2A}{h} - a$

26. $V = \dfrac{1}{3}\pi r^2 h \implies 3V = \pi r^2 h \implies \dfrac{3V}{\pi r^2} = h \implies h = \dfrac{3V}{\pi r^2}$

27. $\dfrac{1}{2}x - \dfrac{3}{4}y = 2 \implies 2x - 3y = 8 \implies -3y = -2x + 8 \implies y = \dfrac{2}{3}x - \dfrac{8}{3} \implies f(x) = \dfrac{2}{3}x - \dfrac{8}{3}$

28. $-7(x - 4y) = y + 1 \implies -7x + 28y = y + 1 \implies 27y = 7x + 1 \implies y = \dfrac{7}{27}x + \dfrac{1}{27} \implies f(x) = \dfrac{7}{27}x + \dfrac{1}{27}$

29. $3x + 4y = 10 - y \implies 5y = -3x + 10 \implies y = -\dfrac{3}{5}x + 2 \implies f(x) = -\dfrac{3}{5}x + 2$

30. $\dfrac{y}{x} = 3 \implies y = 3x \implies f(x) = 3x$

31. $2x + 25 = 19;\ 2x + 25 = 19 \implies 2x = -6 \implies x = \dfrac{-6}{2} \implies x = -3$

32. $2x - 5 = x + 1;\ 2x - 5 = x + 1 \implies x = 6$

33. From the table $f(x) < 5$ when $x > -1; \{x | x > -1\}.$

34. $2(3-x)+4<0 \Rightarrow 6-2x+4<0 \Rightarrow -2x+10<0 \Rightarrow -2x<-10 \Rightarrow x>5; \{x|x>5\}$

35. From the graph $y_1 \geq y_2$ when $x \geq 2; \{x|x \geq 2\}$.

36. (a) $5-4x=-2 \Rightarrow -4x=-7 \Rightarrow x=\dfrac{7}{4}$

     (b) $5-4x<-2 \Rightarrow -4x<-7 \Rightarrow x>\dfrac{7}{4} \Rightarrow \left\{x|x>\dfrac{7}{4}\right\}$

     (c) $5-4x>-2 \Rightarrow -4x>-7 \Rightarrow x<\dfrac{7}{4} \Rightarrow \left\{x|x<\dfrac{7}{4}\right\}$

37. $-2x+1 \leq 3 \Rightarrow -2x \leq 2 \Rightarrow x \geq -1; \{x|x \geq -1\}$

38. $x-5 \geq 2x+3 \Rightarrow -x \geq 8 \Rightarrow x \leq -8; \{x|x \leq -8\}$

39. $\dfrac{3x-1}{4}>\dfrac{1}{2} \Rightarrow 3x-1>2 \Rightarrow 3x>3 \Rightarrow x>1; \{x|x>1\}$

40. $-3.2(x-2)<1.6x \Rightarrow -3.2x+6.4<1.6x \Rightarrow 6.4<4.8x \Rightarrow \dfrac{6.4}{4.8}<x \Rightarrow x>\dfrac{4}{3}; \left\{x|x>\dfrac{4}{3}\right\}$

41. $\dfrac{2}{5}t-(5-t)+3>2\left(\dfrac{3-t}{5}\right) \Rightarrow 2t-25+5t+15>6-2t \Rightarrow 7t-10>6-2t \Rightarrow$

    $9t>16 \Rightarrow t>\dfrac{16}{9}; \left\{t|t>\dfrac{16}{9}\right\}$

42. $\dfrac{2(3x-5)}{5}-3 \geq \dfrac{2-x}{3}+2 \Rightarrow 18x-30-45 \geq 10-5x+30 \Rightarrow 18x-75 \geq -5x+40 \Rightarrow$

    $23x \geq 115 \Rightarrow x \geq 5; \{x|x \geq 5\}$

43. $0.05t-0.15 \leq 0.03-0.75(t-2) \Rightarrow 0.05t-0.15 \leq -0.75t+1.53 \Rightarrow 0.8t \leq 1.68 \Rightarrow t \leq 2.1; \{t|t \leq 2.1\}$

44. $0.6z-1.55(z+5)<1-0.35(2z-1) \Rightarrow 0.6z-1.55z-7.75<1-0.7z+0.35 \Rightarrow$

    $-0.95z-7.75<-0.7z+1.35 \Rightarrow -0.25z<9.1 \Rightarrow z>-36.4; \{z|z>-36.4\}$

45. Graph $Y_1 = 0.72(\pi-1.3X)$ and $Y_2 = 0.54$ in $[-10,10,1]$ by $[-10,10,1]$. The solution is

    $\{x|x \leq 1.840\}$. See Figure 45.

    $[-10,10,1]$ by $[-10,10,1]$ $[-10,10,1]$ by $[-10,10,1]$.

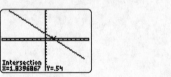

         Figure 45   Figure 46

46. Graph $Y_1 = \sqrt{(2)}-(4-\pi X)$ and $Y_2 = \sqrt{(7)}$ in $[-10,10,1]$ by $[-10,10,1]$. The solution is

    $\{x|x>1.665\}$. See Figure 46.

47. $x+1 \le 3 \Rightarrow x \le 2$ and $x+1 \ge -1 \Rightarrow x \ge -2$; $[-2, 2]$. See Figure 47.

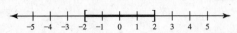

Figure 47

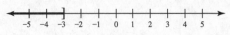

Figure 48

48. $2x+7 < 5 \Rightarrow 2x < -2 \Rightarrow x < -1$ and $-2x \ge 6 \Rightarrow x \le -3$; $(-\infty, -3]$. See Figure 48.

49. $5x-1 \le 3 \Rightarrow 5x \le 4 \Rightarrow x \le \dfrac{4}{5}$ or $1-x < -1 \Rightarrow 2 < x$; $\left(-\infty, \dfrac{4}{5}\right] \cup (2, \infty)$. See Figure 49.

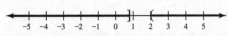

Figure 49

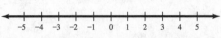

Figure 50

50. $3x+1 > -1 \Rightarrow 3x > -2 \Rightarrow x > -\dfrac{2}{3}$ or $3x+1 < 10 \Rightarrow 3x < 9 \Rightarrow x < 3$; $(-\infty, \infty)$. See Figure 50.

51. $2x+2$ is between -2 and 4 when $-2 \le x \le 1$; $[-2, 1]$

52. (a)   The intersection point of y_1 and y_2 is $(-4, -100)$. The solution is $x = -4$.

    (b)   The intersection point of y_2 and y_3 is $(2, 50)$. The solution is $x = 2$.

    (c) y_2 is between y_1 and y_3 when $-4 \le x \le 2$; $[-4, 2]$

    (d) y_2 is below y_3 when $x < 2$; $(-\infty, 2)$

53. (a)   The intersection point of y_1 and y_2 is $(2, 2)$. The solution is $x = 2$.

    (b) y_1 is below y_2 when $x > 2$; $(2, \infty)$

    (c) y_1 is above y_2 when $x < 2$; $(-\infty, 2)$

54. (a)   The intersection point of $f(x)$ and $g(x)$ is $(4, 20)$. The solution is $x = 4$.

    (b)   The intersection point of $g(x)$ and $h(x)$ is $(2, 40)$. The solution is $x = 2$.

    (c) $g(x)$ is between $f(x)$ and $h(x)$ when $2 < x < 4$; $(2, 4)$

55. $\left[-3, \dfrac{2}{3}\right]$

56. $(-6, 45]$

57. $\left(-\infty, \dfrac{7}{2}\right)$

58. $[1.8, \infty)$

59. $(-3, 4)$

60. $(-\infty, 4) \cup (10, \infty)$

61. $-4 < x+1 < 6 \Rightarrow -5 < x < 5$; The solution set is $(-5, 5)$.

62. $20 \le 2x+4 \le 60 \Rightarrow 16 \le 2x \le 56 \Rightarrow 8 \le x \le 28$; The solution set is $[8, 28]$.

63. $-3 < 4 - \frac{1}{3}x < 7 \Rightarrow -7 < -\frac{1}{3}x < 3 \Rightarrow 21 > x > -9 \Rightarrow -9 < x < 21;$ The solution set is $(-9, 21)$.

64. $2 \le \frac{1}{2}x - 2 \le 12 \Rightarrow 4 \le \frac{1}{2}x \le 14 \Rightarrow 8 \le x \le 28;$ The solution set is $[8, 28]$.

65. $-3 \le \frac{4-5x}{3} - 2 < 3 \Rightarrow -9 \le 4 - 5x - 6 < 9 \Rightarrow -9 \le -5x - 2 < 9 \Rightarrow -7 \le -5x < 11 \Rightarrow$

    $\frac{7}{5} \ge x > -\frac{11}{5} \Rightarrow -\frac{11}{5} < x \le \frac{7}{5}; \left(-\frac{11}{5}, \frac{7}{5}\right]$

66. $30 \le \frac{2x-6}{5} - 4 < 50 \Rightarrow 150 \le 2x - 6 - 20 < 250 \Rightarrow 150 \le 2x - 26 < 250 \Rightarrow$

    $176 \le 2x < 276 \Rightarrow 88 \le x < 138; [88, 138)$

67. No, $|12(-3) - 24| = |-60| = 60 \ne 24.$ No, $|12(2) - 24| = |0| = 0 \ne 24.$

68. No, $\left|5 - 3\left(\frac{4}{3}\right)\right| = |1| = 1 \not> 3.$ Yes, $|5 - 3(0)| = |5| = 5 > 3.$

69. No, $|3(-3) - 6| = |-15| = 15 \not\le 6.$ Yes, $|3(4) - 6| = |6| = 6 \le 6.$

70. No, $|2 + 3(-3)| + 4 = |-7| + 4 = 7 + 4 = 11 \not< 11.$ Yes, $\left|2 + 3\left(\frac{2}{3}\right)\right| + 4 = |4| + 4 = 4 + 4 = 8 < 11.$

71. (a) From the table, $y_1 = 2$ when $x = 0$ or $x = 4$.

    (b) From the table, $y_1 < 2$ when $0 < x < 4;$ $(0, 4)$.

    (c) From the table, $y_1 > 2$ when $x < 0$ or $x > 4;$ $(-\infty, 0) \cup (4, \infty)$.

72. (a) From the graph, $|2x + 2| = 4$ when $x = -3$ or $x = 1$.

    (b) From the graph, $|2x + 2| \le 4$ when $-3 \le x \le 1;$ $[-3, 1]$.

    (c) From the graph, $|2x + 2| \ge 4$ when $x \le -3$ or $x \ge 1;$ $(-\infty, -3] \cup [1, \infty)$.

73. $x = -22$ or $x = 22$

74. $2x - 9 = 7 \Rightarrow 2x = 16 \Rightarrow x = 8$ or $2x - 9 = -7 \Rightarrow 2x = 2 \Rightarrow x = 1$

75. $4 - \frac{1}{2}x = 17 \Rightarrow -\frac{1}{2}x = 13 \Rightarrow x = -26$ or $4 - \frac{1}{2}x = -17 \Rightarrow -\frac{1}{2}x = -21 \Rightarrow x = 42$

76. First note that $\frac{1}{3}|3x - 1| + 1 = 9 \Rightarrow \frac{1}{3}|3x - 1| = 8 \Rightarrow |3x - 1| = 24.$

    $3x - 1 = 24 \Rightarrow 3x = 25 \Rightarrow x = \frac{25}{3}$ or $3x - 1 = -24 \Rightarrow 3x = -23 \Rightarrow x = -\frac{23}{3}$

77. $2x - 5 = 5 - 3x \Rightarrow 5x = 10 \Rightarrow x = 2$ or $2x - 5 = -5 + 3x \Rightarrow -x = 0 \Rightarrow x = 0$

78. $-3 + 3x = -2x + 6 \Rightarrow 5x = 9 \Rightarrow x = \frac{9}{5}$ or $-3 + 3x = 2x - 6 \Rightarrow x = -3$

79. (a) $x + 1 = 7 \Rightarrow x = 6$ or $x + 1 = -7 \Rightarrow x = -8$

(b)   The solutions to $|x+1| \le 7$ satisfy $c \le x \le d$ where c and d are the solutions to $|x+1| = 7$.

From part (a), the interval is $[-8, \ 6]$.

(c)   The solutions to $|x+1| \ge 7$ satisfy $x \le c$ or $x \ge d$ where c and d are the solutions to $|x+1| = 7$.

From part (a), the interval is $(-\infty, \ -8] \cup [6, \ \infty)$.

80. (a) $1 - 2x = 6 \Rightarrow -2x = 5 \Rightarrow x = -\dfrac{5}{2}$ or $1 - 2x = -6 \Rightarrow -2x = -7 \Rightarrow x = \dfrac{7}{2}$

(b)   The solutions to $|1-2x| \le 6$ satisfy $c \le x \le d$ where c and d are the solutions to $|1-2x| = 6$.

From part (a), the interval is $\left[-\dfrac{5}{2}, \ \dfrac{7}{2} \right]$.

(c)   The solutions to $|1-2x| \ge 6$ satisfy $x \le c$ or $x \ge d$ where c and d are

the solutions to $|1-2x| = 6$. From part (a), the interval is $\left(-\infty, \ -\dfrac{5}{2} \right] \cup \left[\dfrac{7}{2}, \ \infty \right)$.

81. The solutions to $|x| > 3$ satisfy $x < c$ or $x > d$ where c and d are the solutions to $|x| = 3$.

$|x| = 3$ is equivalent to $x = -3$ or $x = 3$. The interval is $(-\infty, \ -3) \cup (3, \ \infty)$.

82. The solutions to $|-5x| < 20$ satisfy $c < x < d$ where c and d are the solutions to $|-5x| = 20$.

$|-5x| = 20$ is equivalent to $-5x = -20 \Rightarrow x = 4$ or $-5x = 20 \Rightarrow x = -4$. The interval is $(-4, \ 4)$.

83. The solutions to $|4x-2| \le 14$ satisfy $c \le x \le d$ where c and d are the solutions to $|4x-2| = 14$.

$|4x-2| = 14$ is equivalent to $4x - 2 = -14 \Rightarrow x = -3$ or $4x - 2 = 14 \Rightarrow x = 4$.

The interval is $[-3, \ 4]$.

84. The solutions to $\left| 1 - \dfrac{4}{5}x \right| \ge 3$ satisfy $x \le c$ or $x \ge d$ where c and d are the solutions to $\left| 1 - \dfrac{4}{5}x \right| = 3$.

$\left| 1 - \dfrac{4}{5}x \right| = 3$ is equivalent to $1 - \dfrac{4}{5}x = -3 \Rightarrow x = 5$ or $1 - \dfrac{4}{5}x = 3 \Rightarrow x = -\dfrac{5}{2}$.

The interval is $\left(-\infty, \ -\dfrac{5}{2} \right] \cup [5, \ \infty)$.

85. The solutions to $|t-4.5| \le 0.1$ satisfy $c \le t \le d$ where c and d are the solutions to $|t-4.5| = 0.1$.

$|t-4.5| = 0.1$ is equivalent to $t - 4.5 = -0.1 \Rightarrow t = 4.4$ or $t - 4.5 = 0.1 \Rightarrow t = 4.6$.

The interval is $[4.4, \ 4.6]$.

86. First divide each side of $-2|13t-5| \ge -4$ by -2 to obtain $|13t-5| \le 2$.

The solutions to $|13t-5| \le 2$ satisfy $c \le t \le d$ where c and d are the solutions to $|13t-5| = 2$.

$|13t-5| = 2$ is equivalent to $13t - 5 = -2 \Rightarrow t = \dfrac{3}{13}$ or $13t - 5 = 2 \Rightarrow t = \dfrac{7}{13}$. The interval is $\left[\dfrac{3}{13}, \ \dfrac{7}{13} \right]$.

87. The inequality $|5-4x| > -5$ is true for all values of x because absolute value is never negative.

    The interval is $(-\infty, \infty)$.

88. Since absolute value can never be negative, $|2t-3| \le 0$ is equivalent to $|2t-3| = 0$.

    $|2t-3| = 0$ is equivalent to $2t-3 = 0 \Rightarrow t = \dfrac{3}{2}$. The only solution is $\dfrac{3}{2}$.

89. Graph $Y_1 = \text{abs}(2X)$ and $Y_2 = 3$ in $[-3, 3, 1]$ by $[0, 5, 1]$. See Figures 89a and 89b.

    From the graph, $|2x| \ge 3$ in the interval $(-\infty, -1.5] \cup [1.5, \infty)$.

    $[-3, 3, 1]$ by $[0, 5, 1]$ $[-3, 3, 1]$ by $[0, 5, 1]$

    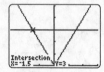   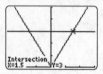

    Figure 89a   Figure 89b

90. Graph $Y_1 = \text{abs}((1/2)X - 1)$ and $Y_2 = 2$ in $[-7, 7, 1]$ by $[0, 4, 1]$. See Figures 90a and 90b.

    From the graph, $\left|\dfrac{1}{2}x - 1\right| \le 2$ in the interval $[-2, 6]$.

    $[-7, 7, 1]$ by $[0, 4, 1]$ $[-7, 7, 1]$ by $[0, 4, 1]$

    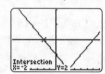   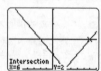

    Figure 90a   Figure 90b

91. $|x| \le 0.05$

92. $|5x-1| > 4$

93. Let x represent the loan amount at 5%. Then $7700 - x$ represents the loan amount at 7%.

    $0.05x + 0.07(7700 - x) = 469 \Rightarrow 0.05x + 539 - 0.07x = 469 \Rightarrow -0.02x = -70 \Rightarrow x = 3500$

    The loan amount at 5% was $3500 and the loan amount at 7% was $7700 - 3500 = \$4200$.

94. Let x represent the amount of time spent running at 8 mph. Then $1.4 - x$ represents the time running

    at 10 mph. $8x + 10(1.4 - x) = 12.8 \Rightarrow 8x + 14 - 10x = 12.8 \Rightarrow -2x = -1.2 \Rightarrow x = 0.6$

    The athlete ran for 0.6 hour at 8 mph and for $1.4 - 0.6 = 0.8$ hour at 10 mph.

95. Let x represent the cost of the house if it was located in San Francisco.

    $0.54x = 415,000 \Rightarrow x = \dfrac{415,000}{0.54} \approx 768,518.5$

    The house would cost about $769,000 in San Francisco.

96. (a) $m = \dfrac{5-10}{2008-1992} = \dfrac{-5}{16} = -0.3125$ Since x is the number of years after 1992 the y-intercept is

10 and the function is $f(x) = -0.3125x + 10$.

   (b)   Let $x = 13 \Rightarrow f(13) = -0.3125(13) + 10 \approx 5.9$ The result is about 5.9 million.

97. (a)   The distance between the riders is zero (they meet) when $x = 3$ hours.

   (b)   The distance between the riders is 20 when $x = 2$ hours and $x = 4$ hours.

   (c)   The riders are less than 20 miles apart between 2 and 4 hours exclusively.

   (d)   The sum of the speeds of the two bicyclists is $\dfrac{60 \text{ miles}}{3 \text{ hours}} = 20$ mph.

98. (a)   Car 1 is traveling faster since its graph has the steeper slope.

   (b)   The cars are both 200 miles from Austin at $x = 3$ hours.

   (c)   Car 1 is closer to Austin than Car 2 at times before $x = 3$ hours.

99. (a) $1.11x - 23.3 = 12 \Rightarrow 1.11x = 35.3 \Rightarrow x \approx 31.8$ inches

   (b)   Walleyes with lengths less than 31.8 inches would weigh less than 12 pounds.

100. (a)   When $x = 0$ (ground level), $T(x) = 60$ so the T-intercept is $b = 60$. Since the temperature

decreases by 19° for each one-mile increase in altitude, the slope is $a = -19$.

The formula is $T(x) = -19x + 60$.

   (b) $20 \le -19x + 60 \le 40 \Rightarrow -40 \le -19x \le -20 \Rightarrow 2.11 \ge x \ge 1.05 \Rightarrow 1.05 \le x \le 2.11$ (approximate)

The temperature is from 40°F to 20°F for altitudes from about 1.05 to 2.11 miles.

101. (a) $30 = 0.238x + 27.9 \Rightarrow 2.1 = 0.238x \Rightarrow x \approx 8.8$ Since x represents the number of years after

1970 the result will be about 1979.

   (b)   The median age increased, on average, by 0.238 year per year. In 1970 ($x = 0$) the median age

was 27.9.

102. Let x be the width of the rectangle. Then its length is $2x + 5$ and its perimeter is $2x + 2(2x + 5)$.

$2x + 2(2x + 5) = 88 \Rightarrow 2x + 4x + 10 = 88 \Rightarrow 6x = 78 \Rightarrow x = 13$ The width is 13 feet and the length is

$2(13) + 5 = 31$ feet. The rectangle is 13 ft by 31 ft.

103. $-48 \le \dfrac{9}{5}C + 32 \le 107 \Rightarrow -80 \le \dfrac{9}{5}C \le 75 \Rightarrow -80\left(\dfrac{5}{9}\right) \le C \le 75\left(\dfrac{5}{9}\right) \Rightarrow -44.\overline{4} \le C \le 41.\overline{6}$

The temperature range in Celsius is $-44.\overline{4}$°C to $41.\overline{6}$°C.

104. $|L - 160| \le 1;\ L - 160 = 1 \Rightarrow L = 161$ or $L - 160 = -1 \Rightarrow L = 159;\ \{L | 159 \le L \le 161\}$

105. (a) $|A - 3.9| \le 1.7$

   (b) $A - 3.9 = 1.7 \Rightarrow A = 5.6$ or $A - 3.9 = -1.7 \Rightarrow A = 2.2;\ \{A | 2.2 \le A \le 5.6\}$

106. The solutions to $\left|\dfrac{T-35}{35}\right| < 0.08$ satisfy $c < T < d$ where c and d are the solutions to $\left|\dfrac{T-35}{35}\right| = 0.08.$

$\left|\dfrac{T-35}{35}\right| = 0.08$ is equivalent to $\dfrac{T-35}{35} = -0.08 \Rightarrow x = 32.2$ and $\dfrac{T-35}{35} = 0.08 \Rightarrow x = 37.8.$

The interval is $(32.2,\ 37.8).$ The values must be between 32.2 and 37.8, exclusively.

Chapter 3 Test

1. Yes; $3 - 4\left(1 - 2 \cdot \dfrac{1}{6}\right) = 3 - 4\left(1 - \dfrac{2}{6}\right) = 3 - 4\left(\dfrac{4}{6}\right) = 3 - \dfrac{16}{6} = \dfrac{18}{6} - \dfrac{16}{6} = \dfrac{2}{6} = \dfrac{1}{3}$ and $2 \cdot \dfrac{1}{6} = \dfrac{2}{6} = \dfrac{1}{3}$

2.
x	-2	-1	0	1	2
$2 - 3x$	8	5	2	-1	-4

The solution set for $2 - 3x \le -1$ is $\{x \mid x \ge 1\}.$

3. $3 - 5x = 18 \Rightarrow -5x = 15 \Rightarrow x = -3.$ Check: $3 - 5(-3) = 3 - (-15) = 3 + 15 = 18$

4. (a)   The intersection point of y_1 and y_2 is $(2,\ 1).$ The solution is $x = 2.$

   (b) y_1 is above y_2 when $x \le 2$; $(-\infty,\ 2]$

   (c) y_1 is below y_2 when $x \ge 2$; $[2,\ \infty)$

5. The graphs of $y_1 = 4 - 2x$ and $y_2 = 1 + x$ (not shown) intersect at the point $(1,\ 2).$

   The solution is $x = 1.$

6. (a) $-\dfrac{2}{3}(3x - 2) + 1 = x \Rightarrow -2x + \dfrac{4}{3} + 1 = x \Rightarrow \dfrac{7}{3} = 3x \Rightarrow \dfrac{1}{3}\left(\dfrac{7}{3}\right) = x \Rightarrow x = \dfrac{7}{9}$

   (b) $\dfrac{2z + 1}{3} = \dfrac{3(1 - z)}{2} \Rightarrow 2(2z + 1) = 3\left[3(1 - z)\right] \Rightarrow 4z + 2 = 9 - 9z \Rightarrow 13z = 7 \Rightarrow z = \dfrac{7}{13}$

   (c) $0.4(2 - 3x) = 0.5x - 0.2 \Rightarrow 4(2 - 3x) = 5x - 2 \Rightarrow 8 - 12x = 5x - 2 \Rightarrow 10 = 17x \Rightarrow x = \dfrac{10}{17}$

7. x-intercept: $-6x + 3(0) = 9 \Rightarrow -6x = 9 \Rightarrow x = -\dfrac{3}{2}$

   y-intercept: $-6(0) + 3y = 9 \Rightarrow 3y = 9 \Rightarrow y = 3$

   $-6x + 3y = 9 \Rightarrow 3y = 6x + 9 \Rightarrow y = 2x + 3$

8. $2 + 5x = x - 4$; $2 + 5x = x - 4 \Rightarrow 4x = -6 \Rightarrow x = \dfrac{-6}{4} \Rightarrow x = -\dfrac{3}{2}$

9. $3x - 2y = 6 \Rightarrow -2y = -3x + 6 \Rightarrow y = \dfrac{3}{2}x - 3$; $f(x) = \dfrac{3}{2}x - 3$

10. $-5x + 1 \le 4 \Rightarrow -5x \le 3 \Rightarrow x \ge -\dfrac{3}{5}$; $\left[-\dfrac{3}{5},\ \infty\right)$

11. $3.1(3-x) < 2.9x \Rightarrow 9.3 - 3.1x < 2.9x \Rightarrow -6x < -9.3 \Rightarrow x > 1.55; \ (1.55, \ \infty)$

12. $2x+6 < 2$ and $-3x \geq 3 \Rightarrow 2x < -4$ and $x \leq -1 \Rightarrow x < -2$ and $x \leq -1;$ See Figure 12.

    Figure 12

13. From the table, $-3x < -3$ when $x > 1$ and $-3x > 6$ when $x < -2$. The interval is $(-\infty, \ -2) \cup (1, \ \infty)$.

14. (a)   The intersection point of y_1 and y_2 is $(-5, \ -300)$. The solution is $x = -5$.

    (b)   The intersection point of y_2 and y_3 is $(5, \ 100)$. The solution is $x = 5$.

    (c) y_2 is between y_1 and y_3 when $-5 \leq x \leq 5$. The interval is $[-5, \ 5]$.

    (d) y_2 is below y_3 when $x < 5$. The interval is $(-\infty, \ 5)$.

15. $-2 < 2 + \dfrac{1}{2}x < 2 \Rightarrow -4 < \dfrac{1}{2}x < 0 \Rightarrow -8 < x < 0; \ (-8, \ 0)$

16. $2 - \dfrac{1}{3}x = 6 \Rightarrow -\dfrac{1}{3}x = 4 \Rightarrow x = -12$ or $2 - \dfrac{1}{3}x = -6 \Rightarrow -\dfrac{1}{3}x = -8 \Rightarrow x = 24$

17. (a) $-5 \leq x \leq 5; [-5, \ 5]$

    (b)   The solution set is all real numbers except 0, or $(-\infty, \ 0) \cup (0, \ \infty)$.

    (c)   There are no solutions for this inequality. There are no values of x for which $|2x+7|$ is

           negative.

    (d) $5x = 10 \Rightarrow x = 2$ or $5x = -10 \Rightarrow x = -2$

           The solution set is $x < -2$ or $x > 2$, or $(-\infty, \ -2) \cup (2, \ \infty)$.

    (e) $-2|2-5x| + 1 \leq 5 \Rightarrow |2-5x| \geq -2$ This is true for all values of x, or $(-\infty, \ \infty)$.

    (f) $-5 < 3 - x < 5 \Rightarrow -8 < -x < 2 \Rightarrow -2 < x < 8; (-2, \ 8)$

18. (a) $1 - 5x = 3 \Rightarrow -5x = 2 \Rightarrow x = -\dfrac{2}{5}$ or $1 - 5x = -3 \Rightarrow -5x = -4 \Rightarrow x = \dfrac{4}{5}$

    (b)   The solutions to $|1-5x| \leq 3$ satisfy $c \leq x \leq d$ where c and d are the solutions to $|1-5x| = 3$.

           From part (a), the interval is $\left[-\dfrac{2}{5}, \ \dfrac{4}{5}\right]$.

    (c)   The solutions to $|1-5x| \geq 3$ satisfy $x \leq c$ or $x \geq d$ where c and d are the solutions to $|1-5x| = 3$.

           From part (a), the interval is $\left(-\infty, \ -\dfrac{2}{5}\right] \cup \left[\dfrac{4}{5}, \ \infty\right)$.

19. (a)   A sports drink with 16 grams of carbohydrates will have 64 calories.

    (b)   A sports drink with more than 16 grams of carbohydrates will have more than 64 calories.

    (c)   A sports drink with less than 16 grams of carbohydrates will have less than 64 calories.

    (d)   Each gram of carbohydrates corresponds to 4 calories.

20. (a) First divide the weight 150 by 2 to get 75. The function is $f(x) = 0.4x + 75$.

    (b) $0.4x + 75 = 89 \Rightarrow 0.4x = 14 \Rightarrow x = \dfrac{14}{0.4} \Rightarrow x = 35$ minutes

21. $d = \dfrac{1}{2}gt^2 \Rightarrow 2d = gt^2 \Rightarrow \dfrac{2d}{t^2} = g \Rightarrow g = \dfrac{2d}{t^2}$

22. Let x represent the loan amount at 4%. Then $5000 - x$ represents the loan amount at 6%.

    $0.04x + 0.06(5000 - x) = 230 \Rightarrow 0.04x + 300 - 0.06x = 230 \Rightarrow -0.02x = -70 \Rightarrow x = 3500$

    The loan amount at 4% was $3500 and the loan amount at 6% was $5000 - 3500 = \$1500$.

23. x: measure of smallest angle, $2x$: measure of largest angle, $x + 16$: measure of third angle

    $x + 2x + x + 16 = 180 \Rightarrow 4x + 16 = 180 \Rightarrow 4x = 164 \Rightarrow x = 41,\ 2x = 82,\ x + 16 = 57$

    The measures of the angles are $41°$, $57°$, and $82°$.

24. x: retail price of jacket, $x - 0.25x$: sale price of jacket, $x - 0.25x = 81 \Rightarrow 0.75x = 81 \Rightarrow x = 108$

    The retail price was $108.

25. (a) $7900 \le 190x + 6000 \le 9800 \Rightarrow 1900 \le 190x \le 3800 \Rightarrow 10 \le x \le 20$ Since x represents the
         number of years after 1980 the result is from 1990 to 2000.

    (b) The number of women increased, on average, by 190 per year. In 1980 ($x = 0$) there were
         6000 women marines.

Chapter 3 Extended and Discovery Exercises

1. (a) The temperature at ground level is $T(0) = 90 - 19(0) = 90\,°\text{F}$.

        The dew point at ground level is $D(0) = 70 - 5.8(0) = 70\,°\text{F}$.

   (b) $90 - 19x = 70 - 5.8x \Rightarrow 20 = 13.2x \Rightarrow \dfrac{20}{13.2} = x \Rightarrow x \approx 1.52$

        The temperature and dew point are equal at an altitude of about 1.52 miles.

   (c) Clouds will not form at altitudes below about 1.52 miles.

   (d) Clouds will form at altitudes above about 1.52 miles.

2. (a) At some altitudes it is likely that the air temperature will reach the dew point.

   (b) Clouds will form at much lower altitudes than usual. In this case, it may become foggy.

3. (a) The scatterplot is shown in Figure 3a.

   (b) The relationship appears to be linear. This seems reasonable if one considers that when the
        number of megabytes doubles, the number of seconds should also double.

(c)   Using the points $(0.129, 6.010)$ and $(1.260, 60.18)$

$$m = \frac{60.18 - 6.010}{1.260 - 0.129} \approx 47.9, \ x_1 = 0.129 \text{ and } y_1 = 6.010.$$

The equation is $y \approx 47.9(x - 0.129) + 6.010 \Rightarrow y \approx 47.9x - 0.1691$.

Each additional megabyte of memory can record approximately 47.9 seconds of music.

(d)   The scatterplot and the line are shown in Figure 3d.

(e) $47.9x - 0.1691 = 120$

(f)   Symbolic: $47.9x - 0.1691 = 120 \Rightarrow 47.9x = 120.1691 \Rightarrow x \approx 2.5$ MB

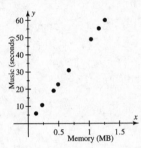
Figure 3a

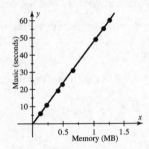
Figure 3d

Chapters 1-3 Cumulative Review Exercises

1. Natural: $\frac{8}{2} = 4$; Whole: $0, \ \frac{8}{2} = 4$; Integer: $-7, \ 0, \ \frac{8}{2} = 4$;

   Rational: $-7, \ -\frac{3}{5}, \ 0, \ \frac{8}{2} = 4, \ 5.\overline{12}$; Irrational: $\sqrt{5}$

2. Natural: $\sqrt{9} = 3$; Whole: $\frac{0}{9}, \ \sqrt{9} = 3$; Integer: $-\frac{6}{3}, \ \frac{0}{9}, \ \sqrt{9} = 3$;

   Rational: $-\frac{6}{3}, \ \frac{0}{9}, \ \sqrt{9} = 3, \ 4.\overline{6}$; Irrational: π

3. Distributive

4. Associative

5. $11 + 26 + (-1) + 14 = 11 + (-1) + 26 + 14 = 10 + 40 = 50$

6. $7 \cdot 99 = 7(100 - 1) = 700 - 17 = 693$

7.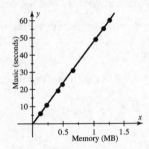

8. $-(-5y - 4) = 5y + 4$

9. $\dfrac{24x^{-4}y^2}{8xy^{-5}} = \dfrac{24}{8} \cdot x^{-4-1}y^{2-(-5)} = 3x^{-5}y^7 = \dfrac{3y^7}{x^5}$

10. $\left(\dfrac{3a^2}{4b^3}\right)^{-3} = \left(\dfrac{4b^3}{3a^2}\right)^{3} = \dfrac{4^3 b^9}{3^3 a^6} = \dfrac{64b^9}{27a^6}$

11. $15 - 2^3 \div 4 = 15 - 8 \div 4 = 15 - 2 = 13$

12. Move the decimal point 5 places to the right: $0.000059 = 5.9 \times 10^{-5}$.

13. $J = \sqrt{38 - (13)} = \sqrt{25} = 5$

14. $r = 3 - (-3)^3 = 3 - (-27) = 3 + 27 = 30$

15. By testing data points in each formula, we find that the best model is (ii).

16. The domain corresponds to the x-coordinates and the range corresponds to the y-coordinates.

   $D = \{-2,\ 0,\ 1,\ 3\};\ R = \{0,\ 2,\ 4\}$

17. When $x = -2$, $y = -2(-2) + 1 = 5$. The other ordered pairs can be found similarly.

   The ordered pairs are $(-2,\ 5)$, $(-1,\ 3)$, $(0,\ 1)$, $(1,\ -1)$, $(2,\ -3)$. See Figure 17.

18. When $x = -2$, $y = \dfrac{(-2)^3 + 4}{2} = \dfrac{-8 + 4}{2} = \dfrac{-4}{2} = -2$. The other ordered pairs can be found similarly.

   The ordered pairs are $(-2,\ -2)$, $\left(-1,\ \dfrac{3}{2}\right)$, $(0,\ 2)$, $\left(1,\ \dfrac{5}{2}\right)$, $(2,\ 6)$. See Figure 18.

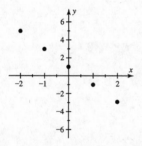
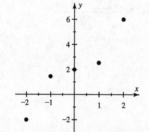
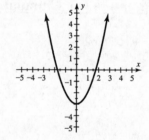

Figure 17   Figure 18   Figure 19

19. See Figure 19.

20. The function is defined for all values of the variable except 0. The domain is $\{x \mid x \neq 0\}$.

21. From the graph, when $x = 0$, $y = 2$. Therefore $f(0) = 2$. Similarly, $f(-2) = -2$.

22. From the table, when $x = 0$, $f(x) = 2$. Therefore $f(0) = 2$. Similarly, $f(-2) = -4$.

23. Yes; $m = -3$, $b = 11$

24. No. A linear function cannot contain a square root.

25. The slope is $m = \dfrac{11 - 7}{2 - 1} = 4$. Since $f(x) = 3$ when $x = 0$ the y-intercept is 3. Here $f(x) = 4x + 3$.

26. See Figure 26.

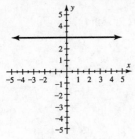

Figure 26

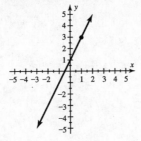

Figure 29

27. The equation is in the form $f(x) = mx + b$. The slope is 5 and the y-intercept is -4.

28. $m = \dfrac{6 - (-2)}{-1 - 3} = \dfrac{8}{-4} = -2$

29. See Figure 29.

30. The graph falls 1 unit for each 2 units of run. The slope is $-\dfrac{1}{2}$. The y-intercept is -1. So

    $y = -\dfrac{1}{2}x - 1$.

31. Since $f(x)$ increases by 3 for each unit increase in x, the slope is 3. From the point $(-1, 0)$, the

    x-intercept is -1. From the point $(0, 3)$, the y-intercept is 3.

32. Vertical lines have equations of the form $x = k$. The equation of the vertical line passing through

    $(5, 7)$ is $x = 5$.

33. Since the line is parallel to $y = 3x - 4$, the slope is $m = 3$. Using the point-slope form gives

    $y = 3(x - 2) + 1 \Rightarrow y = 3x - 6 + 1 \Rightarrow y = 3x - 5$

34. Since the line is perpendicular to $y = -x - 5$, the slope is $m = 1$. Using the point-slope form gives

    $y = 1(x - (-3)) + 0 \Rightarrow y = x + 3$

35. Yes since $\dfrac{1}{2}\left(\dfrac{8}{5}\right) - 4\left(\dfrac{8}{5} - 1\right) = \dfrac{4}{5} - 4\left(\dfrac{3}{5}\right) = \dfrac{4}{5} - \dfrac{12}{5} = -\dfrac{8}{5}$ and $\dfrac{1}{4}\left(\dfrac{8}{5}\right) - 2 = -\dfrac{8}{5}$.

36. The graphs intersect at the point $(1, -2)$, so $y_1 = y_2$ when $x = 1$.

37. $\dfrac{3}{4}(x - 2) + 4 = 2 \Rightarrow \dfrac{3}{4}(x - 2) = -2 \Rightarrow x - 2 = -\dfrac{8}{3} \Rightarrow x = -\dfrac{2}{3}$

38. $\dfrac{2}{3}\left(\dfrac{t - 7}{4}\right) - 2 = \dfrac{1}{3}t - (2t + 3) \Rightarrow \dfrac{t - 7}{6} - 2 = \dfrac{1}{3}t - 2t - 3 \Rightarrow t - 7 - 12 = 2t - 12t - 18 \Rightarrow$

    $t - 19 = -10t - 18 \Rightarrow 11t = 1 \Rightarrow t = \dfrac{1}{11}$

39. To find the x-intercept, let $y = 0$ and solve for x: $x - \dfrac{1}{3}(0) = 2 \Rightarrow x = 2$; to find the

    y-intercept, let $x = 0$ and solve for y: $0 - \dfrac{1}{3}y = 2 \Rightarrow -\dfrac{1}{3}y = 2 \Rightarrow y = -6$;

    $x - \dfrac{1}{3}y = 2 \Rightarrow -\dfrac{1}{3}y = -x + 2 \Rightarrow y = 3x - 6$

40. To find the x-intercept, let $y = 0$ and solve for x: $5x - 4(0) = -10 \Rightarrow 5x = -10 \Rightarrow x = -2$; to find the

    y-intercept, let $x = 0$ and solve for y: $5(0) - 4y = -10 \Rightarrow -4y = -10 \Rightarrow y = \dfrac{5}{2}$;

    $5x - 4y = -10 \Rightarrow -4y = -5x - 10 \Rightarrow y = \dfrac{5}{4}x + \dfrac{5}{2}$

41. Here $y < 2$ when $x > 0$. $\{x \mid x > 0\}$

42. Here y_1 is below y_2 when $x \le -3$. The interval is $(-\infty,\ -3]$.

43. $\dfrac{4x - 9}{6} > \dfrac{1}{2} \Rightarrow 4x - 9 > 3 \Rightarrow 4x > 12 \Rightarrow x > 3$. The interval is $(3,\ \infty)$.

44. $\dfrac{2}{3}z - 2 \le \dfrac{1}{4}z - (2z + 2) \Rightarrow 8z - 24 \le 3z - 24z - 24 \Rightarrow 8z - 24 \le -21z - 24 \Rightarrow$

    $29z \le 0 \Rightarrow z \le 0$. The interval is $(-\infty,\ 0]$.

45. $x + 2 > 1 \Rightarrow x > -1$ and $2x - 1 \le 9 \Rightarrow 2x \le 10 \Rightarrow x \le 5$

    The solutions must satisfy both of these inequalities. The interval is $(-1,\ 5]$. See Figure 45.

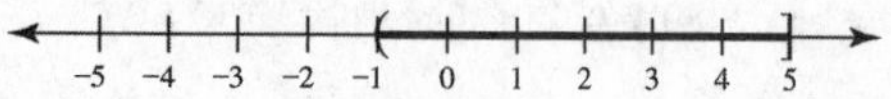

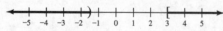

    Figure 45

46. $4x + 7 < 1 \Rightarrow 4x < -6 \Rightarrow x < -\dfrac{3}{2}$ or $3x + 2 \ge 11 \Rightarrow 3x \ge 9 \Rightarrow x \ge 3$. The solutions may satisfy only

    one (or both) of these inequalities. The interval is $\left(-\infty,\ -\dfrac{3}{2}\right) \cup [3,\ \infty)$. See Figure 46.

    Figure 46

47. $-7 \le 2x - 3 \le 5 \Rightarrow -4 \le 2x \le 8 \Rightarrow -2 \le x \le 4$. The interval is $[-2,\ 4]$.

48. $-8 \le -\dfrac{1}{2}x - 3 \le 5 \Rightarrow -5 \le -\dfrac{1}{2}x \le 8 \Rightarrow 10 \ge x \ge -16 \Rightarrow -16 \le x \le 10$. The interval is $[-16,\ 10]$.

49. The solutions to $|3x + 5| > 13$ satisfy $x < c$ or $x > d$ where c and d are the solutions to $|3x + 5| = 13$.

    $|3x + 5| = 13$ is equivalent to $3x + 5 = -13 \Rightarrow x = -6$ or $3x + 5 = 13 \Rightarrow x = \dfrac{8}{3}$.

    The interval is $(-\infty,\ -6) \cup \left(\dfrac{8}{3},\ \infty\right)$.

50. First divide each side of $-3|2t-11| \ge -9$ by -3 to obtain $|2t-11| \le 3$.

   The solutions to $|2t-11| \le 3$ satisfy $c \le t \le d$ where c and d are the solutions to $|2t-11| = 3$.

   $|2t-11| = 3$ is equivalent to $2t-11 = -3 \Rightarrow t = 4$ or $2t-11 = 3 \Rightarrow t = 7$. The interval is $[4, 7]$.

51. (a) $y_1 = 2$ when $x = -3$ or when $x = 1$.

   (b) $y_1 \le 2$ when $-3 \le x \le 1$. $[-3, 1]$

   (c) $y_1 \ge 2$ when $x \le -3$ or $x \ge 1$. $(-\infty, -3] \cup [1, \infty)$

52. $\frac{2}{3}x - 4 = -8 \Rightarrow \frac{2}{3}x = -4 \Rightarrow x = -6$ or $\frac{2}{3}x - 4 = 8 \Rightarrow \frac{2}{3}x = 12 \Rightarrow x = 18$

53. (a) $A = 4300(1.05)^{10} \approx \$7004.25$

   (b) $A = 11{,}000(1.05)^{6} \approx \$14{,}741.05$

54. For each 2-hour increase in time, the car travels an additional 130 miles. The car's speed is 65 mph. $d = 65t$

55. (a) There are 120 milligrams of sodium in a 12-ounce can so there are 10 milligrams of sodium per ounce. The formula is $f(x) = 10x$.

   (b) The slope of this line is 10.

   (c) The soda contains 10 milligrams of sodium per ounce.

56. See Figure 56.

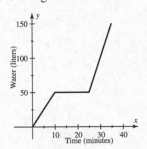

Figure 56

57. (a) The bicyclist is traveling away from Euclid because the slope of the line for the bicyclist is positive.

   (b) The lines intersect at the point (3, 200). When 3 hours has elapsed, they are both 200 miles from Euclid.

   (c) The motorcyclist is farther from Euclid from 0 to 3 hours, not including 3.

58. $600 \le 17(x-1960) + 56 \le 800 \Rightarrow 544 \le 17(x-1960) \le 744 \Rightarrow \frac{544}{17} \le x - 1960 \le \frac{744}{17} \Rightarrow$

   $\frac{544}{17} + 1960 \le x \le \frac{744}{17} + 1960 \Rightarrow 1992 \le x \le 2003.8$ (approximate)..

   The result is from 1992 to about 2004.

Chapter 4: Systems of Linear Equations

4.1: Systems of Linear Equations in Two Variables

1. No, two straight lines cannot have exactly two points of intersection.

3. Numerically and graphically.

5. There are no solutions.

7. Inconsistent

9. $(1, -2)$ since it satisfies both equations. The ordered pair $(4, 4)$ does not satisfy $3x + y = 1$.

11. $\left(-1, \dfrac{13}{3}\right)$ since it satisfies both equations. The ordered pair $(4, 6)$ does not satisfy $4x + 3y = 9$.

13. See Figure 13. The solution is $(1, 1)$.

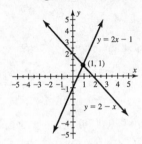
Figure 13

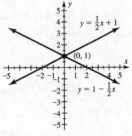
Figure 15

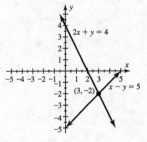
Figure 17

15. See Figure 15. The solution is $(0, 1)$.

17. See Figure 17. The solution is $(3, -2)$.

19. See Figure 19. The solution is $\left(\dfrac{3}{2}, \dfrac{3}{2}\right)$

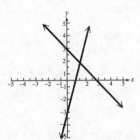
Figure 19

21. The solution is the intersection point $(3, 1)$.

23. The two lines do not intersect. There are no solutions.

25. The system can be solved graphically by solving each equation for y and then graphing.

$-x + y = 1 \Rightarrow y = x + 1$ and $x + y = 3 \Rightarrow y = 3 - x$ The graphs (not shown) intersect at the point

$(1, 2)$. The system is consistent. The equations are independent.

27. The system can be solved graphically by solving each equation for y and then graphing.

    $2x + y = 5 \Rightarrow y = 5 - 2x$ and $-2x + y = -3 \Rightarrow y = 2x - 3$ The graphs (not shown) intersect at the

    point $(2, 1)$. The system is consistent. The equations are independent.

29. The system can be solved graphically by solving each equation for y and then graphing.

    $x + y = 3 \Rightarrow y = 3 - x$ and $2x + 2y = 6 \Rightarrow y = 3 - x$ The graphs (not shown) are identical,

    $\{(x, y) | x + y = 3\}$. The system is consistent. The equations are dependent.

31. The system can be solved graphically by solving each equation for y and then graphing.

    $3x - y = 0 \Rightarrow y = 3x$ and $2x + y = 5 \Rightarrow y = 5 - 2x$ The graphs (not shown) intersect at the point

    $(1, 3)$. The system is consistent. The equations are independent.

33. The system can be solved graphically by solving each equation for y and then graphing.

    $-2x + y = 3 \Rightarrow y = 2x + 3$ and $4x - 2y = 2 \Rightarrow y = 2x - 1$ The graphs (not shown) are

    parallel. There are no solutions. The system is inconsistent.

35. The system can be solved graphically by solving each equation for y and then graphing.

    $x + y = 6 \Rightarrow y = 6 - x$ and $x - y = 2 \Rightarrow y = x - 2$ The graphs (not shown) intersect at the point

    $(4, 2)$. The system is consistent. The equations are independent.

37. The system can be solved graphically by solving each equation for y and then graphing.

    $x - y = 4 \Rightarrow y = x - 4$ and $2x - 2y = 4 \Rightarrow y = x - 2$ The graphs (not shown) are parallel. There are

    no solutions. The system is inconsistent.

39. The system can be solved graphically by solving each equation for y and then graphing.

    $6x - 4y = -2 \Rightarrow y = \dfrac{3x + 1}{2}$ and $-3x + 2y = 1 \Rightarrow y = \dfrac{3x + 1}{2}$ The graphs (not shown) are identical,

    $\{(x, y) | -3x + 2y = 1\}$. The system is consistent. The equations are dependent.

41. The system can be solved graphically by solving each equation for y and then graphing.

    $4x + 3y = 2 \Rightarrow y = \dfrac{2 - 4x}{3}$ and $5x + 2y = 6 \Rightarrow y = \dfrac{6 - 5x}{2}$ The graphs (not shown) intersect at the

    point $(2, -2)$. The system is consistent. The equations are independent.

43. The system can be solved graphically by solving each equation for y and then graphing.

    $2x + 2y = 4 \Rightarrow y = 2 - x$ and $x - 3y = -2 \Rightarrow y = \dfrac{x + 2}{3}$

    The graphs (not shown) intersect at the point $(1, 1)$.

45. The system can be solved graphically by solving each equation for y and then graphing.

    $-\dfrac{1}{2}x - \dfrac{1}{2}y = \dfrac{3}{2} \Rightarrow y = -x - 3$ and $x - \dfrac{1}{2}y = 3 \Rightarrow y = 2x - 6$

    The graphs (not shown) intersect at the point $(1, -4)$.

47. (a) $x + y = 4$: To find the x-intercept, let $y = 0$ and solve for x: $x + 0 = 4 \Rightarrow x = 4$; to find the

   y-intercept, let $x = 0$ and solve for y: $0 + y = 4 \Rightarrow y = 4$.

   $x - y = -2$: To find the x-intercept, let $y = 0$ and solve for x: $x - 0 = -2 \Rightarrow x = -2$; to find the

   y-intercept, let $x = 0$ and solve for y: $0 - y = -2 \Rightarrow -y = -2 \Rightarrow y = 2$.

   (b) See Figure 47b. The solution is (1, 3).

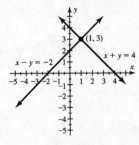

Figure 47b   Figure 49b

49. (a) $-2x + 3y = 12$: To find the x-intercept, let $y = 0$ and solve for x:

   $-2x + 3(0) = 12 \Rightarrow -2x = 12 \Rightarrow x = -6$; to find the y-intercept, let $x = 0$ and solve for y:

   $-2(0) + 3y = 12 \Rightarrow 3y = 12 \Rightarrow y = 4$.

   $-2x + y = 8$: To find the x-intercept, let $y = 0$ and solve for x:

   $-2x + 0 = 8 \Rightarrow -2x = 8 \Rightarrow x = -4$; to find the y-intercept, let $x = 0$ and solve for y:

   $-2(0) + y = 8 \Rightarrow y = 8$.

   (b) See Figure 49b. The solution is (−3, 2).

51. Note that $\frac{1}{4}x + \frac{1}{2}y = \frac{3}{20} \Rightarrow y = -\frac{1}{2}x + \frac{3}{10}$ and $\frac{1}{8}x - y = -\frac{3}{10} \Rightarrow y = \frac{1}{8}x + \frac{3}{10}$.

   Graph $Y_1 = (-1/2)X + (3/10)$ and $Y_2 = (1/8)X + (3/10)$ in [−1, 1, 0.1] by [−1, 1, 0.1]. See Figure 51.

   The unique solution is the intersection point (0, 0.3).

   [−1, 1, 0.1] by [−1, 1, 0.1]   [−4, 4, 1] by [−4, 4, 1]

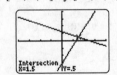

   Figure 51   Figure 53

53. Note that $0.1x + 0.2y = 0.25 \Rightarrow y = -\frac{1}{2}x + \frac{5}{4}$ and $0.7x - 0.3y = 0.9 \Rightarrow y = \frac{7}{3}x - 3$.

   Graph $Y_1 = (-1/2)X + (5/4)$ and $Y_2 = (7/3)X - 3$ in [−4, 4, 1] by [−4, 4, 1]. See Figure 53.

   The unique solution is the intersection point (1.5, 0.5).

55. Note that $0.1x + 0.2y = 50 \Rightarrow y = -0.5x + 250$ and $0.3x - 0.1y = 10 \Rightarrow y = 3x - 100$.

   Graph $Y_1 = -0.5X + 250$ and $Y_2 = 3X - 100$ in [0, 300, 50] by [0, 300, 50]. See Figure 55.

   The unique solution is the intersection point (100, 200).

[0, 300, 50] by [0, 300, 50]   [–10, 10, 1] by [–10, 10, 1]

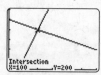   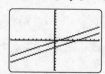

Figure 55   Figure 57

57. Note that $x - 2y = 5 \Rightarrow y = \frac{1}{2}x - \frac{5}{2}$ and $-2x + 4y = -2 \Rightarrow y = \frac{1}{2}x - \frac{1}{2}$.

Graph $Y_1 = (1/2)X - (5/2)$ and $Y_2 = (1/2)X - (1/2)$ in [–10, 10, 1] by [–10, 10, 1]. See Figure 57.

Since the lines are parallel, there is no solution.

59. Since $y_1 = y_2 = 3$ when $x = 2$, the solution is (2, 3).

61. Since $y_1 = y_2 = 1$ when $x = 1$, the solution is (1, 1).

63. Note that $x + y = 3 \Rightarrow y = 3 - x$ and $x - y = 7 \Rightarrow y = x - 7$.

Table $Y_1 = 3 - X$ and $Y_2 = X - 7$ with TblStart = 0 and ΔTbl = 1. See Figure 63.

Since $Y_1 = Y_2 = -2$ when X = 5, the solution is (5, –2).

                      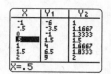

Figure 63   Figure 65   Figure 67

65. Note that $3x + 2y = 5 \Rightarrow y = -\frac{3}{2}x + \frac{5}{2}$ and $-x - y = -5 \Rightarrow y = -x + 5$.

Table $Y_1 = (-3/2)X + (5/2)$ and $Y_2 = -X + 5$ with TblStart = –8 and ΔTbl = 1. See Figure 65.

Since $Y_1 = Y_2 = 10$ when X = –5, the solution is (–5, 10).

67. Note that $0.5x - 0.1y = 0.1 \Rightarrow y = 5x - 1$ and $0.1x - 0.3y = -0.4 \Rightarrow y = \frac{1}{3}x + \frac{4}{3}$.

Table $Y_1 = 5X - 1$ and $Y_2 = (1/3)X + (4/3)$ with TblStart = –1 and ΔTbl = 0.5. See Figure 67.

Since $Y_1 = Y_2 = 1.5$ when X = 0.5, the solution is (0.5, 1.5).

69. $x + ay = 1 \Rightarrow ay = -x + 1 \Rightarrow y = -\frac{1}{a}x + \frac{1}{a}$; $2x + 2ay = 4 \Rightarrow 2ay = -2x + 4 \Rightarrow y = -\frac{1}{a}x + \frac{2}{a}$. So the

graphs of the two lines have the same slope $\left(-\frac{1}{a}\right)$ but different y-intercepts. Therefore, the lines are

parallel and do not intersect, so there are no solutions.

71. (a)   Let x and y represent the two numbers. Then the system needed is $x + y = 18$ and $x - y = 6$.

    (b)   Note that $x + y = 18 \Rightarrow y = -x + 18$ and $x - y = 6 \Rightarrow y = x - 6$. The graphs of these

            equations (not shown) intersect at the point (12, 6). The numbers are 12 and 6.

73. (a)   Let x and y represent the time spent running at 6 mph and 8 mph respectively.

Then the system needed is $x + y = 1$ and $6x + 8y = 7$.

(b)   Note that $x + y = 1 \Rightarrow y = 1 - x$ and $6x + 8y = 7 \Rightarrow y = \dfrac{7 - 6x}{8}$. The graphs of these equations

(not shown) intersect at the point (0.5, 0.5). The athlete ran for 0.5 hour at 6 mph and for 0.5

hour at 8 mph.

75. (a)   Let x and y represent the length and width of the rectangle respectively.

Then the system needed is $2x + 2y = 76$ and $x - y = 4$.

(b)   Note that $2x + 2y = 76 \Rightarrow y = -x + 38$ and $x - y = 4 \Rightarrow y = x - 4$.

The graphs of these equations (not shown) intersect at the point (21, 17).

The length of the rectangle is 21 inches and the width is 17 inches.

77. (a)   Let x and y represent the length and width of the rectangle respectively.

Then the system needed is $2x + 2y = 30$ and $x - 2y = 0$

(b)   Note that $2x + 2y = 30 \Rightarrow y = -x + 15$ and $x - 2y = 0 \Rightarrow y = \dfrac{1}{2}x$

The graphs of these equations (not shown) intersect at the point (10, 5).

The length of the rectangle is 10 cm and the width is 5 cm.

79. (a)   Let x and y represent the measure of the largest angle and the measure of one of the equal

angles respectively. Then the system needed is $x + 2y = 180$ and $x - y = 60$.

(b)   Note that $x + 2y = 180 \Rightarrow y = -\dfrac{1}{2}x + 90$ and $x - y = 60 \Rightarrow y = x - 60$.

The graphs of these equations (not shown) intersect at the point (100, 40).

The measure of the largest angle is 100° and the measure of each of the smaller angles is 40°.

81.   Let x and y represent the highest priced ticket and lowest priced ticket respectively.

Then the system needed is $x + y = 436$ and $x - y = 378$. See Figure 81a for the graphical solution

and Figure 81b for the numerical solution.

[400, 500, 1 by 20, 40, 1]

Figure 81a   Figure 81b

83.   Note that $x + y = 5 \Rightarrow y = -x + 5$ and $x = 2$ will be a vertical line.

The graphs of these equations (not shown) intersect at the point (2, 3). It follows that $x = 2$ and $y = 3$

and the person's age is 23.

85. Note that $3x + y = 9 \Rightarrow y = -3x + 9$ and $y = 3$ will be a horizontal line.

    The graphs of these equations (not shown) intersect at the point $(2, 3)$. It follows that $x = 2$ and $y = 3$ and the person's age is 23.

87. Let x and y represent the 2005 speed and the 2007 speed respectively. Then the system needed is

    $\dfrac{x + y}{2} = 25$ and $x - y = 2$. Note that $\dfrac{x + y}{2} = 25 \Rightarrow x + y = 50$ and $x - y = 2 \Rightarrow y = x - 2$. The

    graphs of these equations (not shown) intersect at the point $(26, 24)$. The 2005 speed is 26 mph and the 2007 speed is 24 mph.

89. Let x and y represent the number of home runs hit by McGwire and Sosa respectively. Then the

    system needed is $x + y = 136$ and $x - y = 4$. By solving each equation for y and graphing each

    (figure not shown) we find the intersection point is $(70, 66)$. McGwire hit 70 home runs and Sosa hit 66.

91. Let x represent the amount invested at 4% interest, and y represent the amount invested at 5%

    interest. Then the system needed is $x + y = 600$ and $y = x + 100$. By solving the first equation for y

    and graphing both equations (figure not shown) we find the intersection point is $(250, 350)$. $250 is

    invested at 4% and $350 is invested at 5%.

Section 4.2 The Substitution and Elimination Methods

1. Substitution and elimination.

3. Elimination yields a contradiction.

5. Solve the first equation for y.

7. Add the equations to eliminate the y variable.

9. Substituting $y = 2x$ into the second equation yields the following:

    $3x + (2x) = 5 \Rightarrow 5x = 5 \Rightarrow x = 1$ and so $y = 2(1) \Rightarrow y = 2$. The solution is $(1, 2)$.

11. Substituting $x = 2y - 1$ into the second equation yields the following:

    $(2y - 1) + 5y = 20 \Rightarrow 7y = 21 \Rightarrow y = 3$ and so $x = 2(3) - 1 \Rightarrow x = 5$. The solution is $(5, 3)$.

13. Note that $x - 2y = 0 \Rightarrow x = 2y$. Substituting $x = 2y$ into the second equation yields the following:

    $3(2y) + y = 7 \Rightarrow 7y = 7 \Rightarrow y = 1$ and so $x = 2(1) \Rightarrow x = 2$. The solution is $(2, 1)$.

    This result is supported by the graph's intersection point of $(2, 1)$.

15. Substituting $y = 3x$ into the second equation yields the following:

    $x + (3x) = 4 \Rightarrow 4x = 4 \Rightarrow x = 1$ and so $y = 3(1) = 3$. The solution is $(1, 3)$.

    By solving each equation for y and graphing each (figure not shown), the intersection point is $(1, 3)$.

17. Note that $x + y = 2 \Rightarrow y = 2 - x$. Substituting $y = 2 - x$ into the second equation yields:

    $2x - (2 - x) = 1 \Rightarrow 3x - 2 = 1 \Rightarrow 3x = 3 \Rightarrow x = 1$ and so $y = 2 - (1) = 1$. The solution is $(1, 1)$.

    By solving each equation for y and graphing each (figure not shown), the intersection point is $(1, 1)$.

19. Note that $x - 2y = -4 \Rightarrow x = 2y - 4$. Substituting $x = 2y - 4$ into the first equation yields:

    $2(2y - 4) + 3y = 6 \Rightarrow 7y = 14 \Rightarrow y = 2$ and so $x = 2(2) - 4 = 0$. The solution is $(0, 2)$. By

    solving each equation for y and graphing each (figure not shown), the intersection point is $(0, 2)$.

21. Note that $4x + y = 3 \Rightarrow y = -4x + 3$. Substituting $y = -4x + 3$ into the second equation yields:

    $2x - 3(-4x + 3) = -2 \Rightarrow 2x + 12x - 9 = -2 \Rightarrow 14x = 7 \Rightarrow x = \dfrac{1}{2}$ and so $y = -4\left(\dfrac{1}{2}\right) + 3 \Rightarrow y = 1$.

    The solution is $\left(\dfrac{1}{2}, 1\right)$.

23. Note that $x - 2y = 0 \Rightarrow x = 2y$. Substituting $x = 2y$ into the second equation yields:

    $-3(2y) + 2y = -1 \Rightarrow -6y + 2y = -1 \Rightarrow -4y = -1 \Rightarrow y = \dfrac{1}{4}$ and so $x = 2\left(\dfrac{1}{4}\right) = \dfrac{1}{2}$.

    The solution is $\left(\dfrac{1}{2}, \dfrac{1}{4}\right)$.

25. Note that $-x + 3y = 0 \Rightarrow x = 3y$. Substituting $x = 3y$ into the first equation yields:

    $2(3y) - 5y = -1 \Rightarrow y = -1$ and so $x = 3(-1) \Rightarrow x = -3$. The solution is $(-3, -1)$.

27. Note that $x - y = 1 \Rightarrow x = y + 1$. Substituting $x = y + 1$ into the second equation yields:

    $2(y + 1) + 6y = -2 \Rightarrow 8y = -4 \Rightarrow y = -\dfrac{1}{2}$ and so $x = \left(-\dfrac{1}{2}\right) + 1 \Rightarrow x = \dfrac{1}{2}$.

    The solution is $\left(\dfrac{1}{2}, -\dfrac{1}{2}\right)$.

29. Note that $2x - y = 6 \Rightarrow y = 2x - 6$. Substituting $y = 2x - 6$ into the first equation yields:

    $\dfrac{1}{2}x - \dfrac{1}{2}(2x - 6) = 1 \Rightarrow -\dfrac{1}{2}x = -2 \Rightarrow x = 4$ and so $y = 2(4) - 6 = 2$. The solution is $(4, 2)$.

31. Note that $\dfrac{1}{6}x - \dfrac{1}{3}y = -1 \Rightarrow x = 2y - 6$. Substituting $x = 2y - 6$ into the second equation yields:

    $\dfrac{1}{3}(2y - 6) + \dfrac{5}{6}y = 7 \Rightarrow \dfrac{3}{2}y = 9 \Rightarrow y = 6$ and so $x = 2(6) - 6 = 6$. The solution is $(6, 6)$.

33. Note that $\dfrac{1}{4}x - \dfrac{1}{3}y = 3 \Rightarrow x = \dfrac{4}{3}y + 12$. Substituting $x = \dfrac{4}{3}y + 12$ into the first equation yields:

    $\dfrac{1}{2}\left(\dfrac{4}{3}y + 12\right) + \dfrac{2}{3}y = -2 \Rightarrow \dfrac{4}{3}y = -8 \Rightarrow y = -6$ and so $x = \dfrac{4}{3}(-6) + 12 \Rightarrow x = 4$.

    The solution is $(4, -6)$.

35. Note that $0.1x + 0.4y = 1.3 \Rightarrow x = -4y + 13$. Substituting $x = -4y + 13$ into the second equation

    yields: $0.3(-4y + 13) - 0.2y = 1.1 \Rightarrow -1.4y = -2.8 \Rightarrow y = 2$ and so $x = -4(2) + 13 \Rightarrow x = 5$.

    The solution is (5, 2).

37. Note that $x - y = 5 \Rightarrow x = y + 5$. Substituting $x = y + 5$ into the second equation yields:

    $2(y + 5) - 2y = 10 \Rightarrow 10 = 10$ and this result is an identity. The system is dependent.

39. Note that $-\frac{5}{3}x - y = 2 \Rightarrow y = -\frac{5}{3}x - 2$. Substituting $y = -\frac{5}{3}x - 2$ into the first equation yields:

    $5x + 3\left(-\frac{5}{3}x - 2\right) = 6 \Rightarrow 5x - 5x - 6 = 6 \Rightarrow -6 = 6$, a contradiction. Therefore the system is

    inconsistent and there are no solutions.

41. Note that $x + 3y = -2 \Rightarrow x = -3y - 2$. Substituting $x = -3y - 2$ into the second equation yields:

    $-\frac{1}{2}(-3y - 2) - \frac{3}{2}y = 1 \Rightarrow \frac{3}{2}y + 1 - \frac{3}{2}y = 1 \Rightarrow 1 = 1$, an identity. Therefore the system is dependent and

    the solution set is $\{(x, y) \mid x + 3y = -2\}$.

43. Adding the two equations will eliminate the variable y.

    $\begin{aligned} x - y &= 5 \\ \underline{x + y} &= \underline{9} \\ 2x &= 14 \end{aligned}$ Thus, $x = 7$. And so $(7) - y = 5 \Rightarrow y = 2$. The solution is (7, 2).

    This result is supported by the graph's intersection point of (7, 2).

45. Adding the two equations will eliminate the variable y.

    $\begin{aligned} x + y &= 3 \\ \underline{x - y} &= \underline{1} \\ 2x &= 4 \end{aligned}$ Thus, $x = 2$. And so $(2) + y = 3 \Rightarrow y = 1$. The solution is $(2, 1)$.

    By solving each equation for y and graphing each (figure not shown), the intersection point is $(2, 1)$.

47. Multiplying the first equation by 2 and adding the two equations will eliminate the variable y.

    $8x - 2y = 8$

    $\dfrac{x + 2y = 1}{9x = 9}$ Thus, $x = 1$. And so $(1) + 2y = 1 \Rightarrow y = 0$. The solution is $(1, 0)$.

    By solving each equation for y and graphing each (figure not shown), the intersection point is $(1, 0)$.

49. Adding the two equations will eliminate the variable y.

    $x + y = 4$

    $\dfrac{x - y = 2}{2x = 6}$ Thus, $x = 3$. And so $3 + y = 4 \Rightarrow y = 1$. The solution is (3, 1).

51. Multiplying the first equation by -1 and adding the two equations will eliminate the variable x.

    $3x + y = -5$

    $\dfrac{-3x + 2y = -1}{3y = -6}$ Thus $y = -2$. And so $-3x - (-2) = 5 \Rightarrow -3x = 3 \Rightarrow x = -1$. The solution is $(-1, -2)$.

53. Adding the two equations will eliminate the variable y.

$$\begin{array}{l} 2x+y=\ \ 4 \\ \underline{2x-y=-2} \\ \ \ \ \ 4x=2 \end{array}$$ Thus, $x=\dfrac{1}{2}$. And so $2\left(\dfrac{1}{2}\right)+y=4 \Rightarrow y=3$. The solution is $\left(\dfrac{1}{2},3\right)$.

55. Multiplying the second equation by –2 and adding the two equations will eliminate the variable x.

$$\begin{array}{l} 6x-4y=12 \\ \underline{-6x-10y=12} \\ \ \ \ \ -14y=24 \end{array}$$ Thus, $y=-\dfrac{12}{7}$. And so $6x-4\left(-\dfrac{12}{7}\right)=12 \Rightarrow 6x=\dfrac{36}{7} \Rightarrow x=\dfrac{6}{7}$.

The solution is $\left(\dfrac{6}{7},-\dfrac{12}{7}\right)$.

57. Multiplying the second equation by 2 and adding the two equations will eliminate both variables.

$$\begin{array}{l} 2x-4y=5 \\ \underline{-2x+4y=18} \\ \ \ \ \ 0=23 \end{array}$$ This is always false and the system is inconsistent. No solutions.

59. Multiplying the first equation by –2 and adding the two equations will eliminate both variables.

$$\begin{array}{l} -4x-2y=-4 \\ \underline{\ \ \ 4x+2y=4} \\ \ \ \ \ \ \ 0=0 \end{array}$$ This is always true and the system is dependent with solutions: $\{(x,\ y)|2x+y=2\}$.

61. Multiply the first equation by $-\dfrac{1}{2}$, the second equation by $\dfrac{2}{15}$ and add the equations to eliminate the

variable y. $\begin{array}{l} -x-2y=11 \\ \underline{10x+2y=-16} \\ \ \ \ \ 9x=-5 \end{array}$ Thus, $x=-\dfrac{5}{9}$. And so $2\left(-\dfrac{5}{9}\right)+4y=-22 \Rightarrow 4y=-\dfrac{188}{9} \Rightarrow y=-\dfrac{47}{9}$.

The solution is $\left(-\dfrac{5}{9},-\dfrac{47}{9}\right)$.

63. Multiply the first equation by 30, the second equation by –20 and add the equations to eliminate the

variable y. $\begin{array}{l} 9x+6y=24 \\ \underline{-8x-6y=-22} \\ \ \ \ \ x=2 \end{array}$ Thus, $x=2$. And so $0.3(2)+0.2y=0.8 \Rightarrow 0.2y=0.2 \Rightarrow y=1$.

The solution is $(2,1)$

65. Note that $2x-y=-13 \Rightarrow y=2x+13$. Substituting $y=2x+13$ into the first equation yields:

$2x+3(2x+13)=7 \Rightarrow 8x=-32 \Rightarrow x=-4$ and so $x=2(-4)+13=5$. The solution is $(-4,5)$.

67. Note that $5u+v=2 \Rightarrow v=2-5u$. Substituting $v=2-5u$ into the first equation yields:

$3u-5(2-5u)=4 \Rightarrow 28u=14 \Rightarrow u=\dfrac{1}{2}$ and so $v=2-5\left(\dfrac{1}{2}\right)=-\dfrac{1}{2}$. The solution is $\left(\dfrac{1}{2},-\dfrac{1}{2}\right)$.

69. Multiplying the first equation by 2 and adding the two equations will eliminate both variables.

$4r - 6t = 14$
$\underline{-4r + 6t = -14}$ This is an identity. The system is dependent with solutions of the
$0 = 0$

form $\{(r,\ t) \mid 2r - 3t = 7\}$.

71. Multiplying the first equation by -1 and adding the two equations will eliminate both variables.

$-m + n = -5$
$\underline{m - n = 7}$ This is a contradiction. The system is inconsistent and has no solutions
$0 = 2$

73. Note that $2x - 3y = 2 \Rightarrow x = \dfrac{3}{2}y + 1$. Substituting $x = \dfrac{3}{2}y + 1$ into the second equation yields:

$3\left(\dfrac{3}{2}y + 1\right) - 5y = 4 \Rightarrow -\dfrac{1}{2}y = 1 \Rightarrow y = -2$ and so $x = \dfrac{3}{2}(-2) + 1 = -2$. The solution is $(-2, -2)$.

75. Note that $0.1x - 0.3y = -5 \Rightarrow x = 3y - 50$. Substituting $x = 3y - 50$ into the second equation yields:

$0.5(3y - 50) + 1.1y = 27 \Rightarrow 2.6y = 52 \Rightarrow y = 20$ and so $x = 3(20) - 50 = 10$.

The solution is $(10,\ 20)$.

77. Multiplying the first equation by -4 and the second equation by 4 will eliminate the variable z when

$-2y + 2z = 4$
adding. $\underline{3y - 2z = 4}$ Thus, $y = 8$. And so $-2(8) + 2z = 4 \Rightarrow z = 10$. The solution is $(8,\ 10)$.
$y = 8$

79. Multiplying the first equation by -5 and adding the two equations will eliminate the variable x.

$-x + 2y = 3$
$\underline{x - y = 1}$ Thus, $y = 4$. And so $x - (4) = 1 \Rightarrow x = 5$. The solution is $(5,\ 4)$.
$y = 4$

81. Note that $x - y = 4 \Rightarrow y = x - 4$ and $x + y = 6 \Rightarrow y = -x + 6$.

Graphical: Graph $Y_1 = X - 4$ and $Y_2 = -X + 6$ in [0, 10, 1] by [0, 2, 1]. See Figure 81a.

The unique solution is the intersection point $(5,\ 1)$.

Numerical: Table $Y_1 = X - 4$ and $Y_2 = -X + 6$ with TblStart = 0 and ΔTbl = 1. See Figure 81b.

Since $Y_1 = Y_2 = 1$ when $X = 5$, the solution is $(5,\ 1)$.

Symbolic: Adding the two equations will eliminate the variable y.

$x - y = 4$
$\underline{x + y = 6}$ Thus, $x = 5$. And so $(5) - y = 4 \Rightarrow -y = -1 \Rightarrow y = 1$. The solution is $(5,\ 1)$.
$2x = 10$

[0, 10, 1] by [0, 2, 1]   [0, 3, 1] by [0, 3, 1]

         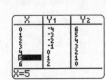   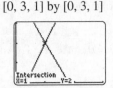   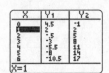

Figure 81a   Figure 81b   Figure 83a   Figure 83b

83. Note that $5x + 2y = 9 \Rightarrow y = -\dfrac{5}{2}x + \dfrac{9}{2}$ and $3x - y = 1 \Rightarrow y = 3x - 1$.

Graphical: Graph $Y_1 = (-5/2)X + (9/2)$ and $Y_2 = 3X - 1$ in [0, 3, 1] by [0, 3, 1]. See Figure 83a.

The unique solution is the intersection point (1, 2).

Numerical: Table $Y_1 = (-5/2)X + (9/2)$ and $Y_2 = 3X - 1$ with TblStart = 0 and ΔTbl = 1.

See Figure 83b. Since $Y_1 = Y_2 = 2$ when X = 1, the solution is (1, 2).

Symbolic: Multiplying the second equation by 2 and adding the two equations will eliminate the

variable y. $\begin{array}{r} 5x + 2y = 9 \\ 6x - 2y = 2 \\ \hline 11x = 11 \end{array}$ Thus, $x = 1$. And so $5(1) + 2y = 9 \Rightarrow 2y = 4 \Rightarrow y = 2$.

The solution is (1, 2).

85. Note that $3x - 2y = 8.5 \Rightarrow y = \dfrac{3x - 8.5}{2}$ and $2x + 4y = 3 \Rightarrow y = \dfrac{-2x + 3}{4}$.

Graphical: Graph $Y_1 = (3X - 8.5)/2$ and $Y_2 = (-2X + 3)/4$ in [−5, 5, 1] by [−5, 5, 1].

See Figure 85a. The unique solution is the intersection point $(2.5, -0.5)$.

Numerical: Table $Y_1 = (3X - 8.5)/2$ and $Y_2 = (-2X + 3)/4$ with TblStart = 0 and ΔTbl = 0.5.

See Figure 85b. Since $Y_1 = Y_2 = -0.5$ when X = 2.5, the solution is $(2.5, -0.5)$.

Symbolic: Multiplying the first equation by 2 and adding the two equations will eliminate the

variable y. $\begin{array}{r} 6x - 4y = 17 \\ 2x + 4y = 3 \\ \hline 8x = 20 \end{array}$ Thus, $x = 2.5$. And so $2(2.5) + 4y = 3 \Rightarrow y = -0.5$.

The solution is $(2.5, -0.5)$.

[−5, 5, 1] by [−5, 5, 1]

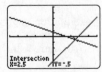   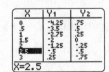

Figure 85a   Figure 85b

87. Multiplying the second equation by –1 and adding the two equations will eliminate the variable y:

$$ax + y = 4$$
$$\underline{-x - y = -4}$$ Thus $x = 0$ since $a \neq 1$. And so $0 + y = 4 \Rightarrow y = 4$. The solution is (0, 4).
$$(a-1)x = 0$$

89. Let x and y represent the time spent on the rowing machine and stair climber respectively. Then the system needed is $x + y = 60$ and $10x + 11.5y = 633$. Multiplying the first equation by –10 and adding the two equations will eliminate the variable x.

$$-10x\ \ -10y = -600$$
$$\underline{\ \ 10x + 11.5y = 633\ \ }$$ Thus, $y = 22$. And so $x + (22) = 60 \Rightarrow x = 38$.
$$1.5y = 33$$

The athlete spent 38 minutes on the rowing machine and 22 minutes on the stair climber.

91. Let x and y represent the amount of 10% and 80% solution respectively. Then the system needed is $x + y = 4$ and $0.10x + 0.80y = 0.50(4)$. Multiplying the second equation by –10 and adding the two

equations will eliminate the variable x.
$$x\ + y = 4$$
$$\underline{-x - 8y = -20}$$ Thus, $y = \dfrac{16}{7}$. And so
$$-7y = -16$$

$x + \left(\dfrac{16}{7}\right) = 4 \Rightarrow x = \dfrac{12}{7}$. Drain and replace $\dfrac{16}{7}$ gallons.

93. Let x and y represent the number of premium and regular rooms respectively. Then the system needed is $x + y = 50$ and $115x + 80y = 4945$. Multiplying the first equation by -80 and adding will

eliminate y.
$$-80x - 80y = -4000$$
$$\underline{\ 115x + 80y =\ \ 4945\ }$$ Thus, $x = 27$. And so $(27) + y = 50 \Rightarrow y = 23$.
$$35x = 945$$

The hotel rented 27 premium rooms and 23 regular rooms.

95. Let x and y represent the larger and smaller angles respectively. Then the system needed is $x + y = 180$ and $x - 2y = 30$. Multiplying the first equation by 2 and adding will eliminate y.

$$2x + 2y = 360$$
$$\underline{x - 2y =\ \ 30\ }$$ Thus, $x = 130$. And so $(130) + y = 180 \Rightarrow y = 50$. The angles are 50° and 130°.
$$3x = 390$$

97. Let x and y represent the numbers of males and females (in millions) in 2010, respectively. Then the system needed is $x + y = 308$ and $y = x + 5$. Substituting $y = x + 5$ into the first equation yields:

$x + x + 5 = 308 \Rightarrow 2x = 303 \Rightarrow x = 151.5$ and so $y = 151.5 + 5 \Rightarrow y = 156.5$. In 2010, there were 151.5 million males and 156.5 million females in the United States.

99. Let x and y represent the average speed of the plane and the jet stream respectively. Using the

formula $d = rt$ and converting the times to minutes, the system is $2400 = (x - y)(250)$ and

$2400 = (x + y)(225)$. These equations may be written as follows: $5x - 5y = 48$ and $3x + 3y = 32$.

Multiplying the first equation by 3, the second by 5 and adding the equations will eliminate the

variable y. $\begin{array}{r} 15x - 15y = 144 \\ 15x + 15y = 160 \\ \hline 30x = 304 \end{array}$ Thus, $x = \dfrac{152}{15}$. And so $3\left(\dfrac{152}{15}\right) + 3y = 32 \Rightarrow 3y = \dfrac{8}{5} \Rightarrow y = \dfrac{8}{15}$.

Plane: $\dfrac{152}{15}$ miles per minute or 608 mph; jet stream: $\dfrac{8}{15}$ miles per minute or 32 mph.

101. Let x and y represent the speed of the boat and the speed of the current respectively. Then the system

needed is $x + y = 30$ and $x - y = 20$. Adding the two equations will eliminate y.

$\begin{array}{r} x + y = 30 \\ x - y = 20 \\ \hline 2x = 50 \end{array}$ Thus, $x = 25$. And so $(25) + y = 30 \Rightarrow y = 5$.

The speed of the boat is 25 mph and the speed of the current is 5 mph.

103. Let x and y represent the amount borrowed at 8% and at 9% respectively. The system needed is

$x + y = 3500$ and $0.08x + 0.09y = 294$. Multiplying the first equation by -8, the second by 100 and

adding the equations will eliminate the variable x. $\begin{array}{r} -8x - 8y = -28,000 \\ 8x + 9y = 29,400 \\ \hline y = 1400 \end{array}$ Thus, $y = 1400$. And so

$x + 1400 = 3500 \Rightarrow x = 2100$. There was $2100 borrowed at 8% and $1400 borrowed at 9%.

105. Let x and y represent the cost of each day credit and each night credit respectively. The system

needed is $12x + 6y = 1800$ and $x - y = 30$. Multiplying the second by 6 and adding the equations

will eliminate y. $\begin{array}{r} 12x + 6y = 1800 \\ 6x - 6y = 180 \\ \hline 18x = 1980 \end{array}$ Thus, $x = 110$. And so $(110) - y = 30 \Rightarrow y = 80$.

Day credits cost $110 each and night credits cost $80 each.

107. Multiplying the second equation by $\dfrac{2}{\sqrt{3}}$ and adding the equations will eliminate W_2.

$\begin{array}{r} W_1 - W_2 = \quad 0 \\ W_1 + W_2 \approx 231 \\ \hline 2W_1 \approx 231 \end{array}$ Thus, $W_1 \approx 115.5$. And so $(115.5) - W_2 = 0 \Rightarrow W_2 = 115.5$.

The weight exerted on each rafter is about 115.5 pounds.

109. Let x and y represent the length and width of the court respectively. The system needed is

$2x + 2y = 296$ and is $x - y = 44$. Multiplying the first equation by 0.5 and adding the equations will

eliminate the variable y.

$x + y = 148$

(a) $x - y = 44$ Thus, $x = 96$. And so $2(96) + 2y = 296 \Rightarrow 2y = 104 \Rightarrow y = 52$.

$\dfrac{}{2x = 192}$

The court has a length 96 feet and a width 52 feet. Note that $2x + 2y = 296 \Rightarrow y = -x + 148$

and $x - y = 44 \Rightarrow y = x - 44$.

(b) Graph $Y_1 = -X + 148$ and $Y_2 = X - 44$ in [0, 150, 10] by [0, 100, 10]. See Figure 109b.

The unique solution is the intersection point (96, 52).

(c) Table $Y_1 = -X + 148$ and $Y_2 = X - 44$ with TblStart = 92 and ΔTbl = 1. See Figure 109c.

Since $Y_1 = Y_2 = 52$ when $X = 96$, the solution is (96, 52).

[0, 150, 10] by [0, 100, 10]

   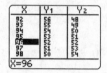

Figure 109b   Figure 109c

111. Let x and y represent the length and width of the rectangle respectively. The system needed is

$2x + 2y = 41$ and is $x - y = 5.5$. Multiplying the second equation by 2 and adding the equations will

eliminate the variable y.

$2x + 2y = 41$

$\dfrac{2x - 2y = 11}{4x = 52}$ Thus, $x = 13$. And so $2(13) + 2y = 41 \Rightarrow 2y = 15 \Rightarrow y = 7.5$.

The rectangle has a length 13 meters and a width 7.5 meters.

Checking Basic Concepts Sections 4.1 and 4.2

1. Note that $2x - y = 5 \Rightarrow y = 2x - 5$ and $-x + 3y = 0 \Rightarrow y = \dfrac{1}{3}x$.

(a) Graph $Y_1 = 2X - 5$ and $Y_2 = (1/3)X$ in [0, 5, 1] by [0, 2, 1]. See Figure 1a.

The unique solution is the intersection point (3, 1).

(b) Table $Y_1 = 2X - 5$ and $Y_2 = (1/3)X$ with TblStart = 0 and ΔTbl = 1. See Figure 1b.

Since $Y_1 = Y_2 = 1$ when $X = 3$, the solution is (3, 1). Yes, the answers agree.

[0, 5, 1] by [0, 2, 1]

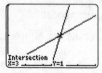   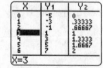

Figure 1a   Figure 1b

2. Note that $x + 4y = 14 \Rightarrow x = 14 - 4y$. Substituting $x = 14 - 4y$ into the first equation yields the following: $4(14 - 4y) - 3y = -1 \Rightarrow -19y = -57 \Rightarrow y = 3$ and so $x = 14 - 4(3) \Rightarrow x = 2$. The solution is (2, 3).

3. Multiply the first equation by 2, the second equation by 3 and add the equations to eliminate the variable y.

$$\begin{array}{r} 8x - 6y = -34 \\ -18x + 6y = 69 \\ \hline -10x = 35 \end{array}$$ Thus, $x = -\dfrac{7}{2}$. And so $4\left(-\dfrac{7}{2}\right) - 3y = -17 \Rightarrow -3y = -3 \Rightarrow y = 1$.

The solution is $\left(-\dfrac{7}{2}, 1\right)$. The system is not dependent. The system is not inconsistent.

4. (a)   Let x and y represent the larger and smaller angles respectively. Then the system needed is $x + y = 90$ and $x - y = 40$.

   (b)   Adding the two equations will eliminate y.

$$\begin{array}{r} x + y = 90 \\ x - y = 40 \\ \hline 2x = 130 \end{array}$$ Thus, $x = 65$. And so $(65) + y = 90 \Rightarrow y = 25$. The angles are 65° and 25°

Section 4.3 Systems of Linear Inequalities

1. Two

3. Yes, since $5(3) - 2(1) = 13 > 8$. No, since $5(2) - 2(1) = 8 \nless 8$.

5. solid

7. See Figure 7.

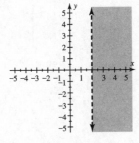

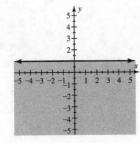

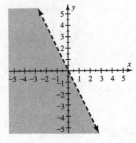

Figure 7   Figure 9   Figure 11

9. See Figure 9.

11. See Figure 11.

13. See Figure 13.

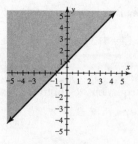

Figure 13

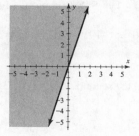

Figure 15

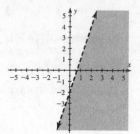

Figure 17

15. See Figure 15.

17. See Figure 17.

19. See Figure 19.

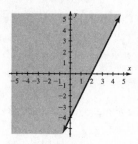

Figure 19

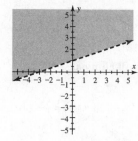

Figure 21

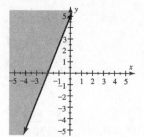

Figure 23

21. See Figure 21.

23. See Figure 23.

25. See Figure 25.

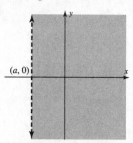

Figure 25

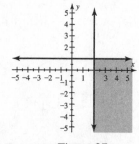

Figure 27

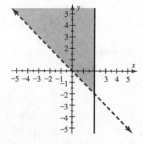

Figure 29

27. See Figure 27.

29. See Figure 29.

31. See Figure 31.

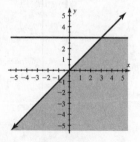

Figure 31

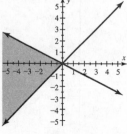

Figure 33

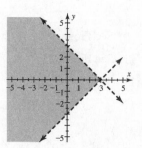

Figure 35

33. See Figure 33.

35. See Figure 35.

37. See Figure 37.

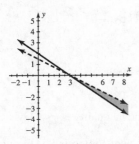

Figure 37

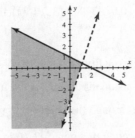

Figure 39

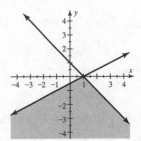

Figure 41

39. See Figure 39.

41. See Figure 41.

43. See Figure 43.

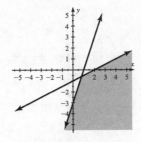

Figure 43

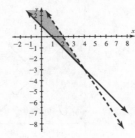

Figure 45

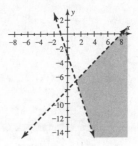

Figure 47

45. See Figure 45.

47. See Figure 47.

49. This region can be shaded using the *Shade* feature of the TI-83, found under the DRAW menu.

Shade $(-3, 5)$ in [–10, 10, 1] by [–10, 10, 1]. See Figure 49.

[–10, 10, 1] by [–10, 10, 1]   [–10, 10, 1] by [–10, 10, 1]   [–10, 10, 1] by [–10, 10, 1]

Figure 49

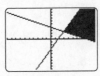

Figure 51

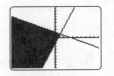

Figure 53

51. This region can be shaded using the *Shade* feature of the TI-83, found under the DRAW menu.

Note that $x + 2y \geq 8 \Rightarrow y \geq -\frac{1}{2}x + 4$ and $6x - 3y \geq 10 \Rightarrow y \leq 2x - \frac{10}{3}$.

Shade $\left((-1/2)X + 4, 2X - (10/3)\right)$ in [–10, 10, 1] by [–10, 10, 1]. See Figure 51.

53. This region can be shaded using the *Shade* feature of the TI-83, found under the DRAW menu.

   Note that $0.9x + 1.7y \le 3.2 \Rightarrow y \le \dfrac{-0.9}{1.7}x + \dfrac{3.2}{1.7}$ and $1.9x - 0.7y \le 1.3 \Rightarrow y \ge \dfrac{1.9}{0.7}x - \dfrac{1.3}{0.7}$.

   Shade $\left((1.9/0.7)X - (1.3/0.7), (-0.9/1.7)X + (3.2/1.7)\right)$ in [–10, 10, 1] by [–10, 10, 1].

   See Figure 53.

55. *b*; The point (1, 8) satisfies this inequality. Only graph *b* contains this point.

57. *a*; The point (1, –7) satisfies this inequality. Only graph *a* contains this point.

59. The equation of a horizontal line through the point (0, 2) is $y = 2$. The inequality is $y \ge 2$.

61. The equation of a vertical line through the point (–2, 0) is $x = -2$. The equation of a line through the

   points (0, 0) and (1, 1) is $y = x$. The system of inequalities is $y \ge x$ and $x \ge -2$.

63. First note that the amount of candy cannot be negative so $x \ge 0$ and $y \ge 0$. Also, if the cost must be

   less than \$15, the inequality $3x + 5y < 15$ must be satisfied. See Figure 63.

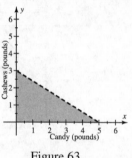

Figure 63

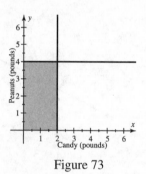

Figure 65   Figure 73

65. First note that the number of CD players and radios cannot be negative so $x \ge 0$ and $y \ge 0$. Then, if

   the business must manufacture at least as many radios as CD players, the inequality $y \ge x$ must be

   satisfied. Also, the total number made each day cannot exceed 40 so the inequality $x + y \le 40$ must

   be satisfied. See Figure 65.

67. (a) The range is approximately 105 to 134 bpm.

   (b) The inequalities are $y \le -0.6(x - 20) + 140$ and $y \ge -0.5(x - 20) + 110$.

69. The person weighs less than recommended.

71. The inequalities are $25h - 7w \le 800$ and $5h - w \ge 170$.

73. First note that the amounts of candy and peanuts cannot be negative so $x \ge 0$ and $y \ge 0$. Also,

   $3x \le 6 \Rightarrow x \le 2$ and $2y \le 8 \Rightarrow y \le 4$. See Figure 73.

75. (a) See Figure 75.

   (b) (7, 60): At mile marker 7, 60 mph is a lawful speed. Answers may vary.

77. (a) See Figure 77.

   (b) The point (387, 283) is not within the graphed boundaries.

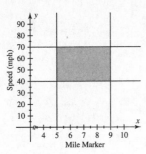

Figure 75

Figure 77

Section 4.4 Introduction to Linear Programming

1. linear programming

3. feasible solutions

5. vertex

7. See Figure 7.

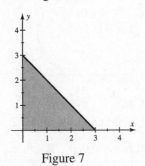

Figure 7

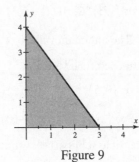

Figure 9

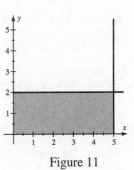

Figure 11

9. See Figure 9.

11. See Figure 11.

13. See Figure 13.

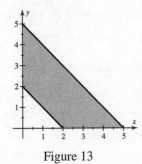

Figure 13

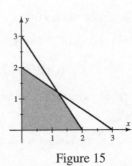

Figure 15

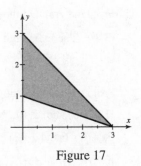

Figure 17

15. See Figure 15.

17. See Figure 17.

19. See Figure 19.

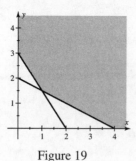

Figure 19

21. The maximum value of R occurs at one of the vertices.

    For $(1,1)$, $R = 4(1)+5(1) = 9$. For $(1,3)$, $R = 4(1)+5(3) = 19$. For $(4,1)$, $R = 4(4)+5(1) = 21$.

    For $(4,3)$, $R = 4(4)+5(3) = 31$. The maximum value is $R = 31$.

23. The maximum value of R occurs at one of the vertices.

    For $(0,2)$, $R = (0)+3(2) = 6$. For $(0,5)$, $R = (0)+3(5) = 15$. For $(3,3)$, $R = (3)+3(3) = 12$.

    For $(5,0)$, $R = (5)+3(0) = 5$. For $(2,0)$, $R = (2)+3(0) = 2$. The maximum value is $R = 15$.

25. The minimum value of C occurs at one of the vertices.

    For $(1,1)$, $C = 2(1)+3(1) = 5$. For $(1,3)$, $C = 2(1)+3(3) = 11$. For $(4,1)$, $C = 2(4)+3(1) = 11$.

    For $(4,3)$, $C = 2(4)+3(3) = 17$. The minimum value is $C = 5$.

27. The minimum value of C occurs at one of the vertices.

    For $(0,2)$, $C = 5(0)+(2) = 2$. For $(0,5)$, $C = 5(0)+(5) = 5$. For $(3,3)$, $C = 5(3)+(3) = 18$.

    For $(5,0)$, $C = 5(5)+(0) = 25$. For $(2,0)$, $C = 5(2)+(0) = 10$. The minimum value is $C = 2$.

29. From the graph of the region of feasible solutions (not shown), the vertices are (0, 0), (0, 150), and
    (150, 0). The maximum value of R occurs at one of the vertices. For $(0,0)$, $R = 3(0)+5(0) = 0$.

    For $(0,150)$, $R = 3(0)+5(150) = 750$. For $(150,0)$, $R = 3(150)+5(0) = 450$.

    The maximum value is $R = 750$.

31. From the graph of the region of feasible solutions (not shown), the vertices are (0, 0), (0, 6), and
    (3, 0). The maximum value of R occurs at one of the vertices. For $(0,0)$, $R = 3(0)+2(0) = 0$.

    For $(0,6)$, $R = 3(0)+2(6) = 12$. For $(3,0)$, $R = 3(3)+2(0) = 9$. The maximum value is $R = 12$.

33. From the graph of the region of feasible solutions (not shown), the vertices are (0, 0), (0, 2),
    (1.5, 1.5) and (2, 0). Note: to find the intersection point (1.5, 1.5), solve the system of equations
    $3x + y = 6$ and $x + 3y = 6$. The maximum value of R occurs at one of the vertices.

    For $(0,0)$, $R = 12(0)+9(0) = 0$. For $(0,2)$, $R = 12(0)+9(2) = 18$.

    For $(1.5,1.5)$, $R = 12(1.5)+9(1.5) = 31.5$. For $(2,0)$, $R = 12(2)+9(0) = 24$.

    The maximum value is $R = 31.5$.

35. From the graph of the region of feasible solutions (not shown), the vertices are $(0, 2)$, $(4, 0)$, and $(2, 0)$. The maximum value of R occurs at one of the vertices. For $(0, 2)$, $R = 4(0) + 5(2) = 10$. For $(4, 0)$, $R = 4(4) + 5(0) = 16$. For $(2, 0)$, $R = 4(2) + 5(0) = 8$. The maximum value is $R = 16$.

37. From the graph of the region of feasible solutions (not shown), the vertices are $(0, 0)$, $(0, 2)$, $(3, 2)$ and $(3, 0)$. The minimum value of C occurs at one of the vertices. For $(0, 0)$, $C = (0) + 2(0) = 0$. For $(0, 2)$, $C = (0) + 2(2) = 4$. For $(3, 2)$, $C = (3) + 2(2) = 7$. For $(3, 0)$, $C = (3) + 2(0) = 3$. The minimum value is $C = 0$.

39. From the graph of the region of feasible solutions (not shown), the vertices are $(0, 4)$, and $(4, 0)$. Note that this region is unbounded. The minimum value of C occurs at one of the vertices. For $(0, 4)$, $C = 8(0) + 15(4) = 60$. For $(4, 0)$, $C = 8(4) + 15(0) = 32$. The minimum value is $C = 32$.

41. From the graph of the region of feasible solutions (not shown), the vertices are $(0, 2)$, $(0, 6)$, $(3, 0)$ and $(2, 0)$. The minimum value of C occurs at one of the vertices. For $(0, 2)$, $C = 30(0) + 40(2) = 80$. For $(0, 6)$, $C = 30(0) + 40(6) = 240$. For $(3, 0)$, $C = 30(3) + 40(0) = 90$. For $(2, 0)$, $C = 30(2) + 40(0) = 60$. The minimum value is $C = 60$.

43. Let x and y represent the amount of candy and coffee respectively. The business can sell no more than a total of 100 pounds so $x + y \leq 100$. Also, because at least 20 pounds of candy must be sold each day $x \geq 20$. Finally, the amount of coffee cannot be negative so $y \geq 0$. Here the revenue function is $R = 4x + 6y$. From the graph of the region of feasible solutions (not shown), the vertices are $(20, 0)$, $(20, 80)$, and $(100, 0)$. The maximum value of R occurs at one of the vertices. For $(20, 0)$, $R = 4(20) + 6(0) = 80$. For $(20, 80)$, $R = 4(20) + 6(80) = 560$. For $(100, 0)$, $R = 4(100) + 6(0) = 400$. To maximize revenue, the business should sell 20 pounds of candy and 80 pounds of coffee.

45. Let x and y represent the number of graphing calculators and scientific calculators respectively. The business must make a total of at least 90 graphing and scientific calculators so $x + y \geq 90$. Also, because at least twice as many scientific calculators as graphing calculators must be manufactured $y \geq 2x$. Finally the quantities cannot be negative so $x \geq 0$ and $y \geq 0$. Here the cost function is $C = 50x + 20y$. From the graph of the region of feasible solutions (not shown), the vertices are $(30, 60)$, and $(0, 90)$. Note that this region is unbounded. The minimum value of C occurs at one of the vertices. For $(30, 60)$, $C = 50(30) + 20(60) = 2700$. For $(0, 90)$, $C = 50(0) + 20(90) = 1800$. To minimize cost, the business should manufacture 0 graphing calculators and 90 scientific calculators.

47. Let x and y represent the amount of Brand X and Brand Y respectively. Since each ounce of Brand X contains 20 units of vitamin A, each ounce of Brand Y contains 10 units of vitamin A and the total

amount of vitamin A must be at least 40 units, $20x+10y \geq 40$. Since each ounce of Brand X

contains 10 units of vitamin C, each ounce of Brand Y contains 10 units of vitamin C and the total

amount of vitamin C must be at least 30 units, $10x+10y \geq 30$. Finally the quantities cannot be

negative so $x \geq 0$ and $y \geq 0$. Here the cost function is $C = 0.90x + 0.60y$. From the graph of the

region of feasible solutions (not shown), the vertices are (0, 4), (1, 2) and (3, 0). Note that this region

is unbounded. To find the intersection point (1, 2) solve the system of equations

$20x + 10y = 40$ and $10x + 10y = 30$. The minimum value of C occurs at one of the vertices.

For $(0, 4)$, $C = 0.90(0) + 0.60(4) = 2.40$. For $(1, 2)$, $C = 0.90(1) + 0.60(2) = 2.10$.

For $(3, 0)$, $C = 0.90(3) + 0.60(0) = 2.70$. To minimize cost, 1 ounce of Brand X and 2 ounces of

Brand Y should be mixed.

49. Let x and y represent the number of hamsters and mice respectively. Since the total number of

animals cannot exceed 50, $x + y \leq 50$. Because no more than 20 hamsters can be raised,

$x \leq 20$. Here the revenue function is $R = 15x + 10y$. From the graph of the region of feasible

solutions (not shown), the vertices are (0, 50), (20, 30) and (20, 0). To find (20, 30) solve the

equations $x + y = 50$ and $x = 20$. The maximum value of R occurs at one of the vertices. For

$(0, 50)$, $R = 15(0) + 10(50) = 500$. For $(20, 30)$, $R = 15(20) + 10(30) = 600$.

For $(20, 0)$, $R = 15(20) + 10(0) = 300$. The maximum revenue is $600.

Checking Basic Concepts Sections 4.3 and 4.4

1. The equation of a line through the points (0, 3) and (1, 1) is $y = -2x + 3$.

   The inequality is $y \leq -2x + 3$.

2. See Figure 2.

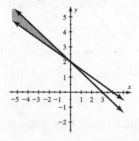

   Figure 2

3. From the graph of the region of feasible solutions (not shown), the vertices are (0, 0), (0, 5), (2, 3)

   and (4, 0). To find the intersection point (2, 3) solve the system of equations

   $3x + 2y = 12$ and $4x + 4y = 20$. The maximum value of R occurs at one of the vertices. For

   $(0, 0)$, $R = 3(0) + 2(0) = 0$. For $(0, 5)$, $R = 3(0) + 2(5) = 10$. For $(2, 3)$, $R = 3(2) + 2(3) = 12$.

   For $(4, 0)$, $R = 3(4) + 2(0) = 12$. The maximum value is $R = 12$.

Section 4.5 Systems of Linear Equations in Three Variables

1. No, three planes cannot intersect at exactly 2 points.

3. Yes, since $1+2+3=6$.

5. Two

7. No

9. $(1, 2, 3)$ satisfies all three equations.

11. $(-1, 1, 2)$ satisfies all three equations.

13. Substitute $z=1$ into the second equation: $2y+(1)=-1 \Rightarrow 2y=-2 \Rightarrow y=-1$

    Substitute $z=1$ and $y=-1$ into the first equation: $x+(-1)-(1)=1 \Rightarrow x-2=1 \Rightarrow x=3$

    The solution is $(3, -1, 1)$.

15. Substitute $z=2$ into the second equation: $2y+3(2)=3 \Rightarrow 2y=-3 \Rightarrow y=-\dfrac{3}{2}$

    Substitute $z=2$ and $y=-\dfrac{3}{2}$ into the first equation: $-x-3\left(-\dfrac{3}{2}\right)+(2)=-2 \Rightarrow -x=-\dfrac{17}{2} \Rightarrow x=\dfrac{17}{2}$

    The solution is $\left(\dfrac{17}{2}, -\dfrac{3}{2}, 2\right)$.

17. Substitute $c=-2$ into the second equation: $-3b+(-2)=4 \Rightarrow -3b=6 \Rightarrow b=-2$

    Substitute $c=-2$ and $b=-2$ into the first equation: $a-(-2)+2(-2)=3 \Rightarrow a-2=3 \Rightarrow a=5$

    The solution is $(5, -2, -2)$.

19. Add the first two equations together to eliminate the variable x.
$$\begin{array}{r} x+y-z=11 \\ -x+2y+3z=-1 \\ \hline 3y+2z=10 \end{array}$$

    From the third equation, $2z=4 \Rightarrow z=2$. And so $3y+2(2)=10 \Rightarrow 3y=6 \Rightarrow y=2$.

    Substitute $z=2$ and $y=2$ into the first equation: $x+(2)-(2)=11 \Rightarrow x=11$

    The solution is $(11, 2, 2)$.

21. Add the first two equations together to eliminate both of the variables x and z.
$$\begin{array}{r} x+y-z=-2 \\ -x+z=1 \\ \hline y=-1 \end{array}$$

    Substitute $y=-1$ into the third equation, $(-1)+2z=3 \Rightarrow 2z=4 \Rightarrow z=2$.

    Substitute $z=2$ and $y=-1$ into the first equation: $x+(-1)-(2)=-2 \Rightarrow x=1$

    The solution is $(1, -1, 2)$.

23. Add the second and third equations together to eliminate the variable y.
$$\begin{array}{r} y+z=-1 \\ -y+3z=9 \\ \hline 4z=8 \end{array}$$

    And so $z=2$. Substitute $z=2$ into the second equation, $y+(2)=-1 \Rightarrow y=-3$.

Substitute $z = 2$ and $y = -3$ into the first equation: $x + (-3) - 2(2) = -7 \Rightarrow x = 0$

The solution is $(0, -3, 2)$.

25. Multiply the second equation by -2 and add the first and second equations to eliminate the variables

y and z.
$$\begin{array}{r} x + 2y + 2z = 1 \\ -2x - 2y - 2z = 0 \\ \hline -x = 1 \end{array}$$
And so $x = -1$. Add the first and third equations together to eliminate

the variables x and y.
$$\begin{array}{r} x + 2y + 2z = 1 \\ -x - 2y + 3z = -11 \\ \hline 5z = -10 \end{array}$$
And so $z = -2$. Substitute $x = -1$ and $z = -2$ into the

second equation: $(-1) + y + (-2) = 0 \Rightarrow y = 3$ The solution is $(-1, 3, -2)$.

27. Multiply the second equation by -1 and add the first and second equations to eliminate the variables

y and z.
$$\begin{array}{r} x + y + z = 5 \\ -y - z = -6 \\ \hline x = -1 \end{array}$$
And so $x = -1$. Substitute $x = -1$ into the third equation:

$(-1) + z = 3 \Rightarrow z = 4$ Substitute $x = -1$ and $z = 4$ into the first equation $(-1) + y + (4) = 5 \Rightarrow y = 2$

The solution is $(-1, 2, 4)$.

29. Add the second and third equations to eliminate the variables x and z.
$$\begin{array}{r} -x + y + 2z = 1 \\ x + y - 2z = 9 \\ \hline 2y = 10 \end{array}$$

And so, $y = 5$. Now add the first and second equations to eliminate the variable x.

$$\begin{array}{r} x + 2y + 3z = 24 \\ -x + y + 2z = 1 \\ \hline 3y + 5z = 25 \end{array}$$
Now substitute $y = 5$ in this new equation: $3(5) + 5z = 25 \Rightarrow 5z = 10 \Rightarrow z = 2$

Finally substitute $y = 5$ and $z = 2$ in the first equation: $x + 2(5) + 3(2) = 24 \Rightarrow x = 8$

The solution is $(8, 5, 2)$.

31. Add the first and second equations to eliminate the variable y.
$$\begin{array}{r} x + y + z = 2 \\ x - y + z = 1 \\ \hline 2x + 2z = 3 \end{array}$$

Multiply the third equation by -2 and add to this new equation to eliminate the variables x and z.
$$\begin{array}{r} 2x + 2z = 3 \\ -2x + -2z = -6 \\ \hline 0 = -3 \end{array}$$
This is a contradiction, so there are no solutions.

33. Add the first two equations to eliminate the variable y.
$$\begin{aligned} x+y+z&=6 \\ x-y+z&=2 \end{aligned}$$
$$\overline{\quad 2x+2z=8 \quad}$$
$\Rightarrow x+z=4 \Rightarrow x=4-z$

Add the second and third equations to eliminate the variables x and z.
$$\begin{aligned} x-y+z&=2 \\ -x+5y-z&=6 \end{aligned}$$
$$\overline{\quad 4y\;\;=8 \quad}$$
$\Rightarrow y=2.$

The system is dependent, and the solutions are all ordered triples of the form $(4-z, 2, z)$.

35. Add the first and second equations to eliminate the variables y and z.
$$\begin{aligned} 2x+y+z&=3 \\ 2x-y-z&=9 \end{aligned}$$
$$\overline{\quad 4x=12 \quad}$$

And so, $x=3$. Add the second and third equations together to eliminate the variable y.

$$\begin{aligned} 2x-y-z&=9 \\ x+y-z&=0 \end{aligned}$$
$$\overline{\quad 3x-2z=9 \quad}$$
Now substitute $x=3$ in this new equation: $3(3)-2z=9 \Rightarrow -2z=0 \Rightarrow z=0$

Finally substitute $x=3$ and $z=0$ in the third equation: $(3)+y-(0)=0 \Rightarrow y=-3$

The solution is $(3, -3, 0)$.

37. Multiply the first equation by -1 and add the first and second equations to eliminate the variable x.

$$\begin{aligned} -2x-6y+2z&=-47 \\ 2x+y+3z&=-28 \end{aligned}$$
$$\overline{\quad -5y+5z=-75 \quad}$$
Multiply the third equation by 2 and add the first and third equations to

eliminate the variables x and z.
$$\begin{aligned} 2x+6y-2z&=47 \\ -2x+2y+2z&=-7 \end{aligned}$$
$$\overline{\quad 8y=40 \quad}$$
And so $y=5$. Substitute $y=5$ into the first

new equation: $-5(5)+5z=-75 \Rightarrow 5z=-50 \Rightarrow z=-10$ Substitute $y=5$ and $z=-10$ into the

original third equation: $-x+(5)+(-10)=-\dfrac{7}{2} \Rightarrow x=-\dfrac{3}{2}$ The solution is $\left(-\dfrac{3}{2}, 5, -10\right)$.

39. Multiply the second equation by -1 and add the first and second equations to eliminate the variable y.

$$\begin{aligned} x+3y-4z&=\frac{13}{2} \\ 2x-3y+z&=-\frac{1}{2} \end{aligned}$$
$$\overline{\quad 3x-3z=6 \quad}$$
Multiply this *new* equation by -1 and add it to the third equation to eliminate the

variable x.
$$\begin{aligned} -3x+3z&=-6 \\ 3x+z&=4 \end{aligned}$$
$$\overline{\quad 4z=-2 \quad}$$
And so $z=-\dfrac{1}{2}$. Substitute $z=-\dfrac{1}{2}$ into the first *new* equation:

$3x-3\left(-\dfrac{1}{2}\right)=6 \Rightarrow 3x=\dfrac{9}{2} \Rightarrow x=\dfrac{3}{2}$ Substitute $x=\dfrac{3}{2}$ and $z=-\dfrac{1}{2}$ into the *original* first equation:

$\left(\dfrac{3}{2}\right)+3y-4\left(-\dfrac{1}{2}\right)=\dfrac{13}{2} \Rightarrow y=1$ The solution is $\left(\dfrac{3}{2}, 1, -\dfrac{1}{2}\right)$.

41. Multiply the second equation by −1 and add to the third equation to eliminate the variables x, y,

and z. $\dfrac{\begin{array}{r} -x+y-z=-1 \\ x-y+z=3 \end{array}}{0=2}$ This is a contradiction. There are no solutions.

43. Add the first two equations to eliminate the variable y. $\dfrac{\begin{array}{r} x+y+z=5 \\ x-y+z=3 \end{array}}{2x \quad +2z=8} \Rightarrow x+z=4 \Rightarrow x=4-z$

    Multiply the first equation by −1 and add the second equation to eliminate the variables x and z.

    $\dfrac{\begin{array}{r} -x-y-z=-5 \\ x-y+z=3 \end{array}}{-2y \quad =-2} \Rightarrow y=1$ The system is dependent, and the solutions are all ordered pairs of the

    form $(4-z, 1, z)$.

45. Multiply the first equation by −1 and add the second equation to eliminate the variables x, y, and z.

    $\dfrac{\begin{array}{r} -x-y-z=-a \\ x+y+z=2a \end{array}}{0=a}$ But $a \neq 0$, so this is a contradiction. There are no solutions.

47. (a) $\begin{array}{l} x+2y+4z=10 \\ x+4y+6z=15 \\ 3y+2z=6 \end{array}$

    (b) Using techniques similar to those used in exercises 19-44, the solution is (2, 1, 1.5).

         A hamburger costs $2.00, fries cost $1.00 and a soft drink costs $1.50.

49. (a) $\begin{array}{l} x+y+z=180 \\ x \quad -z=55 \\ x-y-z=-10 \end{array}$

    (b) Using techniques similar to those used in exercises 19-44, the solution is (85, 65, 30).

         The angles are $x=85°$, $y=65°$, and $z=30°$.

    (c) These values check. $\begin{array}{l} x+y+z=180 \Rightarrow 85+65+30=180 \\ x \quad -z=55 \Rightarrow 85-30=55 \\ x-y-z=-10 \Rightarrow 85-65-30=-10 \end{array}$

51. Add the first two equations to eliminate the variables y and z. $\dfrac{\begin{array}{r} x+y+z=180 \\ x-y-z=40 \end{array}}{2x \quad =220} \Rightarrow x=110$

    $x-z=90 \Rightarrow z=x-90 \Rightarrow z=110-90 \Rightarrow z=20$ Substitute the values for x and z into the first

    equation and solve for y. $110+y+20=180 \Rightarrow y=50$ The angle measures are $110°$, $50°$, and $20°$.

$$a + 600b + 4c = 525$$

53. (a) $a + 400b + 2c = 365$

$$a + 900b + 5c = 805$$

(b) Using techniques similar to those used in exercises 19-44, the solution is (5, 1, –20).

That is, $a = 5$, $b = 1$, and $c = -20$ and so the equation is $F = 5 + A - 20W$.

(c) When $A = 500$ and $W = 3$, $F = 5 + (500) - 20(3) = 445$ fawns.

$$N + P + K = 80$$

55. (a) $N + P - K = 8$

$$9P - K = 0$$

(b) Using techniques similar to those used in exercises 19-44, the solution is (40, 4, 36).

The sample contains 40 pounds of nitrogen, 4 pounds of phosphorus and 36 pounds of potassium.

57. Let x, y and z represent the amounts invested at 4%, 5% and 7.5% respectively. The system needed

is: $\begin{aligned} x \quad + y \quad + z &= 30{,}000 \\ 0.04x + 0.05y + 0.75z &= 1775 \\ x \quad + y \quad - z &= 2000 \end{aligned}$ Using techniques similar to those used in exercises 19-44, the

solution is (7500, 8500, 14,000). There was \$7500 invested at 4%, \$8500 invested at 5% and \$14,000 invested at 7.5%.

59. $x = 2$, $y = 3$, $x + y + z = 6 \Rightarrow 2 + 3 + z = 6 \Rightarrow z = 1$ The solution is (2,3,1) and the pin is 231.

61. $x = 2$, $2x + y + z = 8$, $x + 2y + z = 9 \Rightarrow 4 + y + z = 8$ and $2 + 2y + z = 9$ Rewrite the last two equations

as $y + z = 4$ and $2y + z = 7$ to solve for y and z. Then back substitute to find x. The solution is

(2,3,1) and the pin is 231.

Section 4.6 Matrix Solutions of Linear Systems

1. A rectangular array of numbers

3. $\begin{bmatrix} 1 & 3 & | & 10 \\ 2 & -6 & | & 4 \end{bmatrix}$; 2×3; *Answers may vary.*

5. $\begin{bmatrix} 1 & 0 & | & -3 \\ 0 & 1 & | & 5 \end{bmatrix}$; *Answers may vary.*

7. 3×3

9. 3×2

11. $\begin{bmatrix} 1 & -3 & | & 1 \\ -1 & 3 & | & -1 \end{bmatrix}$

13. $\begin{bmatrix} 2 & -1 & 2 & | & -4 \\ 1 & -2 & 0 & | & 2 \\ -1 & 1 & -2 & | & -6 \end{bmatrix}$

15. $\begin{aligned} x + 2y &= -6 \\ 5x - y &= 4 \end{aligned}$

17. $\begin{aligned} x - y + 2z &= 6 \\ 2x + y - 2z &= 1 \\ -x + 2y - z &= 3 \end{aligned}$

19. $\begin{aligned} x &= 4 \\ y &= -2 \\ z &= 7 \end{aligned}$

21. $\begin{bmatrix} 0 & 1 & 1 & 1 \\ 1 & 0 & 1 & 0 \\ 0 & 0 & 0 & 1 \\ 1 & 0 & 1 & 0 \end{bmatrix}$

23. $\begin{bmatrix} 1 & 1 & | & 4 \\ 1 & 3 & | & 10 \end{bmatrix} R_2 - R_1 \rightarrow \begin{bmatrix} 1 & 1 & | & 4 \\ 0 & 2 & | & 6 \end{bmatrix} (1/2)R_2 \rightarrow \begin{bmatrix} 1 & 1 & | & 4 \\ 0 & 1 & | & 3 \end{bmatrix} R_1 - R_2 \rightarrow \begin{bmatrix} 1 & 0 & | & 1 \\ 0 & 1 & | & 3 \end{bmatrix}$ The solution is $(1, 3)$.

25. $\begin{bmatrix} 2 & 3 & | & 3 \\ -2 & 2 & | & 7 \end{bmatrix} R_2 + R_1 \rightarrow \begin{bmatrix} 2 & 3 & | & 3 \\ 0 & 5 & | & 10 \end{bmatrix} (1/5)R_2 \rightarrow \begin{bmatrix} 2 & 3 & | & 3 \\ 0 & 1 & | & 2 \end{bmatrix} R_1 - 3R_2 \rightarrow \begin{bmatrix} 2 & 0 & | & -3 \\ 0 & 1 & | & 2 \end{bmatrix}$

   $(1/2)R_2 \rightarrow \begin{bmatrix} 1 & 0 & | & -\frac{3}{2} \\ 0 & 1 & | & 2 \end{bmatrix}$ The solution is $\left(-\frac{3}{2}, 2\right)$.

27. $\begin{bmatrix} 1 & -1 & | & 5 \\ 1 & 3 & | & -1 \end{bmatrix} R_2 - R_1 \rightarrow \begin{bmatrix} 1 & -1 & | & 5 \\ 0 & 4 & | & -6 \end{bmatrix} (1/4)R_2 \rightarrow \begin{bmatrix} 1 & -1 & | & 5 \\ 0 & 1 & | & -\frac{3}{2} \end{bmatrix} R_1 + R_2 \rightarrow \begin{bmatrix} 1 & 0 & | & \frac{7}{2} \\ 0 & 1 & | & -\frac{3}{2} \end{bmatrix}$

   The solution is $\left(\frac{7}{2}, -\frac{3}{2}\right)$.

29. $\begin{bmatrix} 4 & -8 & | & -10 \\ 1 & 1 & | & 2 \end{bmatrix} \begin{matrix} Exchange \\ R_2 \leftrightarrow R_1 \end{matrix} \begin{bmatrix} 1 & 1 & | & 2 \\ 4 & -8 & | & -10 \end{bmatrix} (-1/2)R_2 \rightarrow \begin{bmatrix} 1 & 1 & | & 2 \\ -2 & 4 & | & 5 \end{bmatrix} R_2 + 2R_1 \rightarrow \begin{bmatrix} 1 & 1 & | & 2 \\ 0 & 6 & | & 9 \end{bmatrix}$

   $(1/6)R_2 \rightarrow \begin{bmatrix} 1 & 1 & | & 2 \\ 0 & 1 & | & \frac{3}{2} \end{bmatrix} R_1 - R_2 \rightarrow \begin{bmatrix} 1 & 0 & | & \frac{1}{2} \\ 0 & 1 & | & \frac{3}{2} \end{bmatrix}$ The solution is $\left(\frac{1}{2}, \frac{3}{2}\right)$.

31. $\begin{bmatrix} 1 & 1 & 1 & | & 6 \\ 0 & 2 & -1 & | & 1 \\ 0 & 1 & 1 & | & 5 \end{bmatrix} \begin{matrix} R_1 - R_3 \rightarrow \\ R_2 - 2R_3 \rightarrow \end{matrix} \begin{bmatrix} 1 & 0 & 0 & | & 1 \\ 0 & 0 & -3 & | & -9 \\ 0 & 1 & 1 & | & 5 \end{bmatrix} \begin{matrix} Exchange \\ R_2 \leftrightarrow R_3 \end{matrix} \begin{bmatrix} 1 & 0 & 0 & | & 1 \\ 0 & 1 & 1 & | & 5 \\ 0 & 0 & -3 & | & -9 \end{bmatrix}$

   $(-1/3)R_3 \rightarrow \begin{bmatrix} 1 & 0 & 0 & | & 1 \\ 0 & 1 & 1 & | & 5 \\ 0 & 0 & 1 & | & 3 \end{bmatrix} R_2 - R_3 \rightarrow \begin{bmatrix} 1 & 0 & 0 & | & 1 \\ 0 & 1 & 0 & | & 2 \\ 0 & 0 & 1 & | & 3 \end{bmatrix}$ The solution is $(1, 2, 3)$.

33. $\begin{bmatrix} 1 & 2 & 3 & | & 6 \\ -1 & 3 & 4 & | & 0 \\ 1 & 1 & -2 & | & -6 \end{bmatrix} \begin{matrix} R_2 + R_1 \rightarrow \\ R_3 - R_1 \rightarrow \end{matrix} \begin{bmatrix} 1 & 2 & 3 & | & 6 \\ 0 & 5 & 7 & | & 6 \\ 0 & -1 & -5 & | & -12 \end{bmatrix} (-1)R_3 \rightarrow \begin{bmatrix} 1 & 2 & 3 & | & 6 \\ 0 & 5 & 7 & | & 6 \\ 0 & 1 & 5 & | & 12 \end{bmatrix}$

   $\begin{matrix} R_1 - 2R_3 \rightarrow \\ R_2 - 5R_3 \rightarrow \end{matrix} \begin{bmatrix} 1 & 0 & -7 & | & -18 \\ 0 & 0 & -18 & | & -54 \\ 0 & 1 & 5 & | & 12 \end{bmatrix} \begin{matrix} Exchange \\ R_2 \leftrightarrow R_3 \end{matrix} \begin{bmatrix} 1 & 0 & -7 & | & -18 \\ 0 & 1 & 5 & | & 12 \\ 0 & 0 & -18 & | & -54 \end{bmatrix} (-1/18)R_3 \rightarrow \begin{bmatrix} 1 & 0 & -7 & | & -18 \\ 0 & 1 & 5 & | & 12 \\ 0 & 0 & 1 & | & 3 \end{bmatrix}$

$$\begin{matrix} R_1 + 7R_3 \to \\ R_2 - 5R_3 \to \\ \ \end{matrix}\begin{bmatrix} 1 & 0 & 0 & 3 \\ 0 & 1 & 0 & -3 \\ 0 & 0 & 1 & 3 \end{bmatrix}$$ The solution is $(3, -3, 3)$.

35. $\begin{bmatrix} 1 & 1 & 1 & 0 \\ 2 & 1 & 2 & -1 \\ 1 & 1 & 0 & 0 \end{bmatrix} \begin{matrix} \\ R_2 - 2R_1 \to \\ R_3 - R_1 \to \end{matrix} \begin{bmatrix} 1 & 1 & 1 & 0 \\ 0 & -1 & 0 & -1 \\ 0 & 0 & -1 & 0 \end{bmatrix} \begin{matrix} \\ (-1)R_2 \to \\ (-1)R_3 \to \end{matrix} \begin{bmatrix} 1 & 1 & 1 & 0 \\ 0 & 1 & 0 & 1 \\ 0 & 0 & 1 & 0 \end{bmatrix} R_1 - R_2 \to \begin{bmatrix} 1 & 0 & 1 & -1 \\ 0 & 1 & 0 & 1 \\ 0 & 0 & 1 & 0 \end{bmatrix}$

$R_1 - R_3 \to \begin{bmatrix} 1 & 0 & 0 & -1 \\ 0 & 1 & 0 & 1 \\ 0 & 0 & 1 & 0 \end{bmatrix}$ The solution is $(-1, 1, 0)$.

37. $\begin{bmatrix} 1 & 1 & 1 & 3 \\ -1 & 0 & -1 & -2 \\ 1 & 1 & 2 & 4 \end{bmatrix} \begin{matrix} \\ R_2 + R_1 \to \\ R_3 - R_1 \to \end{matrix} \begin{bmatrix} 1 & 1 & 1 & 3 \\ 0 & 1 & 0 & 1 \\ 0 & 0 & 1 & 1 \end{bmatrix} R_1 - R_2 \to \begin{bmatrix} 1 & 0 & 1 & 2 \\ 0 & 1 & 0 & 1 \\ 0 & 0 & 1 & 1 \end{bmatrix} R_1 - R_3 \to \begin{bmatrix} 1 & 0 & 0 & 1 \\ 0 & 1 & 0 & 1 \\ 0 & 0 & 1 & 1 \end{bmatrix}$

The solution is $(1, 1, 1)$.

39. $\begin{bmatrix} 1 & 2 & 1 & 3 \\ 2 & 1 & -1 & -6 \\ -1 & -1 & 2 & 5 \end{bmatrix} \begin{matrix} \\ R_2 - 2R_1 \to \\ R_3 + R_1 \to \end{matrix} \begin{bmatrix} 1 & 2 & 1 & 3 \\ 0 & -3 & -3 & -12 \\ 0 & 1 & 3 & 8 \end{bmatrix} (-1/3)R_2 \to \begin{bmatrix} 1 & 2 & 1 & 3 \\ 0 & 1 & 1 & 4 \\ 0 & 1 & 3 & 8 \end{bmatrix}$

$\begin{matrix} \\ \\ R_3 - R_2 \to \end{matrix} \begin{bmatrix} 1 & 2 & 1 & 3 \\ 0 & 1 & 1 & 4 \\ 0 & 0 & 2 & 4 \end{bmatrix} \begin{matrix} R_1 - 2R_2 \to \\ \\ (1/2)R_3 \to \end{matrix} \begin{bmatrix} 1 & 0 & -1 & -5 \\ 0 & 1 & 1 & 4 \\ 0 & 0 & 1 & 2 \end{bmatrix} \begin{matrix} R_1 + R_3 \to \\ R_2 - R_3 \to \\ \ \end{matrix} \begin{bmatrix} 1 & 0 & 0 & -3 \\ 0 & 1 & 0 & 2 \\ 0 & 0 & 1 & 2 \end{bmatrix}$

The solution is $(-3, 2, 2)$.

41. See example 7 in the text for graphing calculator instructions.

$[A] = \begin{bmatrix} 1 & 4 & 13 \\ 5 & -3 & -50 \end{bmatrix}$; $\text{rref}([A]) = \begin{bmatrix} 1 & 0 & -7 \\ 0 & 1 & 5 \end{bmatrix}$; The solution is $(-7, 5)$.

43. See example 7 in the text for graphing calculator instructions.

$[A] = \begin{bmatrix} 2 & -1 & 3 & 9 \\ -4 & 5 & 2 & 12 \\ 2 & 0 & 7 & 23 \end{bmatrix}$; $\text{rref}([A]) = \begin{bmatrix} 1 & 0 & 0 & 1 \\ 0 & 1 & 0 & 2 \\ 0 & 0 & 1 & 3 \end{bmatrix}$; The solution is $(1, 2, 3)$.

45. See example 7 in the text for graphing calculator instructions.

$[A] = \begin{bmatrix} 6 & 2 & 1 & 4 \\ -2 & 4 & 1 & -3 \\ 2 & -8 & 0 & -2 \end{bmatrix}$; $\text{rref}([A]) = \begin{bmatrix} 1 & 0 & 0 & 1 \\ 0 & 1 & 0 & 0.5 \\ 0 & 0 & 1 & -3 \end{bmatrix}$; The solution is $(1, 0.5, -3)$.

47. See example 7 in the text for graphing calculator instructions.

$[A] = \begin{bmatrix} 4 & 3 & 12 & -9.25 \\ -1 & 15 & 8 & -4.75 \\ 0 & 6 & 7 & -5.5 \end{bmatrix}$; $\text{rref}([A]) = \begin{bmatrix} 1 & 0 & 0 & 0.5 \\ 0 & 1 & 0 & 0.25 \\ 0 & 0 & 1 & -1 \end{bmatrix}$; The solution is $(0.5, 0.25, -1)$.

49. See example 7 in the text for graphing calculator instructions.

$[A] = \begin{bmatrix} 1.2 & -0.9 & 2.7 & 5.37 \\ 3.1 & -5.1 & 7.2 & 14.81 \\ 0.2 & 1.8 & -3.6 & -6.38 \end{bmatrix}$; $\text{rref}([A]) = \begin{bmatrix} 1 & 0 & 0 & 0.5 \\ 0 & 1 & 0 & -0.2 \\ 0 & 0 & 1 & 1.7 \end{bmatrix}$; The solution is $(0.5, -0.2, 1.7)$.

51. $\begin{bmatrix} 1 & 2 & 4 \\ -2 & -4 & -8 \end{bmatrix} R_2 + 2R_1 \to \begin{bmatrix} 1 & 2 & 4 \\ 0 & 0 & 0 \end{bmatrix}$

Row 2 represents the equation $0 = 0$. The system is dependent.

53. $\begin{bmatrix} 1 & 1 & 1 & | & 3 \\ 1 & 1 & -1 & | & 1 \\ 1 & 1 & 0 & | & 3 \end{bmatrix} \begin{matrix} \\ R_2 - R_1 \to \\ R_3 - R_1 \to \end{matrix} \begin{bmatrix} 1 & 1 & 1 & | & 3 \\ 0 & 0 & -2 & | & -2 \\ 0 & 0 & -1 & | & 0 \end{bmatrix} R_2 - 2R_3 \to \begin{bmatrix} 1 & 1 & 1 & | & 3 \\ 0 & 0 & 0 & | & -2 \\ 0 & 0 & -1 & | & 0 \end{bmatrix}$

Row 2 represents the equation $0 = -2$. The system is inconsistent.

55. $\begin{bmatrix} 1 & 2 & 3 & | & 14 \\ 2 & -3 & -2 & | & -10 \\ 3 & -1 & 1 & | & 4 \end{bmatrix} \begin{matrix} \\ R_2 - 2R_1 \to \\ R_3 - 3R_1 \to \end{matrix} \begin{bmatrix} 1 & 2 & 3 & | & 14 \\ 0 & -7 & -8 & | & -38 \\ 0 & -7 & -8 & | & -38 \end{bmatrix} R_3 - R_2 \to \begin{bmatrix} 1 & 2 & 3 & | & 14 \\ 0 & -7 & -8 & | & -38 \\ 0 & 0 & 0 & | & 0 \end{bmatrix}$

Row 3 represents the equation $0 = 0$. The system is dependent.

57. $\begin{bmatrix} a & 0 & 0 & | & 1 \\ 0 & b & 0 & | & 1 \\ 0 & 0 & ab & | & 2 \end{bmatrix} \begin{matrix} (1/a)R_1 \to \\ (1/b)R_2 \to \\ (1/(ab))R_3 \to \end{matrix} \begin{bmatrix} 1 & 0 & 0 & | & \frac{1}{a} \\ 0 & 1 & 0 & | & \frac{1}{b} \\ 0 & 0 & 1 & | & \frac{2}{ab} \end{bmatrix}$ The solution is $\left(\frac{1}{a}, \frac{1}{b}, \frac{2}{ab} \right)$.

59. The equation found in example 8 is $W = -374 + 19H + 6L$. When $H = 12$ and $L = 60$,

$$W = -374 + 19(12) + 6(60) = 214 \text{ lb.}$$

61. (a) $\begin{aligned} a + 16b + 26c &= 80 \\ a + 28b + 45c &= 344 \\ a + 31b + 54c &= 416 \end{aligned}$

   (b)   Using the graphing calculator to solve the system, the solution is $a \approx -272.9$, $b \approx 19.8$, $c \approx 1.4$.

   So the equation is $W \approx -272.9 + 19.8N + 1.4C$.

   (c)   When $N = 22$ and $C = 38$, $W \approx -272.9 + 19.8(22) + 1.4(38) \approx 216$ lb.

63. Let x, y and z represent the time spent running at 5, 6 and 8 mph respectively. The system needed is:

   $\begin{aligned} x + y + z &= 2 \\ 5x + 6y + 8z &= 12.5 \\ x - z &= 0 \end{aligned}$ and so $[A] = \begin{bmatrix} 1 & 1 & 1 & | & 2 \\ 5 & 6 & 8 & | & 12.5 \\ 1 & 0 & -1 & | & 0 \end{bmatrix}$; $\text{rref}([A]) = \begin{bmatrix} 1 & 0 & 0 & | & 0.5 \\ 0 & 1 & 0 & | & 1 \\ 0 & 0 & 1 & | & 0.5 \end{bmatrix}$

   The solution is (0.5, 1, 0.5). The runner ran 0.5 hr at 5 mph, 1 hr at 6 mph and 0.5 hr at 8 mph.

65. Let x, y and z represent the amount invested at 5%, 8% and 12% respectively. The system needed is:

   $\begin{aligned} x + y + z &= 3000 \\ 0.03x + 0.04y + 0.06z &= 145 \\ 3x - z &= 0 \end{aligned}$ and so $[A] = \begin{bmatrix} 1 & 1 & 1 & | & 3000 \\ 0.03 & 0.04 & 0.06 & | & 145 \\ 3 & 0 & -1 & | & 0 \end{bmatrix}$; $\text{rref}([A]) = \begin{bmatrix} 1 & 0 & 0 & | & 500 \\ 0 & 1 & 0 & | & 1000 \\ 0 & 0 & 1 & | & 1500 \end{bmatrix}$

   The solution is (500, 1000, 1500). There was $500 invested at 3%, $1000 at 4% and $1500 at 6%.

67. (a)   See example below. *Answers may vary.*

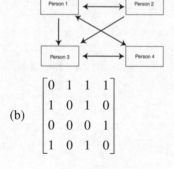

   (b) $\begin{bmatrix} 0 & 1 & 1 & 1 \\ 1 & 0 & 1 & 0 \\ 0 & 0 & 0 & 1 \\ 1 & 0 & 1 & 0 \end{bmatrix}$

Checking Basic Concepts Section 4.5 and 4.6

1. $(1, 3, -1)$ satisfies all three equations.

2. Multiply the first equation by -2 and add the first and second equations to eliminate the variable x.

$-2x + 2y - 2z = -4$

$\underline{2x - 3y\ + z = -1}$ Add the first and third equations to eliminate the variables x and y.

$-y - z = -5$

$x - y + z = 2$

$\underline{-x + y + z = 4}$ And so $z = 3$. Substitute $z = 3$ into the first *new* equation:

$2z = 6$

$-y - (3) = -5 \Rightarrow -y = -2 \Rightarrow y = 2$ Substitute $y = 2$ and $z = 3$ into the *original* first

equation: $x - (2) + (3) = 2 \Rightarrow x = 1$ The solution is $(1, 2, 3)$.

3. (a) $\begin{bmatrix} 1 & 2 & 1 & | & 1 \\ 1 & 1 & 1 & | & -1 \\ 0 & 1 & 1 & | & 1 \end{bmatrix} \begin{matrix} \\ R_2 - R_1 \rightarrow \\ \\ \end{matrix} \begin{bmatrix} 1 & 2 & 1 & | & 1 \\ 0 & -1 & 0 & | & -2 \\ 0 & 1 & 1 & | & 1 \end{bmatrix} \begin{matrix} R_1 + 2R_2 \rightarrow \\ -1R_2 \rightarrow \\ R_3 + R_2 \rightarrow \end{matrix} \begin{bmatrix} 1 & 0 & 1 & | & -3 \\ 0 & 1 & 0 & | & 2 \\ 0 & 0 & 1 & | & -1 \end{bmatrix}$

$R_1 - R_3 \rightarrow \begin{bmatrix} 1 & 0 & 0 & | & -2 \\ 0 & 1 & 0 & | & 2 \\ 0 & 0 & 1 & | & -1 \end{bmatrix}$; The solution is $(-2, 2, -1)$.

   (b) $[A] = \begin{bmatrix} 1 & 2 & 1 & | & 1 \\ 1 & 1 & 1 & | & -1 \\ 0 & 1 & 1 & | & 1 \end{bmatrix}$; $\text{rref}([A]) = \begin{bmatrix} 1 & 0 & 0 & | & -2 \\ 0 & 1 & 0 & | & 2 \\ 0 & 0 & 1 & | & -1 \end{bmatrix}$; The solution is $(-2, 2, -1)$.

4. Let x, y and z represent the amount invested at 1%, 2% and 4% respectively. The system needed is:

$x + \quad y \quad + z = 1500$

$0.01x + 0.02y + 0.04z = 46$ and so $[A] = \begin{bmatrix} 1 & 1 & 1 & | & 1500 \\ 0.01 & 0.02 & 0.04 & | & 46 \\ 2 & -1 & 0 & | & 0 \end{bmatrix}$; $\text{rref}([A]) = \begin{bmatrix} 1 & 0 & 0 & | & 200 \\ 0 & 1 & 0 & | & 400 \\ 0 & 0 & 1 & | & 900 \end{bmatrix}$

$2x - y \qquad\quad = 0$

The solution is $(200, 400, 900)$. There was \$200 invested at 1%, \$400 at 2% and \$900 at 4%.

Section 4.7 Determinants

1. square

3. system of linear equations

5. $\det A = 1(-8) - 3(-2) = -8 + 6 = -2$

7. $\det A = -3(-1) - 8(7) = 3 - 56 = -53$

9. $\det A = 23(-13) - 6(4) = -299 - 24 = -323$

11. $\det A = 1\left[(1)(7) - (-4)(-3)\right] - 0\left[(-1)(7) - (-4)(2)\right] + 0\left[(-1)(-3) - (1)(2)\right] = -5 - 0 + 0 = -5$

13. $\det A = 2\left[(-2)(8) - (1)(6)\right] - 1\left[(-1)(8) - (1)(0)\right] + 0\left[(-1)(6) - (-2)(0)\right] = -44 + 8 + 0 = -36$

15. $\det A = (-1)\big[(-3)(7)-(-3)(5)\big]-3\big[(3)(7)-(-3)(5)\big]+2\big[(3)(5)-(-3)(5)\big]=6-108+60=-42$

17. $\det A = 5\big[(-2)(5)-(0)(0)\big]-0\big[(0)(5)-(0)(0)\big]+0\big[(0)(0)-(-2)(0)\big]=-50-0+0=-50$

19. $\det A = 0\big[(3)(9)-(5)(-9)\big]-0\big[(2)(9)-(5)(-3)\big]+0\big[(2)(-9)-(3)(-3)\big]=0-0+0=0$

21. Using the calculator we find $\det([A])=-3555$.

23. Using the calculator we find $\det([A])=-7466.5$.

25. $\det A = a\,(bc-0)-0\,(0\cdot c-0)+0\,(0-0\cdot b)=abc$

27. The triangle has vertices $(3, 2)$, $(5, 8)$ and $(9, 5)$. The matrix needed is $A=\begin{bmatrix}3&5&9\\2&8&5\\1&1&1\end{bmatrix}$.

   The area is $D=\left|\dfrac{1}{2}\det([A])\right|=15\text{ ft}^2$.

29. The triangle has vertices $(-6, -4)$, $(2, 6)$ and $(6, -2)$. The matrix needed is $A=\begin{bmatrix}-6&2&6\\-4&6&-2\\1&1&1\end{bmatrix}$.

   The area is $D=\left|\dfrac{1}{2}\det([A])\right|=52\text{ ft}^2$.

31. Split the figure into two triangles with vertices $(2, 1)$, $(3, 6)$, $(9, 3)$ and vertices $(3, 6)$, $(7, 7)$, $(9, 3)$.

   The matrices needed are $A=\begin{bmatrix}2&3&9\\1&6&3\\1&1&1\end{bmatrix}$ and $B=\begin{bmatrix}3&7&9\\6&7&3\\1&1&1\end{bmatrix}$.

   The area is $D=\left|\dfrac{1}{2}\det([A])\right|+\left|\dfrac{1}{2}\det([B])\right|=16.5+9=25.5\text{ ft}^2$.

33. $E=\det\begin{bmatrix}4&3\\20&-4\end{bmatrix}=-16-60=-76;\quad F=\det\begin{bmatrix}5&4\\6&20\end{bmatrix}=100-24=76$

   $D=\det\begin{bmatrix}5&3\\6&-4\end{bmatrix}=-20-18=-38;\ $ The solution is $x=\dfrac{E}{D}=\dfrac{-76}{-38}=2$ and $y=\dfrac{F}{D}=\dfrac{76}{-38}=-2$.

35. $E=\det\begin{bmatrix}-3&-5\\-8&6\end{bmatrix}=-18-40=-58;\quad F=\det\begin{bmatrix}7&-3\\-4&-8\end{bmatrix}=-56-12=-68$

   $D=\det\begin{bmatrix}7&-5\\-4&6\end{bmatrix}=42-20=22;\ $ The solution is $x=\dfrac{E}{D}=\dfrac{-58}{22}=-\dfrac{29}{11}$ and $y=\dfrac{F}{D}=\dfrac{-68}{22}=-\dfrac{34}{11}$.

37. $E=\det\begin{bmatrix}-61&-3\\-23&-4\end{bmatrix}=244-69=175;\quad F=\det\begin{bmatrix}8&-61\\-1&-23\end{bmatrix}=-184-61=-245$

   $D=\det\begin{bmatrix}8&-3\\-1&-4\end{bmatrix}=-32-3=-35;\ $ The solution is $x=\dfrac{E}{D}=\dfrac{175}{-35}=-5$ and $y=\dfrac{F}{D}=\dfrac{-245}{-35}=7$.

Checking Basic Concepts Section 4.7

1. (a) $\det A = -3(3) - (-2)(4) = -9 + 8 = -1$

   (b) $\det A = 1\big[(1)(-1) - (2)(1)\big] - 5\big[(-2)(-1) - (2)(3)\big] + 0\big[(-2)(1) - (1)(3)\big] = -3 + 20 + 0 = 17$

2. $E = \det \begin{bmatrix} -14 & -1 \\ -36 & -4 \end{bmatrix} = 56 - 36 = 20;$ $F = \det \begin{bmatrix} 2 & -14 \\ 3 & -36 \end{bmatrix} = -72 - (-42) = -30$

   $D = \det \begin{bmatrix} 2 & -1 \\ 3 & -4 \end{bmatrix} = -8 - (-3) = -5;$ The solution is $x = \dfrac{E}{D} = \dfrac{20}{-5} = -4$ and $y = \dfrac{F}{D} = \dfrac{-30}{-5} = 6.$

3. The triangle has vertices $(-1, 2)$, $(5, 6)$ and $(2, -3)$. The matrix needed is $A = \begin{bmatrix} -1 & 5 & 2 \\ 2 & 6 & -3 \\ 1 & 1 & 1 \end{bmatrix}$.

   The area is $D = \left| \dfrac{1}{2} \det([A]) \right| = 21$ square units.

Chapter 4 Review

1. $(3, 2)$ is a solution since it satisfies both equations.

2. $(4, -3)$ is a solution since it satisfies both equations.

3. The unique solution is the intersection point $(-1, -3)$.

4. Since $Y_1 = Y_2 = 8.5$ when $x = 3.5,$ the solution is $(3.5, 8.5)$.

5. Note that $x + y = 6 \Rightarrow y = -x + 6$ and $x - y = -4 \Rightarrow y = x + 4.$

   Graph $Y_1 = -X + 6$ and $Y_2 = X + 4$ in $[0, 5, 1]$ by $[0, 8, 1]$. See Figure 5.

   The intersection point is $(1, 5)$. Since there is one unique intersection point, the system is consistent. The equations are independent.

   [0, 5, 1] by [0, 8, 1]   [–10, 10, 1] by [–10, 10, 1]

      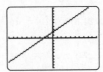

   Figure 5   Figure 6

6. Note that $x - y = -2 \Rightarrow y = x + 2$ and $-2x + 2y = 4 \Rightarrow y = x + 2.$

   Graph $Y_1 = X + 2$ and $Y_2 = X + 2$ in $[-10, 10, 1]$ by $[-10, 10, 1]$. See Figure 6. The graphs coincide. The solution is $\{(x, y) | x - y = -2\}$. The system is consistent, and the equations are dependent.

7. Note that $4x + 2y = 1 \Rightarrow y = -2x + \dfrac{1}{2}$ and $2x + y = 5 \Rightarrow y = -2x + 5.$

   Graph $Y_1 = -2X + (1/2)$ and $Y_2 = -2X + 5$ in $[-10, 10, 1]$ by $[-10, 10, 1]$. See Figure 7.

   Since the two lines are parallel, the system has no solutions. The system is inconsistent.

8. Note that $x - 3y = 5 \Rightarrow y = \frac{1}{3}x - \frac{5}{3}$ and $x + 5y = -3 \Rightarrow y = -\frac{1}{5}x - \frac{3}{5}$. Graph

   $Y_1 = (1/3)X - (5/3)$ and $Y_2 = (-1/5)X - (3/5)$ in [0, 3, 1] by [–3, 0, 1]. See Figure 8.

   The intersection point is $(2, -1)$. Since there is one unique intersection point, the system is

   consistent. The equations are independent.

   [–10, 10, 1] by [–10, 10, 1]   [0, 3, 1] by [–3, 0, 1]

        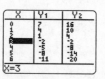   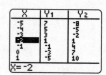

   Figure 7   Figure 8   Figure 9   Figure 10

9. Note that $3x + y = 7 \Rightarrow y = -3x + 7$ and $6x + y = 16 \Rightarrow y = -6x + 16$.

   Table $Y_1 = -3X + 7$ and $Y_2 = -6X + 16$ with TblStart = 0 and ΔTbl = 1. See Figure 9.

   Since $Y_1 = Y_2 = -2$ when X = 3, the solution is (3, –2).

10. Note that $4x + 2y = -6 \Rightarrow y = -2x - 3$ and $3x - y = -7 \Rightarrow y = 3x + 7$.

    Table $Y_1 = -2X - 3$ and $Y_2 = 3X + 7$ with TblStart = –5 and ΔTbl = 1. See Figure 10.

    Since $Y_1 = Y_2 = 1$ when X = –2, the solution is (–2, 1).

11. (a) Let x and y represent the two numbers. Then the system needed is $x + y = 25$ and $x - y = 10$.

    (b) Note that $x + y = 25 \Rightarrow y = -x + 25$ and $x - y = 10 \Rightarrow y = x - 10$.

         The graphs of these equations (not shown) intersect at the point (17.5, 7.5). The numbers are

         17.5 and 7.5.

12. (a) Let x and y represent the two numbers. Then the system needed is $3x - 2y = 19$ and

         $x + y = 18$.

    (b) Note that $3x - 2y = 19 \Rightarrow y = \frac{3}{2}x - \frac{19}{2}$ and $x + y = 18 \Rightarrow y = -x + 18$. The graphs of these

         equations (not shown) intersect at the point (11, 7). The numbers are 11 and 7.

13. Note that $\pi x - 2.1y = \sqrt{2} \Rightarrow -2.1y = -\pi x + \sqrt{2} \Rightarrow y = -\frac{1}{2.1}(-\pi x + \sqrt{2})$ and

    $\sqrt{3}x + y = \frac{5}{6} \Rightarrow y = -\sqrt{3}x + \frac{5}{6}$. Graph $Y_1 = -1/2.1(-\pi X + \sqrt{(2)})$ and $Y_2 = -\sqrt{(3)}X + 5/6$ in

    [–10, 10, 1] by [–10, 10, 1]. See Figure 13. The solution is approximately (0.467, 0.025).

    [–10, 10, 1] by [–10, 10, 1]   [–10, 10, 1] by [–10, 10, 1].

    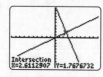

    Figure 13   Figure 14

14. Note that $\sqrt{5}x - \pi y = \dfrac{2}{7} \Rightarrow -\pi y = -\sqrt{5}x + \dfrac{2}{7} \Rightarrow y = -\dfrac{1}{\pi}\left(-\sqrt{5}x + \dfrac{2}{7}\right)$ and

$x + 0.3y = \pi \Rightarrow 0.3y = -x + \pi \Rightarrow y = \dfrac{1}{0.3}(-x + \pi)$. Graph $Y_1 = -1/\pi\,(-\sqrt{(5)}X + 2/7)$

and $Y_2 = 1/0.3\,(-X + \pi)$ in $[-10, 10, 1]$ by $[-10, 10, 1]$. See Figure 14.

The solution is approximately $(2.611, 1.768)$.

15. Note that $x + 2y = -1 \Rightarrow x = -2y - 1$. Substituting $x = -2y - 1$ into the first equation yields the

following: $2(-2y - 1) + 5y = -1 \Rightarrow y = 1$ and so $x = -2(1) - 1 \Rightarrow x = -3$. The solution is $(-3, 1)$.

16. Note that $3x + y = 6 \Rightarrow y = -3x + 6$. Substituting $y = -3x + 6$ into the second equation yields the

following: $4x + 5(-3x + 6) = 8 \Rightarrow -11x = -22 \Rightarrow x = 2$ and so $y = -3(2) + 6 \Rightarrow y = 0$.

The solution is $(2, 0)$.

17. Note that $4x + 2y = 0 \Rightarrow y = -2x$. Substituting $y = -2x$ into the first equation yields the following:

$2x - 3(-2x) = -8 \Rightarrow 8x = -8 \Rightarrow x = -1$ and so $y = -2(-1) = 2$. The solution is $(-1, 2)$.

18. Note that $5x + 3y = -1 \Rightarrow y = \dfrac{-5x - 1}{3}$. Substituting $y = \dfrac{-5x - 1}{3}$ in the second equation yields the

following: $3x - 5\left(\dfrac{-5x - 1}{3}\right) = -21 \Rightarrow 9x + 25x + 5 = -63 \Rightarrow 34x = -68 \Rightarrow x = -2$ and so

$y = \dfrac{-5(-2) - 1}{3} = 3$. The solution is $(-2, 3)$.

19. Adding the two equations will eliminate the variable y. $\begin{array}{r} 3x + y = 4 \\ 2x - y = -2 \\ \hline 5x = 2 \end{array}$ Thus, $x = \dfrac{2}{5}$.

And so $3\left(\dfrac{2}{5}\right) + y = 4 \Rightarrow y = 4 - \dfrac{6}{5} \Rightarrow y = \dfrac{14}{5}$. The solution is $\left(\dfrac{2}{5}, \dfrac{14}{5}\right)$.

20. Multiply the first equation by 2, the second equation by 3 and add the equations to eliminate the

variable y. $\begin{array}{r} 4x + 6y = -26 \\ 9x - 6y = 0 \\ \hline 13x = -26 \end{array}$ Thus, $x = -2$. And so $2(-2) + 3y = -13 \Rightarrow 3y = -9 \Rightarrow y = -3$.

The solution is $(-2, -3)$.

21. Multiplying the first equation by 2 and adding the two equations will eliminate both variables.

$\begin{array}{r} 6x - 2y = 10 \\ -6x + 2y = -10 \\ \hline 0 = 0 \end{array}$ This is always true and the system is dependent with solutions: $\{(x, y) | 3x - y = 5\}$.

22. Multiplying the second equation by 2 and adding the two equations will eliminate both variables.

$\begin{array}{r} 8x - 6y = 7 \\ -8x + 6y = 22 \\ \hline 0 = 29 \end{array}$ This is always false and the system is inconsistent. No solution.

23. See Figure 23.

24. See Figure 24.

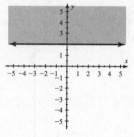

Figure 23

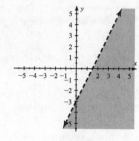

Figure 24

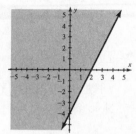

Figure 25

25. See Figure 25.

26. See Figure 26.

27. See Figure 27.

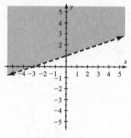

Figure 26

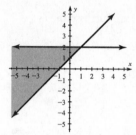

Figure 27

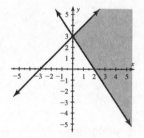

Figure 28

28. See Figure 28.

29. See Figure 29.

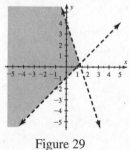

Figure 29

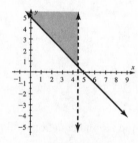

Figure 30

30. See Figure 30.

31. The equation of a vertical line through the point $(1, 0)$ is $x = 1$. The equation of a horizontal line through the point $(0, -1)$ is $y = -1$. The system of inequalities is $y < -1$ and $x > 1$.

32. The equation of a line through the points $(-1, -1)$ and $(0, 1)$ is $y = 2x + 1$. The equation of a line through the points $(0, 4)$ and $(4, 0)$ is $y = -x + 4$. The system of inequalities is
$y \leq -x + 4$ and $y \geq 2x + 1$.

33. The maximum value of R occurs at one of the vertices. For $(1, 1)$, $R = 7(1) + 8(1) = 15$.

For $(1, 3)$, $R = 7(1) + 8(3) = 31$. For $(3, 1)$, $R = 7(3) + 8(1) = 29$. The maximum value is $R = 31$.

34. The minimum value of R occurs at one of the vertices. For $(1, 1)$, $C = (1) + 2(1) = 3$.

For $(1, 3)$, $C = (1) + 2(3) = 7$. For $(3, 1)$, $C = (3) + 2(1) = 5$. The minimum value is $C = 3$.

35. From the graph of the region of feasible solutions (not shown), the vertices are $(0, 1)$, $(0, 3)$, $(3, 0)$, and $(1, 0)$. The maximum value of R occurs at one of the vertices. For $(0, 1)$, $R = 2(0) + (1) = 1$.

    For $(0, 3)$, $R = 2(0) + (3) = 3$. For $(3, 0)$, $R = 2(3) + (0) = 6$. For $(1, 0)$, $R = 2(1) + (0) = 2$.

    The maximum value is $R = 6$.

36. From the graph of the region of feasible solutions (not shown), the vertices are $(0, 0)$, $(0, 4)$, $(3, 3)$, and $(4, 0)$. Note: to find the intersection point $(3, 3)$, solve the system of equations

    $3x + y = 12$ and $x + 3y = 12$. The maximum value of R occurs at one of the vertices.

    For $(0, 0)$, $R = 6(0) + 9(0) = 0$. For $(0, 4)$, $R = 6(0) + 9(4) = 36$.

    For $(3, 3)$, $R = 6(3) + 9(3) = 45$. For $(4, 0)$, $R = 6(4) + 9(0) = 24$. The maximum value is $R = 45$.

37. Yes, since $3 + (-4) + 5 = 4$.

38. $(1, -1, 2)$ is a solution since it satisfies all three of the equations.

39. Add the first two equations together to eliminate the variable x.
$$\begin{array}{r} x - y - 2z = -11 \\ -x + 2y + 3z = 16 \\ \hline y + z = 5 \end{array}$$

    From the third equation, $3z = 6 \Rightarrow z = 2$. And so $y + (2) = 5 \Rightarrow y = 3$.

    Substitute $z = 2$ and $y = 3$ into the first equation: $x - (3) - 2(2) = -11 \Rightarrow x = -4$

    The solution is $(-4, 3, 2)$.

40. Multiply the second equation by -5 and the third equation by 3. Add these equations to eliminate the

    $$\begin{array}{r} 10x - 5y - 15z = 10 \\ \text{variable } z. \quad 3x - 6y + 15z = -78 \\ \hline 13x - 11y = -68 \end{array}$$ Multiply the first equation by 11 and add it to this *new* equation to

    eliminate the variable y. $\begin{array}{r} 11x + 11y = 44 \\ 13x - 11y = -68 \\ \hline 24x = -24 \end{array}$ And so $x = -1$. Substitute $x = -1$ into the first equation:

    $(-1) + y = 4 \Rightarrow y = 5$ Substitute $x = -1$ and $y = 5$ into the second equation:

    $-2(-1) + (5) + 3z = -2 \Rightarrow 3z = -9 \Rightarrow z = -3$ The solution is $(-1, 5, -3)$.

41. Multiply the second equation by -1 and add it to the third equation to eliminate the variable z.

    $$\begin{array}{r} -x - 2y - z = -7 \\ -2x + y + z = 7 \\ \hline -3x - y = 0 \end{array}$$ Multiply the first equation by -1 and add it to this *new* equation to eliminate the

    variable y. $\begin{array}{r} -2x + y = 5 \\ -3x - y = 0 \\ \hline -5x = 5 \end{array}$ And so $x = -1$. Substitute $x = -1$ into the first equation:

    $2(-1) - y = -5 \Rightarrow y = 3$ Substitute $x = -1$ and $y = 3$ into the third equation:

$-2(-1)+(3)+z=7 \Rightarrow z=2$　The solution is $(-1, 3, 2)$.

42. Multiply the second equation by 2 and add the first and second equations to eliminate the variable x.

$$2x+3y\ +z=6$$
$$\underline{-2x+4y+4z=6}$$ Add the second and third equations together to eliminate the variable x.
$$7y+5z=12$$

$$-x+2y+2z=3$$
$$\underline{x\ +y+2z=4}$$ Multiply the first *new* equation by 4 and the second *new* equation by -5. Add
$$3y+4z=7$$

$$28y+20z=48$$
these to eliminate the variable z.　$\underline{-15y-20z=-35}$ And so $y=1$. Substitute $y=1$ into the second
$$13y=13$$

new equation: $3(1)+4z=7 \Rightarrow 4z=4 \Rightarrow z=1$ Substitute $y=1$ and $z=1$ into the *original* third

equation: $x+(1)+2(1)=4 \Rightarrow x=1$ The solution is $(1, 1, 1)$.

$$x-y+3z=2$$
43. Add the first two equations to eliminate the variable y.　$\underline{2x+y+4z=3}$ Multiply the second
$$3x\ \ \ \ +7z=5$$

$$-4x-2y-8z=-6$$
equation by -2 and add to the third equation to eliminate the variable y.　$\underline{x+2y+\ z=5}$
$$-3x\ \ \ \ -7z=-1$$

$$3x+7z=5$$
Add this new equation to the result of the first sum to eliminate the variables x and z.　$\underline{-3x-7z=-1}$
$$0=4$$

This is a contradiction. There are no solutions.

44. Solve the third equation for x: $x+z=2 \Rightarrow x=2-z$. Substitute $x=2-z$ into the second equation:

$2-z+y-z=1 \Rightarrow y-2z=-1 \Rightarrow y=2z-1$. The system is dependent, and all solutions are of the

form of the ordered triple $(2-z, 2z-1, z)$.

45. $\begin{bmatrix} 1 & 1 & 1 & | & -6 \\ 1 & 2 & 1 & | & -8 \\ 0 & 1 & 1 & | & -5 \end{bmatrix} R_2 - R_1 \rightarrow \begin{bmatrix} 1 & 1 & 1 & | & -6 \\ 0 & 1 & 0 & | & -2 \\ 0 & 1 & 1 & | & -5 \end{bmatrix} \begin{matrix} R_1 - R_3 \rightarrow \\ \\ R_3 - R_2 \rightarrow \end{matrix} \begin{bmatrix} 1 & 0 & 0 & | & -1 \\ 0 & 1 & 0 & | & -2 \\ 0 & 0 & 1 & | & -3 \end{bmatrix}$;　The solution is $(-1, -2, -3)$.

46. $\begin{bmatrix} 1 & 1 & 1 & | & -3 \\ -1 & 1 & 0 & | & 5 \\ 0 & 1 & 1 & | & -1 \end{bmatrix} \begin{matrix} R_1 - R_3 \rightarrow \\ R_2 + R_1 \rightarrow \end{matrix} \begin{bmatrix} 1 & 0 & 0 & | & -2 \\ 0 & 2 & 1 & | & 2 \\ 0 & 1 & 1 & | & -1 \end{bmatrix} R_2 - R_3 \rightarrow \begin{bmatrix} 1 & 0 & 0 & | & -2 \\ 0 & 1 & 0 & | & 3 \\ 0 & 1 & 1 & | & -1 \end{bmatrix} R_3 - R_2 \rightarrow \begin{bmatrix} 1 & 0 & 0 & | & -2 \\ 0 & 1 & 0 & | & 3 \\ 0 & 0 & 1 & | & -4 \end{bmatrix}$

The solution is $(-2, 3, -4)$.

47. $\begin{bmatrix} 1 & 2 & -1 & | & 1 \\ -1 & 1 & -2 & | & 5 \\ 0 & 2 & 1 & | & 10 \end{bmatrix} \begin{matrix} R_1 - R_3 \to \\ R_2 + R_1 \to \\ \end{matrix} \begin{bmatrix} 1 & 0 & -2 & | & -9 \\ 0 & 3 & -3 & | & 6 \\ 0 & 2 & 1 & | & 10 \end{bmatrix} (1/3)R_2 \to \begin{bmatrix} 1 & 0 & -2 & | & -9 \\ 0 & 1 & -1 & | & 2 \\ 0 & 2 & 1 & | & 10 \end{bmatrix}$

$\begin{matrix} \\ \\ R_3 - 2R_2 \to \end{matrix} \begin{bmatrix} 1 & 0 & -2 & | & -9 \\ 0 & 1 & -1 & | & 2 \\ 0 & 0 & 3 & | & 6 \end{bmatrix} (1/3)R_3 \to \begin{bmatrix} 1 & 0 & -2 & | & -9 \\ 0 & 1 & -1 & | & 2 \\ 0 & 0 & 1 & | & 2 \end{bmatrix} \begin{matrix} R_1 + 2R_3 \to \\ R_2 + R_3 \to \\ \end{matrix} \begin{bmatrix} 1 & 0 & 0 & | & -5 \\ 0 & 1 & 0 & | & 4 \\ 0 & 0 & 1 & | & 2 \end{bmatrix}$

The solution is (–5, 4, 2).

48. $\begin{bmatrix} 2 & 2 & -2 & | & -14 \\ -2 & -3 & 2 & | & 12 \\ 1 & 1 & -4 & | & -22 \end{bmatrix} \begin{matrix} (1/2)R_1 \to \\ R_2 + R_1 \to \\ R_3 - (1/2)R_1 \to \end{matrix} \begin{bmatrix} 1 & 1 & -1 & | & -7 \\ 0 & -1 & 0 & | & -2 \\ 0 & 0 & -3 & | & -15 \end{bmatrix} \begin{matrix} R_1 + R_2 \to \\ -1R_2 \to \\ (-1/3)R_3 \to \end{matrix} \begin{bmatrix} 1 & 0 & -1 & | & -9 \\ 0 & 1 & 0 & | & 2 \\ 0 & 0 & 1 & | & 5 \end{bmatrix}$

$R_1 + R_3 \to \begin{bmatrix} 1 & 0 & 0 & | & -4 \\ 0 & 1 & 0 & | & 2 \\ 0 & 0 & 1 & | & 5 \end{bmatrix}$; The solution is (–4, 2, 5).

49. See example 7 in section 4.6 in the text for graphing calculator instructions.

$[A] = \begin{bmatrix} 3 & -2 & 6 & | & -17 \\ -2 & -1 & 5 & | & 20 \\ 0 & 4 & 7 & | & 30 \end{bmatrix}$; rref$([A]) = \begin{bmatrix} 1 & 0 & 0 & | & -7 \\ 0 & 1 & 0 & | & 4 \\ 0 & 0 & 1 & | & 2 \end{bmatrix}$; The solution is (–7, 4, 2).

50. See example 7 in section 4.6 in the text for graphing calculator instructions.

$[A] = \begin{bmatrix} 19 & -13 & -7 & | & 7.4 \\ 22 & 33 & -8 & | & 110.5 \\ 10 & -56 & 9 & | & 23.7 \end{bmatrix}$; rref$([A]) = \begin{bmatrix} 1 & 0 & 0 & | & 5.4 \\ 0 & 1 & 0 & | & 2.1 \\ 0 & 0 & 1 & | & 9.7 \end{bmatrix}$; The solution is (5.4, 2.1, 9.7)

51. $\det A = 6(2) - (-4)(-5) = 12 - 20 = -8$

52. $\det A = 0(9) - 5(-6) = 0 + 30 = 30$

53. $\det A = 3\left[(4)(1) - (-3)(7)\right] - 1\left[(-5)(1) - (-3)(-3)\right] + 0\left[(-5)(7) - (4)(-3)\right] = 75 - (-14) + 0 = 89$

54. $\det A = -2\left[(1)(8) - (-5)(-3)\right] - 2\left[(-1)(8) - (-5)(-7)\right] + 3\left[(-1)(-3) - (1)(-7)\right] =$

$14 - (-86) + 30 = 130$

55. Using the calculator we find $\det\left([A]\right) = 181,845$

56. Using the calculator we find $\det\left([A]\right) = 67.688$

57. The triangle has vertices (–4, 6), (–2, –4) and (6, 2). The matrix needed is $A = \begin{bmatrix} -4 & -2 & 6 \\ 6 & -4 & 2 \\ 1 & 1 & 1 \end{bmatrix}$.

The area is $D = \left|\frac{1}{2}\det([A])\right| = 46 \text{ ft}^2$.

58. The triangle has vertices (–12, –8), (4, 8) and (8, –4). The matrix needed is $A = \begin{bmatrix} -12 & 4 & 8 \\ -8 & 8 & -4 \\ 1 & 1 & 1 \end{bmatrix}$.

The area is $D = \left|\frac{1}{2}\det\left([A]\right)\right| = 128 \text{ ft}^2$.

59. $E = \det \begin{bmatrix} 8 & 6 \\ 18 & -8 \end{bmatrix} = -64 - 108 = -172;$ $F = \det \begin{bmatrix} 7 & 8 \\ 5 & 18 \end{bmatrix} = 126 - 40 = 86$

$D = \det \begin{bmatrix} 7 & 6 \\ 5 & -8 \end{bmatrix} = -56 - 30 = -86;$ The solution is $x = \dfrac{E}{D} = \dfrac{-172}{-86} = 2$ and $y = \dfrac{F}{D} = \dfrac{86}{-86} = -1.$

60. $E = \det \begin{bmatrix} 25 & 5 \\ -3 & 4 \end{bmatrix} = 100 + 15 = 115;$ $F = \det \begin{bmatrix} -2 & 25 \\ 3 & -3 \end{bmatrix} = 6 - 75 = -69$

$D = \det \begin{bmatrix} -2 & 5 \\ 3 & 4 \end{bmatrix} = -8 - 15 = -23;$ The solution is $x = \dfrac{E}{D} = \dfrac{115}{-23} = -5$ and $y = \dfrac{F}{D} = \dfrac{-69}{-23} = 3.$

61. $E = \det \begin{bmatrix} 1.5 & -6 \\ 8 & -5 \end{bmatrix} = -7.5 + 48 = 40.5;$ $F = \det \begin{bmatrix} 3 & 1.5 \\ 7 & 8 \end{bmatrix} = 24 - 10.5 = 13.5$

$D = \det \begin{bmatrix} 3 & -6 \\ 7 & -5 \end{bmatrix} = -15 + 42 = 27;$ The solution is $x = \dfrac{E}{D} = \dfrac{40.5}{27} = \dfrac{3}{2}$ and $y = \dfrac{F}{D} = \dfrac{13.5}{27} = \dfrac{1}{2}.$

62. $E = \det \begin{bmatrix} -47 & 4 \\ 63 & -7 \end{bmatrix} = 329 - 252 = 77;$ $F = \det \begin{bmatrix} -5 & -47 \\ 6 & 63 \end{bmatrix} = -315 + 282 = -33$

$D = \det \begin{bmatrix} -5 & 4 \\ 6 & -7 \end{bmatrix} = 35 - 24 = 11;$ The solution is $x = \dfrac{E}{D} = \dfrac{77}{11} = 7$ and $y = \dfrac{F}{D} = \dfrac{-33}{11} = -3.$

63. Let x and y represent pedestrian fatalities for 1994 and 2004 respectively. Then the system needed is $x + y = 10,130$ and $x - y = 848$. Adding the two equations will eliminate the variable y.

$x + y = 10,130$

$\dfrac{x - y = 848}{2x = 10,978}$ Thus, $x = 5489$. And so $(5489) + y = 10,130 \Rightarrow y = 4641.$

There were 5489 pedestrian fatalities in 1994 and 4641 in 2004.

64. Let x and y represent the time spent on the stair climber and the bicycle respectively. Then the system needed is $x + y = 30$ and $11.5x + 9y = 290$. Multiplying the first equation by –9 and adding

the two equations will eliminate the variable y. $\dfrac{\begin{aligned} -9x - 9y &= -270 \\ 11.5x + 9y &= 290 \end{aligned}}{2.5x = 20}$ Thus, $x = 8$. And so

$(8) + y = 30 \Rightarrow y = 22$. The athlete spent 8 minutes on the stair climber and 22 minutes on the stationary bicycle.

65. First note that the amounts of candy and cashews cannot be negative, so $x \geq 0$ and $y \geq 0$. Also, if the total cost must be less than or equal to $20 then the inequality $4x + 5y \leq 20$ must be satisfied. See Figure 65.

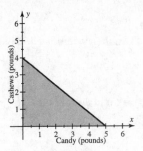

Figure 65

66. Let x and y represent the amounts of the unsubsidized and subsidized loans, respectively. Then the system of equations is $x + y = 4000$ and $x = y - 500$. Substitute $x = y - 500$ into the first equation:

$y - 500 + y = 4000 \Rightarrow 2y = 4500 \Rightarrow y = 2250$, and so $x = 2250 - 500 \Rightarrow x = 1750$. The unsubsidized loan is for \$1750 and the subsidized loan is for \$2250.

67. Let x and y represent the number of shirts and pants respectively. Since a shirt requires 20 minutes of cutting, pants require 10 minutes of cutting and the cutting machine is only available for 360 minutes each day, $20x + 10y \le 360$. Since a shirt requires 10 minutes of sewing, pants require 20 minutes of sewing and the sewing machine is only available for 480 minutes each day, $10x + 20y \le 480$. Since the number of shirts and pants cannot be negative $x \ge 0$ and $y \ge 0$. Here the profit function is

$P = 20x + 25y$. From the graph of the region of feasible solutions (not shown), the vertices are

(0, 0), (0, 24), (8, 20) and (18, 0). To find (8, 20) solve the system of equations

$20x + 10y = 360$ and $10x + 20y = 480$. The maximum value of P occurs at one of the vertices.

For $(0, 0)$, $P = 20(0) + 25(0) = 0$. For $(0, 24)$, $P = 20(0) + 25(24) = 600$.

For $(8, 20)$, $P = 20(8) + 25(20) = 660$. For $(18, 0)$, $P = 20(18) + 25(0) = 360$.

The maximum profit of \$660 is attained when 8 shirts and 20 pants are sold.

68. Let x and y represent the number of gallons of 30% and 55% solution respectively. Then the system needed is $x + y = 4$ and $0.30x + 0.55y = 1.6$. Multiplying the first equation by –3, the second equation by 10 and adding the two equations will eliminate the variable x.

$$\begin{array}{r} -3x - 3y = -12 \\ 3x + 5.5y = \ 16 \\ \hline 2.5y = 4 \end{array}$$

Thus, $y = 1.6$. And so $x + (1.6) = 4 \Rightarrow x = 2.4$.

The mechanic should add 2.4 gallons of 30% solution and 1.6 gallons of 55% solution.

69. Let x and y represent the speed of the boat and the current respectively. Using the formula $d = rt$

the system is $18 = (x + y)(1)$ and $18 = (x - y)(1.5)$. These equations may be written: $x + y = 18$

and $1.5x - 1.5y = 18$. Multiplying the first equation by 1.5 and adding the equations together will

eliminate the variable y. $\begin{array}{r} 1.5x + 1.5y = 27 \\ 1.5x - 1.5y = 18 \\ \hline 3x = 45 \end{array}$ Thus, $x = 15$. And so $(15) + y = 18 \Rightarrow y = 3$.

The boat travels at 15 mph and the river flows at 3 mph.

70. Let x and y represent the number of \$8 and \$12 tickets respectively. Then the system needed is

$x + y = 480$ and $8x + 12y = 4620$. Note that $x + y = 480 \Rightarrow y = -x + 480$ and

$8x + 12y = 4620 \Rightarrow y = -\dfrac{2}{3}x + 385$. Multiplying the first equation by –8 and adding the two

equations will eliminate the variable x.
$$\begin{array}{r} -8x - 8y = -3840 \\ \underline{8x + 12y = 4620} \\ 4y = 780 \end{array}$$
Thus, $y = 195$. And so

$x + (195) = 480 \Rightarrow x = 285$. The solution is (285, 195).

There were 285 tickets sold costing \$8 each and 195 tickets sold costing \$12 each.

71. (a) $\begin{array}{l} m + 3c + 5b = 14 \\ m + 2c + 4b = 11 \\ c + 3b = 5 \end{array}$

(b)   Using a graphing calculator to solve the system, the solution is (3, 2, 1).

A malt costs \$3.00, cones cost \$2.00 and an ice cream bar costs \$1.00.

72. Let x, y and z represent the measure of the largest, middle and smallest angle respectively. The

system needed is: $\begin{array}{l} x + y + z = 180 \\ x - y - z = 20 \\ x - z = 85 \end{array}$ Using a graphing calculator to solve the system, the solution

is (100, 65, 15). The measures of the three angles are 100°, 65° and 15°.

73. Let x, y and z represent the amount of \$1.50, \$2.00 and \$2.50 candy respectively. The system needed

is: $\begin{array}{l} x + y + z = 12 \\ 1.50x + 2.00y + 2.50z = 26 \\ -y + z = 2 \end{array}$ Using a graphing calculator to solve the system, the solution

is (2, 4, 6). There should be 2 lb of \$1.50 candy, 4 lb of \$2.00 candy and 6 lb of \$2.50 candy.

74. (a) $\begin{array}{l} a + 202b + 63c = 40 \\ a + 365b + 70c = 50 \\ a + 446b + 77c = 55 \end{array}$

(b)   Using the graphing calculator to solve the system, the solution is

$a \approx 27.134$, $b \approx 0.061$, $c \approx 0.009$. So the equation is $C \approx 27.134 + 0.061W + 0.009L$.

(c)   When $W = 300$ and $L = 68$, $C \approx 27.134 + 0.061(300) + 0.009(68) = 46.046 \approx 46$ inches.

Chapter 4 Test

1. Note that $2x + y = 7 \Rightarrow y = 7 - 2x$ and $3x - 2y = 7 \Rightarrow y = \dfrac{3x - 7}{2}$.

Graph $Y_1 = 7 - 2X$ and $Y_2 = (3X - 7)/2$ in [–5, 5, 1] by [–5, 5, 1]. See Figure 1.

The graphs intersect at (3, 1). The system is consistent. The equations are independent.

2. Note that $8x - 4y = 3 \Rightarrow y = 2x - \frac{3}{4}$ and $-4x + 2y = 6 \Rightarrow y = 2x + 3$.

   Graph $Y_1 = 2X - (3/4)$ and $Y_2 = 2X + 3$ in $[-5, 5, 1]$ by $[-5, 5, 1]$. See Figure 2.

   The system is inconsistent since the lines are parallel. No solutions.

   $[-5, 5, 1]$ by $[-5, 5, 1]$ $[-5, 5, 1]$ by $[-5, 5, 1]$ $[-5, 5, 1]$ by $[-5, 5, 1]$ $[-10, 10, 1]$ by $[-10, 10, 1]$.

      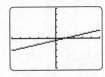   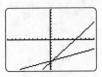

   Figure 1   Figure 2   Figure 3   Figure 4

3. Note that $2x - 6y = 2 \Rightarrow y = \frac{-2x + 2}{-6}$ and $-3x + 9y = -3 \Rightarrow y = \frac{3x - 3}{9}$. Graph $Y_1 = (-2X + 2)/-6$

   and $Y_2 = (3X - 3)/9$ in $[-5, 5, 1]$ by $[-5, 5, 1]$. The lines are identical, so the system is consistent

   and the equations are dependent. See Figure 3. The solution is $\{(x, y) \mid x - 3y = 1\}$.

4. Note that $2x - 5y = 36 \Rightarrow y = \frac{-2x + 36}{-5}$ and $-4x + 3y = -23 \Rightarrow y = \frac{4x - 23}{3}$. Graph

   $Y_1 = (-2X + 36)/-5$ and $Y_2 = (4X - 23)/3$ in $[-10, 10, 1]$ by $[-10, 10, 1]$. The graphs intersect at

   $\left(\frac{1}{2}, -7\right)$. See Figure 4. The system is consistent. The equations are independent.

5. Note that $2x + 5y = -1 \Rightarrow y = -\frac{2}{5}x - \frac{1}{5}$. Substituting $y = -\frac{2}{5}x - \frac{1}{5}$ into the second equation yields:

   $3x + 2\left(-\frac{2}{5}x - \frac{1}{5}\right) = -7 \Rightarrow \frac{11}{5}x = -\frac{33}{5} \Rightarrow x = -3$ and so $y = -\frac{2}{5}(-3) - \frac{1}{5} \Rightarrow y = 1$.

   The solution is $(-3, 1)$.

6. (a) Let x and y represent the first and second number respectively. The system is $x - y = 34$ and

   $x - 2y = 0$.

   (b) Multiplying the first equation by -1 and adding the two equations will eliminate the variable x.

   $\begin{aligned} -x \ \ + y &= -34 \\ \underline{x - 2y} &= 0 \\ -y &= -34 \end{aligned}$ Thus, $y = 34$. And so $x - (34) = 34 \Rightarrow x = 68$. The solution is $(68, 34)$.

7. Note that $-\pi x + \sqrt{3}y = 3.3 \Rightarrow y = \frac{\pi x + 3.3}{\sqrt{3}}$ and $\sqrt{5}x + (1 + \sqrt{2})y = 2.1 \Rightarrow y = \frac{-\sqrt{5}x + 2.1}{1 + \sqrt{2}}$.

   Graph $Y_1 = (\pi X + 3.3)/\sqrt{3}$ and $Y_2 = (-\sqrt{5}X + 2.1)/(1 + \sqrt{2})$ in $[-5, 5, 1]$ by $[-5, 5, 1]$.

   The graphs intersect at approximately $(-0.378, 1.220)$.

8. Note that $3x + y = 17 \Rightarrow y = -3x + 17$ and $2x - 3y = 37 \Rightarrow y = \dfrac{-2x + 37}{-3}$. The solution is $(8, -7)$.

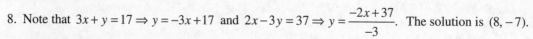

9. See Figure 9.

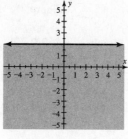

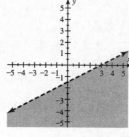

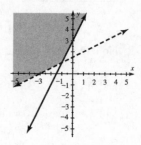

Figure 9   Figure 10   Figure 11

10. See Figure 10.

11. See Figure 11.

12. Multiply the third equation by –1 and add the second and third equations to eliminate the variables y

and z.
$$\begin{array}{r} -2x + y + z = 5 \\ -y - z = 3 \\ \hline -2x = 8 \end{array}$$
And so $x = -4$. Substitute $x = -4$ into the first equation:

$(-4) + 3y = 2 \Rightarrow 3y = 6 \Rightarrow y = 2$ Substitute $x = -4$ and $y = 2$ into the second equation:

$-2(-4) + (2) + z = 5 \Rightarrow z = -5$ The solution is $(-4, 2, -5)$.

13. Add the first and second equations to eliminate the variable z.
$$\begin{array}{r} x + y - z = 1 \\ 2x - 3y + z = 0 \\ \hline 3x - 2y = 1 \end{array}$$

Multiply the first equation by 2 and add the first and third equations to eliminate the variable z.
$$\begin{array}{r} 2x + 2y - 2z = 2 \\ x - 4y + 2z = 2 \\ \hline 3x - 2y = 4 \end{array}$$
So the two new equations are $3x - 2y = 1$ and $3x - 2y = 4$. This is a contradiction,

so there are no solutions.

14. Add the first and second equation and eliminate both the y and z variable:
$$\begin{array}{r} x - y + z = 1 \\ x + y - z = 1 \\ \hline 2x = 2 \end{array} \Rightarrow x = 1$$

Substituting $x = 1$ into equation 2 we have $1 + y - z = 1 \Rightarrow y - z = 0 \Rightarrow y = z$. The system is

dependent and the solution is $(1, z, z)$.

15. The equation of a line through the points $(-2, 0)$ and $(0, 2)$ is $y = x + 2$. The equation of a line

   through the points $(0, -1)$ and $(1, 2)$ is $y = 3x - 1$. The system of inequalities is

   $y \leq x + 2$ and $y \geq 3x - 1$.

16. (a) $\begin{bmatrix} 1 & 1 & 1 & | & 2 \\ 1 & -1 & -1 & | & 3 \\ 2 & 2 & 1 & | & 6 \end{bmatrix}$

   (b) $\begin{bmatrix} 1 & 1 & 1 & | & 2 \\ 1 & -1 & -1 & | & 3 \\ 2 & 2 & 1 & | & 6 \end{bmatrix} \begin{matrix} \\ R_2 - R_1 \rightarrow \\ R_3 - 2R_1 \rightarrow \end{matrix} \begin{bmatrix} 1 & 1 & 1 & | & 2 \\ 0 & -2 & -2 & | & 1 \\ 0 & 0 & -1 & | & 2 \end{bmatrix} \begin{matrix} R_1 + (1/2)R_2 \rightarrow \\ (-1/2)R_2 \rightarrow \\ -1R_3 \rightarrow \end{matrix} \begin{bmatrix} 1 & 0 & 0 & | & \frac{5}{2} \\ 0 & 1 & 1 & | & -\frac{1}{2} \\ 0 & 0 & 1 & | & -2 \end{bmatrix}$

   $R_2 - R_3 \rightarrow \begin{bmatrix} 1 & 0 & 0 & | & \frac{5}{2} \\ 0 & 1 & 0 & | & \frac{3}{2} \\ 0 & 0 & 1 & | & -2 \end{bmatrix}$ The solution is $\left(\frac{5}{2}, \frac{3}{2}, -2 \right)$.

17. $\det A = 3\big[(2)(-3) - (8)(-6)\big] - 6\big[(2)(-3) - (8)(-1)\big] + 0\big[(2)(-6) - (2)(-1)\big] = 126 - 12 + 0 = 114$

18. $E = \det \begin{bmatrix} 7 & -3 \\ 11 & 2 \end{bmatrix} = 14 - (-33) = 47; \quad F = \det \begin{bmatrix} 5 & 7 \\ -4 & 11 \end{bmatrix} = 55 - (-28) = 83$

   $D = \det \begin{bmatrix} 5 & -3 \\ -4 & 2 \end{bmatrix} = 10 - 12 = -2;$ The solution is $x = \dfrac{E}{D} = \dfrac{47}{-2} = -\dfrac{47}{2}$ and $y = \dfrac{F}{D} = \dfrac{83}{-2} = -\dfrac{83}{2}$.

19. (a) Let x and y represent private and public tuition respectively. The system needed is

   $\begin{aligned} x \quad - y &= 19,700 \\ x - 3.6y &= 0 \end{aligned}$

   (b) Note that $x - y = 19,700 \Rightarrow y = x - 19,700$. Substituting $y = x - 19,700$ into the second

   equation gives: $x - 3.6(x - 19,700) = 0 \Rightarrow -2.6x = -7092 \Rightarrow x \approx 27,277$ and so

   $y = 27,277 - 19,700 = 7577$. Private tuition was \$27,277 and public tuition was \$7577.

20. Let x and y represent the time spent running and on the rowing machine respectively.

   The system needed is $x + y = 60$ and $12x + 9y = 669$. Multiplying the first equation by -9 and

   $$-9x - 9y = -540$$

   adding the two equations will eliminate the variable y. $\dfrac{12x + 9y = 669}{3x = 129}$ Thus, $x = 43$. And so

   $(43) + y = 60 \Rightarrow y = 17$. The solution is $(43, 17)$. The athlete spent 43 minutes running and 17

   minutes on the rowing machine.

21. Let x and y represent the average speed of the plane and the wind respectively. Using the formula

   $d = rt$, the system is $600 = (x - y)(2.5)$ and $600 = (x + y)(2)$. These equations may be written as

   follows: $x - y = 240$ and $x + y = 300$. Adding the two equations together will eliminate the variable

   y. $\dfrac{\begin{aligned} x - y &= 240 \\ x + y &= 300 \end{aligned}}{2x = 540}$ Thus, $x = 270$. And so $(270) - y = 240 \Rightarrow -y = -30 \Rightarrow y = 30$. The solution is

   $(270, 30)$. The speed of the plane is 270 mph and the speed of the wind is 30 mph.

22. Let x, y and z represent the measure of the largest, middle and smallest angle respectively. The

system needed is: $\begin{array}{rrrr} x+y+z &=& 180 \\ x & -z &=& 50 \\ -x+y+z &=& 10 \end{array}$ and so $[A] = \begin{bmatrix} 1 & 1 & 1 & | & 180 \\ 1 & 0 & -1 & | & 50 \\ -1 & 1 & 1 & | & 10 \end{bmatrix}$; $\operatorname{rref}([A]) = \begin{bmatrix} 1 & 0 & 0 & | & 85 \\ 0 & 1 & 0 & | & 60 \\ 0 & 0 & 1 & | & 35 \end{bmatrix}$

The solution is $(85, 60, 35)$. The angles are $85°$, $60°$, and $35°$.

23. From the graph of the region of feasible solutions (not shown), the vertices are $(0, 0)$, $(0, 2)$, and

$(3, 0)$. The maximum value of R occurs at one of the vertices. For $(0, 0)$, $R = (0) + 2(0) = 0$.

For $(0, 2)$, $R = (0) + 2(2) = 4$. For $(3, 0)$, $R = (3) + 2(0) = 3$. The maximum value is $R = 4$.

24. The constraint is $3x + 5y < 30$. See Figure 24.

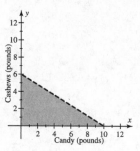

Figure 24

25. x: amount of subsidized loan $x - 700$: amount of unsubsidized loan

$$x + x - 700 = 4400 \Rightarrow 2x - 700 = 4400 \Rightarrow x = 2550, \; x - 700 = 1850$$

The unsubsidized loan is for $2550 and the subsidized loan is for $1850.

Extended and Discovery Exercises

1. $D = 1\big[(1)(3) - (1)(2)\big] - 2\big[(1)(3) - (1)(1)\big] + 0\big[(1)(2) - (1)(1)\big] = 1 - 4 + 0 = -3$

$E = 6\big[(1)(3) - (1)(2)\big] - 9\big[(1)(3) - (1)(1)\big] + 9\big[(1)(2) - (1)(1)\big] = 6 - 18 + 9 = -3$

$F = 1\big[(9)(3) - (9)(2)\big] - 2\big[(6)(3) - (9)(1)\big] + 0\big[(6)(1) - (9)(1)\big] = 9 - 18 + 0 = -9$

$G = 1\big[(1)(9) - (1)(9)\big] - 2\big[(1)(9) - (1)(6)\big] + 0\big[(1)(9) - (1)(6)\big] = 0 - 6 + 0 = -6$

$x = \dfrac{E}{D} = \dfrac{-3}{-3} = 1$, $y = \dfrac{F}{D} = \dfrac{-9}{-3} = 3$, $z = \dfrac{G}{D} = \dfrac{-6}{-3} = 2$. The solution is $(1, 3, 2)$.

2. $D = 0\big[(-1)(-1) - (1)(-1)\big] - 2\big[(1)(-1) - (1)(1)\big] + 1\big[(1)(-1) - (-1)(1)\big] = 0 + 4 + 0 = 4$

$E = 1\big[(-1)(-1) - (1)(-1)\big] - (-1)\big[(1)(-1) - (1)(1)\big] + 3\big[(1)(-1) - (-1)(1)\big] = 2 - 2 + 0 = 0$

$F = 0\big[(-1)(-1) - (3)(-1)\big] - 2\big[(1)(-1) - (3)(1)\big] + 1\big[(1)(-1) - (-1)(1)\big] = 0 + 8 + 0 = 8$

$G = 0[(-1)(3) - (1)(-1)] - 2[(1)(3) - (1)(1)] + 1[(1)(-1) - (-1)(1)] = 0 - 4 + 0 = -4$

$x = \dfrac{E}{D} = \dfrac{0}{4} = 0$, $y = \dfrac{F}{D} = \dfrac{8}{4} = 2$, $z = \dfrac{G}{D} = \dfrac{-4}{4} = -1$. The solution is $(0, 2, -1)$.

3. $D = 1\big[(1)(2) - (1)(0)\big] - 1\big[(0)(2) - (1)(1)\big] + 0\big[(0)(0) - (1)(1)\big] = 2 + 1 + 0 = 3$

$E = 2\big[(1)(2) - (1)(0)\big] - 0\big[(0)(2) - (1)(1)\big] + 1\big[(0)(0) - (1)(1)\big] = 4 + 0 - 1 = 3$

$F = 1\big[(0)(2) - (1)(0)\big] - 1\big[(2)(2) - (1)(1)\big] + 0\big[(0)(2) - (1)(0)\big] = 0 - 3 + 0 = -3$

$G = 1\big[(1)(1) - (1)(0)\big] - 1\big[(0)(1) - (1)(2)\big] + 0\big[(0)(0) - (1)(2)\big] = 1 + 2 + 0 = 3$

$x = \dfrac{E}{D} = \dfrac{3}{3} = 1, \quad y = \dfrac{F}{D} = \dfrac{-3}{3} = -1, \quad z = \dfrac{G}{D} = \dfrac{3}{3} = 1.$ The solution is $(1, -1, 1)$.

4. $D = 1\big[(-2)(-3) - (1)(-3)\big] - (-1)\big[(1)(-3) - (1)(2)\big] + 0\big[(1)(-3) - (-2)(2)\big] = 9 - 5 + 0 = 4$

$E = 1\big[(-2)(-3) - (1)(-3)\big] - (-2)\big[(1)(-3) - (1)(2)\big] + 5\big[(1)(-3) - (-2)(2)\big] = 9 - 10 + 5 = 4$

$F = 1\big[(-2)(-3) - (5)(-3)\big] - (-1)\big[(1)(-3) - (5)(2)\big] + 0\big[(1)(-3) - (-2)(2)\big] = 21 - 13 + 0 = 8$

$G = 1[(-2)(5) - (1)(-2)] - (-1)[(1)(5) - (1)(1)] + 0[(1)(-2) - (-2)(1)] = -8 + 4 + 0 = -4$

$x = \dfrac{E}{D} = \dfrac{4}{4} = 1, \quad y = \dfrac{F}{D} = \dfrac{8}{4} = 2, \quad z = \dfrac{G}{D} = \dfrac{-4}{4} = -1.$ The solution is $(1, 2, -1)$.

5. $D = 1\big[(1)(2) - (-1)(1)\big] - (-1)\big[(0)(2) - (-1)(2)\big] + 2\big[(0)(1) - (1)(2)\big] = 3 + 2 - 4 = 1$

$E = 7\big[(1)(2) - (-1)(1)\big] - 5\big[(0)(2) - (-1)(2)\big] + 6\big[(0)(1) - (1)(2)\big] = 21 - 10 - 12 = -1$

$F = 1\big[(5)(2) - (6)(1)\big] - (-1)\big[(7)(2) - (6)(2)\big] + 2\big[(7)(1) - (5)(2)\big] = 4 + 2 - 6 = 0$

$G = 1\big[(1)(6) - (-1)(5)\big] - (-1)\big[(0)(6) - (-1)(7)\big] + 2\big[(0)(5) - (1)(7)\big] = 11 + 7 - 14 = 4$

$x = \dfrac{E}{D} = \dfrac{-1}{1} = -1, \quad y = \dfrac{F}{D} = \dfrac{0}{1} = 0, \quad z = \dfrac{G}{D} = \dfrac{4}{1} = 4.$ The solution is $(-1, 0, 4)$.

6. $D = 1\big[(-3)(-2) - (4)(-1)\big] - 2\big[(2)(-2) - (4)(3)\big] + 1\big[(2)(-1) - (-3)(3)\big] = 10 + 32 + 7 = 49$

$E = -1\big[(-3)(-2) - (4)(-1)\big] - 12\big[(2)(-2) - (4)(3)\big] + (-12)\big[(2)(-1) - (-3)(3)\big] =$

$-10 + 192 - 84 = 98$

$F = 1\big[(12)(-2) - (-12)(-1)\big] - 2\big[(-1)(-2) - (-12)(3)\big] + 1\big[(-1)(-1) - (12)(3)\big] =$

$-36 - 76 - 35 = -147$

$G = 1\big[(-3)(-12) - (4)(12)\big] - 2\big[(2)(-12) - (4)(-1)\big] + 1\big[(2)(12) - (-3)(-1)\big] = -12 + 40 + 21 = 49$

$x = \dfrac{E}{D} = \dfrac{98}{49} = 2, \quad y = \dfrac{F}{D} = \dfrac{-147}{49} = -3, \quad z = \dfrac{G}{D} = \dfrac{49}{49} = 1.$ The solution is $(2, -3, 1)$.

7. Since Denver is city 1 and Las Vegas is city 4, we look at either entry a_{14} or a_{41}. The distance is 760 miles.

8. Add entry a_{12} to entry a_{23}. The distance is 360 miles. 9. The dimension would be 20×20 and the matrix would contain 400 elements.

10. The elements on the main diagonal represent the distance from a city to itself, which is always zero.

11. $a_{14} + a_{41} = 760 + 760 = 1520$ miles

12. $a_{11} + a_{44} = 0 + 0 = 0$ miles

13. See Figure 13.

$$\begin{bmatrix} 0 & 130 & 95 & 75 \\ 130 & 0 & 186 & \star \\ 95 & 186 & 0 & 57 \\ 75 & \star & 57 & 0 \end{bmatrix}$$

$$\begin{bmatrix} 0 & 97 & \star & \star & 59 \\ 97 & 0 & 113 & 118 & \star \\ \star & 113 & 0 & 94 & \star \\ \star & 118 & 94 & 0 & 177 \\ 59 & \star & \star & 177 & 0 \end{bmatrix}$$

Figure 13   Figure 14

14. See Figure 14.

15. All maps for adjacency matrix A must have the same distances between cities, but the location of each city may vary. The solution is not unique. One possible solution is shown in Figure 15.

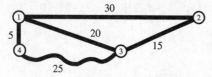

Figure 15

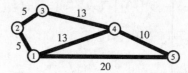

Figure 16

16. All maps for adjacency matrix A must have the same distances between cities, but the location of each city may vary. The solution is not unique. One possible solution is shown in Figure 16.

17. (a) $$\begin{bmatrix} 1 & 19 & 57.5 & 32 & | & 125 \\ 1 & 26 & 65 & 42 & | & 316 \\ 1 & 30 & 72 & 48 & | & 436 \\ 1 & 30.5 & 75 & 54 & | & 514 \end{bmatrix}$$

(b) $a \approx -552.272$, $b \approx 8.733$, $c \approx 2.859$, $d \approx 10.843$

(c) $N = 24$, $L = 63$ and $C = 39$, $W \approx -552.272 + 8.733(24) + 2.859(63) + 10.843(39) \approx 260$

A bear with a 24-inch neck, 63-inch length and 39-inch chest weighs approximately 260 pounds.

Chapters 1-4 Cumulative Review Exercises

1. Identity

2. Associative

3. $\dfrac{1+2}{3-1} - 2^2 \cdot 4 = \dfrac{3}{2} - 4 \cdot 4 = 1.5 - 16 = -14.5$

4. $-1 + 8 - 19 + 12 = 0$

5. $-\dfrac{a}{b}$

6. (a) $\left(\dfrac{4^2}{3}\right)^{-2} = \left(\dfrac{3}{4^2}\right)^2 = \dfrac{3^2}{4^4} = \dfrac{9}{256}$

    (b) $\dfrac{(2ab^{-4})^2}{a^7(b^2)^{-3}} = \dfrac{4a^2b^{-8}}{a^7b^{-6}} = 4a^{2-7}b^{-8-(-6)} = 4a^{-5}b^{-2} = \dfrac{4}{a^5b^2}$

    (c) $(x^2y^3)^{-2}(xy^2)^4 = x^{-4}y^{-6}x^4y^8 = x^0y^2 = y^2$

7. $0.0056 = 5.6 \times 10^{-3}$

8. $f(-2) = 5 - 4(-2) = 5 + 8 = 13$

9. $V = \dfrac{1}{3}\pi(3)(2)^2 = \dfrac{1}{3}\pi(3)(4) = 4\pi$

10. The denominator cannot equal zero, so $x \neq -\dfrac{1}{2}$.

11. See Figure 11.

12. See Figure 12.

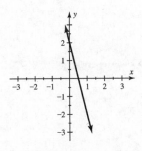

          Figure 11   Figure 12

13. (a) Yes. The graph passes the vertical line test.

     (b) D: all real numbers; R: all real numbers

     (c) $f(-1) = -1.5;\ f(2) = 0$

     (d) x-intercept: 2; y-intercept: -1

     (e) Two points on the line are $(2, 0)$ and $(0, -1)$. $m = \dfrac{y_2 - y_1}{x_2 - x_1} = \dfrac{-1-0}{0-2} = \dfrac{-1}{-2} = \dfrac{1}{2}$

     (f) $f(x) = \dfrac{1}{2}x - 1$

14. The slope of a line perpendicular to $y = -\dfrac{3}{4}x$ is $m = \dfrac{4}{3}$.

     $y - (3) = \dfrac{4}{3}(x - (-6)) \Rightarrow y = \dfrac{4}{3}(x + 6) + 3 \Rightarrow y = \dfrac{4}{3}x + 11$

15. Two points on the graph are $(0, 2)$ and $(2, 10)$. $m = \dfrac{y_2 - y_1}{x_2 - x_1} = \dfrac{10-2}{2-0} = \dfrac{8}{2} = 4$; From the table, the y-intercept is 2, so an equation of the line is $y = 4x + 2$. To find the x-intercept, let $y = 0$ and solve for

$$x: 0 = 4x + 2 \Rightarrow -2 = 4x \Rightarrow x = -\frac{1}{2}.$$

16. $\frac{1}{2}(1-x) + 2x = 3x + 1 \Rightarrow \frac{1}{2} - \frac{1}{2}x + 2x = 3x + 1$

$$\Rightarrow -\frac{1}{2}x + 2x - 3x = -\frac{1}{2} + 1 \Rightarrow -\frac{3}{2}x = \frac{1}{2} \Rightarrow x = -\frac{2}{3} \cdot \frac{1}{2} \Rightarrow x = -\frac{1}{3}$$

17. $2(1-x) > 4x - (2+x) \Rightarrow 2 - 2x > 4x - 2 - x \Rightarrow -2x - 4x + x > -2 - 2 \Rightarrow -5x > -4 \Rightarrow x < \frac{4}{5}$; the

solutions are $\left(-\infty, \frac{4}{5} \right)$.

18. $x + 1 < 2 \Rightarrow x < 1$ or $x + 1 > 4 \Rightarrow x > 3 \Rightarrow (-\infty, 1) \cup (3, \infty)$

19. $-3 \le 2 - 4x < 5 \Rightarrow -5 \le -4x < 3 \Rightarrow \frac{5}{4} \ge x > -\frac{3}{4} \Rightarrow -\frac{3}{4} < x \le \frac{5}{4} \Rightarrow \left(-\frac{3}{4}, \frac{5}{4} \right]$

20. (a) $|3x - 2| = 4 \Rightarrow 3x - 2 = 4$ or $3x - 2 = -4 \Rightarrow 3x = 6$ or $3x = -2 \Rightarrow x = 2$ or $x = -\frac{2}{3}$

(b) $|3x - 2| < 4 \Rightarrow -4 < 3x - 2 < 4 \Rightarrow -2 < 3x < 6 \Rightarrow -\frac{2}{3} < x < 2 \Rightarrow \left(-\frac{2}{3}, 2 \right)$

(c) $|3x - 2| > 4 \Rightarrow 3x - 2 < -4$ or $3x - 2 > 4 \Rightarrow 3x < -2$ or $3x > 6 \Rightarrow x < -\frac{2}{3}$ or

$$x > 2 \Rightarrow \left(-\infty, -\frac{2}{3} \right) \cup (2, \infty)$$

21. (a) Note that $2x + y = 3 \Rightarrow y = -2x + 3$ and $x - 2y = -1 \Rightarrow -2y = -x - 1 \Rightarrow y = \frac{1}{2}x + \frac{1}{2}$. Graph

$Y_1 = -2X + 3$ and $Y_2 = 1/2X + 1/2$ in $[-5, 5, 1]$ by $[-5, 5, 1]$. The solution is $(1, 1)$.

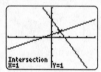

(b) Note that $4x + 3y = 8 \Rightarrow 3y = -4x + 8 \Rightarrow y = -\frac{4}{3}x + \frac{8}{3}$ and

$2x + 7y = 4 \Rightarrow 7y = -2x + 4 \Rightarrow y = -\frac{2}{7}x + \frac{4}{7}$. Table $Y_1 = -4/3X + 8/3$ and $Y_2 = -2/7X + 4/7$ with

TblStart $= 0$ and ΔTbl $= 1$. The solution is $(2, 0)$.

X	Y₁	Y₂
0	2.6667	.57143
1	1.3333	.28571
2	0	0
3	-1.333	-.2857
4	-2.667	-.5714
5	-4	-.8571
6	-5.333	-1.143

X=2

22. (a) Add the two equations to eliminate the variable y.
$$\begin{array}{r} x+2y=4 \\ 3x-2y=-2 \\ \hline 4x\quad\ \ =2 \end{array} \Rightarrow x=\frac{1}{2}. \text{ Substitute } x=\frac{1}{2}$$

in the first equation and solve for $y: \frac{1}{2}+2y=4 \Rightarrow 2y=\frac{7}{2} \Rightarrow y=\frac{7}{4}$. The solution is $\left(\frac{1}{2},\frac{7}{4}\right)$.

(b) Multiply the first equation by 2 and add the equations to eliminate the variables x and y.

$$\begin{array}{r} -2x+2y=4 \\ 2x-2y=5 \\ \hline 0=9 \end{array}$$ This is a contradiction. There are no solutions.

23. (a) Note that $x+2y<3 \Rightarrow y<-\frac{1}{2}x+\frac{3}{2}$ See Figure 23a.

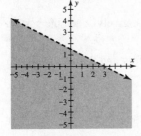   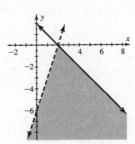

Figure 23a   Figure 23b

(b) Note that $x+y\le 2 \Rightarrow y\le -x+2$ and $3x-y>6 \Rightarrow y<3x-6$ See Figure 23b.

24. Add the second and third equations to eliminate the variable y.
$$\begin{array}{r} 3x+y+\ z=12 \\ x-y-2z=-1 \\ \hline 4x\quad -z=11 \end{array}$$

Multiply the second equation by -2 and add to the first equation to eliminate the variable y.

$$\begin{array}{r} -x+2y\ +z=2 \\ -6x-2y-2z=-24 \\ \hline -7x\qquad -z=-22 \end{array}$$ Multiply the result of the first sum by -1 and add to this new equation to

eliminate the variable z. $\begin{array}{r} -4x+z=-11 \\ -7x-z=-22 \\ \hline -11x\quad =-33 \end{array} \Rightarrow x=3.$ Then $4(3)-z=11 \Rightarrow 12-z=11 \Rightarrow z=1.$

Substitute the values $x=3$ and $z=1$ into the second equation and solve for y.

$3(3)+y+(1)=12 \Rightarrow 9+y+1=12 \Rightarrow y=2.$ The solution is (3, 2, 1).

25. $\begin{bmatrix} 1 & 1 & 1 & | & -3 \\ 2 & -1 & 1 & | & -11 \\ -1 & -1 & 1 & | & -5 \end{bmatrix} \begin{array}{l} -2R_1+R_2 \to \\ R_1+R_3 \to \end{array} \begin{bmatrix} 1 & 1 & 1 & | & -3 \\ 0 & -3 & -1 & | & -5 \\ 0 & 0 & 2 & | & -8 \end{bmatrix} \frac{1}{2}R_3 \to \begin{bmatrix} 1 & 1 & 1 & | & -3 \\ 0 & -3 & -1 & | & -5 \\ 0 & 0 & 1 & | & -4 \end{bmatrix} \begin{array}{l} -R_3+R_1 \to \\ R_3+R_2 \to \end{array}$

$\begin{bmatrix} 1 & 1 & 0 & | & 1 \\ 0 & -3 & 0 & | & -9 \\ 0 & 0 & 1 & | & -4 \end{bmatrix} -\frac{1}{3}R_2 \to \begin{bmatrix} 1 & 1 & 0 & | & 1 \\ 0 & 1 & 0 & | & 3 \\ 0 & 0 & 1 & | & -4 \end{bmatrix} -R_2+R_1 \to \begin{bmatrix} 1 & 0 & 0 & | & -2 \\ 0 & 1 & 0 & | & 3 \\ 0 & 0 & 1 & | & -4 \end{bmatrix};$ The solution is $(-2, 3, -4)$.

26. $1(0\cdot3-4\cdot4)-2(-1\cdot3-2\cdot4)+0(-1\cdot4-2\cdot0)=1(0-16)-2(-3-8)+0(-4-0)$

$=-16-2(-11)+0=-16+22=6$

27. (a) $I(x)=0.07x$

(b) $I(500)=0.07(500)=\$35;$ the interest after one year for $500 at 7% is $35.

28. (a) $T(1990)=572.3(1990-1980)+3617=572.3(10)+3617=5723+3617=9340;$ average tuition

and fees in 1990 were $9340.

(b)   slope $=572.3;$ tuition and fees increased, on average, by $572.30 per year from 1980 to 2000.

29. Let x represent the width. Then $2(x)+2(x+7)=74\Rightarrow2x+2x+14=74\Rightarrow4x=60\Rightarrow x=15;$ the

dimensions are 15 inches by 22 inches.

30. $|P-12.4|\le0.2$

31. Let x, y, and z represent the amounts invested at 5%, 6%, and 8%, respectively. Then the system of

equations is $\begin{array}{l}x+y+z=5000\\z=y+500\\y=x+z-1000\end{array}\quad\Rightarrow\quad\begin{array}{l}x+y+z=5000\\-y+z=500\\-x+y-z=-1000\end{array}$ Add the first and third equations to

eliminate the variables x and z. $\dfrac{\begin{array}{l}x+y+z=5000\\-x+y-z=-1000\end{array}}{2y\qquad=4000}\Rightarrow y=2000.$ Then $z=2000+500=2500.$

Substitute the values $y=2000$ and $z=2500$ into the first equation and solve for x:

$x+2000+2500=5000\Rightarrow x=500.$ So $500 is invested at 5%, $2000 is invested at 6% and $2500 is

invested at 8%.

32. Let x and y represent the populations of Texas and Florida in 2010, respectively. Then the system

needed is $\begin{array}{l}x+y=43\\x=y+7\end{array}$ Substitute $x=y+7$ into the first equation and solve for y:

$y+7+y=43\Rightarrow2y=36\Rightarrow y=18$ and so $x=18+7=25.$ In 2010, the population of

Texas was 25 million and the population of Florida was 18 million.

Chapter 5: Polynomial Expressions and Functions

5.1: Polynomial Functions

1. $3x^2$; *Answers may vary.*

3. No, the powers must match for each variable.

5. No, the opposite is $-x^2-1$.

7. No; a function has only one output for each input.

9. $f(2)=(2)^2+2=4+2=6$

11. $(f-g)(x)=f(x)-g(x)=3x^2+7-(2x^2-5)=3x^2+7-2x^2+5=x^2+12$

13. Yes

15. No, there are two terms.

17. Yes

19. No, the variable is in a denominator.

21. No, the exponent is negative.

23. x^2

25. πr^2

27. xy

29. The degree is 7. The coefficient is 3.

31. The degree is 7. The coefficient is –3.

33. The degree is 6. The coefficient is –1.

35. The degree is 2. The leading coefficient is 5.

37. The degree is 3. The leading coefficient is $-\dfrac{2}{5}$.

39. The degree is 5. The leading coefficient is 1.

41. $x^2+4x^2=5x^2$

43. $6y^4-3y^4=3y^4$

45. Not possible

47. $9x^2-x+4x-6x^2=3x^2+3x$

49. $x^2+9xy-y^2+4x^2+y^2=5x^2+9xy$

51. $4x+7x^3y^7-\dfrac{1}{2}x^3y^7+9x-\dfrac{3}{2}x^3y^7=5x^3y^7+13x$

53. $(3x+1)+(-x+1)=2x+2$

55. $(x^2-2x+15)+(-3x^2+5x-7)=-2x^2+3x+8$

57. $\left(3x^3 - 4x + 3\right) + \left(5x^2 + 4x + 12\right) = 3x^3 + 5x^2 + 15$

59. $\left(x^4 - 3x^2 - 4\right) + \left(-8x^4 + x - \dfrac{1}{2}\right) = -7x^4 - 3x^2 + x - \dfrac{9}{2}$

61. $\left(4r^4 - r + 2\right) + \left(r^3 - 5r\right) = 4r^4 + r^3 - 6r + 2$

63. $\left(4xy - x^2 + y^2\right) + \left(4y^2 - 8xy - x^2\right) = 5y^2 - 4xy - 2x^2$

65. $-6x^5$

67. $-19x^5 + 5x^3 - 3x$

69. $7z^4 - z^2 + 8$

71. $(5x - 3) - (2x + 4) = 5x - 3 - 2x - 4 = 3x - 7$

73. $\left(x^2 - 3x + 1\right) - \left(-5x^2 + 2x - 4\right) = x^2 - 3x + 1 + 5x^2 - 2x + 4 = 6x^2 - 5x + 5$

75. $3\left(4x^4 + 2x^2 - 9\right) - 4\left(x^4 - 2x^2 - 5\right) = 12x^4 + 6x^2 - 27 - 4x^4 + 8x^2 + 20 = 8x^4 + 14x^2 - 7$

77. $4\left(x^4 - 1\right) - \left(4x^4 + 3x + 7\right) = 4x^4 - 4 - 4x^4 - 3x - 7 = -3x - 11$

79. Yes. This is a fourth degree polynomial.

81. No, the exponent on the variable is not positive.

83. No, the variable is located in the denominator.

85. Yes. The degree of the polynomial is 2. It is a quadratic polynomial.

87. $-4(2)^2 = -4(4) = -16$

89. $2(2)^2(3) = 2(4)(3) = 24$

91. $(-3)^2(4) - (-3)(4)^2 = 9(4) + 3(16) = 36 + 48 = 84$

93. When $x = 1$, $y = -1$ and so $f(1) = -1$. When $x = -2$, $y = 2$ and so $f(-2) = 2$.

95. When $x = -1$, $y = -4$ and so $f(-1) = -4$. When $x = 2$, $y = 2$ and so $f(2) = 2$.

97. $f(-2) = 3(-2)^2 = 3(4) = 12$

99. $f\left(-\dfrac{1}{2}\right) = 5 - 4\left(-\dfrac{1}{2}\right) = 5 + 2 = 7$

101. $f(-1) = 0.5(-1)^4 - 0.3(-1)^3 + 5 = 0.5 + 0.3 + 5 = 5.8$

103. $f(-1) = -(-1)^3 = -(-1) = 1$

105. $f(-3) = -(-3)^2 - 3(-3) = -9 + 9 = 0$

107. $f(2.4) = 1 - 2(2.4) + (2.4)^2 = 1 - 4.8 + 5.76 = 1.96$

109. $f(-1) = (-1)^5 - 5 = -1 - 5 = -6$

111. $f(a) = a^2 - 2a$

113. (a) $(f+g)(2) = (3(2)-1) + (5-(2)) = 6-1+5-2 = 8$

   (b) $(f-g)(-1) = (3(-1)-1) - (5-(-1)) = -3-1-5-1 = -10$

   (c) $(f+g)(x) = (3x-1) + (5-x) = 2x+4$

   (d) $(f-g)(x) = (3x-1) - (5-x) = 3x-1-5+x = 4x-6$

115. (a) $(f+g)(2) = \left(-3(2)^2\right) + \left((2)^2+1\right) = -3(4)+4+1 = -7$

   (b) $(f-g)(-1) = \left(-3(-1)^2\right) - \left((-1)^2+1\right) = -3-1-1 = -5$

   (c) $(f+g)(x) = \left(-3x^2\right) + \left(x^2+1\right) = -2x^2+1$

   (d) $(f-g)(x) = \left(-3x^2\right) - \left(x^2+1\right) = -4x^2-1$

117. (a) $(f+g)(2) = \left(-(2)^3\right) + \left(3(2)^3\right) = -8+3(8) = 16$

   (b) $(f-g)(-1) = \left(-(-1)^3\right) - \left(3(-1)^3\right) = 1-(-3) = 4$

   (c) $(f+g)(x) = \left(-x^3\right) + \left(3x^3\right) = 2x^3$

   (d) $(f-g)(x) = \left(-x^3\right) - \left(3x^3\right) = -4x^3$

119. (a) $(f+g)(2) = \left(2^2 - 2(2)+1\right) + \left(4(2)^2+3(2)\right) = 4-4+1+16+6 = 23$

   (b) $(f-g)(-1) = \left((-1)^2 - 2(-1)+1\right) - \left(4(-1)^2+3(-1)\right) = 1+2+1-4+3 = 3$

   (c) $(f+g)(x) = \left(x^2 - 2x+1\right) + \left(4x^2+3x\right) = 5x^2+x+1$

   (d) $(f-g)(x) = \left(x^2 - 2x+1\right) - \left(4x^2+3x\right) = -3x^2-5x+1$

121. (a) When $x = 1980$ the y-value is about 1300. When $x = 2000$ the y-value is about 10,000. In 1980 there were about 1300 women running. In 2000 there were about 10,000.

   (b) In 1980 $x = 0$, and so $3.56(0)^2 + 360(0)+1300 = 1300$.

   In 2000 $x = 20$, and so $3.56(20)^2 + 360(20)+1300 = 9924$. The answers are similar.

   (c) $9924-1300 = 8624$. There was an increase of 8624 runners.

123. (a) $f(10) = 2.4(10)^2 - 14(10)+23 = 240-140+23 = 123$ thousand, which is close to the actual value.

   (b) $f(17) = 2.4(17)^2 - 14(17)+23 = 693.6-238+23 = 478.6$ thousand, which is too high. AIDS deaths did not continue to rise as rapidly as the model predicts.

125. The volume of one cube is s^3, so the volume of five identical cubes is $5s^3$.

127. A square with side x has area x^2 and a circle with radius x has area πx^2.

The polynomial is $x^2 + \pi x^2$. When $x = 10$ we have $10^2 + \pi \cdot 10^2 = 100 + 100\pi \approx 414.2 \text{ in}^2$.

129. The table is shown in Figure 129. The athlete's heart rate was between 80 and 110 beats per minute from 4 to 8 minutes after exercise had stopped.

t	3	4	5	6	7	8	9
$y = f(t)$	126.875	110	96.875	87.5	81.875	80	81.875

Figure 129

131. The best model is $f(x)$. This can be verified numerically using a table.

133. (a) $R(x) = 16x$

(b) $P(x) = R(x) - C(x) = 16x - (4x + 2000) = 12x - 2000$

(c) $P(3000) = 12(3000) - 2000 = 36,000 - 2000 = 34,000$; the profit is \$34,000 for

selling 3000 DVDs.

135. (a) $C(100) = 0.3(100) + 100 = 130$. It costs \$130 thousand to make 100 notebook computers.

(b)   The y-intercept for C is 100. The company has \$100 thousand in fixed costs even if they make 0 computers. The y-intercept for R is 0. If the company sells 0 computers, their revenue is \$0.

(c) $P(x) = R(x) - C(x) = 0.75x - (0.3x + 100) = 0.45x - 100$

(d)   A profit will occur when revenue is greater than cost.

$0.75x > 0.3x + 100 \Rightarrow 0.45x > 100 \Rightarrow x > 222.22$. So 223 or more computers need to be sold.

Section 5.2 Multiplication of Polynomials

1. Distributive

3. $x^{3+5} = x^8$

5. $a^2 - b^2$

7. $x^2 - x$

9. False

11. d

13. $x^4 \cdot x^8 = x^{4+8} = x^{12}$

15. $-5y^7 \cdot 4y = -20y^{7+1} = -20y^8$

17. $(-xy)(4x^3 y^5) = -4x^{1+3} \cdot y^{1+5} = -4x^4 y^6$

19. $(5y^2 z)(4x^2 yz^5) = 20x^2 \cdot y^{2+1} \cdot z^{1+5} = 20x^2 y^3 z^6$

21. $5(y + 2) = 5y + 10$

23. $-2(5x+9) = -10x-18$

25. $-6y(y-3) = -6y^2+18y$

27. $(9-4x)3 = 27-12x$

29. $-ab(a^2-b^2) = -a^3b+ab^3$

31. $-5m(n^3+m) = -5mn^3-5m^2$

33. $x^2+x+2x+2 = x^2+3x+2$; When $x=5, (5)^2+3(5)+2 = 25+15+2 = 42$ in^2.

35. $2x^2+2x+x+1 = 2x^2+3x+1$; When $x=5, 2(5)^2+3(5)+1 = 50+15+1 = 66$ in^2.

37. $(x+5)(x+6) = x^2+6x+5x+30 = x^2+11x+30$

39. $(2x+1)(2x+1) = 4x^2+2x+2x+1 = 4x^2+4x+1$

41. No, a square of a sum does not equal the sum of the squares.

43. No, $(x-2)^2 = x^2-4x+4 \neq x^2-4$.

45. Yes, $(x+3)(x-2) = x^2+x-6$.

47. No, $2z(3z+1) = 6z^2+2z \neq 6z^2+1$.

49. $(x+5)(x+10) = x^2+10x+5x+50 = x^2+15x+50$

51. $(x-3)(x-4) = x^2-4x-3x+12 = x^2-7x+12$

53. $(2z-1)(z+2) = 2z^2+4z-z-2 = 2z^2+3z-2$

55. $(y+3)(y-4) = y^2-4y+3y-12 = y^2-y-12$

57. $(4x-3)(4-9x) = 16x-36x^2-12+27x = -36x^2+43x-12$

59. $(-2z+3)(z-2) = -2z^2+4z+3z-6 = -2z^2+7z-6$

61. $\left(z-\dfrac{1}{2}\right)\left(z+\dfrac{1}{4}\right) = z^2+\dfrac{1}{4}z-\dfrac{1}{2}z-\dfrac{1}{8} = z^2-\dfrac{1}{4}z-\dfrac{1}{8}$

63. $(x^2+1)(2x^2-1) = 2x^4-x^2+2x^2-1 = 2x^4+x^2-1$

65. $(x+y)(x-2y) = x^2-2xy+xy-2y^2 = x^2-xy-2y^2$

67. $4x(x^2-2x-3) = 4x^3-8x^2-12x$

69. $-x(x^4-3x^2+1) = -x^5+3x^3-x$

71. $(2n^2-4n+1)(3n^2) = 6n^4-12n^3+3n^2$

73. $(x+1)(x^2+2x-3) = x^3+2x^2-3x+x^2+2x-3 = x^3+3x^2-x-3$

75. $z(2+z)(1-z-z^2) = (2z+z^2)(1-z-z^2) = 2z-2z^2-2z^3+z^2-z^3-z^4 = -z^4-3z^3-z^2+2z$

77. $2ab^2(2a^2-ab+3b^2) = 4a^3b^2-2a^2b^3+6ab^4$

79. $(2r-4t)(3r^2+rt-t^2) = 6r^3+2r^2t-2rt^2-12r^2t-4rt^2+4t^3 = 6r^3-10r^2t-6rt^2+4t^3$

81. $(2^3)^2 = 2^{3\cdot2} = 2^6 = 64$

83. $2(z^3)^6 = 2z^{3\cdot6} = 2z^{18}$

85. $(-5x)^2 = (-5)^2 \cdot x^2 = 25x^2$

87. $(-2xy^2)^3 = (-2)^3 \cdot x^3 \cdot y^{2\cdot3} = -8x^3y^6$

89. $(-4a^2b^3)^2 = (-4)^2 \cdot a^{2\cdot2} \cdot b^{3\cdot2} = 16a^4b^6$

91. $(x-4)(x+4) = x^2-16$

93. $(3-2x)(3+2x) = 3^2-(2x)^2 = 9-4x^2$

95. $(x-y)(x+y) = x^2-y^2$

97. $(x+2)^2 = x^2+2(2x)+4 = x^2+4x+4$

99. $(2x+1)^2 = 4x^2+2(2x)+1 = 4x^2+4x+1$

101. $(x-1)^2 = x^2+2(-x)+1 = x^2-2x+1$

103. $(3x-2)^2 = 9x^2+2(-6x)+4 = 9x^2-12x+4$

105. $3x(x+1)(x-1) = 3x(x^2-1) = 3x^3-3x$

107. $3rt(t-2r)(t+2r) = 3rt(t^2-4r^2) = 3rt^3-12r^3t$

109. $(a^2+2b^2)(a^2-2b^2) = (a^2)^2-(2b^2)^2 = a^4-4b^4$

111. $(3m^3+5n^2)^2 = (3m^3)^2+2(3m^3)(5n^2)+(5n^2)^2 = 9m^6+30m^3n^2+25n^4$

113. $(x^3-2y^3)^2 = (x^3)^2-2(x^3)(2y^3)+(2y^3)^2 = x^6-4x^3y^3+4y^6$

115. $((x-3)+y)((x-3)-y) = (x-3)^2-y^2 = x^2-6x+9-y^2 = x^2-6x-y^2+9$

117. $(r-(t+2))(r+(t+2)) = r^2-(t+2)^2 = r^2-(t^2+4t+4) = r^2-t^2-4t-4$

119. $(102)(98) = (100+2)(100-2) = 100^2-2^2 = 10,000-4 = 9996$

121. $(fg)(2) = (2+1)(2-2) = (3)(0) = 0; \quad (fg)(x) = (x+1)(x-2) = x^2-2x+x-2 = x^2-x-2$

123. $(fg)(2) = \left((2)^2\right)(3-5(2)) = (4)(-7) = -28;$ $(fg)(x) = \left(x^2\right)(3-5x) = 3x^2 - 5x^3 = -5x^3 + 3x^2$

125. $(fg)(2) = \left(2(2)^2 + 1\right)\left((2)^2 + 2 - 3\right) = (9)(3) = 27;$

$(fg)(x) = \left(2x^2 + 1\right)\left(x^2 + x - 3\right) = 2x^4 + 2x^3 - 6x^2 + x^2 + x - 3 = 2x^4 + 2x^3 - 5x^2 + x - 3$

127. $x(x+4);$ When $x = 20, (20)(20+4) = 20(24) = 480$ ft^2.

129. $(x+3)^2;$ When $x = 20, (20+3)^2 = 23^2 = 529$ ft^2.

131. (a) $D = 40 - \dfrac{1}{2}(20) = 40 - 10 = 30;$ demand is 30 thousand units.

(b) $R = Dp = \left(40 - \dfrac{1}{2}p\right)(p) = 40p - \dfrac{1}{2}p^2$

(c) $R = 40(30) - \dfrac{1}{2}(30)^2 = 1200 - \dfrac{1}{2}(900) = 1200 - 450 = 750;$ revenue is \$750 thousand.

133. (a) $100\left(1 - \dfrac{x}{100}\right)^2 = 100\left(1 - \dfrac{x}{100}\right)\left(1 - \dfrac{x}{100}\right) = 100\left(1 - \dfrac{2x}{100} + \dfrac{x^2}{10,000}\right) = 100 - 2x + \dfrac{x^2}{100}$

(b) $100\left(1 - \dfrac{34}{100}\right)^2 = 100(.66)^2 = 43.56;$ $100 - 2(34) + \dfrac{34^2}{100} = 43.56$ The answers are the same.

(c) If one serve is out of bounds 34% of the time, then two consecutive serves will be in bounds 43.56% of the time.

135. (a) The average American used 1900 gallons of water daily in 1980, and this average is decreasing by 24 gallons each year.

(b) $T(x) = (-24x + 1900)(2.6x + 230) = -62.4x^2 - 5520x + 4940x + 437,000 = $
$-62.4x^2 - 580x + 437,000$

(c) $T(30) = -62.4(30)^2 - 580(30) + 437,000 = 363,440.$ The daily usage in 2010 was

363,440,000,000 gallons.

137. Let x be the smaller integer. Then $x+1$ is the next consecutive integer.

The product of the two consecutive integers is $x(x+1) = x^2 + x$.

139. Let l represent the length of the rectangle. Then since the perimeter is 100 feet,

$2l + 2x = 100 \Rightarrow l = 50 - x.$ Since the area is given by $A = xl$, we may write the area as follows:

$x(50 - x) = 50x - x^2$.

Checking Basic Concepts Sections 5.1 and 5.2

1. (a) $8x^2 + 4x - 5x^2 + 3x = 3x^2 + 7x$

(b) $\left(5x^2 - 3x + 2\right) - \left(3x^2 - 5x^3 + 1\right) = 5x^2 - 3x + 2 - 3x^2 + 5x^3 - 1 = 5x^3 + 2x^2 - 3x + 1$

2. The polynomial is $x(x+120)$. When tickets cost \$10 each, the revenue will be

   $10(10+120) = \$1300$.

3. (a)   When exercise is stopped, $t = 0$, and so $f(0) = 2(0)^2 - 25(0) + 160 = 0 - 0 + 160 = 160$ bpm.

   (b)   After 4 minutes, $t = 4$, and so $f(4) = 2(4)^2 - 25(4) + 160 = 32 - 100 + 160 = 92$ bpm.

4. (a) $-5(x-6) = -5x + 30$

   (b) $4x^3\left(3x^2 - 5x\right) = 12x^5 - 20x^4$

   (c) $(2x-1)(x+3) = 2x^2 + 6x - x - 3 = 2x^2 + 5x - 3$

5. (a) $(5x-6)(5x+6) = 25x^2 - 36$

   (b) $(3x-4)^2 = 9x^2 - 24x + 16$

6. See Figure 6.

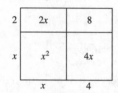

Figure 6

Section 5.3 Factoring Polynomials

1. To solve equations

3. Yes, $2x$ is a common factor of each term.

5. No, since $\dfrac{1}{2}(4) = 2$, but $\dfrac{1}{2} \neq 2$ and $4 \neq 2$.

7. d

9. $10x - 15 = 5(2x - 3)$

11. $4x + 6y = 2(2x + 3y)$

13. $9r - 15t = 3(3r - 5t)$

15. $2x^3 - 5x = x\left(2x^2 - 5\right)$

17. $8a^3 + 10a = 2a\left(4a^2 + 5\right)$

19. $6r^3 - 18r^5 = 6r^3\left(1 - 3r^2\right)$

21. $8x^3 - 4x^2 + 16x = 4x\left(2x^2 - x + 4\right)$

23. $9n^4 - 6n^2 + 3n = 3n\left(3n^3 - 2n + 1\right)$

25. $6t^6 - 4t^4 + 2t^2 = 2t^2\left(3t^4 - 2t^2 + 1\right)$

27. $5x^2y^2 - 15x^2y^3 = 5x^2y^2\left(1 - 3y\right)$

29. $6a^3b^2 - 15a^2b^3 = 3a^2b^2\left(2a - 5b\right)$

31. $18mn^2 + 12m^2n^3 = 6mn^2\left(3 + 2mn\right)$

33. $15x^2y + 10xy - 25x^2y^2 = 5xy\left(3x + 2 - 5xy\right)$

35. $4a^2 - 2ab + 6ab^2 = 2a\left(2a - b + 3b^2\right)$

37. $-2x^2 + 4x - 6 = -2\left(x^2 - 2x + 3\right)$

39. $-8z^4 - 16z^3 = -8z^3\left(z + 2\right)$

41. $-4m^2n^3 - 6mn^2 - 8mn = -2mn\left(2mn^2 + 3n + 4\right)$

43. $m = 0$ or $n = 0$

45. $3z = 0 \Rightarrow z = 0$ or $z + 4 = 0 \Rightarrow z = -4$

47. $r - 1 = 0 \Rightarrow r = 1$ or $r + 3 = 0 \Rightarrow r = -3$

49. $x + 2 = 0 \Rightarrow x = -2$ or $3x - 1 = 0 \Rightarrow 3x = 1 \Rightarrow x = \dfrac{1}{3}$

51. $3x = 0 \Rightarrow x = 0$ or $y - 6 = 0 \Rightarrow y = 6$

53. (a)   The graph crosses the x-axis at -3 and 0.

   (b)   From the graph, $y = P(x) = 0$ when $x = 0$ or $x = -3$.

   (c)   The zeros of $P(x)$ are -3 and 0.

55. (a)   The graph crosses the x-axis at 0 and 2.

   (b)   From the graph $y = P(x) = 0$ when $x = 0$ or $x = 2$.

   (c)   The zeros of $P(x)$ are 0 and 2.

57. (a)   The graph crosses the x-axis at -1 and 1.

   (b)   From the graph $y = P(x) = 0$ when $x = -1$ or $x = 1$.

   (c)   The zeros of $P(x)$ are -1 and 1.

59. The graph of $y = x^2 - 2x$ (not shown) crosses the x-axis at 0 and 2. The solutions are 0 and 2.

61. The graph of $y = x - x^2$ (not shown) crosses the x-axis at 0 and 1. The solutions are 0 and 1.

63. The graph of $y = x^2 + 4x$ (not shown) crosses the x-axis at -4 and 0. The solutions are -4 and 0.

65. $x^2 - x = 0 \Rightarrow x(x-1) = 0 \Rightarrow$ Either $x = 0$ or $x - 1 = 0 \Rightarrow x = 1.$

67. $5x^2 - x = 0 \Rightarrow x(5x-1) = 0 \Rightarrow$ Either $x = 0$ or $5x - 1 = 0 \Rightarrow x = \dfrac{1}{5}.$

69. $10x^2 + 5x = 0 \Rightarrow 5x(2x+1) = 0 \Rightarrow$ Either $5x = 0 \Rightarrow x = 0$ or $2x + 1 = 0 \Rightarrow x = -\dfrac{1}{2}.$

71. $15x^2 = 10x \Rightarrow 15x^2 - 10x = 0 \Rightarrow 5x(3x-2) = 0 \Rightarrow$ Either $5x = 0 \Rightarrow x = 0$ or $3x - 2 = 0 \Rightarrow x = \dfrac{2}{3}.$

73. $25x = 10x^2 \Rightarrow 25x - 10x^2 = 0 \Rightarrow 5x(5-2x) = 0 \Rightarrow$ Either $5x = 0 \Rightarrow x = 0$ or $5 - 2x = 0 \Rightarrow x = \dfrac{5}{2}.$

75. $32x^4 - 16x^3 = 0 \Rightarrow 16x^3(2x-1) = 0 \Rightarrow$ Either $16x^3 = 0 \Rightarrow x = 0$ or $2x - 1 = 0 \Rightarrow x = \dfrac{1}{2}.$

77. Numerically: Table $Y_1 = X^2 + 2X$ with Tblstart $= -5$ and ΔTbl $= 1$. See Figure 77a.

Graphically: Graph $Y_1 = X^2 + 2X$ in $[-5,\ 5,\ 1]$ by $[-5, 5, 1]$. See Figure 77b.

$x^2 + 2x = 0 \Rightarrow x(x+2) = 0 \Rightarrow x = 0$ or $x + 2 = 0 \Rightarrow x = -2$. The solutions are $-2,\ 0$.

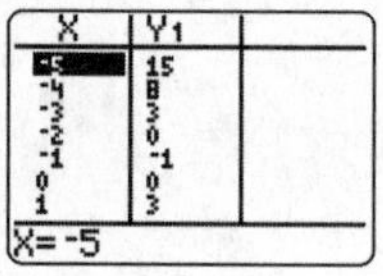

Figure 77a 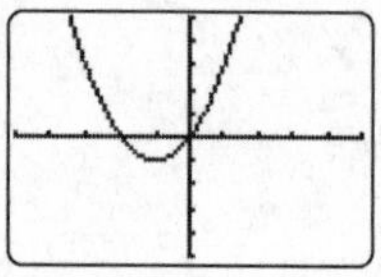   Figure 77b

79. Numerically: Table $Y_1 = 2X^2 - 3X$ with Tblstart $= -1$ and ΔTbl $= 0.5$. See Figure 79a.

Graphically: Graph $Y_1 = 2X^2 - 3X$ in $[-5,\ 5,\ 0.5]$ by $[-5, 5, 1]$. See Figure 79b.

$2x^2 - 3x = 0 \Rightarrow x(2x-3) = 0 \Rightarrow x = 0$ or $2x - 3 = 0 \Rightarrow x = \dfrac{3}{2}$. The solutions are $0,\ \dfrac{3}{2}$.

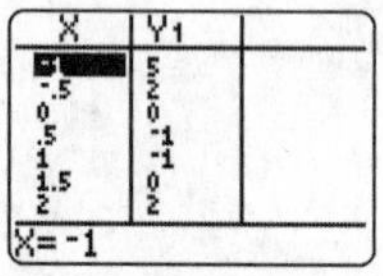

Figure 79a 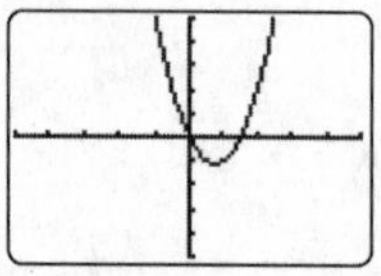   Figure 79b

81. $2x(x+2) + 3(x+2) = (2x+3)(x+2)$

83. $(x-5)x^2 - (x-5)2 = (x-5)(x^2-2)$

85. $x^3 + 3x^2 + 2x + 6 = x^2(x+3) + 2(x+3) = (x+3)(x^2+2)$

87. $6x^3 - 4x^2 + 9x - 6 = 2x^2(3x-2) + 3(3x-2) = (3x-2)(2x^2+3)$

89. $2x^3 - 3x^2 + 2x - 3 = x^2(2x-3) + 1(2x-3) = (x^2+1)(2x-3)$

91. $x^3 - 7x^2 - 3x + 21 = x^2(x-7) - 3(x-7) = (x^2-3)(x-7)$

93. $3x^3 - 15x^2 + 5x - 25 = 3x^2(x-5) + 5(x-5) = (x-5)(3x^2+5)$

95. $xy + x + 3y + 3 = x(y+1) + 3(y+1) = (y+1)(x+3)$

97. $ab - 3a + 2b - 6 = a(b-3) + 2(b-3) = (a+2)(b-3)$

99. Golf ball: graph b, because the height starts at 0, then increases until it reaches its highest point and then decreases until the height is 0 again.

Yo – yo: graph a, because the height decreases until its lowest point, then it increases and returns to its starting position.

101. (a) $-16t^2 + 128t = 0 \Rightarrow -16t(t-8) = 0 \Rightarrow$ Either $-16t = 0 \Rightarrow t = 0$ or $t - 8 = 0 \Rightarrow t = 8$.

The golf ball hits the ground after 8 seconds.

(b) The graph crosses the t-axis at about 8 seconds.

(c) See Figure 101. Yes, the answer is the same.

(d) The maximum altitude given by either the graph or the table is 256 feet which is attained after 4 seconds.

t	3	4	5	6	7	8
$y = h(t)$	240	256	240	192	112	0

Figure 101

103. $f(52) = 0.0148(52)^2 + 0.686(52) + 315 \approx 391$; CO_2 concentration in 2010 was 391 ppm.

105. $8y = \pi y^2 \Rightarrow 8y - \pi y^2 = 0 \Rightarrow y(8 - \pi y) = 0 \Rightarrow y = 0$ or $8 - \pi y = 0$, that is $y = 0$ or $y = \dfrac{8}{\pi}$.

The positive value is $\dfrac{8}{\pi}$.

107. (a) See Figure 107. It models the data quite well.

(b) From the table, the height is about 54,000 feet.

(c) $11t^2 + 6t = 0 \Rightarrow t(11t + 6) = 0 \Rightarrow$ Either $t = 0$ or $11t + 6 = 0 \Rightarrow t = -\dfrac{6}{11}$.

When $t = 0$ the shuttle has not yet left the ground. The value $t = -\dfrac{6}{11}$ has no physical meaning.

t	20	30	40	50	60	70	80
$y = f(t)$	4520	10,080	17,840	27,800	39,960	54,320	70,880

Figure 107

109. No, there are many possibilities such as 9×24, 8×27 or 13.5×16.

111. One example is a satellite dish. *Answers may vary.*

Section 5.4 Factoring Trinomials

1. A polynomial with 3 terms: $x^2 - x + 1$; *Answers may vary.*

3. $a = 3$, $b = -1$, $c = -3$

5. $(x+3)(x-1)$

7. c

9. Yes, it checks using FOIL.

11. No, $(x+5)(x-4)=x^2+x-20$.

13. Yes, it checks using FOIL.

15. No, $(5m+2)(2m-5)=10m^2-21m-10$.

17. $x^2+7x+10=(x+2)(x+5)$

19. $x^2+8x+12=(x+2)(x+6)$

21. $x^2-13x+36=(x-9)(x-4)$

23. $x^2-7x-8=(x-8)(x+1)$

25. $z^2+z-72=(z-8)(z+9)$

27. $t^2-15t+56=(t-8)(t-7)$

29. $y^2-18y+72=(y-12)(y-6)$

31. $m^2-18m-40=(m-20)(m+2)$

33. $n^2-20n-300=(n-30)(n+10)$

35. $2x^2+7x+3=(x+3)(2x+1)$

37. $6x^2-7x-5=(2x+1)(3x-5)$

39. $4z^2+19z+12=(z+4)(4z+3)$

41. $6t^2-17t+12=(2t-3)(3t-4)$

43. $10y^2+13y-3=(2y+3)(5y-1)$

45. $6m^2-m-12=(2m-3)(3m+4)$

47. $42n^2+5n-25=(6n+5)(7n-5)$

49. $1+x-2x^2=(1-x)(1+2x)$

51. $20+7x-6x^2=(5-2x)(4+3x)$

53. $5y^2+5y-30=5\left(y^2+y-6\right)=5(y-2)(y+3)$

55. $2z^2+12z+16=2\left(z^2+6z+8\right)=2(z+2)(z+4)$

57. $z^3+9z^2+14z=z\left(z^2+9z+14\right)=z(z+2)(z+7)$

59. $t^3 - 10t^2 + 21t = t\left(t^2 - 10t + 21\right) = t(t-7)(t-3)$

61. $m^4 + 6m^3 + 5m^2 = m^2\left(m^2 + 6m + 5\right) = m^2(m+1)(m+5)$

63. $5x^3 + x^2 - 6x = x\left(5x^2 + x - 6\right) = x(x-1)(5x+6)$

65. $6x^3 + 21x^2 + 9x = 3x\left(2x^2 + 7x + 3\right) = 3x(x+3)(2x+1)$

67. $2x^3 - 14x^2 + 20x = 2x\left(x^2 - 7x + 10\right) = 2x(x-5)(x-2)$

69. $60z^3 + 230z^2 - 40z = 10z\left(6z^2 + 23z - 4\right) = 10z(z+4)(6z-1)$

71. $4x^4 + 10x^3 - 6x^2 = 2x^2\left(2x^2 + 5x - 3\right) = 2x^2(x+3)(2x-1)$

73. $x^2 + (2+3)x + 2\cdot3 = x^2 + 2x + 3x + 2\cdot3 = x(x+2) + 3(x+2) = (x+2)(x+3)$

75. $x^2 + (a+b)x + ab = x^2 + ax + bx + ab = x(x+a) + b(x+a) = (x+a)(x+b)$

77. $x = 1,\ x = 2,\ x^2 + bx + c = 0 \Rightarrow (x-1)(x-2) = 0 \Rightarrow x^2 - 2x - x + 2 = 0$

$\Rightarrow x^2 - 3x + 2 = 0 \Rightarrow b = -3,\ c = 2$

79. $h(x) = x^2 + 3x + 2 = (x+1)(x+2) \Rightarrow f(x) = x+1,\ g(x) = x+2$

81. $h(x) = x^2 - 2x - 8 = (x+2)(x-4) \Rightarrow f(x) = x+2,\ g(x) = x-4$

83. Since $x = 2$ and $x = 4$ are zeros of the polynomial, the factors are $(x-2)(x-4)$.

This checks using FOIL.

85. Since $x = -1$ and $x = 2$ are zeros of the polynomial, the factors are $2(x+1)(x-2)$.

This checks using FOIL.

87. Since $x = -1$ and $x = 2$ are zeros of the polynomial, the factors are $-(x-2)(x+1)$.

This checks using FOIL.

89. Since $Y_1 = 0$ when $x = -1$ and $x = 4$, the factors are $(x+1)(x-4)$. This checks using FOIL.

91. Since $Y_1 = 0$ when $x = -1$ and $x = 2$, the factors are $2(x+1)(x-2)$. This checks using FOIL.

93. Table $Y_1 = X^2 + 3X - 10$ with TblStart $= -6.4$ and ΔTbl $= 1.4$. See Figure 93.

Since $Y_1 = 0$ when $x = -5$ and $x = 2$, the factors are $(x-2)(x+5)$.

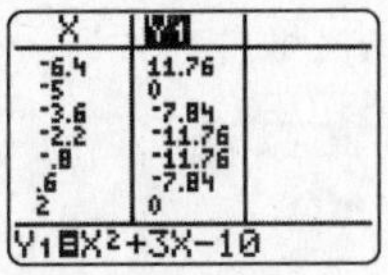

Figure 93

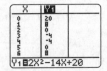
Figure 95

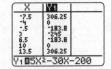
Figure 97

Figure 99

95. Table $Y_1 = X^2 - 3X - 28$ with TblStart $= -6.2$ and ΔTbl $= 2.2$. See Figure 95.

Since $Y_1 = 0$ when $x = -4$ and $x = 7$, the factors are $(x-7)(x+4)$.

97. Table $Y_1 = 2X^2 - 14X + 20$ with TblStart = 0 and ΔTbl = 1. See Figure 97.

Since $Y_1 = 0$ when $x = 2$ and $x = 5$, the factors are $2(x-5)(x-2)$.

99. Table $Y_1 = 5X^2 - 30X - 200$ with TblStart = -7.5 and ΔTbl = 3.5. See Figure 99.

Since $Y_1 = 0$ when $x = -4$ and $x = 10$, the factors are $5(x-10)(x+4)$.

101. Table $Y_1 = 8X^2 - 44X + 20$ with TblStart = -1 and ΔTbl = 1.5. See Figure 101.

Since $Y_1 = 0$ when $x = 0.5$ and $x = 5$, the factors are $8(x-5)(x-0.5) = 4(x-5)(2x-1)$.

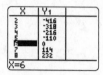

Figure 101   Figure 107

103. (a) $(35-x)(1200+100x)$

(b) $(35-x)(1200+100x) = 54{,}000 \Rightarrow 42{,}000 + 2300x - 100x^2 = 54{,}000 \Rightarrow x^2 - 23x + 120 = 0 \Rightarrow$

$(x-15)(x-8) = 0 \Rightarrow$ Either $x - 15 = 0 \Rightarrow x = 15$ or $x - 8 = 0 \Rightarrow x = 8$

When $x = 15$ the ticket price is $35 - 15 = \$20$. When $x = 8$ the ticket price is $35 - 8 = \$27$.

105. $x(x+6) = 91 \Rightarrow x^2 + 6x - 91 = 0 \Rightarrow (x+13)(x-7) = 0 \Rightarrow$ Either $x = -13$ or $x = 7$.

Since the width cannot be negative, the width is 7 feet and the length is $7 + 6 = 13$ feet.

107. (a) $2t^2 + 88t = 600 \Rightarrow t^2 + 44t - 300 = 0 \Rightarrow (t+50)(t-6) = 0 \Rightarrow$ Either $t = -50$ or $t = 6$.

Since -50 seconds has no meaning, the car travels 600 feet in 6 seconds.

(b)   Table $Y_1 = 2X^2 + 88X - 600$ with TblStart = 2 and ΔTbl = 1. See Figure 107.

Checking Basic Concepts Sections 5.3 and 5.4

1. (a) $3x^2 - 6x = 3x(x-2)$

(b) $16x^3 - 8x^2 + 4x = 4x(4x^2 - 2x + 1)$

2. (a) $x^2 - 2x = 0 \Rightarrow x(x-2) = 0 \Rightarrow$ Either $x = 0$ or $x - 2 = 0 \Rightarrow x = 2$.

(b) $9x^2 = 81x \Rightarrow 9x(x-9) = 0 \Rightarrow$ Either $9x = 0 \Rightarrow x = 0$ or $x - 9 = 0 \Rightarrow x = 9$.

3. (a) $x^2 + 3x - 10 = (x-2)(x+5)$

(b) $x^2 - 3x - 10 = (x-5)(x+2)$

(c) $8x^2 + 14x + 3 = (2x+3)(4x+1)$

4. $x^2 + 3x + 2 = 0 \Rightarrow (x+1)(x+2) = 0 \Rightarrow x+1 = 0 \Rightarrow x = -1 \text{ or } x+2 = 0 \Rightarrow x = -2$

5. $-16t^2 + 64t + 8 = 56 \Rightarrow -16t^2 + 64t - 48 = 0 \Rightarrow -16(t-1)(t-3) = 0 \Rightarrow t = 1 \text{ or } t = 3$

   After 1 and 3 seconds.

Section 5.5 Special Types of Factoring

1. $x^2 - 9$; *Answers may vary.*

3. $x^3 + 8$; *Answers may vary.*

5. $a^2 + 2ab + b^2 = (a+b)^2$

7. c

9. Yes; $x^2 - 25 = (x-5)(x+5)$

11. No; $x^3 + y^3 = (x+y)(x^2 - xy + y^2)$

13. $x^2 - 36 = (x-6)(x+6)$

15. $25 - z^2 = (5-z)(5+z)$

17. The sum of squares cannot be factored.

19. $36x^2 - 100 = 4(9x^2 - 25) = 4(3x-5)(3x+5)$

21. $49a^2 - 64b^2 = (7a-8b)(7a+8b)$

23. $64z^2 - 25z^4 = z^2(64 - 25z^2) = z^2(8-5z)(8+5z)$

25. $5x^3 - 125x = 5x(x^2 - 25) = 5x(x-5)(x+5)$

27. The sum of squares cannot be factored.

29. $16t^4 - r^2 = (4t^2 - r)(4t^2 + r)$

31. $(x+1)^2 - 25 = ((x+1)-5)((x+1)+5) = (x-4)(x+6)$

33. $100 - (n-4)^2 = (10-(n-4))(10+(n-4)) = (14-n)(6+n)$

35. $y^4 - 16 = (y^2 - 4)(y^2 + 4) = (y-2)(y+2)(y^2 + 4)$

37. $16x^4 - y^4 = (4x^2 - y^2)(4x^2 + y^2) = (2x-y)(2x+y)(4x^2 + y^2)$

39. $x^3 + x^2 - x - 1 = x^2(x+1) - 1(x+1) = (x^2 - 1)(x+1) = (x-1)(x+1)(x+1) = (x-1)(x+1)^2$

41. $4x^3 - 8x^2 - x + 2 = 4x^2(x-2) - 1(x-2) = (4x^2 - 1)(x-2) = (2x-1)(2x+1)(x-2)$

43. No, the middle term would need to be $8x$.

45. Yes, $x^2 + 8x + 16 = (x+4)^2$

47. Yes, $4z^2 - 4z + 1 = (2z-1)^2$

49. No, the middle term would need to be $-24t$.

51. $x^2 + 2x + 1 = (x+1)^2$

53. $4x^2 + 20x + 25 = (2x+5)^2$

55. $x^2 - 12x + 36 = (x-6)^2$

57. $9z^2 - 24z + 16 = (3z-4)^2$

59. $4y^4 + 4y^3 + y^2 = y^2 \left(4y^2 + 4y + 1\right) = y^2 (2y+1)^2$

61. $9z^3 - 6z^2 + z = z\left(9z^2 - 6z + 1\right) = z(3z-1)^2$

63. $9x^2 + 6xy + y^2 = (3x+y)^2$

65. $49a^2 - 28ab + 4b^2 = (7a-2b)^2$

67. $4x^4 - 4x^3 y + x^2 y^2 = x^2 \left(4x^2 - 4xy + y^2\right) = x^2 (2x-y)^2$

69. $x^3 - 8 = (x-2)\left(x^2 + 2x + 4\right)$

71. $y^3 + z^3 = (y+z)\left(y^2 - yz + z^2\right)$

73. $27x^3 - 8 = (3x-2)\left(9x^2 + 6x + 4\right)$

75. $64z^3 + 27t^3 = (4z+3t)\left(16z^2 - 12zt + 9t^2\right)$

77. $8x^4 + 125x = x\left(8x^3 + 125\right) = x(2x+5)\left(4x^2 - 10x + 25\right)$

79. $27y - 8x^3 y = y\left(27 - 8x^3\right) = y(3-2x)\left(9 + 6x + 4x^2\right)$

81. $z^6 - 27y^3 = \left(z^2 - 3y\right)\left(z^4 + 3z^2 y + 9y^2\right)$

83. $125z^6 + 8y^9 = \left(5z^2 + 2y^3\right)\left(25z^4 - 10z^2 y^3 + 4y^6\right)$

85. $5m^6 + 40n^3 = 5\left(m^6 + 8n^3\right) = 5\left(m^2 + 2n\right)\left(m^4 - 2m^2 n + 4n^2\right)$

87. $25x^2 - 64 = (5x-8)(5x+8)$

89. $x^3 + 27 = (x+3)\left(x^2 - 3x + 9\right)$

91. $64x^2 + 16x + 1 = (8x+1)^2$

93. $3x^2 + 14x + 8 = (x+4)(3x+2)$

95. $x^4 + 8x = x(x^3 + 8) = x(x+2)(x^2 - 2x + 4)$

97. $64x^3 + 8y^3 = 8(8x^3 + y^3) = 8(2x+y)(4x^2 - 2xy + y^2)$

99. $2r^2 - 8t^2 = 2(r^2 - 4t^2) = 2(r - 2t)(r + 2t)$

101. $a^3 + 4a^2b + 4ab^2 = a(a^2 + 4ab + 4b^2) = a(a+2b)^2$

103. $x^2 - 3x + 2 = (x-2)(x-1)$

105. $4z^2 - 25 = (2z-5)(2z+5)$

107. $x^4 + 16x^3 + 64x^2 = x^2(x^2 + 16x + 64) = x^2(x+8)^2$

109. $z^3 - 1 = (z-1)(z^2 + z + 1)$

111. $3t^2 - 5t - 8 = (t+1)(3t-8)$

113. $7a^3 + 20a^2 - 3a = a(7a^2 + 20a - 3) = a(a+3)(7a-1)$

115. $x^6 - y^6 = (x^3 - y^3)(x^3 + y^3) = (x-y)(x^2 + xy + y^2)(x+y)(x^2 - xy + y^2)$

117. $100x^2 - 1 = (10x - 1)(10x + 1)$

119. $p^3q^3 - 27 = (pq - 3)(p^2q^2 + 3pq + 9)$

121. Area of larger square: $A = s^2 = x^2$, Area of smaller square: $A = s^2 = 5^2 = 25$

   Difference of areas: $x^2 - 25 = (x-5)(x+5)$

Section 5.6 Summary of Factoring

1. Factor out the GCF.

3. No; the sum of two squares cannot be factored.

5. $a^2 - a = a(a-1)$

7. $a^2 - 9 = (a-3)(a+3)$

9. $x^2 - 2x + 1 = (x-1)^2$

11. $x^3 - a^3 = (x-a)(x^2 + ax + a^2)$

13. $a^2 + 4$ cannot be factored because it is a sum of squares.

15. $x(x+2) - 3(x+2) = (x-3)(x+2)$

17. $x^3 + 2x^2 + x + 2 = x^2(x+2) + 1(x+2) = (x^2 + 1)(x+2)$

19. $6x^2 - 14x = 2x(3x-7)$

21. $2x^3 - 18x = 2x\left(x^2 - 9\right) = 2x(x-3)(x+3)$

23. $4a^4 - 64 = 4\left(a^4 - 16\right) = 4\left(a^2 - 4\right)\left(a^2 + 4\right) = 4(a-2)(a+2)\left(a^2 + 4\right)$

25. $6x^3 - 13x^2 - 15x = x\left(6x^2 - 13x - 15\right) = x(x-3)(6x+5)$

27. $2x^4 - 5x^3 - 25x^2 = x^2\left(2x^2 - 5x - 25\right) = x^2(x-5)(2x+5)$

29. $2x^4 + 5x^2 + 3 = \left(x^2 + 1\right)\left(2x^2 + 3\right)$

31. $x^3 + 3x^2 + x + 3 = x^2(x+3) + 1(x+3) = \left(x^2 + 1\right)(x+3)$

33. $5x^3 - 5x^2 + 10x - 10 = 5\left(x^3 - x^2 + 2x - 2\right) = 5\left(x^2(x-1) + 2(x-1)\right) = 5\left(x^2 + 2\right)(x-1)$

35. $ax + bx - ay - by = x(a+b) - y(a+b) = (x-y)(a+b)$

37. $18x^2 + 12x + 2 = 2\left(9x^2 + 6x + 1\right) = 2(3x+1)^2$

39. $-4x^3 + 24x^2 - 36x = -4x\left(x^2 - 6x + 9\right) = -4x(x-3)^2$

41. $8x^3 - 27 = (2x-3)\left(4x^2 + 6x + 9\right)$

43. $-x^4 - 8x = -x\left(x^3 + 8\right) = -x(x+2)\left(x^2 - 2x + 4\right)$

45. $x^4 - 2x^3 - x + 2 = x^3(x-2) - 1(x-2) = \left(x^3 - 1\right)(x-2) = (x-1)\left(x^2 + x + 1\right)(x-2)$

47. $r^4 - 16 = \left(r^2 - 4\right)\left(r^2 + 4\right) = (r-2)(r+2)\left(r^2 + 4\right)$

49. $25x^2 - 4a^2 = (5x - 2a)(5x + 2a)$

51. $2x^4 - 2y^4 = 2\left(x^4 - y^4\right) = 2\left(x^2 - y^2\right)\left(x^2 + y^2\right) = 2(x-y)(x+y)\left(x^2 + y^2\right)$

53. $9x^3 + 6x^2 - 3x = 3x\left(3x^2 + 2x - 1\right) = 3x(3x-1)(x+1)$

55. $(z-2)^2 - 9 = ((z-2) - 3)((z-2) + 3) = (z-5)(z+1)$

57. $3x^5 - 27x^3 + 3x^2 - 27 = 3\left(x^5 - 9x^3 + x^2 - 9\right) = 3\left(x^3\left(x^2 - 9\right) + 1\left(x^2 - 9\right)\right) = 3\left(x^3 + 1\right)\left(x^2 - 9\right)$

     $= 3(x+1)\left(x^2 - x + 1\right)(x-3)(x+3)$

59. (a) $x^3 - 390x^2 + 2x - 780 = x^2(x-390) + 2(x-390) = \left(x^2 + 2\right)(x-390)$

      (b) $\left(x^2 + 2\right)(x-390) = 0 \Rightarrow x^2 = -2$ (this result is not possible) or $x = 390$; the record high

         amount was 390 ppm.

61. (a)   Volume $= x(x-1)(x-1-1) = x(x-1)(x-2) = x\left(x^2 - 3x + 2\right) = x^3 - 3x^2 + 2x$

      (b) $x^3 - 3x^2 + 2x = 6$

(c) $x^3 - 3x^2 + 2x = 6 \Rightarrow x^3 - 3x^2 + 2x - 6 = 0 \Rightarrow x^2(x-3) + 2(x-3) = 0 \Rightarrow (x^2 + 2)(x-3) = 0$

$\Rightarrow x^2 = -2$ (this result is not possible) or $x = 3$; the height is 3 feet.

Checking Basic Concepts Sections 5.5 and 5.6

1. (a) $25x^2 - 16 = (5x-4)(5x+4)$

   (b) $x^2 + 12x + 36 = (x+6)^2$

   (c) $9x^2 - 30x + 25 = (3x-5)^2$

   (d) $x^3 - 27 = (x-3)(x^2 + 3x + 9)$

   (e) $81x^4 - 16 = (9x^2 - 4)(9x^2 + 4) = (3x-2)(3x+2)(9x^2 + 4)$

2. (a) $x^3 - 2x^2 + 3x - 6 = x^2(x-2) + 3(x-2) = (x^2 + 3)(x-2)$

   (b) $x^2 - 4x - 21 = (x+3)(x-7)$

   (c) $12x^3 + 2x^2 - 4x = 2x(6x^2 + x - 2) = 2x(2x-1)(3x+2)$

Section 5.7 Polynomial Equations

1. Factoring is important so that we can solve polynomial equations. *Answers may vary.*

3. Subtract 16 from each side.

5. Since $x - 1$ must equal zero, the solution is 1.

7. The graph of $y = x^2 - 1$ (not shown) crosses the x-axis at $x = -1$ and $x = 1$.

   The solutions are -1 and 1.

9. The graph of $y = \dfrac{1}{4}x^2 - 1$ (not shown) crosses the x-axis at $x = -2$ and $x = 2$.

   The solutions are -2 and 2.

11. The graph of $y = x^2 - x - 2$ (not shown) crosses the x-axis at $x = -1$ and $x = 2$.

    The solutions are -1 and 2.

13. $z^2 - 64 = 0 \Rightarrow (z-8)(z+8) = 0 \Rightarrow z - 8 = 0$ or $z + 8 = 0 \Rightarrow z = 8$ or $z = -8$.

    The solutions are -8 and 8.

15. $4y^2 - 1 = 0 \Rightarrow (2y-1)(2y+1) = 0 \Rightarrow 2y - 1 = 0$ or $2y + 1 = 0 \Rightarrow y = \dfrac{1}{2}$ or $y = -\dfrac{1}{2}$.

The solutions are $-\dfrac{1}{2}$ and $\dfrac{1}{2}$.

17. $x^2 - 3x - 4 = 0 \Rightarrow (x-4)(x+1) = 0 \Rightarrow x - 4 = 0$ or $x + 1 = 0 \Rightarrow x = 4$ or $x = -1$.

The solutions are -1 and 4.

19. $x^2 + 4x - 12 = 0 \Rightarrow (x+6)(x-2) = 0 \Rightarrow x + 6 = 0$ or $x - 2 = 0 \Rightarrow x = -6$ or $x = 2$.

The solutions are -6 and 2.

21. $2x^2 + 5x - 3 = 0 \Rightarrow (2x-1)(x+3) = 0 \Rightarrow 2x - 1 = 0$ or $x + 3 = 0 \Rightarrow x = \dfrac{1}{2}$ or $x = -3$.

The solutions are -3 and $\dfrac{1}{2}$.

23. $2x^2 = 32 \Rightarrow 2x^2 - 32 = 0 \Rightarrow x^2 - 16 = 0 \Rightarrow (x-4)(x+4) = 0 \Rightarrow x - 4 = 0$ or $x + 4 = 0 \Rightarrow$

$x = 4$ or $x = -4$.   The solutions are -4 and 4.

25. $z^2 + 14z + 49 = 0 \Rightarrow (z+7)^2 = 0 \Rightarrow z + 7 = 0 \Rightarrow z = -7$.   The solution is -7.

27. $9t^2 + 1 = 6t \Rightarrow 9t^2 - 6t + 1 = 0 \Rightarrow (3t-1)^2 = 0 \Rightarrow 3t - 1 = 0 \Rightarrow t = \dfrac{1}{3}$.   The solution is $\dfrac{1}{3}$.

29. $15n^2 = 7n + 2 \Rightarrow 15n^2 - 7n - 2 = 0 \Rightarrow (5n+1)(3n-2) = 0 \Rightarrow 5n + 1 = 0$ or $3n - 2 = 0 \Rightarrow$

$n = -\dfrac{1}{5}$ or $n = \dfrac{2}{3}$.   The solutions are $-\dfrac{1}{5}$ and $\dfrac{2}{3}$.

31. $24m^2 + 23m = 12 \Rightarrow 24m^2 + 23m - 12 = 0 \Rightarrow (3m+4)(8m-3) = 0 \Rightarrow$

$3m + 4 = 0$ or $8m - 3 = 0 \Rightarrow m = -\dfrac{4}{3}$ or $m = \dfrac{3}{8}$.   The solutions are $-\dfrac{4}{3}$ and $\dfrac{3}{8}$.

33. $x^2 + 12 = 0 \Rightarrow x^2 = -12$.   There are no solutions.

35.   From the graph $x = -3$ or $x = 2$. $x^2 + x - 6 = 0 \Rightarrow (x+3)(x-2) = 0 \Rightarrow$ Either $x = -3$ or $x = 2$

37.   From the graph $x = -2$ or $x = 4$. $2x^2 - 4x - 16 = 0 \Rightarrow 2(x+2)(x-4) = 0 \Rightarrow$ Either $x = -2$ or $x = 4$

39.   The parabola should not cross the x-axis.   See Figure 39.

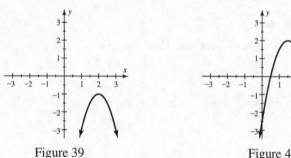

Figure 39   Figure 41

41.   The parabola should cross the x-axis two times.   See Figure 41.

43. $z^3 = 9z \Rightarrow z^3 - 9z = 0 \Rightarrow z(z^2 - 9) = 0 \Rightarrow z(z-3)(z+3) = 0 \Rightarrow z = 0$ or $z = 3$ or $z = -3$

The solutions are -3, 0, and 3.

45. $x^3 + x = 0 \Rightarrow x(x^2 + 1) = 0 \Rightarrow x = 0$, $\left(x^2 = -1 \text{ is not possible}\right)$ The solution is 0.

47. $2x^3 - 6x^2 = 20x \Rightarrow 2x^3 - 6x^2 - 20x = 0 \Rightarrow 2x(x^2 - 3x - 10) = 0 \Rightarrow 2x(x - 5)(x + 2) = 0 \Rightarrow$

   $x = 0$ or $x = 5$ or $x = -2$. The solutions are -2, 0, and 5.

49. $t^4 - 4t^3 - 5t^2 = 0 \Rightarrow t^2(t^2 - 4t - 5) = 0 \Rightarrow t^2(t - 5)(t + 1) = 0 \Rightarrow t = 0$ or $t = 5$ or $t = -1$

   The solutions are -1, 0, and 5.

51. $x^3 + 7x^2 - 4x - 28 = 0 \Rightarrow x^2(x + 7) - 4(x + 7) = 0 \Rightarrow (x^2 - 4)(x + 7) = 0 \Rightarrow$

   $(x - 2)(x + 2)(x + 7) = 0 \Rightarrow x = 2$ or $x = -2$ or $x = -7$. The solutions are -7, -2, and 2.

53. $n^3 - 2n^2 - 36n + 72 = 0 \Rightarrow n^2(n - 2) - 36(n - 2) = 0 \Rightarrow (n^2 - 36)(n - 2) = 0 \Rightarrow$

   $(n - 6)(n + 6)(n - 2) = 0 \Rightarrow n = 6$ or $n = -6$ or $n = 2$. The solutions are -6, 2, and 6.

55. (a) $x^2 - 2x - 15 = 0 \Rightarrow (x + 3)(x - 5) = 0 \Rightarrow$ Either $x = -3$ or $x = 5$

   (b)   Graph $Y_1 = X^2 - 2X - 15$ in $[-5, 7, 1]$ by $[-20, 10, 2]$. See Figures 55a & 55b.

         Either $x = -3$ or $x = 5$.

   (c)   Table $Y_1 = X^2 - 2X - 15$ with TblStart $= -5$ and ΔTbl $= 2$. See Figure 55c.

         Either $x = -3$ or $x = 5$.

$[-5, 7, 1]$ by $[-20, 10, 2]$ $[-5, 7, 1]$ by $[-20, 10, 2]$

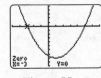   

Figure 55a   Figure 55b   Figure 55c

57. (a) $2x^2 - 3x = 2 \Rightarrow 2x^2 - 3x - 2 = 0 \Rightarrow (2x + 1)(x - 2) = 0 \Rightarrow$ Either $x = -\dfrac{1}{2}$ or $x = 2$

   (b)   Graph $Y_1 = 2X^2 - 3X - 2$ in $[-4, 4, 1]$ by $[-4, 4, 1]$. See Figures 57a & 57b.

         Either $x = -\dfrac{1}{2}$ or $x = 2$.

   (c)   Table $Y_1 = 2X^2 - 3X - 2$ with TblStart $= -1$ and ΔTbl $= 0.5$. See Figure 57c.

         Either $x = -\dfrac{1}{2}$ or $x = 2$.

$[-4, 4, 1]$ by $[-4, 4, 1]$ $[-4, 4, 1]$ by $[-4, 4, 1]$

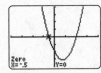   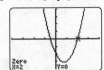   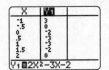

Figure 57a   Figure 57b   Figure 57c

59. (a) $3t^2 + 18t + 15 = 0 \Rightarrow 3(t+5)(t+1) = 0 \Rightarrow$ Either $t = -5$ or $t = -1$

    (b)   Graph $Y_1 = 3X^2 + 18X + 15$ in $[-6, 0, 1]$ by $[-15, 5, 1]$. See Figures 59a & 59b.

    Either $t = -5$ or $t = -1$.

    (c)   Table $Y_1 = 3X^2 + 18X + 15$ with TblStart $= -6$ and ΔTbl $= 1$. See Figure 59c.

    Either $t = -5$ or $t = -1$.

$[-6, 0, 1]$ by $[-15, 5, 1]$ $[-6, 0, 1]$ by $[-15, 5, 1]$

        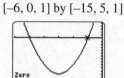   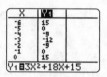

Figure 59a   Figure 59b   Figure 59c

61. (a) $4x^4 + 16x^2 = 16x^3 \Rightarrow 4x^4 - 16x^3 + 16x^2 = 0 \Rightarrow 4x^2(x-2)^2 = 0 \Rightarrow x = 0$ or $x = 2$

    (b)   Graph $Y_1 = 4X^4 - 16X^3 + 16X^2$ in $[-2, 4, 1]$ by $[-2, 5, 1]$. See Figures 61a & 61b.

    Either $x = 0$ or $x = 2$.

    (c)   Table $Y_1 = 4X^4 - 16X^3 + 16X^2$ with TblStart $= -2$ and ΔTbl $= 1$. See Figure 61c.

    Either $x = 0$ or $x = 2$.

$[-2, 4, 1]$ by $[-2, 5, 1]$ $[-2, 4, 1]$ by $[-2, 5, 1]$

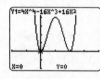   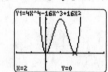   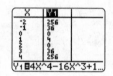

Figure 61a   Figure 61b   Figure 61c

63. $x^4 - 2x^2 - 8 = 0 \Rightarrow (x^2 - 4)(x^2 + 2) = 0 \Rightarrow (x - 2)(x + 2)(x^2 + 2) = 0 \Rightarrow$

    $x = 2$ or $x = -2$, $(x^2 = -2$ is not possible$)$. The solutions are -2 and 2.

65. $x^4 - 26x^2 + 25 = 0 \Rightarrow (x^2 - 25)(x^2 - 1) = 0 \Rightarrow (x - 5)(x + 5)(x - 1)(x + 1) = 0 \Rightarrow$

    $x = 5$ or $x = -5$ or $x = 1$ or $x = -1$. The solutions are -5, -1, 1, and 5.

67. $x^4 - 13x^2 + 36 = 0 \Rightarrow (x^2 - 9)(x^2 - 4) = 0 \Rightarrow (x - 3)(x + 3)(x - 2)(x + 2) = 0 \Rightarrow$

    $x = 3$ or $x = -3$ or $x = 2$ or $x = -2$. The solutions are -3, -2, 2, and 3.

69. $x^4 - 16 = 0 \Rightarrow (x^2 - 4)(x^2 + 4) = 0 \Rightarrow (x - 2)(x + 2)(x^2 + 4) = 0 \Rightarrow x = 2$ or $x = -2$,

    $(x^2 = -4$ is not possible$)$. The solutions are -2 and 2.

71. $9x^4 - 13x^2 + 4 = 0 \Rightarrow (x^2 - 1)(9x^2 - 4) = 0 \Rightarrow (x-1)(x+1)(3x-2)(3x+2) = 0 \Rightarrow$

    $x = 1$ or $x = -1$ or $x = \dfrac{2}{3}$ or $x = -\dfrac{2}{3}$. The solutions are -1, $-\dfrac{2}{3}$, $\dfrac{2}{3}$, and 1.

73. Let x be the height of the picture and $x + 4$ be its width. The overall area is given by $(x+4)(x+8)$.

    $(x+4)(x+8) = 525 \Rightarrow x^2 + 12x + 32 = 525 \Rightarrow x^2 + 12x - 493 = 0 \Rightarrow (x-17)(x+29) = 0 \Rightarrow$

    $x = 17$ or $x = -29$. The only valid solution is 17. The dimensions are 17 inches by 21 inches.

75. $-0.0001x^2 + 500 = 400 \Rightarrow 0.0001x^2 - 100 = 0 \Rightarrow (0.01x - 10)(0.01x + 10) = 0 \Rightarrow$

    $x = 1000$ or $x = -1000$. The values are -1000 and 1000.

77. Let x be the width of the pool. The area of the pool is given by $x(x+20)$ whereas the total area of

    the pool and the sidewalk is $(x+10)(x+30)$. Since the area of just the sidewalk portion is 900

    square feet, we have $(x+10)(x+30) - x(x+20) = 900 \Rightarrow x^2 + 40x + 300 - x^2 - 20x = 900 \Rightarrow$

    $20x = 600 \Rightarrow x = 30$ The pool is 30 feet by 50 feet.

79. (a)   The ball is 2 feet above the ground when it is hit.

    (b)   According the graph the ball is 70 feet in the air after about 2 seconds.

    (c) $-16t^2 + 66t + 2 = 70 \Rightarrow 16t^2 - 66t + 68 = 0 \Rightarrow 8t^2 - 33t + 34 = 0 \Rightarrow (8t - 17)(t-2) = 0 \Rightarrow$

           That is $t = \dfrac{17}{8}$ or $t = 2$. The baseball is 70 feet in the air at 2 seconds and $\dfrac{17}{8} = 2.125$ seconds.

    (d)   Symbolic solutions can be more accurate than reading a graph.

81. $\dfrac{1}{11}x^2 + \dfrac{11}{3}x = 220 \Rightarrow 3x^2 + 121x = 7260 \Rightarrow 3x^2 + 121x - 7260 = 0 \Rightarrow (x-33)(3x+220) = 0 \Rightarrow$

    $x = 33$ or $x = -\dfrac{220}{3}$. The only valid solution is 33 mph.

83. Let x represent the thickness of the iPod. Then $5.904 = x(x+1.8)(x+3.5)$.

    Graph $Y_1 = 5.904$ and $Y_2 = X(X+1.8)(X+3.5)$ in $[0, 2.5, 0.5]$ by $[0, 10, 1]$. See Figure 83.

    The solution is $x = 0.6$, so the dimensions of the iPod are 0.6 inches by 2.4 inches by 4.1 inches.

    $[0, 2.5, 0.5]$ by $[0, 10, 1]$ $[0, 5, 1]$ by $[0, 300, 50]$.

x	0	300	600	900	1200	1500
$y = E(x)$	500	428	392	392	428	500

    Figure 83   Figure 85

85. (a)   The table is shown in Figure 85. The elevation begins at 500 feet, decreases, and then

           increases back to 500 feet.

(b) $0.0002x^2 - 0.3x + 500 = 400 \Rightarrow 0.0002x^2 - 0.3x + 100 = 0 \Rightarrow x^2 - 1500x + 500,000 = 0 \Rightarrow$

$(x - 500)(x - 1000) = 0 \Rightarrow x = 500$ or $x = 1000$.   The elevation is 400 ft at 500 ft and 1000 ft.

Checking Basic Concepts Section 5.7

1. The graph crosses the x-axis at $x = -3$ and $x = -1$. The solutions are -3 and -1.

   $x^2 + 4x + 3 = 0 \Rightarrow (x + 3)(x + 1) = 0 \Rightarrow x = -3$ or $x = -1$. The solutions are -3 and -1.

2. The graph of $y = x^2 - 9$ (not shown) crosses the x-axis at $x = -3$ and $x = 3$. The solutions are

   -3 and 3. $x^2 - 9 = 0 \Rightarrow (x - 3)(x + 3) = 0 \Rightarrow x = 3$ or $x = -3$. The solutions are -3 and 3.

3. (a) $x^2 + 9 = 6x \Rightarrow x^2 - 6x + 9 = 0 \Rightarrow (x - 3)^2 = 0 \Rightarrow x - 3 = 0 \Rightarrow x = 3$

   (b) $x^2 + 9 = 0 \Rightarrow x^2 = -9$. There is no solution.

   (c) $12x^2 + 7x - 10 = 0 \Rightarrow (4x + 5)(3x - 2) = 0 \Rightarrow x = -\dfrac{5}{4}$ or $x = \dfrac{2}{3}$

4. (a) $x^3 + 5x = 0 \Rightarrow x(x^2 + 5) = 0 \Rightarrow x = 0$, $\left(x^2 = -5 \text{ is not possible}\right)$. The only solution is 0.

   (b) $x^4 - 81 = 0 \Rightarrow (x^2 - 9)(x^2 + 9) = 0 \Rightarrow (x - 3)(x + 3)(x^2 + 9) = 0 \Rightarrow x = 3$ or $x = -3$

   $\left(x^2 = -9 \text{ is not possible}\right)$. The only solutions are -3 and 3.

   (c) $z^4 + z^2 = 20 \Rightarrow z^4 + z^2 - 20 = 0 \Rightarrow (z^2 - 4)(z^2 + 5) = 0 \Rightarrow (z - 2)(z + 2)(z^2 + 5) = 0 \Rightarrow$

   $z = 2$ or $z = -2$, $\left(z^2 = -5 \text{ is not possible}\right)$. The only solutions are -2 and 2.

   (d) $x^3 + 2x^2 - 49x = 98 \Rightarrow x^3 + 2x^2 - 49x - 98 = 0 \Rightarrow x^2(x + 2) - 49(x + 2) = 0 \Rightarrow$

   $(x^2 - 49)(x + 2) = 0 \Rightarrow (x - 7)(x + 7)(x + 2) = 0 \Rightarrow x = 7, -7, \text{ or } -2.$

   The solutions are $-7, -2,$ and 7.

Chapter 5 Review

1. $x + 5$; $x^2 - 3x + 1$; *Answers may vary.*

2. $10xy^2$; When $x = 5$ and $y = 8$, $10(5)(8)^2 = 3200$ in^3.

3. The degree is 5. The coefficient is -4.

4. The degree is 3. The coefficient is 1.

5. The degree is 7. The coefficient is 5.

6. The degree is 10. The coefficient is -9.

7. $5x - 4x + 10x = 11x$

8. $9x^3 - 5x^3 + x^2 = 4x^3 + x^2$

9. $6x^3y - 4x^3 + 8x^3y + 5x^3 = 14x^3y + x^3$

10. $2(2)^3(-1)^2 - 3(-1) = 16 - (-3) = 16 + 3 = 19$

11. The degree is 2. The leading coefficient is 5.

12. The degree is 4. The leading coefficient is 2.

13. $\left(3x^2 - x + 7\right) + \left(5x^2 + 4x - 8\right) = 8x^2 + 3x - 1$

14. $\left(6z^3 + z\right) + \left(17z^3 - 4z^2\right) = 23z^3 - 4z^2 + z$

15. $\left(-4x^2 - 6x + 1\right) - \left(-3x^2 - 7x + 1\right) = -4x^2 - 6x + 1 + 3x^2 + 7x - 1 = -x^2 + x$

16. $\left(3x^3 - 5x + 7\right) - \left(8x^3 + x^2 - 2x + 1\right) = 3x^3 - 5x + 7 - 8x^3 - x^2 + 2x - 1 = -5x^3 - x^2 - 3x + 6$

17. $f(-1) = 2(-1)^2 - 3(-1) + 2 = 2 + 3 + 2 = 7$

18. $f(4) = 1 - (4) - 4(4)^3 = 1 - 4 - 256 = -259$

19. When $x = 2$, $y = -4$ and so $f(2) = -4$.

20. $(f + g)(-2) = \left((-2)^2 - 1\right) + \left((-2) - 3(-2)^2\right) = 4 - 1 + (-2) - 3(4) = -11;$

    $(f - g)(x) = \left(x^2 - 1\right) - \left(x - 3x^2\right) = x^2 - 1 - x + 3x^2 = 4x^2 - x - 1$

21. $5(3x - 4) = 15x - 20$

22. $-2x\left(1 + x - 4x^2\right) = -2x - 2x^2 + 8x^3$

23. $x^3 \cdot x^5 = x^{3+5} = x^8$

24. $-2x^3 \cdot 3x = -6x^{3+1} = -6x^4$

25. $-7xy^7 \cdot 6xy = -42x^{1+1} \cdot y^{7+1} = -42x^2y^8$

26. $12xy^4 \cdot 5x^2y = 60x^{1+2} \cdot y^{4+1} = 60x^3y^5$

27. $(x + 4)(x + 5) = x^2 + 5x + 4x + 20 = x^2 + 9x + 20$

28. $(x - 7)(x - 8) = x^2 - 8x - 7x + 56 = x^2 - 15x + 56$

29. $(6x + 3)(2x - 9) = 12x^2 - 54x + 6x - 27 = 12x^2 - 48x - 27$

30. $\left(y - \dfrac{1}{3}\right)\left(y + \dfrac{1}{3}\right) = y^2 - \left(\dfrac{1}{3}\right)^2 = y^2 - \dfrac{1}{9}$

31. $4x^2\left(2x^2 - 3x - 1\right) = 8x^4 - 12x^3 - 4x^2$

32. $-x\left(4+5x-7x^2\right)=-4x-5x^2+7x^3$

33. $(4x+y)(4x-y)=16x^2-y^2$

34. $(x+3)^2=x^2+6x+9$

35. $(2y-5)^2=4y^2-20y+25$

36. $(a-b)\left(a^2+ab+b^2\right)=a^3-b^3$

37. $\left(5m-2n^4\right)^2=25m^2-20mn^4+4n^8$

38. $((r-1)+t)((r-1)-t)=(r-1)^2-t^2=r^2-2r+1-t^2$

39. $25x^2-30x=5x(5x-6)$

40. $12x^4+8x^3-16x^2=4x^2\left(3x^2+2x-4\right)$

41. $x^2+3x=0\Rightarrow x(x+3)=0\Rightarrow$ Either $x=0$ or $x+3=0\Rightarrow x=-3$

42. $7x^4=28x^2\Rightarrow 7x^4-28x^2=0\Rightarrow 7x^2\left(x^2-4\right)=0\Rightarrow 7x^2(x-2)(x+2)\Rightarrow x=0$ or $x=2$ or $x=-2$

43. $2t^2-3t+1=0\Rightarrow(2t-1)(t-1)=0\Rightarrow$ Either $2t-1=0\Rightarrow t=\dfrac{1}{2}$ or $t-1=0\Rightarrow t=1$

44. $4z(z-3)+4(z-3)=0\Rightarrow(z-3)(4z+4)=0\Rightarrow$ Either $z-3=0\Rightarrow z=3$ or $4z+4=0\Rightarrow z=-1$

45. $2x^3+2x^2-3x-3=2x^2(x+1)-3(x+1)=(x+1)\left(2x^2-3\right)$

46. $z^3+z^2+z+1=z^2(z+1)+1(z+1)=(z+1)\left(z^2+1\right)$

47. $ax-bx+ay-by=x(a-b)+y(a-b)=(a-b)(x+y)$

48. $(fg)(4)=(4+1)(4-3)=(5)(1)=5;\ (fg)(x)=(x+1)(x-3)=x^2-3x+x-3=x^2-2x-3$

49. (a) The graph crosses the x-axis at 0 and 3.

    (b) The solutions to $P(x)=0$ are 0 and 3.

    (c) The zeros of $P(x)$ are 0 and 3.

50. (a) The graph crosses the x-axis at -1 and 2.

    (b) The solutions to $P(x)=0$ are -1 and 2.

    (c) The zeros of $P(x)$ are -1 and 2.

51. $x^2+8x+12=(x+2)(x+6)$

52. $x^2-5x-50=(x-10)(x+5)$

53. $9x^2+25x-6=(x+3)(9x-2)$

54. $4x^2 - 22x + 10 = 2\left(2x^2 - 11x + 5\right) = 2(x-5)(2x-1)$

55. $x^3 - 4x^2 + 3x = x\left(x^2 - 4x + 3\right) = x(x-3)(x-1)$

56. $2x^4 + 14x^3 + 20x^2 = 2x^2\left(x^2 + 7x + 10\right) = 2x^2(x+2)(x+5)$

57. $5x^4 + 15x^3 - 90x^2 = 5x^2\left(x^2 + 3x - 18\right) = 5x^2(x-3)(x+6)$

58. $10x^3 - 90x^2 + 200x = 10x\left(x^2 - 9x + 20\right) = 10x(x-5)(x-4)$

59. Since $Y_1 = 0$ when $x = -3$ and $x = 5$, the factors are $(x+3)(x-5)$. This checks using FOIL.

60. Since $Y_1 = 0$ when $x = 11$ and $x = 13$, the factors are $(x-11)(x-13)$. This checks using FOIL.

61. Since $x = -7$ and $x = 4$ are zeros of the polynomial, the factors are $(x+7)(x-4)$. This checks using FOIL.

62. Since $x = 8$ and $x = 13$ are zeros of the polynomial, the solutions are 8 and 13. These solutions check.

63. $t^2 - 49 = (t-7)(t+7)$

64. $4y^2 - 9x^2 = (2y-3x)(2y+3x)$

65. $x^2 + 4x + 4 = (x+2)^2$

66. $16x^2 - 8x + 1 = (4x-1)^2$

67. $x^3 - 27 = (x-3)\left(x^2 + 3x + 9\right)$

68. $64x^3 + 27y^3 = (4x+3y)\left(16x^2 - 12xy + 9y^2\right)$

69. $10y^3 - 10y = 10y\left(y^2 - 1\right) = 10y(y-1)(y+1)$

70. $4r^4 - t^6 = \left(2r^2 - t^3\right)\left(2r^2 + t^3\right)$

71. $m^4 - 16n^4 = \left(m^2 - 4n^2\right)\left(m^2 + 4n^2\right) = (m-2n)(m+2n)\left(m^2 + 4n^2\right)$

72. $n^3 - 2n^2 - n + 2 = n^2(n-2) - 1(n-2) = \left(n^2 - 1\right)(n-2) = (n-1)(n+1)(n-2)$

73. $25a^2 - 30ab + 9b^2 = (5a-3b)^2$

74. $2r^3 - 12r^2t + 18rt^2 = 2r\left(r^2 - 6rt + 9t^2\right) = 2r(r-3t)^2$

75. $a^6 + 27b^3 = \left(a^2 + 3b\right)\left(a^4 - 3a^2b + 9b^2\right)$

76. $8p^6 - q^3 = \left(2p^2 - q\right)\left(4p^4 + 2p^2q + q^2\right)$

77. $5x^3 - 10x^2 = 5x^2(x-2)$

78. $-2x^3 + 32x = -2x(x^2 - 16) = -2x(x-4)(x+4)$

79. $x^4 - 16y^4 = (x^2 - 4y^2)(x^2 + 4y^2) = (x-2y)(x+2y)(x^2 + 4y^2)$

80. $4x^3 + 8x^2 - 12x = 4x(x^2 + 2x - 3) = 4x(x+3)(x-1)$

81. $-2x^3 + 11x^2 - 12x = -x(2x^2 - 11x + 12) = -x(2x-3)(x-4)$

82. $x^4 - 8x^2 - 9 = (x^2 + 1)(x^2 - 9) = (x^2 + 1)(x-3)(x+3)$

83. $64a^3 + b^3 = (4a+b)(16a^2 - 4ab + b^2)$

84. $8 - y^3 = (2-y)(4 + 2y + y^2)$

85. $(z+3)^2 - 16 = ((z+3)-4)((z+3)+4) = (z-1)(z+7)$

86. $x^4 - 5x^3 - 4x^2 + 20x = x(x^3 - 5x^2 - 4x + 20) = x(x^2(x-5) - 4(x-5))$

    $= x(x^2 - 4)(x-5) = x(x-2)(x+2)(x-5)$

87. $x^2 - 16 = 0 \Rightarrow (x-4)(x+4) = 0 \Rightarrow x = 4$ or $x = -4$. The solutions are -4 and 4. The graph of

    $y = x^2 - 16$ (not shown) crosses the x-axis at $x = -4$ and $x = 4$. The solutions are -4 and 4.

88. $x^2 - 2x - 3 = 0 \Rightarrow (x-3)(x+1) = 0 \Rightarrow x = 3$ or $x = -1$. The solutions are -1 and 3. The graph of

    $y = x^2 - 2x - 3$ (not shown) crosses the x-axis at $x = -1$ and $x = 3$. The solutions are -1 and 3.

89. $4x^2 - 28x + 49 = 0 \Rightarrow (2x-7)^2 = 0 \Rightarrow 2x - 7 = 0 \Rightarrow x = \dfrac{7}{2}$

90. $x^2 + 8 = 0 \Rightarrow x^2 = -8$. There are no solutions.

91. $3x^2 = 2x + 5 \Rightarrow 3x^2 - 2x - 5 = 0 \Rightarrow (x+1)(3x-5) = 0 \Rightarrow x = -1$ or $x = \dfrac{5}{3}$

92. $4x^2 + 5x = 6 \Rightarrow 4x^2 + 5x - 6 = 0 \Rightarrow (x+2)(4x-3) = 0 \Rightarrow x = -2$ or $x = \dfrac{3}{4}$

93. $x^3 = x \Rightarrow x^3 - x = 0 \Rightarrow x(x^2 - 1) = 0 \Rightarrow x(x-1)(x+1) = 0 \Rightarrow x = 0$ or $x = 1$ or $x = -1$

94. $x^3 - 6x^2 + 11x = 6 \Rightarrow x^3 - 6x^2 + 11x - 6 = 0$; the graph of $y = x^3 - 6x^2 + 11x - 6$ (not shown) crosses

    the x-axis at $x = 1$, $x = 2$ and $x = 3$. The solutions are 1, 2 and 3.

95. $x^3 + x^2 - 72x = 0 \Rightarrow x(x^2 + x - 72) = 0 \Rightarrow x(x-8)(x+9) = 0 \Rightarrow x = 0$ or $x = 8$ or $x = -9$

96. $x^4 - 15x^3 + 56x^2 = 0 \Rightarrow x^2(x^2 - 15x + 56) = 0 \Rightarrow x^2(x-8)(x-7) = 0 \Rightarrow x = 0$ or $x = 8$ or $x = 7$

97. $x^4 = 16 \Rightarrow x^4 - 16 = 0 \Rightarrow (x^2 - 4)(x^2 + 4) = 0 \Rightarrow (x-2)(x+2)(x^2 + 4) = 0 \Rightarrow$

    $x = 2$ or $x = -2$, $(x^2 = -4$ is not possible$)$. The only solutions are -2 and 2.

98. $x^4 + 5x^2 = 36 \Rightarrow x^4 + 5x^2 - 36 = 0 \Rightarrow \left(x^2 - 4\right)\left(x^2 + 9\right) = 0 \Rightarrow (x-2)(x+2)\left(x^2 + 9\right) = 0 \Rightarrow$

$x = 2$ or $x = -2$, $\left(x^2 = -9 \text{ is not possible}\right)$. The only solutions are -2 and 2.

99. (a) $R(x) = 15x$

(b) $P(x) = R(x) - C(x) = 15x - (3x + 9000) = 12x - 9000$

(c) $P(4000) = 12(4000) - 9000 = 48,000 - 9000 = 39,000$; the profit is $39,000 for selling 4000 DVDs.

100. (a) Height $= x$, width $= x + 5$, length $= x + 5 + 5 = x + 10$; volume $= x(x+5)(x+10)$

(b) $x(x+5)(x+10) = 168$

(c) $x\left(x^2 + 15x + 50\right) - 168 = 0 \Rightarrow x^3 + 15x^2 + 50x - 168 = 0 \Rightarrow (x-2)\left(x^2 + 17x + 84\right) = 0 \Rightarrow x = 2$,

since $x^2 + 17x + 84 > 0$ for all x. The dimensions are 2 inches by 7 inches by 12 inches.

101. Let x represent the height of the picture, then the overall area of the picture and frame is

$(x+6)(x+2+6) = (x+6)(x+8) = 224 \Rightarrow x^2 + 14x + 48 - 224 = 0 \Rightarrow x^2 + 14x - 176 = 0$

$\Rightarrow (x-8)(x+22) = 0 \Rightarrow x = 8$, since $x = -22$ has no meaning in this context. The dimensions of the

picture are 8 inches by 10 inches.

102. $\dfrac{1}{2}(x+2)(x-3) = \dfrac{1}{2}\left(x^2 - x - 6\right) = \dfrac{1}{2}x^2 - \dfrac{1}{2}x - 3$

103. (a) In May, $x = 5$; $f(5) = -1.466(5)^2 + 20.25(5) + 9 = 73.6°F$.

(b) Table $Y_1 = -1.466X^2 + 20.25X + 9$ with TblStart $= 1$ and ΔTbl $= 1$. See Figure 103b. It is

greatest in July.

(c) Graph $Y_1 = -1.466X^2 + 20.25X + 9$ in [1, 12, 1] by [30, 90, 10]. See Figure 103c. The

temperature increases from January to July and then decreases from July to December.

[1, 12, 1] by [30, 90, 10]

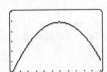

Figure 103b   Figure 103c

104. (a) $100\left(1 - \dfrac{x}{100}\right)^2 = 100\left(1 - 2(1)\left(\dfrac{x}{100}\right) + \left(\dfrac{x}{100}\right)^2\right) = 100\left(1 - \dfrac{x}{50} + \dfrac{x^2}{10,000}\right) = 100 - 2x + \dfrac{x^2}{100}$

$= \dfrac{x^2}{100} - 2x + 100$

(b) $100\left(1 - \dfrac{70}{100}\right)^2 = 100(0.3)^2 = 9\%$ and $\dfrac{1}{100}(70)^2 - 2(70) + 100 = 49 - 140 + 100 = 9\%$

105. If x is the largest of three consecutive integers then the smallest is $x-2$ and the middle integer is

 $x-1$. The product is given by $(x-2)(x-1)x = (x^2 - 3x + 2)x = x^3 - 3x^2 + 2x$.

106. See Figure 106.

$$[-25, 20, 5] \text{ by } [-160, 100, 10]$$

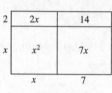   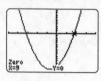   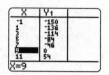

Figure 106   Figure 107a   Figure 107b

107. Let x represent the width of the building. Then $x+7$ represents the length of the building. The

 equation needed is given by $x(x+7) = 144 \Rightarrow x^2 + 7x - 144 = 0$.

 (a)   Graph $Y_1 = X^2 + 7X - 144$ in $[-25, 20, 5]$ by $[-160, 100, 10]$. See Figure 107a.

 Here $Y_1 = 0$ when $X = 9$. The dimensions are 9 feet by 16 feet.

 (b)   Table $Y_1 = X^2 + 7X - 144$ with TblStart $= -1$ and ΔTbl $= 2$. See Figure 107b.

 Here $Y_1 = 0$ when $X = 9$. The dimensions are 9 feet by 16 feet.

 (c) $x^2 + 7x - 144 = 0 \Rightarrow (x-9)(x+16) = 0 \Rightarrow$ Either $x - 9 = 0 \Rightarrow x = 9$ or $x + 16 = 0 \Rightarrow x = -16$.

 Since $x = -16$ has no meaning in this problem, we know that $x = 9$.

 The dimensions are 9 feet by 16 feet.

108. Let l represent the length of the rectangle. Then since the perimeter is 50 feet,

 $2l + 2x = 50 \Rightarrow l = 25 - x$. The area is $A = (25 - x)x = 25x - x^2$.

109. (a) $-16t^2 + 66t = 0 \Rightarrow 8t^2 - 33t = 0 \Rightarrow t(8t - 33) = 0 \Rightarrow t = 0$ or $8t - 33 = 0 \Rightarrow t = 0$ or $t = \dfrac{33}{8}$

 The ball strikes the ground after $\dfrac{33}{8} = 4.125$ seconds.

 (b) $-16t^2 + 66t = 50 \Rightarrow -16t^2 + 66t - 50 = 0 \Rightarrow 8t^2 - 33t + 25 = 0 \Rightarrow (t-1)(8t-25) \Rightarrow$

 $t = 1$ or $t = \dfrac{25}{8}$. The ball is 50 feet high after 1 second and $\dfrac{25}{8} = 3.125$ seconds.

110. (a) $R(x) = (50 - x)(600 + 20x)$

 (b) $(50 - x)(600 + 20x) = 32{,}000 \Rightarrow 30{,}000 + 400x - 20x^2 = 32{,}000 \Rightarrow$

 $x^2 - 20x + 100 = 0 \Rightarrow (x-10)^2 = 0 \Rightarrow x - 10 = 0 \Rightarrow x = 10$.

 When $x = 10$ the ticket price is $50 - 10 = \$40$.

111. $-0.0001x^2 + 500 = 400 \Rightarrow 0.0001x^2 - 100 = 0 \Rightarrow (0.01x - 10)(0.01x + 10) = 0 \Rightarrow$

 $0.01x - 10 = 0$ or $0.01 + 10 = 0 \Rightarrow x = 1000$ or $x = -1000$. The values are -1000 and 1000.

Chapter 5 Test

1. $x^2y^2 - 4x + 9x - 5x^2y^2 = 5x - 4x^2y^2$

2. $\left(-2x^3 - 6x + 1\right) - \left(5x^3 - x^2 + x - 10\right) = -2x^3 - 6x + 1 - 5x^3 + x^2 - x + 10 = -7x^3 + x^2 - 7x + 11$

3. $f(-2) = 2(-2)^3 - (-2)^2 - 5(-2) + 2 = -16 - 4 + 10 + 2 = -8$

4. Since $y = -2$ when $x = 2$, $f(2) = -2$.

5. The degree is $3 + 1 = 4$. The coefficient is 3.

6. $8xyz$

7. $-(-2)^2(3) + 3(-2)(3)^2 = -(4)(3) + 3(-2)(9) = -12 - 54 = -66$

8. $(f - g)(4) = 2(4) - 5 - \left(1 - 4^3\right) = 8 - 5 - (1 - 64) = 8 - 5 + 63 = 66$

   $(f + g)(x) = 2x - 5 + 1 - x^3 = -x^3 + 2x - 4$

9. $-\dfrac{2}{5}x^2(10x - 5) = -4x^3 + 2x^2$

10. $2xy^7 \cdot 7xy = 14 \cdot x^{1+1} \cdot y^{7+1} = 14x^2y^8$

11. $(2x + 1)(5x - 7) = 10x^2 - 14x + 5x - 7 = 10x^2 - 9x - 7$

12. $(5 - 3x)^2 = 25 - 30x + 9x^2 = 9x^2 - 30x + 25$

13. $(5x - 4y)(5x + 4y) = (5x)^2 - (4y)^2 = 25x^2 - 16y^2$

14. $-2x^2\left(x^2 - 3x + 2\right) = -2x^2 \cdot x^2 + 2x^2 \cdot 3x - 2x^2 \cdot 2 = -2x^4 + 6x^3 - 4x^2$

15. $(x - 2y)\left(x^2 + 2xy + 4y^2\right) = x \cdot x^2 + x \cdot 2xy + x \cdot 4y^2 - 2y \cdot x^2 - 2y \cdot 2xy - 2y \cdot 4y^2 =$

    $x^3 + 2x^2y + 4xy^2 - 2xy^2 - 4xy^2 - 8y^3 = x^3 - 8y^3$

16. $2x^2(x - 1)(x + 1) = 2x^2\left(x^2 - 1\right) = 2x^2 \cdot x^2 - 2x^2 \cdot 1 = 2x^4 - 2x^2$

17. $x^2 - 3x - 10 = (x - 5)(x + 2)$

18. $2x^3 + 6x = 2x\left(x^2 + 3\right)$

19. $3x^2 + 7x - 20 = (x + 4)(3x - 5)$

20. $5x^4 - 5x^2 = 5x^2\left(x^2 - 1\right) = 5x^2(x - 1)(x + 1)$

21. $2x^3 + x^2 - 10x - 5 = x^2(2x + 1) - 5(2x + 1) = (2x + 1)\left(x^2 - 5\right)$

22. $49x^2 - 14x + 1 = (7x - 1)^2$

23. $x^3 + 8 = (x + 2)\left(x^2 - 2x + 4\right)$

24. $4x^2y^4 + 8x^4y^2 = 4x^2y^2\left(y^2 + 2x^2\right)$

25. $a^2 - 3ab + 2b^2 = (a-b)(a-2b)$

26. The degree is 3. The leading coefficient is 1.

27. c $\left(2m^3 - 4n^2\right)^2 = 4m^6 - 16m^3n^2 + 16n^4$

28. Since $x = -8$ and $x = 6$ are zeros of the polynomial, the factors are $(x+8)(x-6)$.

29. If x is an even integer, the next consecutive even integer is $x+2$. The product is $x(x+2) = x^2 + 2x$

30. $5x^2 = 15x \Rightarrow 5x^2 - 15x = 0 \Rightarrow 5x(x-3) = 0 \Rightarrow$ Either $5x = 0 \Rightarrow x = 0$ or $x - 3 = 0 \Rightarrow x = 3$

31. $4t^2 + 19t - 5 = 0 \Rightarrow (4t-1)(t+5) = 0 \Rightarrow$ Either $4t - 1 = 0 \Rightarrow t = \dfrac{1}{4}$ or $t + 5 = 0 \Rightarrow t = -5$

32. $2z^4 - 8z^2 = 0 \Rightarrow 2z^2\left(z^2 - 4\right) = 0 \Rightarrow 2z^2(z-2)(z+2) = 0 \Rightarrow z = -2, \ 0, \ \text{or } 2$

33. $x^4 - 2x^2 + 1 = 0 \Rightarrow \left(x^2 - 1\right)\left(x^2 - 1\right) = 0 \Rightarrow x^2 - 1 = 0 \Rightarrow (x+1)(x-1) = 0 \Rightarrow$ Either $x = -1$ or $x = 1$

34. $x(x-3) + (x-3) = 0 \Rightarrow (x+1)(x-3) = 0 \Rightarrow x = -1$ or $x = 3$

35. Let x represent the height of the frame. Then $x + 4$ represents the width of the frame.

    (a)   The equation needed is given by $x(x+4) = 221$.

    (b) $x(x+4) = 221 \Rightarrow x^2 + 4x - 221 = 0 \Rightarrow (x-13)(x+17) = 0 \Rightarrow x = 13$ or $x = -17$.

    Since $x = -17$ has no meaning in this problem, the height is 13 inches and the width is 17 inches.

    (c)   The perimeter is $P = 2(13) + 2(17) = 26 + 34 = 60$ inches.

36. (a)   For May, $x = 5$; $f(5) = -0.091(5)^3 + 0.66(5)^2 + 5.78(5) + 23.5 \approx 57.5°\text{F}$

    (b)   The dew point starts at about 30°F in January and increases to a maximum of about 65°F in July. Then it decreases to 30°F by the end of December.

37. $(x+2)(x+3) = x^2 + 2x + 3x + 6 = x^2 + 5x + 6$

2	2x	6
x	x^2	$3x$
	x	3

38. $-16t^2 + 96t + 3 = 131 \Rightarrow -16t^2 + 96t - 128 = 0 \Rightarrow -16\left(t^2 - 6t + 8\right) = 0 \Rightarrow (t-2)(t-4) = 0 \Rightarrow t = 2$ or

    $t = 4$. The baseball was 131 feet in the air after 2 seconds and 4 seconds.

Extended and Discovery Exercises

1. (a)   Plot the data in [1, 10, 1] by [0, 30, 3]. See Figure 1a. The data appear to be nonlinear.

   (b)   Graph $Y_1 = X^2.5$ in [1, 10, 1] by [0, 30, 3]. See Figure 1b. The model is slightly accurate only for the first 3 planets.

   (c)   By trial and error, the model $y = x^{1.5}$ fits quite well.

   (d)   The orbit for Neptune is $y = (30.1)^{1.5} \approx 165.1$ years. The orbit for Pluto is

           $y = (39.4)^{1.5} \approx 247.3$ years.

[1, 10, 1] by [0, 30, 3]   [1, 10, 1] by [0, 30, 3]   [1993, 2005, 1] by [45, 115, 10]

       Figure 1a   Figure 1b   Figure 2

2. (a)   Plot the data in [1993, 2005, 1] by [45, 115, 10]. See Figure 2.

   (b)   By trial and error the value is $k \approx 0.7$.

   (c)   In 2006 the number of Americans over 100 will be $y = 0.7(2006 - 1994)^2 + 50 \approx 150.8$

         thousand.

3. (a) $\begin{bmatrix} 10,000^2 & 10,000 & 1 & 412 \\ 40,000^2 & 40,000 & 1 & 843 \\ 100,000^2 & 100,000 & 1 & 2550 \end{bmatrix}$

   (b)   The results from a graphing calculator are $a \approx 1.56 \times 10^{-7}$, $b \approx 0.007$, and $c \approx 330.9$.

   (c) $1.56 \times 10^{-7} (20,000)^2 + 0.007(20,000) + 330.9 \approx \$533.30$. This is close to the actual

        value. Note that unrounded values for a, b, and c give \$524.37.

Chapters 1-5 Cumulative Review Exercises

1. $A = \dfrac{1}{2}(5)\left(\dfrac{3}{2}\right) = \dfrac{15}{4}$

2. $\dfrac{b-a}{a}$

3. (a) $\left(\dfrac{x^{-3}}{y}\right)^2 = \dfrac{x^{-6}}{y^2} = \dfrac{1}{x^6 y^2}$

   (b) $\dfrac{(3r^{-1}t)^{-4}}{r^2(t^2)^{-2}} = \dfrac{3^{-4} r^4 t^{-4}}{r^2 t^{-4}} = \dfrac{1}{3^4} \cdot r^{4-2} \cdot t^{-4-(-4)} = \dfrac{1}{81} \cdot r^2 \cdot t^0 = \dfrac{r^2}{81}$

   (c) $(ab^{-2})^4 (ab^3)^{-1} = a^4 b^{-8} a^{-1} b^{-3} = a^3 b^{-11} = \dfrac{a^3}{b^{11}}$

4. $5.859 \times 10^4 = 58,590$

5. $f(-3) = -2(-3)^2 + 3(-3) = -2(9) + 3(-3) = -18 - 9 = -27$

6. The radicand must be greater than or equal to zero. $D : x \geq 4$

7. See Figure 7.

8. See Figure 8.

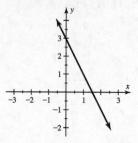

Figure 7   Figure 8

9. (a)   Yes. It passes the vertical line test.

   (b) D: all real numbers; $R : y \geq -4$

   (c) $f(1) = -4$; $f(2) = -3$

   (d)   The graph crosses the x-axis at -1 and 3.

   (e)   The function has the value 0 where the graph crosses the x-axis at -1 and 3; $x = -1, 3$.

   (f) $f(x) = (x+1)(x-3)$

10. $2x - y = 4 \Rightarrow y = 2x - 4$; so a line parallel to this line has slope $m = 2$. The line through $(-2, 5)$

    with slope $m = 2$ has equation $y - 5 = 2(x - (-2)) \Rightarrow y = 2x + 4 + 5 \Rightarrow y = 2x + 9$.

11. Two points on the graph of f are $(-2, 12)$ and $(-1, 7)$.

    $m = \dfrac{y_2 - y_1}{x_2 - x_1} = \dfrac{7 - 12}{-1 - (-2)} = \dfrac{-5}{1} = -5$; $y - 12 = -5(x - (-2)) \Rightarrow y = -5x - 10 + 12 \Rightarrow y = -5x + 2$

12. $-2(5x - 1) = 1 - (5 - x) \Rightarrow -10x + 2 = 1 - 5 + x \Rightarrow -11x = -6 \Rightarrow x = \dfrac{6}{11}$

13. $3x + 4 \leq x - 1 \Rightarrow 2x \leq -5 \Rightarrow x \leq -\dfrac{5}{2} \Rightarrow \left(-\infty, -\dfrac{5}{2}\right]$

14. $|5x - 10| > 5 \Rightarrow 5x - 10 < -5$ or $5x - 10 > 5 \Rightarrow 5x < 5$ or $5x > 15 \Rightarrow x < 1$ or

    $x > 3 \Rightarrow (-\infty, 1) \cup (3, \infty)$

15. $-2 < 2 - 7x \leq 2 \Rightarrow -4 < -7x \leq 0 \Rightarrow \dfrac{4}{7} > x \geq 0 \Rightarrow 0 \leq x < \dfrac{4}{7} \Rightarrow \left[0, \dfrac{4}{7}\right)$

16. Note that $4x + 2y = 10 \Rightarrow 2y = -4x + 10 \Rightarrow y = -2x + 5$ and $-x + 5y = 3 \Rightarrow 5y = x + 3 \Rightarrow y = \dfrac{1}{5}x + \dfrac{3}{5}$.

    Graph $Y_1 = -2X + 5$ and $Y_2 = 1/5\,X + 3/5$ in $[-5, 5, 1]$ by $[-5, 5, 1]$. See Figure 16.

    The solution is $(2, 1)$.

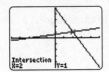

Figure 16

17. (a) $-x - 2y = 5 \Rightarrow -2y - 5 = x.$ Substitute $x = -2y - 5$ into the second equation and solve for y:

$2(-2y - 5) + 4y = -10 \Rightarrow -4y - 10 + 4y = -10 \Rightarrow -10 = -10.$ This is an identity, so the

equations are dependent. The solutions are $\{(x,\ y)|x + 2y = -5\}.$

(b)   Add the two equations to eliminate the variable y. $\begin{array}{l} 3x + 2y = 7 \\ \underline{2x - 2y = 3} \\ 5x \quad\quad = 10 \end{array} \Rightarrow x = 2.$

Substitute $x = 2$ into the first equation and solve for $y : 3(2) + 2y = 7 \Rightarrow 2y = 1 \Rightarrow y = \dfrac{1}{2}.$

The solution is $\left(2,\ \dfrac{1}{2}\right).$

18. Note that $2x + y \le 4 \Rightarrow y \le -2x + 4$ and $x - 2y < -2 \Rightarrow -2y < -x - 2 \Rightarrow y > \dfrac{1}{2}x + 1$

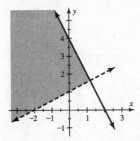

19. Add the first two equations to eliminate the variables y and z. $\begin{array}{l} x + y - z = -2 \\ \underline{x - y + z = -6} \\ 2x \quad\quad\quad = -8 \end{array} \Rightarrow x = -4$

Multiply the second equation by -1 and add to the third equation to eliminate the variables x and y.

$\begin{array}{l} -x + y\ -z = 6 \\ \underline{x - y - 2z = 3} \\ \quad\quad -3z = 9 \end{array}$ $z = -3$ Substitute $x = -4$ and $z = -3$ into the first equation and solve

for $y : -4 + y - (-3) = -2 \Rightarrow y = -1.$ The solution is $(-4,\ -1,\ -3).$

20. $\begin{bmatrix} 1 & -1 & -1 & | & -2 \\ 2 & 1 & 1 & | & 8 \\ -1 & -1 & 2 & | & 0 \end{bmatrix} \begin{array}{l} R_2 + R_1 \to \\ \\ 2R_3 + R_2 \to \end{array} \begin{bmatrix} 3 & 0 & 0 & | & 6 \\ 2 & 1 & 1 & | & 8 \\ 0 & -1 & 5 & | & 8 \end{bmatrix} \tfrac{1}{3}R_1 \to \begin{bmatrix} 1 & 0 & 0 & | & 2 \\ 2 & 1 & 1 & | & 8 \\ 0 & -1 & 5 & | & 8 \end{bmatrix} -2R_1 + R_2 \to$

$\begin{bmatrix} 1 & 0 & 0 & | & 2 \\ 0 & 1 & 1 & | & 4 \\ 0 & -1 & 5 & | & 8 \end{bmatrix} R_2 + R_3 \to \begin{bmatrix} 1 & 0 & 0 & | & 2 \\ 0 & 1 & 1 & | & 4 \\ 0 & 0 & 6 & | & 12 \end{bmatrix} \tfrac{1}{6}R_3 \to \begin{bmatrix} 1 & 0 & 0 & | & 2 \\ 0 & 1 & 1 & | & 4 \\ 0 & 0 & 1 & | & 2 \end{bmatrix} -R_3 + R_2 \to$

$\begin{bmatrix} 1 & 0 & 0 & | & 2 \\ 0 & 1 & 0 & | & 2 \\ 0 & 0 & 1 & | & 2 \end{bmatrix} \Rightarrow (2, 2, 2)$

21. $0(0 \cdot 0 - 1(-1)) - 1(1 \cdot 0 - 3(-1)) + (-3)(1 \cdot 1 - 3 \cdot 0) = 0 - (0 + 3) - 3(1 - 0) = -3 - 3(1) = -3 - 3 = -6$

22. $x^2 + 4 = 0 \Rightarrow x^2 = -4$, which is not possible for any real number x. There are no solutions.

23. $-2x\left(x^2 - 2x + 5\right) = -2x^3 + 4x^2 - 10x$

24. $(2a + b)(2a - b) = 4a^2 - 2ab + 2ab - b^2 = 4a^2 - b^2$

25. $(x-1)(x+1)(x+3) = \left(x^2 - 1\right)(x+3) = x^3 + 3x^2 - x - 3$

26. $(4x + 9)(2x - 1) = 8x^2 - 4x + 18x - 9 = 8x^2 + 14x - 9$

27. $\left(x^2 + 3y^3\right)^2 = \left(x^2 + 3y^3\right)\left(x^2 + 3y^3\right) = x^4 + 3x^2 y^3 + 3x^2 y^3 + 9y^6 = x^4 + 6x^2 y^3 + 9y^6$

28. $-2x\left(1 - x^2\right) = -2x + 2x^3 = 2x^3 - 2x$

29. $x^2 - 8x - 33 = (x + 3)(x - 11)$

30. $10x^3 + 65x^2 - 35x = 5x\left(2x^2 + 13x - 7\right) = 5x(2x - 1)(x + 7)$

31. $4x^2 - 100 = 4\left(x^2 - 25\right) = 4(x - 5)(x + 5)$

32. $49x^2 - 70x + 25 = (7x - 5)^2$

33. $r^4 - r = r\left(r^3 - 1\right) = r(r - 1)\left(r^2 + r + 1\right)$

34. $x^3 + 2x^2 + x + 2 = x^2(x + 2) + 1(x + 2) = \left(x^2 + 1\right)(x + 2)$

35. $4x^2 - 1 = 0 \Rightarrow (2x - 1)(2x + 1) = 0 \Rightarrow 2x - 1 = 0$ or $2x + 1 = 0 \Rightarrow x = \dfrac{1}{2}$ or $x = -\dfrac{1}{2}$.

    The solutions are $-\dfrac{1}{2}$ and $\dfrac{1}{2}$.

36. $3x^2 + 14x - 5 = 0 \Rightarrow (3x - 1)(x + 5) = 0 \Rightarrow 3x - 1 = 0$ or $x + 5 = 0 \Rightarrow x = \dfrac{1}{3}$ or $x = -5$.

    The solutions are -5 and $\dfrac{1}{3}$.

37. $x^3 + 4x = 4x^2 \Rightarrow x^3 - 4x^2 + 4x = 0 \Rightarrow x\left(x^2 - 4x + 4\right) = 0 \Rightarrow x(x - 2)^2 = 0 \Rightarrow x = 0$ or

    $x - 2 = 0 \Rightarrow x = 0$ or $x = 2$. The solutions are 0 and 2.

38. $x^4 = x^2 \Rightarrow x^4 - x^2 = 0 \Rightarrow x^2\left(x^2 - 1\right) = 0 \Rightarrow x^2(x - 1)(x + 1) = 0 \Rightarrow x^2 = 0$ or $x - 1 = 0$ or

    $x + 1 = 0 \Rightarrow x = 0$ or $x = 1$ or $x = -1$. The solutions are $-1, 0$, and 1.

39. Graph $Y_1 = \sqrt{(2)}X - 1.1(X - \pi)$ and $Y_2 = 1 - 2X$ in $[-5, 5, 1]$ by $[-5, 5, 1]$. The solution is

    approximately $x = -1.1$. See Figure 39.

    $[-5, 5, 1]$ by $[-5, 5, 1]$ $[-10, 10, 1]$ by $[-10, 10, 1]$ $[-10, 10, 1]$ by $[-10, 10, 1]$

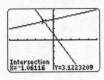

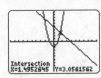

Figure 39   Figure 40a   Figure 40b

40. Graph $Y_1 = (\pi - 1) X \wedge 2 - \sqrt{(3)}$ and $Y_2 = 5 - 1.3X$ in $[-10, 10, 1]$ by $[-10, 10, 1]$. The solutions

   are approximately $x = -2.1$ and $x = 1.5$. See Figures 40a and 40b.

41. Let x represent the amount invested at 6% interest. Then

   $0.06x + 0.07(4000 - x) = 257 \Rightarrow 0.06x + 280 - 0.07x = 257 \Rightarrow -0.01x = -23 \Rightarrow x = 2300.$ So $2300

   is invested at 6% and $1700 is invested at 7%.

42. (a) $C(1995) = 750(1995 - 1990) + 6800 = 750(5) + 6800 = 10,550;$ In 1995 the car cost $10,550.

   (b)   Slope $= 750;$ The cost increased, on average, by $750 per year from 1990 to 2010.

43. Let x represent the length of a side of the square base. $x^2(x - 5) = 1008 \Rightarrow x^3 - 5x^2 - 1008 = 0$

   $\Rightarrow (x - 12)(x^2 + 7x + 84) = 0 \Rightarrow x = 12,$ since $x^2 + 7x + 84$ is never zero. The dimensions are 12

   inches by 12 inches by 7 inches.

44. Let x, y, and z represent the measures of the angles of the triangle from smallest to largest. Then the

   system needed is $\begin{array}{ll} x + y + z = 180 & x + y + z = 180 \\ z = x + y - 20 & \Rightarrow -x - y + z = -20 \\ y + z = x + 100 & -x + y + z = 100 \end{array}$

   Add the first two equations to eliminate the variables x and y. $\begin{array}{l} x + y + z = 180 \\ \underline{-x - y + z = -20} \\ 2z = 160 \end{array} \Rightarrow z = 80$

   Add the first and third equations to eliminate the variable x.

   $\begin{array}{l} x + y + z = 180 \\ \underline{-x + y + z = 100} \\ 2y + 2z = 280 \end{array} \Rightarrow 2y + 2(80) = 280 \Rightarrow 2y = 120 \Rightarrow y = 60$

   Then $x + y + z = 180 \Rightarrow x + 60 + 80 = 180 \Rightarrow x = 40.$ The angles are $40°, 60°,$ and $80°.$

Chapter 6: Rational Expressions and Functions

6.1: Introduction to Rational Functions and Equations

1. A rational expression is a polynomial divided by a nonzero polynomial. Example: $\dfrac{x^2+1}{3x}$

   Answers may vary.

3. Multiply each side by $x+7$.

5. No, it simplifies to $5+x$.

7. a

9. c

11. Yes

13. Yes

15. No, $\sqrt{x}$ is not a polynomial.

17. $f(x)=\dfrac{x}{x+1}$

19. $f(x)=\dfrac{x^2}{x-2}$

21. The denominator cannot equal zero, so $D=\{x\,|\,x\neq -2\}$.

23. The denominator cannot equal zero, so $D=\left\{x\,|\,x\neq \frac{1}{3}\right\}$.

25. The denominator cannot equal zero, and

   $t^2-4=0 \Rightarrow (t-2)(t+2)=0 \Rightarrow t=2$ or $t=-2$, so $D=\{t\,|\,t\neq -2, t\neq 2\}$.

27. The denominator cannot equal zero, and

   $x^2-3x+2=0 \Rightarrow (x-2)(x-1)=0 \Rightarrow x=2$ or $x=1$, so $D=\{x\,|\,x\neq 1, x\neq 2\}$.

29. The denominator cannot equal zero, and

   $x^3-4x=0 \Rightarrow x(x^2-4)=0 \Rightarrow x(x-2)(x+2)=0 \Rightarrow x=0$ or $x=2$ or $x=-2$, so

   $D=\{x\,|\,x\neq -2, x\neq 0, x\neq 2\}$.

31. See Figure 31. The denominator cannot equal zero, so $x-1\neq 0$. The domain is $\{x\,|\,x\neq 1\}$.

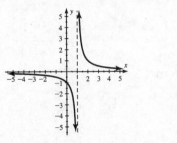

Figure 31

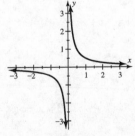

Figure 33

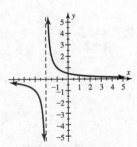

Figure 35

33. See Figure 33. The denominator cannot equal zero, so $2x \neq 0$. The domain is $\{x \mid x \neq 0\}$.

35. See Figure 35. The denominator cannot equal zero, so $x + 2 \neq 0$. The domain is $\{x \mid x \neq -2\}$.

37. See Figure 37. This denominator will never equal zero. The domain is $\{x \mid -\infty < x < \infty\}$.

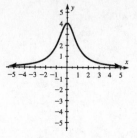
Figure 37

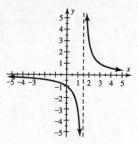
Figure 39

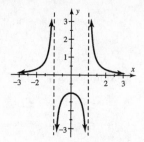
Figure 41

39. See Figure 39. The denominator cannot equal zero, so $2x - 3 \neq 0$. The domain is $\left\{x \mid x \neq \dfrac{3}{2}\right\}$.

41. See Figure 41. The denominator cannot equal zero, so $x^2 - 1 = 0 \Rightarrow (x-1)(x+1) = 0 \Rightarrow x = 1$ or $x = -1$.

The domain is $\{x \mid x \neq -1, x \neq 1\}$.

43. $f(-2) = \dfrac{1}{(-2)-1} = -\dfrac{1}{3}$.

45. $f(-3) = \dfrac{(-3)+1}{(-3)-1} = \dfrac{-2}{-4} = \dfrac{1}{2}$.

47. $f(-2) = \dfrac{(-2)^2 - 3(-2) + 5}{(-2)^2 + 1} = \dfrac{15}{5} = 3$.

49. $f(-3) = 2$ and $f(1) = 0$; The equation of the vertical asymptote is $x = -1$.

51. $f(-1) = 0$ and $f(2)$ is undefined; The equations of the vertical asymptotes are $x = -2$ and $x = 2$.

53.

x	-2	-1	0	1	2
$f(x) = \dfrac{1}{x-1}$	$-\dfrac{1}{3}$	$-\dfrac{1}{2}$	-1	$-$	1

When $x = 1$, $\dfrac{1}{x-1} = \dfrac{1}{1-1} = \dfrac{1}{0}$, so $f(1)$ is undefined.

55. $\dfrac{3}{x} = 5 \Rightarrow 3 = 5x \Rightarrow \dfrac{3}{5} = x \Rightarrow x = \dfrac{3}{5}$

57. $\dfrac{1}{x-2} = -1 \Rightarrow 1 = -1(x-2) \Rightarrow 1 = -x + 2 \Rightarrow -1 = -x \Rightarrow x = 1$

59. $\dfrac{x}{x+1} = 2 \Rightarrow x = 2(x+1) \Rightarrow x = 2x + 2 \Rightarrow -x = 2 \Rightarrow x = -2$

61. $\dfrac{2x+1}{3x-2} = 1 \Rightarrow 2x + 1 = 3x - 2 \Rightarrow 1 = x - 2 \Rightarrow 3 = x \Rightarrow x = 3$

63. $\dfrac{3}{x+2} = x \Rightarrow 3 = x(x+2) \Rightarrow x^2 + 2x - 3 = 0 \Rightarrow (x+3)(x-1) = 0 \Rightarrow x = -3$ or $x = 1$

65. $\dfrac{6}{x+1}=3x \Rightarrow \dfrac{2}{x+1}=x \Rightarrow 2=x(x+1) \Rightarrow x^2+x-2=0 \Rightarrow (x-1)(x+2) \Rightarrow x=1 \text{ or } x=-2$

67. $\dfrac{1}{x^2-1}=-1 \Rightarrow 1=-1(x^2-1) \Rightarrow 1=-x^2+1 \Rightarrow 0=-x^2 \Rightarrow 0=x^2 \Rightarrow 0=x \Rightarrow x=0$

69. $\dfrac{x}{x-5}=\dfrac{2x-5}{x-5} \Rightarrow x=2x-5 \Rightarrow -x=-5 \Rightarrow x=5,$ but this solution does not check. No solution.

71. $\dfrac{4x}{x+2}=\dfrac{-8}{2x+4} \Rightarrow \dfrac{4x}{x+2}=\dfrac{2(-4)}{2(x+2)} \Rightarrow \dfrac{4x}{x+2}=\dfrac{-4}{x+2} \Rightarrow 4x=-4 \Rightarrow x=-1$

73. $\dfrac{2x}{x+2}=\dfrac{x-4}{x+2} \Rightarrow 2x=x-4 \Rightarrow x=-4$

75. Graph $Y_1=(4+X)/(2X)$ and $Y_2=-0.5$ in $[-4.7, 4.7, 1]$ by $[-3.1, 3.1, 1]$. See Figure 75. $x=-2$

$[-4.7, 4.7, 1]$ by $[-3.1, 3.1, 1]$

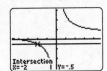

Figure 75

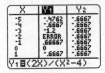

Figure 77

Figure 79

77. Table $Y_1=(2X)/(X^2-4)$ and $Y_2=-2/3$ with TblStart $=-5$ and ΔTbl $=1$.

See Figure 77. $x=-4$ or $x=1$

79. Table $Y_1=1/(X-1)$ and $Y_2=X-1$ with TblStart $=-3$ and ΔTbl $=1$. See Figure 79. $x=0$ or $x=2$

81. (a) Symbolic: $\dfrac{1}{x+2}=1 \Rightarrow 1=x+2 \Rightarrow -1=x \Rightarrow x=-1$

    (b) Graphical: Graph $Y_1=1/(X+2)$ and $Y_2=1$ in $[-4.7, 4.7, 1]$ by $[-3.1, 3.1, 1]$.

    See Figure 81b. $x=-1$

    (c) Numerical: Table $Y_1=1/(X+2)$ and $Y_2=1$ with TblStart $=-3$ and ΔTbl $=1$.

    See Figure 81c. $x=-1$

$[-4.7, 4.7, 1]$ by $[-3.1, 3.1, 1]$

Figure 81b

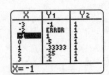

Figure 81c

83. (a) Symbolic: $\dfrac{x}{2x+1}=\dfrac{2}{5} \Rightarrow 5x=2(2x+1) \Rightarrow 5x=4x+2 \Rightarrow x=2$

    (b) Graphical: Graph $Y_1=X/(2X+1)$ and $\dot{Y}_2=2/5$ in $[-2.35, 2.35, 1]$ by $[-1.55, 1.55, 1]$.

    See Figure 83b. $x=2$

(c)   Numerical: Table $Y_1 = X/(2X+1)$ and $Y_2 = 2/5$ with TblStart = 0 and ΔTbl = 1.

See Figure 83c. $x = 2$

[−2.35, 2.35, 1] by [−1.55, 1.55, 1]

Figure 83b

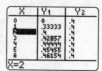

Figure 83c

85. Graph $Y_1 = X^3/(X-1)$ and $Y_2 = -1/(X+1)$ in [−5, 5, 1] by [−5, 5, 1]. The solutions are

approximately $x = -1.62$ and $x = 0.62$. See Figures 85a and 85b.

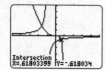

Figure 85a

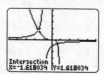

Figure 85b

87. $a = \dfrac{x}{b} \Rightarrow ba = b \cdot \dfrac{x}{b} \Rightarrow ab = x$

89. (a) $(f+g)(3) = 5(3) + (3+1) = 15 + 4 = 19$

(b) $(f-g)(-2) = 5(-2) - (-2+1) = -10 - (-1) = -9$

(c) $(fg)(5) = (5 \cdot 5)(5+1) = 25(6) = 150$

(d) $(f/g)(0) = (5 \cdot 0)/(0+1) = \dfrac{0}{1} = 0$

91. (a) $(f+g)(3) = (2 \cdot 3 - 1) + (4 \cdot 3^2) = 6 - 1 + 4 \cdot 9 = 5 + 36 = 41$

(b) $(f-g)(-2) = (2(-2) - 1) - (4(-2)^2) = -4 - 1 - 4(4) = -5 - 16 = -21$

(c) $(fg)(5) = (2 \cdot 5 - 1)(4 \cdot 5^2) = (9)(100) = 900$

(d) $(f/g)(0) = (2 \cdot 0 - 1)/(4 \cdot 0^2) = \dfrac{-1}{0}$, which is undefined

93. (a) $(f+g)(x) = (x+1) + (x+2) = 2x + 3$

(b) $(f-g)(x) = (x+1) - (x+2) = -1$

(c) $(fg)(x) = (x+1)(x+2) = x^2 + 3x + 2$

(d) $(f/g)(x) = \dfrac{x+1}{x+2}$

95. (a) $(f+g)(x) = (1-x) + (x^2) = x^2 - x + 1$

(b) $(f-g)(x) = (1-x) - (x^2) = 1 - x - x^2$

(c) $(fg)(x) = (1-x)(x^2) = x^2 - x^3$

(d) $(f/g)(x) = \dfrac{1-x}{x^2}$

97. The graph should increase quickly at first and then continue to increase at a slower rate. The answer is graph c.

99. The graph should increase slowly at first and then continue to increase at a faster rate. The answer is graph d.

101. (a) $f(400) = \dfrac{2540}{400} = 6.35;$ A curve with a radius of 400 feet will have an outer rail elevation of 6.35 inches.

(b)   Table $Y_1 = 2540/X$ with TblStart $= 100$ and ΔTbl $= 50$. See Figure 101.

(c)   If the radius of the curve doubles, the outer rail elevation is halved.

(d) $\dfrac{2540}{r} = 5 \Rightarrow 2540 = 5r \Rightarrow r = \dfrac{2540}{5} = 508$ feet

Figure 101

103. (a) $D(0.05) = \dfrac{900}{10.5 + 30(0.05)} = \dfrac{900}{12} = 75;$ The braking distance is 75 feet when the uphill grade is 0.05.

(b) $60 = \dfrac{900}{10.5 + 30x} \Rightarrow 60(10.5 + 30x) = 900 \Rightarrow 630 + 1800x = 900 \Rightarrow 1800x = 270 \Rightarrow x = 0.15$

105. (a) $T(4) = \dfrac{1}{5-(4)} = \dfrac{1}{1} = 1;$ When cars leave the ramp at a rate of 4 vehicles per minute, the wait is 1 minute.

(b)   As more cars try to exit, the waiting time increases. This agrees with intuition.

(c) $3 = \dfrac{1}{5-x} \Rightarrow 3(5-x) = 1 \Rightarrow 15 - 3x = 1 \Rightarrow -3x = -14 \Rightarrow x = 4.\overline{6}$ vehicles per minute.

107. (a) $P(1) = \dfrac{(1)-1}{(1)} = \dfrac{0}{1} = 0$ and $P(50) = \dfrac{(50)-1}{(50)} = \dfrac{49}{50} = 0.98$

When there is only 1 ball there is no chance of losing. With 50 balls there is a 98% chance of losing.

(b)   Graph $Y_1 = (X-1)/X$ in [0, 100, 10] by [0, 1, 0.1]. See Figure 107.

(c)   It increases. This agrees with intuition since there are more balls without the winning number.

(d) $0.975 = \dfrac{x-1}{x} \Rightarrow 0.975x = x-1 \Rightarrow -0.025x = -1 \Rightarrow x = 40$ balls

[0, 100, 10] by [0, 1, 0.1]

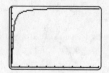

Figure 107

109. $R(1) = \dfrac{100}{1.2(1)+1} \approx 45.45$, $R(3) = \dfrac{100}{1.2(3)+1} \approx 21.7$ After 1 day (3 days) students remember about

45% (22%) of what they learned.

111. (a) $R(x) = \dfrac{350(x-3)^2 + 50}{206(x-3)^2 + 150}$

(b) $R(6) = \dfrac{350(6-3)^2 + 50}{206(6-3)^2 + 150} = \dfrac{350(3)^2 + 50}{206(3)^2 + 150} = \dfrac{3200}{2004} \approx 1.6$ Six years after startup, Google's

revenues were about 1.6 times Facebook's revenues.

Section 6.2 Multiplication and Division of Rational Expressions

1. 1

3. No, $\dfrac{x^2-1}{x-1} = \dfrac{(x-1)(x+1)}{x-1} = x+1.$

5. We multiply $\dfrac{2}{3}$ by $\dfrac{7}{5}$.

7. $\dfrac{ad}{bc}$

9. c, $\dfrac{x^2-1}{x^2+2x+1} = \dfrac{(x-1)(x+1)}{(x+1)(x+1)} = \dfrac{x-1}{x+1}$

11. $\dfrac{1}{2} \cdot \dfrac{4}{5} = \dfrac{1}{1} \cdot \dfrac{2}{5} = \dfrac{2}{5}$

13. $\dfrac{7}{8} \cdot \dfrac{4}{3} \cdot (-3) = \dfrac{7}{2} \cdot \dfrac{1}{3} \cdot \dfrac{-3}{1} = \dfrac{7}{2} \cdot \dfrac{1}{1} \cdot \dfrac{-1}{1} = -\dfrac{7}{2}$

15. $\dfrac{3}{8} \cdot 2 = \dfrac{3}{8} \cdot \dfrac{2}{1} = \dfrac{3}{4} \cdot \dfrac{1}{1} = \dfrac{3}{4}$

17. $-\dfrac{7}{11} \div 14 = -\dfrac{7}{11} \cdot \dfrac{1}{14} = -\dfrac{1}{11} \cdot \dfrac{1}{2} = -\dfrac{1}{22}$

19. $\dfrac{5}{7} \div \dfrac{15}{14} = \dfrac{5}{7} \cdot \dfrac{14}{15} = \dfrac{1}{1} \cdot \dfrac{2}{3} = \dfrac{2}{3}$

21. $6 \div \left(-\dfrac{1}{3}\right) = 6 \cdot (-3) = -18$

23. $\dfrac{5x}{x^2} = \dfrac{5}{x}$

25. $\dfrac{3z+6}{z+2} = \dfrac{3(z+2)}{z+2} = 3$

27. $\dfrac{2z+2}{3z+3} = \dfrac{2(z+1)}{3(z+1)} = \dfrac{2}{3}$

29. $\dfrac{(x-1)(x+1)}{x-1} = x+1$

31. $\dfrac{x^2-4}{x+2} = \dfrac{(x-2)(x+2)}{x+2} = x-2$

33. $\dfrac{x(x-1)}{(x+1)(x-1)} = \dfrac{x}{x+1}$

35. $\dfrac{(3x+1)(x+2)}{(x+2)(5x-2)} = \dfrac{3x+1}{5x-2}$

37. $\dfrac{x+5}{x^2+2x-15} = \dfrac{x+5}{(x+5)(x-3)} = \dfrac{1}{x-3}$

39. $\dfrac{x^2+2x}{x^2+3x+2} = \dfrac{x(x+2)}{(x+1)(x+2)} = \dfrac{x}{x+1}$

41. $\dfrac{6x^2+7x-5}{2x^2-11x+5} = \dfrac{(3x+5)(2x-1)}{(x-5)(2x-1)} = \dfrac{3x+5}{x-5}$

43. $\dfrac{a^2-b^2}{a-b} = \dfrac{(a-b)(a+b)}{a-b} = a+b$

45. $\dfrac{m^3+n^3}{m+n} = \dfrac{(m+n)(m^2-mn+n^2)}{m+n} = m^2-mn+n^2$

47. $-\dfrac{4-t}{t-4} = \dfrac{-1(4-t)}{t-4} = \dfrac{t-4}{t-4} = 1$

49. $\dfrac{4m-n}{-4m+n} = \dfrac{4m-n}{-1(4m-n)} = \dfrac{1}{-1} = -1$

51. $\dfrac{5-y}{y-5} = \dfrac{-1(y-5)}{y-5} = -1$

53. $\dfrac{1}{4x}$

55. $\dfrac{5b}{2a}$

57. $\dfrac{5-x}{3-x}$

59. $\dfrac{x^2+1}{1}=x^2+1$

61. $\dfrac{2}{x}\cdot\dfrac{x-1}{3x}=\dfrac{2(x-1)}{3x^2}$

63. $\dfrac{x-2}{x}\cdot\dfrac{x-3}{x+4}=\dfrac{(x-2)(x-3)}{x(x+4)}$

65. $\dfrac{1}{2x}\cdot\dfrac{4x}{2}=\dfrac{1}{2x}\cdot\dfrac{2x}{1}=\dfrac{2x}{2x}=1$

67. $\dfrac{5a}{4}\cdot\dfrac{12}{5a}=\dfrac{1}{1}\cdot\dfrac{3}{1}=3$

69. $\dfrac{9x^2y^4}{8xy^6}\cdot\dfrac{\left(2xy^2\right)^3}{3(xy)^4}=\dfrac{9x^2y^4}{8xy^6}\cdot\dfrac{8x^3y^6}{3x^4y^4}=\dfrac{9\cdot 8\cdot x^5y^{10}}{8\cdot 3\cdot x^5y^{10}}=\dfrac{9}{3}=3$

71. $\dfrac{x+1}{2x-5}\cdot\dfrac{2x-5}{x}=\dfrac{x+1}{1}\cdot\dfrac{1}{x}=\dfrac{x+1}{x}$

73. $\dfrac{b^2+1}{b^2-1}\cdot\dfrac{b-1}{b+1}=\dfrac{b^2+1}{(b-1)(b+1)}\cdot\dfrac{b-1}{b+1}=\dfrac{b^2+1}{b+1}\cdot\dfrac{1}{b+1}=\dfrac{b^2+1}{(b+1)^2}$

75. $\dfrac{x^2-2x-35}{2x^3-3x^2}\cdot\dfrac{x^3-x^2}{2x-14}=\dfrac{(x-7)(x+5)}{x^2(2x-3)}\cdot\dfrac{x^2(x-1)}{2(x-7)}=\dfrac{x+5}{2x-3}\cdot\dfrac{x-1}{2}=\dfrac{(x-1)(x+5)}{2(2x-3)}$

77. $\dfrac{3n-9}{n^2-9}\cdot\left(n^3+27\right)=\dfrac{3(n-3)\cdot(n+3)\left(n^2-3n+9\right)}{(n+3)(n-3)}=\dfrac{3\left(n^2-3n+9\right)}{1}=3\left(n^2-3n+9\right)$

79. $\dfrac{3n-9}{n^2-9}\cdot\dfrac{n^3+27}{12}=\dfrac{3(n-3)\cdot(n+3)\left(n^2-3n+9\right)}{(n+3)(n-3)\cdot 12}=\dfrac{n^2-3n+9}{4}$

81. $\dfrac{x}{y}\cdot\dfrac{2y}{x}\cdot\dfrac{2}{xy}=\dfrac{x\cdot 2y\cdot 2}{y\cdot x\cdot xy}=\dfrac{4}{xy}$

83. $\dfrac{x-1}{y}\cdot\dfrac{y(x+y)}{2}\cdot\dfrac{y}{x+y}=\dfrac{(x-1)\cdot y^2(x+y)}{2y\cdot(x+y)}=\dfrac{y(x-1)}{2}$

85. $\dfrac{3x}{2}\div\dfrac{2x}{5}=\dfrac{3x}{2}\cdot\dfrac{5}{2x}=\dfrac{3}{2}\cdot\dfrac{5}{2}=\dfrac{15}{4}$

87. $\dfrac{8a^4}{3b} \div \dfrac{a^5}{9b^2} = \dfrac{8a^4}{3b} \cdot \dfrac{9b^2}{a^5} = \dfrac{8}{1} \cdot \dfrac{3b}{a} = \dfrac{24b}{a}$

89. $(2n+4) \div \dfrac{n+2}{n-1} = \dfrac{2(n+2)}{1} \cdot \dfrac{n-1}{n+2} = 2(n-1)$

91. $\dfrac{6b}{b+2} \div \dfrac{3b^4}{2b+4} = \dfrac{6b}{b+2} \cdot \dfrac{2(b+2)}{3b^4} = \dfrac{2}{1} \cdot \dfrac{2}{b^3} = \dfrac{4}{b^3}$

93. $\dfrac{3a+1}{a^7} \div \dfrac{a+1}{3a^8} = \dfrac{3a+1}{a^7} \cdot \dfrac{3a^8}{a+1} = \dfrac{3a+1}{1} \cdot \dfrac{3a}{a+1} = \dfrac{3a(3a+1)}{a+1}$

95. $\dfrac{x+5}{x-x^3} \div \dfrac{25-x^2}{x^3} = \dfrac{x+5}{x(1-x^2)} \cdot \dfrac{x^3}{(5-x)(5+x)} = \dfrac{1}{1-x^2} \cdot \dfrac{x^2}{5-x} = \dfrac{x^2}{(x-5)(x^2-1)}$

97. $\dfrac{x^2-3x+2}{x^2+5x+6} \div \dfrac{x^2+x-2}{x^2+2x-3} = \dfrac{(x-2)(x-1)}{(x+2)(x+3)} \cdot \dfrac{(x+3)(x-1)}{(x+2)(x-1)} = \dfrac{(x-2)(x-1)}{(x+2)^2}$

99. $\dfrac{x^2-4}{x^2+x-2} \div \dfrac{x-2}{x-1} = \dfrac{(x-2)(x+2)}{(x+2)(x-1)} \cdot \dfrac{x-1}{x-2} = 1$

101. $\dfrac{3y}{x^2} \div \dfrac{y^2}{x} \div \dfrac{y}{5x} = \dfrac{3y}{x^2} \cdot \dfrac{x}{y^2} \cdot \dfrac{5x}{y} = \dfrac{15x^2 y}{x^2 y^3} = \dfrac{15}{y^2}$

103. $\dfrac{x-3}{x-1} \div \dfrac{x^2}{x-1} \div \dfrac{x-3}{x} = \dfrac{x-3}{x-1} \cdot \dfrac{x-1}{x^2} \cdot \dfrac{x}{x-3} = \dfrac{1}{x}$

105. $(fg)(x) = \dfrac{x+2}{x^2-3x+2} \cdot \dfrac{x-1}{x+2} = \dfrac{(x+2)(x-1)}{(x-2)(x-1)(x+2)} = \dfrac{1}{x-2}$

107. $(f/g)(x) = \dfrac{x^2-1}{x+1} = \dfrac{(x-1)(x+1)}{x+1} = x-1$

109. (a) $C(5000) = \dfrac{50(5000)+20,000}{5000} = \dfrac{270,000}{5000} = 54$. It costs $54, on average, to make each camera

   when 5000 are produced.

   (b) $T(x) = C(x) \cdot x = \dfrac{50x+20,000}{x} \cdot x = 50x + 20,000$.

   (c) $T(5000) = 50(5000) + 20,000 = 270,000$. It costs $270,000 to make 5000 cameras.

111. (a) $\dfrac{5x^2+12x+4}{x+2} = \dfrac{(5x+2)(x+2)}{x+2} = 5x+2$

   (b) If the width is 8 feet, $x+2 = 8 \Rightarrow x = 6$. The length is $5(6)+2 = 32$ feet.

113. (a) $\dfrac{4x^3+4x^2+x}{x} = 4x^2 + 4x + 1$

   (b) Since $4x^2 + 4x + 1 = (2x+1)^2$, a side of the bottom is $2(10)+1 = 21$. The dimensions are 21

   by 21 by 10.

Checking Basic Concepts Sections 6.1 and 6.2

1. (a) $f(2) = \dfrac{(2)}{(2)-1} = \dfrac{2}{1} = 2$

   (b) The denominator cannot equal zero, so $x - 1 \neq 0$. The domain is $\{x \mid x \neq 1\}$.

   (c) See Figure 1. Vertical asymptote: $x = 1$.

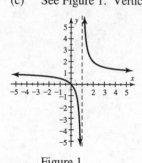

   Figure 1

2. (a) $\dfrac{6}{2x+3} = 3 \Rightarrow 6 = 3(2x+3) \Rightarrow 6 = 6x + 9 \Rightarrow 6x = -3 \Rightarrow x = -\dfrac{1}{2}$

   (b) $\dfrac{2}{x-1} = x \Rightarrow 2 = x(x-1) \Rightarrow x^2 - x - 2 = 0 \Rightarrow (x+1)(x-2) = 0 \Rightarrow x = -1 \text{ or } x = 2$

3. $\dfrac{x^2 - 6x - 7}{x^2 - 1} = \dfrac{(x-7)(x+1)}{(x-1)(x+1)} = \dfrac{x-7}{x-1}$

4. (a) $\dfrac{2x^2}{x^2 - 1} \cdot \dfrac{x+1}{4x} = \dfrac{2x^2}{(x-1)(x+1)} \cdot \dfrac{x+1}{4x} = \dfrac{x}{x-1} \cdot \dfrac{1}{2} = \dfrac{x}{2(x-1)}$

   (b) $\dfrac{1}{x-2} \div \dfrac{3}{(x-2)(x+3)} = \dfrac{1}{x-2} \cdot \dfrac{(x-2)(x+3)}{3} = \dfrac{1}{1} \cdot \dfrac{x+3}{3} = \dfrac{x+3}{3}$

Section 6.3 Addition and Subtraction of Rational Expressions

1. 18

3. $(x-5)(x+5)$

5. A common denominator

7. $\dfrac{a+b}{c}$

9. b, $\dfrac{1-x}{2+x} + \dfrac{x}{2+x} = \dfrac{1-x+x}{2+x} = \dfrac{1}{2+x}$

11. $10 = 2 \cdot 5$ and $15 = 3 \cdot 5$, the LCM is $2 \cdot 3 \cdot 5 = 30$.

13. $34 = 2 \cdot 17$ and $51 = 3 \cdot 17$, the LCM is $2 \cdot 3 \cdot 17 = 102$.

15. $6a = 2 \cdot 3 \cdot a$ and $9a^2 = 3 \cdot 3 \cdot a \cdot a$, the LCM is $2 \cdot 3 \cdot 3 \cdot a \cdot a = 18a^2$.

17. $10x^2 = 2 \cdot 5 \cdot x \cdot x$ and $25(x^2 - x) = 5 \cdot 5 \cdot x \cdot (x-1)$, the LCM is $2 \cdot 5 \cdot 5 \cdot x \cdot x \cdot (x-1) = 50x^2(x-1)$.

19. $x^2 + 2x + 1 = (x+1) \cdot (x+1)$ and $x^2 - 4x - 5 = (x-5) \cdot (x+1)$, the LCM is $(x-5)(x+1)^2$.

21. The LCM is $(x+y)(x-y)$.

23. $\dfrac{1}{7} + \dfrac{4}{7} = \dfrac{5}{7}$

25. $\dfrac{2}{3} + \dfrac{5}{6} + \dfrac{1}{4} = \dfrac{8}{12} + \dfrac{10}{12} + \dfrac{3}{12} = \dfrac{21}{12} = \dfrac{7}{4}$

27. $\dfrac{1}{10} - \dfrac{3}{10} = \dfrac{-2}{10} = -\dfrac{1}{5}$

29. $\dfrac{3}{2} - \dfrac{1}{8} = \dfrac{12}{8} - \dfrac{1}{8} = \dfrac{11}{8}$

31. $\dfrac{1}{x} + \dfrac{3}{x} = \dfrac{4}{x}$

33. $\dfrac{2}{x^2 - 4} - \dfrac{x+1}{x^2 - 4} = \dfrac{2 - (x+1)}{x^2 - 4} = \dfrac{1-x}{x^2 - 4}$

35. $\dfrac{4}{x^2} + \dfrac{5}{x^2} = \dfrac{9}{x^2}$

37. $\dfrac{4}{xy} - \dfrac{7}{xy} = -\dfrac{3}{xy}$

39. $\dfrac{x}{x+1} + \dfrac{1}{x+1} = \dfrac{x+1}{x+1} = 1$

41. $\dfrac{2z}{4-z} - \dfrac{3z-4}{4-z} = \dfrac{2z - (3z-4)}{4-z} = \dfrac{-z+4}{4-z} = \dfrac{4-z}{4-z} = 1$

43. $\dfrac{4r}{5t^2} + \dfrac{r}{5t^2} = \dfrac{4r+r}{5t^2} = \dfrac{5r}{5t^2} = \dfrac{r}{t^2}$

45. $\dfrac{3t}{t^2 - t - 6} + \dfrac{2-2t}{t^2 - t - 6} = \dfrac{3t + 2 - 2t}{t^2 - t - 6} = \dfrac{t+2}{(t-3)(t+2)} = \dfrac{1}{t-3}$

47. $\dfrac{5b}{3a} - \dfrac{7b}{5a} = \dfrac{5b \cdot 5}{15a} - \dfrac{7b \cdot 3}{15a} = \dfrac{25b - 21b}{15a} = \dfrac{4b}{15a}$

49. $\dfrac{4}{n-4} + \dfrac{3}{2-n} = \dfrac{4}{n-4} - \dfrac{3}{n-2} = \dfrac{4(n-2)}{(n-4)(n-2)} - \dfrac{3(n-4)}{(n-4)(n-2)} = \dfrac{4n-8-3n+12}{(n-4)(n-2)} = \dfrac{n+4}{(n-4)(n-2)}$

51. $\dfrac{x}{x+4} - \dfrac{x+1}{x} = \dfrac{x^2}{x(x+4)} - \dfrac{(x+1)(x+4)}{x(x+4)} = \dfrac{x^2 - (x^2 + 5x + 4)}{x(x+4)} = \dfrac{-5x - 4}{x(x+4)}$

53. $\dfrac{2}{x^2} - \dfrac{4x-1}{x} = \dfrac{2}{x^2} - \dfrac{x(4x-1)}{x^2} = \dfrac{2 - (4x^2 - x)}{x^2} = \dfrac{-4x^2 + x + 2}{x^2}$

55. $\dfrac{x+3}{x-5}+\dfrac{5}{x-3}=\dfrac{(x+3)(x-3)}{(x-5)(x-3)}+\dfrac{5(x-5)}{(x-5)(x-3)}=\dfrac{x^2-9+5x-25}{(x-5)(x-3)}=\dfrac{x^2+5x-34}{(x-5)(x-3)}$

57. $\dfrac{4n}{n^2-9}-\dfrac{8}{n-3}=\dfrac{4n}{(n-3)(n+3)}-\dfrac{8(n+3)}{(n-3)(n+3)}=\dfrac{4n-8n-24}{(n-3)(n+3)}=-\dfrac{4(n+6)}{(n-3)(n+3)}$

59. $\dfrac{x}{x^2-9}+\dfrac{5x}{x-3}=\dfrac{x}{(x-3)(x+3)}+\dfrac{5x(x+3)}{(x-3)(x+3)}=\dfrac{x+5x^2+15x}{(x-3)(x+3)}=\dfrac{x(5x+16)}{(x-3)(x+3)}$

61. $\dfrac{b}{2b-4}-\dfrac{b-1}{b-2}=\dfrac{b}{2(b-2)}-\dfrac{2(b-1)}{2(b-2)}=\dfrac{b-(2b-2)}{2(b-2)}=\dfrac{-(b-2)}{2(b-2)}=-\dfrac{1}{2}$

63. $\dfrac{2x}{x-5}+\dfrac{2x-1}{3x^2-16x+5}=\dfrac{2x(3x-1)}{(x-5)(3x-1)}+\dfrac{2x-1}{(x-5)(3x-1)}=\dfrac{6x^2-2x+2x-1}{(x-5)(3x-1)}=\dfrac{6x^2-1}{(x-5)(3x-1)}$

65. $\dfrac{4x}{x-y}-\dfrac{9}{x+y}=\dfrac{4x(x+y)}{(x-y)(x+y)}-\dfrac{9(x-y)}{(x-y)(x+y)}=\dfrac{4x^2+4xy-(9x-9y)}{(x-y)(x+y)}=\dfrac{4x^2+4xy-9x+9y}{(x-y)(x+y)}$

67. $\dfrac{3}{(x-1)(x-2)}+\dfrac{4x}{(x+1)(x-2)}=\dfrac{3(x+1)}{(x-2)(x-1)(x+1)}+\dfrac{4x(x-1)}{(x-2)(x-1)(x+1)}$

$=\dfrac{3x+3+4x^2-4x}{(x-2)(x-1)(x+1)}=\dfrac{4x^2-x+3}{(x-2)(x-1)(x+1)}$

69. $\dfrac{3}{x^2-x-6}-\dfrac{2}{x^2+5x+6}=\dfrac{3}{(x-3)(x+2)}-\dfrac{2}{(x+3)(x+2)}$

$=\dfrac{3(x+3)}{(x-3)(x+2)(x+3)}-\dfrac{2(x-3)}{(x-3)(x+2)(x+3)}=\dfrac{3x+9-2x+6}{(x-3)(x+2)(x+3)}=\dfrac{x+15}{(x-3)(x+2)(x+3)}$

71. $\dfrac{3}{x^2-2x+1}+\dfrac{1}{x^2-3x+2}=\dfrac{3(x-2)}{(x-2)(x-1)^2}+\dfrac{x-1}{(x-2)(x-1)^2}=\dfrac{3x-6+x-1}{(x-2)(x-1)^2}=\dfrac{4x-7}{(x-2)(x-1)^2}$

73. $\dfrac{3x}{x^2+2x-3}+\dfrac{1}{x^2-2x+1}=\dfrac{3x(x-1)}{(x+3)(x-1)(x-1)}+\dfrac{1(x+3)}{(x+3)(x-1)(x-1)}=\dfrac{3x^2-2x+3}{(x+3)(x-1)^2}$

75. $\dfrac{4c}{ab}+\dfrac{3b}{ac}-\dfrac{2a}{bc}=\dfrac{4c^2}{abc}+\dfrac{3b^2}{abc}-\dfrac{2a^2}{abc}=\dfrac{4c^2+3b^2-2a^2}{abc}=\dfrac{-2a^2+3b^2+4c^2}{abc}$

77. $5-\dfrac{6}{n^2-36}+\dfrac{3}{n-6}=\dfrac{5(n-6)(n+6)}{(n-6)(n+6)}-\dfrac{6}{(n-6)(n+6)}+\dfrac{3(n+6)}{(n-6)(n+6)}$

$=\dfrac{5n^2-180-6+3n+18}{(n-6)(n+6)}=\dfrac{5n^2+3n-168}{(n-6)(n+6)}$

79. $\dfrac{3}{x-5}-\dfrac{1}{x-3}-\dfrac{2x}{x-5}=\dfrac{3(x-3)}{(x-5)(x-3)}-\dfrac{x-5}{(x-5)(x-3)}-\dfrac{2x(x-3)}{(x-5)(x-3)}=\dfrac{-2(x^2-4x+2)}{(x-5)(x-3)}$

81. $\dfrac{5}{2x-3}+\dfrac{x}{x+1}-\dfrac{x}{2x-3}=\dfrac{5(x+1)}{(x+1)(2x-3)}+\dfrac{x(2x-3)}{(x+1)(2x-3)}-\dfrac{x(x+1)}{(x+1)(2x-3)}$

$\qquad =\dfrac{5x+5+2x^2-3x-x^2-x}{(x+1)(2x-3)}=\dfrac{x^2+x+5}{(x+1)(2x-3)}$

83. $\dfrac{1}{x-1}-\dfrac{2}{x+1}+\dfrac{x}{x^2-1}=\dfrac{1(x+1)}{(x-1)(x+1)}-\dfrac{2(x-1)}{(x-1)(x+1)}+\dfrac{x}{(x-1)(x+1)}=\dfrac{x+1-2x+2+x}{(x-1)(x+1)}=\dfrac{3}{(x-1)(x+1)}$

85. $\dfrac{a-b}{a+b}+\dfrac{a+b}{a-b}=\dfrac{(a-b)(a-b)}{(a+b)(a-b)}+\dfrac{(a+b)(a+b)}{(a+b)(a-b)}=\dfrac{a^2-2ab+b^2+a^2+2ab+b^2}{(a+b)(a-b)}=\dfrac{2a^2+2b^2}{(a+b)(a-b)}$

87. $(f+g)(x)=\dfrac{1}{x+1}+\dfrac{x}{x+1}=\dfrac{1+x}{x+1}=1,\ \ (f-g)(x)=\dfrac{1}{x+1}-\dfrac{x}{x+1}=\dfrac{1-x}{x+1}$

89. $(f+g)(x)=\dfrac{1}{x+2}+\dfrac{1}{x-2}=\dfrac{x-2}{(x-2)(x+2)}+\dfrac{x+2}{(x-2)(x+2)}=\dfrac{2x}{(x-2)(x+2)}$

$\qquad (f-g)(x)=\dfrac{1}{x+2}-\dfrac{1}{x-2}=\dfrac{x-2}{(x-2)(x+2)}-\dfrac{x+2}{(x-2)(x+2)}=\dfrac{-4}{(x-2)(x+2)}$

91. $C(x)=M(x)+P(x)=\dfrac{5x+1,000,000}{x}+\dfrac{2x+500,000}{x}=\dfrac{7x+1,500,000}{x}$

93. Find the LCM of 12 and 30: $12=2\cdot2\cdot3$ and $30=2\cdot3\cdot5,$ so the LCM is $2\cdot2\cdot3\cdot5=60.$ The planets will be in alignment in 60 years.

95. $\dfrac{1}{R}=\dfrac{1}{80}+\dfrac{1}{300}=\dfrac{30}{2400}+\dfrac{8}{2400}=\dfrac{38}{2400}=\dfrac{19}{1200}$ and so $R=\dfrac{1200}{19}\approx63.2$ ohms

97. $\dfrac{1}{S}=\dfrac{1}{0.1}-\dfrac{1}{10}\ \Rightarrow\ \dfrac{1}{S}=10-0.1\ \Rightarrow\ \dfrac{1}{S}=9.9\ \Rightarrow\ S=\dfrac{1}{9.9}\approx0.101$ feet, or about 1.2 inches.

99. $\dfrac{8}{d^2}+\dfrac{8}{d^2}=\dfrac{16}{d^2}\ \mathrm{W/m^2}$

Section 6.4 Rational Equations

1. Multiply each side by $x+2$.

3. $\dfrac{1}{2}$

5. LCD

7. False, 2 is an extraneous solution.

9. The factored denominators are x and 5. The LCD is $5x$.

11. The factored denominators are $x-1$ and $(x-1)(x+1)$. The LCD is $(x-1)(x+1)=x^2-1$.

13. The factored denominators are 2 and $2x+1$ and $2(x-2)$. The LCD is $2(2x+1)(x-2)$.

15. The LCD is 6. The first step is to multiply both sides of the equation by the LCD.

$$6 \cdot \left(\frac{x}{3} + \frac{1}{2} \right) = \frac{5}{6} \cdot 6 \Rightarrow 2x + 3 = 5 \Rightarrow 2x = 2 \Rightarrow x = \frac{2}{2} = 1$$

17. The LCD is $15x$. The first step is to multiply both sides of the equation by the LCD.

$$15x \cdot \left(\frac{2}{x} - \frac{7}{3} \right) = -\frac{29}{15} \cdot 15x \Rightarrow 30 - 35x = -29x \Rightarrow 30 = 6x \Rightarrow x = \frac{30}{6} = 5$$

19. The LCD is $3(x-1)$. The first step is to multiply both sides of the equation by the LCD.

$$3(x-1) \cdot \frac{x}{x-1} = \frac{4}{3} \cdot 3(x-1) \Rightarrow 3x = 4x - 4 \Rightarrow x = 4$$

21. The LCD is $6x$. The first step is to multiply both sides of the equation by the LCD.

$$6x \cdot \frac{1}{2x} - 6x \cdot \frac{5}{3x} = 1 \cdot 6x \Rightarrow 3 - 10 = 6x \Rightarrow -7 = 6x \Rightarrow x = -\frac{7}{6}$$

23. The LCD is $x+1$. The first step is to multiply both sides of the equation by the LCD.

$$(x+1) \cdot \frac{1}{x+1} - (x+1)(1) = \frac{3}{x+1} \cdot (x+1) \Rightarrow 1 - x - 1 = 3 \Rightarrow -x = 3 \Rightarrow x = -3$$

25. The LCD is $x-3$. The first step is to multiply both sides of the equation by the LCD.

$$(x-3) \cdot \frac{1}{x-3} + (x-3) \cdot \frac{x}{x-3} = \frac{2x}{x-3} \cdot (x-3) \Rightarrow 1 + x = 2x \Rightarrow x = 1$$

27. The LCD is $(x-1)(x+4)$. The first step is to multiply both sides of the equation by the LCD.

$$(x-1)(x+4) \cdot \frac{3}{x-1} = \frac{6}{x+4} \cdot (x-1)(x+4) \Rightarrow 3x + 12 = 6x - 6 \Rightarrow 18 = 3x \Rightarrow x = \frac{18}{3} = 6$$

29. The LCD is $(3z+4)(2z-5)$. The first step is to multiply both sides of the equation by the LCD.

$$(3z+4)(2z-5) \cdot \frac{6}{3z+4} = \frac{4}{2z-5} \cdot (3z+4)(2z-5) \Rightarrow 12z - 30 = 12z + 16 \Rightarrow -30 = 16 \Rightarrow$$

No solutions.

31. The LCD is $(t-1)(t+2)$. The first step is to multiply both sides of the equation by the LCD.

$$(t-1)(t+2) \cdot \frac{5}{t-1} + (t-1)(t+2) \cdot \frac{2}{t+2} = \frac{15}{t^2+t-2} \cdot (t-1)(t+2) \Rightarrow$$

$$5t + 10 + 2t - 2 = 15 \Rightarrow 7t + 8 = 15 \Rightarrow 7t = 7 \Rightarrow t = 1$$

Some expressions in the original equation are not defined when $t = 1$. No solutions.

33. The LCD is $6n$. The first step is to multiply both sides of the equation by the LCD.

$$6n \cdot \frac{1}{3n} - 6n \cdot 2 = \frac{1}{2n} \cdot 6n \Rightarrow 2 - 12n = 3 \Rightarrow -12n = 1 \Rightarrow n = -\frac{1}{12}$$

35. The LCD is x^2. The first step is to multiply both sides of the equation by the LCD.

$$x^2 \cdot \left(\frac{1}{x} + \frac{1}{x^2} \right) = 2 \cdot x^2 \Rightarrow x + 1 = 2x^2 \Rightarrow 2x^2 - x - 1 = 0 \Rightarrow (2x+1)(x-1) = 0 \Rightarrow x = -\frac{1}{2} \text{ or } 1$$

37. The LCD is $(x+2)(x-3)$. The first step is to multiply both sides of the equation by the LCD.

$$(x+2)(x-3)\cdot\frac{x}{x+2}=\frac{4}{x-3}\cdot(x+2)(x-3)\Rightarrow x^2-3x=4x+8\Rightarrow x^2-7x-8=0\Rightarrow$$

$$(x+1)(x-8)=0\Rightarrow x=-1 \text{ or } 8$$

39. $\dfrac{2w+1}{3w}-\dfrac{4w-3}{w}=0\Rightarrow 3w\cdot\dfrac{2w+1}{3w}-3w\cdot\dfrac{(4w-3)}{w}=0\cdot 3w\Rightarrow 2w+1-12w+9=0\Rightarrow$

$$-10w+10=0\Rightarrow -10w=-10\Rightarrow w=1$$

41. $\dfrac{3}{2y}+\dfrac{2y}{y-4}=-\dfrac{11}{2}\Rightarrow 2y(y-4)\cdot\dfrac{3}{2y}+2y(y-4)\cdot\dfrac{2y}{y-4}=-\dfrac{11}{2}\cdot 2y(y-4)\Rightarrow$

$$3y-12+4y^2=-11y^2+44y\Rightarrow 15y^2-41y-12=0\Rightarrow (15y+4)(y-3)=0\Rightarrow y=-\frac{4}{15}\text{ or } y=3$$

43. $\dfrac{1}{(x-1)^2}+\dfrac{3}{x^2-1}=\dfrac{5}{x^2-1}\Rightarrow 1(x+1)+3(x-1)=5(x-1)\Rightarrow x+1+3x-3=5x-5\Rightarrow$

$$4x-2=5x-5\Rightarrow x=3$$

45. $\dfrac{x^2+x-2}{x-2}-\dfrac{4}{x-2}=0\Rightarrow \dfrac{x^2+x-2-4}{x-2}=0\Rightarrow \dfrac{x^2+x-6}{x-2}=0$

$$\Rightarrow\frac{(x+3)(x-2)}{x-2}=0\Rightarrow x+3=0\Rightarrow x=-3$$

47. $\dfrac{4}{x+3}-\dfrac{x}{3-x}=\dfrac{18}{x^2-9}\Rightarrow (x-3)(x+3)\cdot\dfrac{4}{x+3}-(x-3)(x+3)\dfrac{-x}{x-3}$

$$=\frac{18}{(x-3)(x+3)}\cdot(x-3)(x+3)\Rightarrow 4(x-3)-(-x)(x+3)=18$$

$$\Rightarrow 4x-12+x^2+3x=18\Rightarrow x^2+7x-30=0\Rightarrow (x+10)(x-3)=0\Rightarrow x=-10$$

($x=3$ makes expressions in the original equation undefined)

49. $\dfrac{1}{x}-4x=0\Rightarrow x\cdot\dfrac{1}{x}-x\cdot 4x=0\cdot x\Rightarrow 1-4x^2=0\Rightarrow (1-2x)(1+2x)=0\Rightarrow x=\dfrac{1}{2}\text{ or } x=-\dfrac{1}{2}$

51. $\dfrac{a}{x}-\dfrac{b}{x}=c\Rightarrow \dfrac{a-b}{x}=c\Rightarrow x\cdot\dfrac{a-b}{x}=c\cdot x\Rightarrow a-b=cx\Rightarrow \dfrac{a-b}{c}=x$

53. $y_1=y_2$ when $x=1$. By substitution of $x=1$, this answer checks.

55. $y_1=y_2$ when $x=2$. By substitution of $x=2$, this answer checks

57. Graph $Y_1=1/(X-1)$ and $Y_2=1/2$ in $[-4.7, 4.7, 1]$ by $[-3.1, 3.1, 1]$. See Figure 57.

The solution is the x-coordinate of the intersection point, $x=3$.

[–4.7, 4.7, 1] by [–3.1, 3.1, 1]

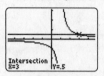

Figure 57

59. Graph $Y_1 = 3/(X+2)$ and $Y_2 = X$ in [–4.7, 4.7, 1] by [–6.2, 6.2, 1]. See Figures 59a & 59b.

The solutions are the x-coordinates of the intersection points, $x = -3$ or $x = 1$.

[–4.7, 4.7, 1] by [–6.2, 6.2, 1]   [–4.7, 4.7, 1] by [–6.2, 6.2, 1]

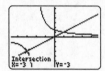   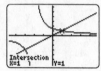

Figure 59a   Figure 59b

61. Graph $Y_1 = 1/(X-2)$ and $Y_2 = 2$ in [–4.7, 4.7, 1] by [–3.1, 3.1, 1]. See Figure 61.

The solution is the x-coordinate of the intersection point, $x = 2.5$.

[–4.7, 4.7, 1] by [–3.1, 3.1, 1]

Figure 61

63. Graph $Y_1 = 1/X + 1/X^2$ and $Y_2 = 15/4$ in [–4.7, 4.7, 1] by [–6.2, 6.2, 1]. See Figures 63a & 63b.

The solutions are the x-coordinates of the intersection points, $x = -0.4$ or $x \approx 0.67$.

[–4.7, 4.7, 1] by [–6.2, 6.2, 1]   [–4.7, 4.7, 1] by [–6.2, 6.2, 1]

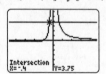   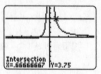

Figure 63a   Figure 63b

65. Graph $Y_1 = 1/(X+2) - 1/(X-2)$ and $Y_2 = 4/3$ in [–4.7, 4.7, 1] by [–3.1, 3.1, 1]. See Figures 65a &

65b. The solutions are the x-coordinates of the intersection points, $x = -1$ or $x = 1$.

[–4.7, 4.7, 1] by [–3.1, 3.1, 1]   [–4.7, 4.7, 1] by [–3.1, 3.1, 1]

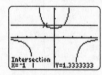   

Figure 65a   Figure 65b

67. Symbolic: The LCD is $3x$. The first step is to multiply both sides of the equation by the LCD.

$$3x \cdot \left(\frac{1}{x} + \frac{2}{3} \right) = 1 \cdot 3x \Rightarrow 3 + 2x = 3x \Rightarrow 3 = x \Rightarrow x = 3$$

Graphical: Graph $Y_1 = 1/X + 2/3$ and $Y_2 = 1$ in $[-4.7, 4.7, 1]$ by $[-1.55, 1.55, 1]$. See Figure 67a.

The solution is the x-coordinate of the intersection point, $x = 3$.

Numerical: Table $Y_1 = 1/X + 2/3$ and $Y_2 = 1$ with TblStart = 0 and ΔTbl = 1. See Figure 67b.

The solution is the x-value when $Y_1 = Y_2$, $x = 3$.

$[-4.7, 4.7, 1]$ by $[-1.55, 1.55, 1]$

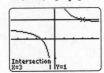   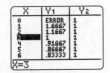

Figure 67a   Figure 67b

69. Symbolic: The LCD is $3x(x+2)$. The first step is to multiply both sides of the equation by the

LCD. $3x(x+2) \cdot \left(\frac{1}{x} + \frac{1}{x+2} \right) = \frac{4}{3} \cdot 3x(x+2) \Rightarrow 3x + 6 + 3x = 4x^2 + 8x \Rightarrow 4x^2 + 2x - 6 = 0 \Rightarrow$

$2(2x+3)(x-1) = 0 \Rightarrow x = -\frac{3}{2}$ or $x = 1$

Graphical: Graph $Y_1 = 1/X + 1/(X+2)$ and $Y_2 = 4/3$ in $[-4.7, 4.7, 1]$ by $[-3.1, 3.1, 1]$.

See Figures 69a & 69b. The solutions are the x-coordinates of the intersection points,

$x = -\frac{3}{2}$ or $x = 1$.

Numerical: Table $Y_1 = 1/X + 1/(X+2)$ and $Y_2 = 4/3$ with TblStart = -2 and ΔTbl = 0.5.

See Figure 69c. The solutions are the x-values when $Y_1 = Y_2$, $x = -\frac{3}{2}$ or $x = 1$.

$[-4.7, 4.7, 1]$ by $[-3.1, 3.1, 1]$ $[-4.7, 4.7, 1]$ by $[-3.1, 3.1, 1]$

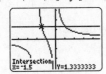   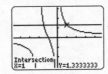   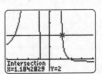

Figure 69a   Figure 69b   Figure 69c

71. Graph $Y_1 = X/(X^2-1) - 3/(X+2)$ and $Y_2 = 2$ in $[-4, 4, 1]$ by $[-1, 4, 1]$. The solutions are the x

coordinates of the intersection points, approximately $x = -3.28$, $x = -0.90$, or $x = 1.18$. See Figures

71a, 71b, and 7c.

   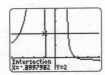   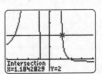

Figure 71a   Figure 71b   Figure 71c

73. $t = \dfrac{d}{r} \Rightarrow rt = d \Rightarrow r = \dfrac{d}{t}$

75. $h = \dfrac{2A}{b} \Rightarrow bh = 2A \Rightarrow b = \dfrac{2A}{h}$

77. $\dfrac{1}{2a} = \dfrac{1}{b} \Rightarrow 2a = b \Rightarrow a = \dfrac{b}{2}$

79. $\dfrac{1}{R} = \dfrac{1}{R_1} + \dfrac{1}{R_2} \Rightarrow R_1 R_2 = RR_2 + RR_1 \Rightarrow R_1 R_2 - RR_1 = RR_2 \Rightarrow R_1(R_2 - R) = RR_2 \Rightarrow R_1 = \dfrac{RR_2}{R_2 - R}$

81. $T = \dfrac{1}{15 - x} \Rightarrow T(15 - x) = 1 \Rightarrow 15 - x = \dfrac{1}{T} \Rightarrow -x = \dfrac{1}{T} - 15 \Rightarrow x = 15 - \dfrac{1}{T}$

83. $\dfrac{1}{r} = \dfrac{1}{t+1} \Rightarrow t + 1 = r \Rightarrow t = r - 1$

85. $\dfrac{1}{r} = \dfrac{a}{a+b} \Rightarrow a + b = ar \Rightarrow b = ar - a \Rightarrow b = a(r-1)$

87. (a) $T = 0.5 \Rightarrow 0.5 = \dfrac{1}{x - 80}$. Graph $Y_1 = 0.5$ and $Y_2 = 1/(X - 80)$ in $[0, 100, 10]$ by $[0, 2, 0.5]$. The

solution is the x-coordinate of the intersection point, $x = 82$ per hour. See Figure 87.

(b)   The wait gets very long.

$[0, 100, 10]$ by $[0, 2, 0.5]$. $[-4.7, 4.7, 1]$ by $[-3.1, 3.1, 1]$

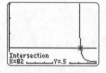

Figure 87   Figure 89

89. (a) $\dfrac{x}{5} + \dfrac{x}{2} = 1$

(b)   The LCD is 10. The first step is to multiply both sides of the equation by the LCD.

$$10 \cdot \left(\dfrac{x}{5} + \dfrac{x}{2} \right) = 1 \cdot 10 \Rightarrow 2x + 5x = 10 \Rightarrow 7x = 10 \Rightarrow x = \dfrac{10}{7} \text{ hours.}$$

(c)   Graph $Y_1 = X/5 + X/2$ and $Y_2 = 1$ in $[-4.7, 4.7, 1]$ by $[-3.1, 3.1, 1]$. See Figure 89.

The solution is the x-coordinate of the intersection point, $x \approx 1.43$ hours.

91. (a) $f(3) = \dfrac{100(3)}{3+1} = \dfrac{300}{4} = 75$. Three testers found 75% of the glitches.

(b) $95 = \dfrac{100x}{x+1} \Rightarrow 100x = 95x + 95 \Rightarrow 5x = 95 \Rightarrow x = 19$ The result is 19 testers.

(c)   As the number of testers becomes large, the percentage of the total number of glitches found

slowly nears 100%, but never actually gets there.

93. (a) $S(36) = \dfrac{500}{36+4} = \dfrac{500}{40} = 12.5$. The number 36 pick receives 12.5% of the number one pick's

salary.

(b) $20 = \dfrac{500}{x+4} \Rightarrow 20x + 80 = 500 \Rightarrow 20x = 420 \Rightarrow x = 21$

95. (a) $\dfrac{x}{40} + \dfrac{x}{70} = 1$

(b) The LCD is 280. The first step is to multiply both sides of the equation by the LCD.

$$280 \cdot \left(\dfrac{x}{40} + \dfrac{x}{70} \right) = 1 \cdot 280 \Rightarrow 7x + 4x = 280 \Rightarrow 11x = 280 \Rightarrow x = \dfrac{280}{11} = 25.\overline{45} \text{ hours.}$$

(c) Graph $Y_1 = X/40 + X/70$ and $Y_2 = 1$ in $[-37.6, 37.6, 4]$ by $[-3.1, 3.1, 1]$. See Figure 95.

The solution is the x-coordinate of the intersection point, $x = 25.\overline{45}$ hours.

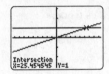

Figure 95

97. Let x represent the speed of the winner. Then $x - 2$ represents the speed of the second-place runner.

Since $t = \dfrac{d}{r}$, the time needed for the winner to finish is $\dfrac{5}{x}$ and the time for the second-place runner

is $\dfrac{5}{x-2}$. Since the winner finished 7.5 minutes $\left(\dfrac{1}{8} \text{ hour} \right)$ ahead of second place, the equation is

$\dfrac{5}{x-2} - \dfrac{5}{x} = \dfrac{1}{8}$. The LCD is $8x(x-2)$. The first step is to multiply both sides of the equation by the

LCD. $8x(x-2) \cdot \left(\dfrac{5}{x-2} - \dfrac{5}{x} \right) = \dfrac{1}{8} \cdot 8x(x-2) \Rightarrow 40x - 40x + 80 = x^2 - 2x \Rightarrow x^2 - 2x - 80 = 0 \Rightarrow$

$(x-10)(x+8) = 0 \Rightarrow x = 10 \text{ or } x = -8$ Since the value $x = -8$ has no meaning, the solutions are

winner: 10 mph, second-place: $10 - 2 = 8$ mph.

99. Let x represent the speed of the current. Then $15 - x$ represents the boat's upstream speed and

$15 + x$ represents the boat's downstream speed. Since $t = \dfrac{d}{r}$, the equation is $\dfrac{36}{15-x} + \dfrac{36}{15+x} = 5$.

The LCD is $(15-x)(15+x)$. The first step is to multiply both sides of the equation by the LCD.

$(15-x)(15+x) \cdot \left(\dfrac{36}{15-x} + \dfrac{36}{15+x} \right) = 5 \cdot (15-x)(15+x) \Rightarrow 540 + 36x + 540 - 36x = 1125 - 5x^2$

$\Rightarrow 5x^2 - 45 = 0 \Rightarrow 5x^2 = 45 \Rightarrow x^2 = 9 \Rightarrow x = \pm\sqrt{9} \Rightarrow x = 3$ mph (the value $x = -3$ has no meaning)

101. Let x represent the plane's speed with no wind. Then $x+50$ represents the plane's speed with the wind and $x-50$ represents the plane's speed into the wind. Since $t = \dfrac{d}{r}$, the equation is

$\dfrac{675}{x+50} = \dfrac{450}{x-50}$. The LCD is $(x+50)(x-50)$. The first step is to multiply both sides of the

equation by the LCD. $(x+50)(x-50) \cdot \dfrac{675}{x+50} = \dfrac{450}{x-50} \cdot (x+50)(x-50) \Rightarrow$

$675x - 33,750 = 450x + 22,500 \Rightarrow 225x = 56,250 \Rightarrow x = 250$ mph

103. Let x and $x+3$ represent the time needed by the faster employee and the slower employee

respectively. $\dfrac{1}{x} + \dfrac{1}{x+3} = \dfrac{1}{2} \Rightarrow 2(x+3) + 2x = x(x+3) \Rightarrow 2x + 6 + 2x = x^2 + 3x \Rightarrow x^2 - x - 6 = 0 \Rightarrow$

$(x-3)(x+2) = 0 \Rightarrow x = 3$ or $x = -2$. The value $x = -2$ has no physical meaning.

The faster employee can mow the field in 3 hours. The slower employee takes 6 hours.

105. Let x and $x-1$ represent the speed of the faster person and the slower person respectively.

$\dfrac{12}{x} = \dfrac{9}{x-1} \Rightarrow 12x - 12 = 9x \Rightarrow 3x = 12 \Rightarrow x = 4$ The faster person walks 4 mph. The slower person

walks 3 mph.

107. Let x represent the time needed to empty the pool.

$\dfrac{x}{40} - \dfrac{x}{60} = 1 \Rightarrow 3x - 2x = 120 \Rightarrow x = 120$ hours. The pool is empty in 120 hours.

109. Yes; the value is slightly greater than 2.

Checking Basic Concepts Sections 6.3 and 6.4

1. (a) $\dfrac{x}{x^2-1} + \dfrac{1}{x^2-1} = \dfrac{x+1}{(x-1)(x+1)} = \dfrac{1}{x-1}$

   (b) $\dfrac{1}{x-2} - \dfrac{3}{x} = \dfrac{x}{x(x-2)} - \dfrac{3(x-2)}{x(x-2)} = \dfrac{x-(3x-6)}{x(x-2)} = \dfrac{-2x+6}{x(x-2)} = \dfrac{-2(x-3)}{x(x-2)} = -\dfrac{2(x-3)}{x(x-2)}$

   (c) $\dfrac{1}{x(x-1)} + \dfrac{1}{x^2-1} - \dfrac{2}{x(x+1)} = \dfrac{1(x+1) + 1x - 2(x-1)}{x(x-1)(x+1)} = \dfrac{3}{x(x-1)(x+1)}$

2. (a) $\dfrac{6}{x} - \dfrac{1}{2} = 1 \Rightarrow 12 - x = 2x \Rightarrow 12 = 3x \Rightarrow x = 4$

   (b) $\dfrac{3}{2x-1} = \dfrac{2}{x+1} \Rightarrow 3(x+1) = 2(2x-1) \Rightarrow 3x + 3 = 4x - 2 \Rightarrow x = 5$

   (c) $\dfrac{2}{x-1} + \dfrac{3}{x+2} = \dfrac{x}{x^2+x-2} \Rightarrow 2(x+2) + 3(x-1) = x \Rightarrow 5x + 1 = x \Rightarrow 4x = -1 \Rightarrow x = -\dfrac{1}{4}$

3. $\dfrac{x}{12} + \dfrac{x}{28} = 1 \Rightarrow 7x + 3x = 84 \Rightarrow 10x = 84 \Rightarrow x = 8.4 \text{ min}$

4. $\dfrac{1}{S} = \dfrac{1}{F} - \dfrac{1}{D} \Rightarrow DF = SD - SF \Rightarrow DF = S(D - F) \Rightarrow \dfrac{DF}{D - F} = S \Rightarrow S = \dfrac{DF}{D - F}$

Section 6.5 Complex Fractions

1. $\dfrac{5}{7} \cdot \dfrac{11}{3} = \dfrac{55}{21}$

3. Multiply both the numerator and the denominator by $x - 1$. *Answers may vary.*

5. $\dfrac{z + \dfrac{3}{4}}{z - \dfrac{3}{4}}$

7. $\dfrac{\dfrac{1}{5}}{\dfrac{4}{7}} = \dfrac{1}{5} \cdot \dfrac{7}{4} = \dfrac{7}{20}$

9. $\dfrac{1 + \dfrac{1}{3}}{1 - \dfrac{1}{3}} = \dfrac{1 + \dfrac{1}{3}}{1 - \dfrac{1}{3}} \cdot \dfrac{3}{3} = \dfrac{3 + 1}{3 - 1} = \dfrac{4}{2} = 2$

11. $\dfrac{2 + \dfrac{2}{3}}{2 - \dfrac{1}{4}} = \dfrac{2 + \dfrac{2}{3}}{2 - \dfrac{1}{4}} \cdot \dfrac{12}{12} = \dfrac{24 + 8}{24 - 3} = \dfrac{32}{21}$

13. $\dfrac{\dfrac{a}{b}}{\dfrac{3a}{2b^2}} = \dfrac{a}{b} \cdot \dfrac{2b^2}{3a} = \dfrac{2b}{3}$

15. $\dfrac{\dfrac{x}{2y}}{\dfrac{2x}{3y}} = \dfrac{x}{2y} \cdot \dfrac{3y}{2x} = \dfrac{3}{4}$

17. $\dfrac{\dfrac{8}{n+1}}{\dfrac{4}{n-1}} = \dfrac{8}{n+1} \cdot \dfrac{n-1}{4} = \dfrac{2(n-1)}{n+1}$

19. $\dfrac{\dfrac{2k+3}{k}}{\dfrac{k-4}{k}} = \dfrac{2k+3}{k} \cdot \dfrac{k}{k-4} = \dfrac{2k+3}{k-4}$

21. $\dfrac{\dfrac{3}{z^2-4}}{\dfrac{z}{z^2-4}} = \dfrac{3}{z^2-4}\cdot\dfrac{z^2-4}{z} = \dfrac{3}{z}$

23. $\dfrac{\dfrac{x}{x^2-16}}{\dfrac{1}{x-4}} = \dfrac{x}{x^2-16}\cdot\dfrac{x-4}{1} = \dfrac{x(x-4)}{x^2-16} = \dfrac{x(x-4)}{(x-4)(x+4)} = \dfrac{x}{x+4}$

25. $\dfrac{1+\dfrac{1}{x}}{x+1} = \dfrac{1+\dfrac{1}{x}}{x+1}\cdot\dfrac{x}{x} = \dfrac{x+1}{x(x+1)} = \dfrac{1}{x}$

27. $\dfrac{\dfrac{1}{x-3}}{\dfrac{1}{x}-\dfrac{3}{x-3}} = \dfrac{\dfrac{1}{x-3}}{\dfrac{1}{x}-\dfrac{3}{x-3}}\cdot\dfrac{x(x-3)}{x(x-3)} = \dfrac{x}{x-3-3x} = \dfrac{x}{-2x-3} = -\dfrac{x}{2x+3}$

29. $\dfrac{\dfrac{1}{x}+\dfrac{2}{x^2}}{\dfrac{3}{x}-\dfrac{1}{x^2}} = \dfrac{\dfrac{1}{x}+\dfrac{2}{x^2}}{\dfrac{3}{x}-\dfrac{1}{x^2}}\cdot\dfrac{x^2}{x^2} = \dfrac{x+2}{3x-1}$

31. $\dfrac{\dfrac{1}{x+3}+\dfrac{2}{x-3}}{2-\dfrac{1}{x-3}} = \dfrac{\dfrac{1}{x+3}+\dfrac{2}{x-3}}{2-\dfrac{1}{x-3}}\cdot\dfrac{(x-3)(x+3)}{(x-3)(x+3)} = \dfrac{x-3+2(x+3)}{2(x-3)(x+3)-(x+3)} = \dfrac{3(x+1)}{(x+3)(2x-7)}$

33. $\dfrac{\dfrac{4}{x-5}}{\dfrac{1}{x+5}+\dfrac{1}{x}} = \dfrac{\dfrac{4}{x-5}}{\dfrac{1}{x+5}+\dfrac{1}{x}}\cdot\dfrac{x(x-5)(x+5)}{x(x-5)(x+5)} = \dfrac{4x(x+5)}{x(x-5)+(x-5)(x+5)} = \dfrac{4x(x+5)}{(x-5)(2x+5)}$

35. $\dfrac{\dfrac{1}{p^2q}+\dfrac{1}{pq^2}}{\dfrac{1}{p^2q}-\dfrac{1}{pq^2}} = \dfrac{\dfrac{1}{p^2q}+\dfrac{1}{pq^2}}{\dfrac{1}{p^2q}-\dfrac{1}{pq^2}}\cdot\dfrac{p^2q^2}{p^2q^2} = \dfrac{q+p}{q-p} = \dfrac{p+q}{q-p}$

37. $\dfrac{\dfrac{1}{a}+\dfrac{1}{b}}{\dfrac{1}{b}-\dfrac{1}{a}} = \dfrac{\dfrac{1}{a}+\dfrac{1}{b}}{\dfrac{1}{b}-\dfrac{1}{a}}\cdot\dfrac{ab}{ab} = \dfrac{b+a}{a-b} = \dfrac{a+b}{a-b}$

39. $\dfrac{\dfrac{1}{x}+\dfrac{1}{x+1}}{\dfrac{2}{x+1}-\dfrac{1}{x+1}} = \dfrac{\dfrac{1}{x}+\dfrac{1}{x+1}}{\dfrac{1}{x+1}} = \dfrac{\dfrac{1}{x}+\dfrac{1}{x+1}}{\dfrac{1}{x+1}}\cdot\dfrac{x(x+1)}{x(x+1)} = \dfrac{x+1+x}{x} = \dfrac{2x+1}{x}$

41. $\dfrac{3^{-1}-4^{-1}}{5^{-1}+4^{-1}} = \dfrac{\dfrac{1}{3}-\dfrac{1}{4}}{\dfrac{1}{5}+\dfrac{1}{4}} = \dfrac{\dfrac{4}{12}-\dfrac{3}{12}}{\dfrac{4}{20}+\dfrac{5}{20}} = \dfrac{\dfrac{1}{12}}{\dfrac{9}{20}} = \dfrac{1}{12}\cdot\dfrac{20}{9} = \dfrac{5}{27}$

43. $\dfrac{m^{-1}-2n^{-2}}{1+(mn)^{-2}} = \dfrac{\dfrac{1}{m}-\dfrac{2}{n^2}}{1+\dfrac{1}{m^2 n^2}} = \dfrac{\dfrac{1}{m}-\dfrac{2}{n^2}}{1+\dfrac{1}{m^2 n^2}} \cdot \dfrac{m^2 n^2}{m^2 n^2} = \dfrac{mn^2-2m^2}{m^2 n^2+1}$

45. $\dfrac{1-(2n+1)^{-1}}{1+(2n+1)^{-1}} = \dfrac{1-\dfrac{1}{2n+1}}{1+\dfrac{1}{2n+1}} = \dfrac{1-\dfrac{1}{2n+1}}{1+\dfrac{1}{2n+1}} \cdot \dfrac{2n+1}{2n+1} = \dfrac{2n+1-1}{2n+1+1} = \dfrac{2n}{2(n+1)} = \dfrac{n}{n+1}$

47. $\dfrac{\dfrac{x}{x^2-4}-\dfrac{1}{x^2-4}}{\dfrac{1}{x+4}} = \dfrac{\dfrac{x-1}{x^2-4}}{\dfrac{1}{x+4}} = \dfrac{x-1}{x^2-4}\cdot\dfrac{x+4}{1} = \dfrac{(x-1)(x+4)}{x^2-4}$

49. $\dfrac{P\left(1+\dfrac{r}{12}\right)^{24}-P}{\dfrac{r}{12}}$

51. $R = \dfrac{1}{\dfrac{1}{R_1}+\dfrac{1}{R_2}} = \dfrac{1}{\dfrac{1}{R_1}+\dfrac{1}{R_2}} \cdot \dfrac{R_1 R_2}{R_1 R_2} = \dfrac{R_1 R_2}{R_2+R_1} \Rightarrow R = \dfrac{R_1 R_2}{R_1+R_2}$

53. (a) $T(11) = \dfrac{10}{\dfrac{1}{10}(11)+\dfrac{9}{10}} = \dfrac{10}{\dfrac{11+9}{10}} = \dfrac{10}{\dfrac{20}{10}} = \dfrac{100}{20} = 5$. After 11 practice trials this task can be

completed in 5 seconds.

(b) $T(x) = \dfrac{10}{\dfrac{1}{10}x+\dfrac{9}{10}} = \dfrac{10\cdot 10}{10\left(\dfrac{1}{10}x+\dfrac{9}{10}\right)} = \dfrac{100}{x+9}$

Section 6.6 Modeling with Proportions and Variation

1. A proportion is a statement that two ratios are equal.

3. It doubles.

5. constant

7. *kxy*

9. It would vary directly. If the number of people doubled, the food bill would double.

11. $\dfrac{x}{14} = \dfrac{5}{7} \Rightarrow 7x = 70 \Rightarrow x = \dfrac{70}{10} = 10$

13. $\dfrac{8}{x} = \dfrac{2}{3} \Rightarrow 2x = 24 \Rightarrow x = \dfrac{24}{2} = 12$

15. $\dfrac{6}{13} = \dfrac{h}{156} \Rightarrow 13h = 936 \Rightarrow h = \dfrac{936}{13} = 72$

17. $\dfrac{7}{4z} = \dfrac{5}{3} \Rightarrow 20z = 21 \Rightarrow z = \dfrac{21}{20}$

19. $\dfrac{2}{3x+1} = \dfrac{5}{x} \Rightarrow 2x = 5(3x+1) \Rightarrow 2x = 15x+5 \Rightarrow -13x = 5 \Rightarrow x = -\dfrac{5}{13}$

21. $\dfrac{4}{x} = \dfrac{x}{9} \Rightarrow 36 = x^2 \Rightarrow x^2 - 36 = 0 \Rightarrow (x-6)(x+6) = 0 \Rightarrow x = 6$ or $x = -6$

23. (a) $\dfrac{7}{9} = \dfrac{10}{x}$

    (b) $\dfrac{7}{9} = \dfrac{10}{x} \Rightarrow 7x = 90 \Rightarrow x = \dfrac{90}{7}$

25. (a) $\dfrac{5}{3} = \dfrac{x}{6}$

    (b) $\dfrac{5}{3} = \dfrac{x}{6} \Rightarrow 3x = 30 \Rightarrow x = \dfrac{30}{3} = 10$

27. (a) $\dfrac{78}{6} = \dfrac{x}{8}$

    (b) $\dfrac{78}{6} = \dfrac{x}{8} \Rightarrow 6x = 624 \Rightarrow x = \dfrac{624}{6} = \$104$

29. (a) $\dfrac{2}{120} = \dfrac{5}{x}$

    (b) $\dfrac{2}{120} = \dfrac{5}{x} \Rightarrow 2x = 600 \Rightarrow x = \dfrac{600}{2} = 300$ min

31. (a) $y = kx \Rightarrow 6 = 3k \Rightarrow k = 2$

    (b) $y = 2x \Rightarrow y = 2(7) = 14$

33. (a) $y = kx \Rightarrow 5 = 2k \Rightarrow k = 2.5$

    (b) $y = 2.5x \Rightarrow y = 2.5(7) \Rightarrow 17.5$

35. (a) $y = kx \Rightarrow -120 = 16k \Rightarrow k = -7.5$

    (b) $y = -7.5x \Rightarrow y = -7.5(7) = -52.5$

37. (a) $y = \dfrac{k}{x} \Rightarrow 5 = \dfrac{k}{4} \Rightarrow k = 20$

    (b) $y = \dfrac{20}{x} \Rightarrow y = \dfrac{20}{10} = 2$

39. (a) $y = \dfrac{k}{x} \Rightarrow 100 = \dfrac{k}{\frac{1}{2}} \Rightarrow k = 50$

    (b) $y = \dfrac{50}{x} \Rightarrow y = \dfrac{50}{10} = 5$

41. (a) $y = \dfrac{k}{x} \Rightarrow 20 = \dfrac{k}{20} \Rightarrow k = 400$

    (b) $y = \dfrac{400}{x} \Rightarrow y = \dfrac{400}{10} = 40$

43. (a) $z = kxy \Rightarrow 6 = 3(8)k \Rightarrow 6 = 24k \Rightarrow k = 0.25$

    (b) $z = 0.25xy \Rightarrow z = 0.25(5)(7) = 8.75$

45. (a) $z = kxy \Rightarrow 5775 = 25(21)k \Rightarrow 5775 = 525k \Rightarrow k = 11$

    (b) $z = 11xy \Rightarrow z = 11(5)(7) = 385$

47. (a) $z = kxy \Rightarrow 25 = \dfrac{1}{2}(5)k \Rightarrow 25 = \dfrac{5}{2}k \Rightarrow k = 10$

    (b) $z = 10xy \Rightarrow z = 10(5)(7) = 350$

49. (a)   Direct. The ratios $\dfrac{y}{x}$ always equal 1.5.

    (b) $y = kx \Rightarrow 3 = 2k \Rightarrow k = 1.5;$ The equation is $y = 1.5x.$

    (c)   See Figure 49.

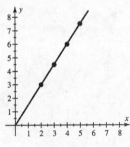

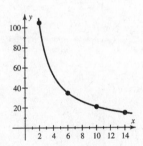

Figure 49   Figure 53

51. (a)   Neither. The products xy do not remain constant.

    (b)   N/A

    (c)   N/A

53. (a)   Inverse. The products xy always equal 210.

    (b) $y = \dfrac{k}{x} \Rightarrow 105 = \dfrac{k}{2} \Rightarrow k = 210;$ The equation is $y = \dfrac{210}{x}.$

    (c)   See Figure 53.

55. (a)   Direct. The ratios $\dfrac{y}{x}$ always equal $-2b.$

    (b) $y = kx \Rightarrow -2b = 1k \Rightarrow k = -2b.$ The equation is $y = -2bx.$

57. Direct. The ratios $\dfrac{y}{x}$ always equal 1. $y = kx \Rightarrow 1 = 1k \Rightarrow k = 1$

59. Neither. The ratios $\dfrac{y}{x}$ do not remain constant.

61. Direct. The ratios $\dfrac{y}{x}$ always equal 2. $\quad y = kx \Rightarrow 2 = 1k \Rightarrow k = 2$

63. Example: $\dfrac{x}{307,000,000} = \dfrac{1}{5} \Rightarrow \; \Rightarrow 5x = 307,000,000 = 61,400,000$ *Answers may vary.*

65. $\dfrac{6}{7} = \dfrac{x}{27} \Rightarrow 7x = 162 \Rightarrow x = \dfrac{162}{7} \approx 23.1$ feet

67. $\dfrac{8}{1} = \dfrac{11}{x} \Rightarrow 8x = 11 \Rightarrow x = \dfrac{11}{8} = 1.375$ inches

69. $\dfrac{618}{2131} = \dfrac{x}{231 \text{ million}} \Rightarrow 2131x = 142,578 \Rightarrow x \approx 67$ million

71. Let x be the total number of largemouth bass in the population.

$\dfrac{300}{x} = \dfrac{17}{112} \Rightarrow 17x = 33,600 \Rightarrow x \approx 2000$

73. $d = kx^2; \quad 300 = k(60)^2 \Rightarrow 300 = 3600k \Rightarrow k = \dfrac{1}{12}; \quad d = \dfrac{1}{12}x^2, \, x = 30 \Rightarrow d = \dfrac{1}{12}(30)^2 = 75$ feet

75. (a) Direct; The ratios $\dfrac{R}{W}$ always equal 0.012.

    (b) $R = 0.012W$. See Figure 75.

    (c) $R = 0.012(3200) = 38.4$ lb

77. (a) Direct; The ratios $\dfrac{D}{W}$ always equal 0.75.

    (b) $D = 0.75W$

    (c) $D = 0.75(11) = 8.25$ inches

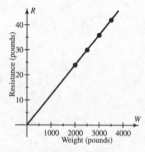

Figure 75

79. (a) $F = \dfrac{k}{L} \Rightarrow 150 = \dfrac{k}{8} \Rightarrow k = 1200$ and so $F = \dfrac{1200}{L}$

    (b) $F = \dfrac{1200}{20} = 60$ lb

81. $\dfrac{6}{435} = \dfrac{11}{x} \Rightarrow 6x = 4785 \Rightarrow \$797.50$;   In this problem k represents the tuition cost per credit.

83. $y = kx \Rightarrow 800 = 1800k \Rightarrow k = \dfrac{4}{9}$ and so $y = \dfrac{4}{9} \cdot 2700 = 1200$ lb per square inch

85. (a) Direct

    (b) $y = kx \Rightarrow -95 = 5k \Rightarrow k = -19$ and so $y = -19x$

    (c) It is negative. For each one mile increase in altitude the temperature decreases by 19°F.

    (d) $y = -19(3.5) = -66.5$; The temperature decreases by 66.5°F.

87. $\dfrac{35}{2} = \dfrac{25}{x} \Rightarrow 35x = 50 \Rightarrow x \approx 1.43$ ohms

89. $z = kx^2 y^3 \Rightarrow 31.9 = (2)^2 (2.5)^3 k \Rightarrow 31.9 = 62.5k \Rightarrow k = 0.5104$ and so $z = 0.5104x^2 y^3$

91. $S = kwt^2 \Rightarrow 300 = k(5)(3)^2 \Rightarrow 300 = 45k \Rightarrow k = \dfrac{20}{3}$ and so $S = \dfrac{20}{3} wt^2$.

    When $w = 5$ and $t = 2$, $S = \dfrac{20}{3}(5)(2)^2 \approx 133$ lb.

93. $W_m = kW_E \Rightarrow 28 = 175k \Rightarrow k = 0.16$; so $W_m = 0.16W_E$

    A 220-pound person would weigh $W_m = 0.16(220) = 35.2$ pounds on the moon.

95. $V = kIR \Rightarrow 220 = (10)(22)k \Rightarrow k = 1$; so $V = IR$

    When $I = 15$ and $R = 50$ the voltage is $V = (15)(50) = 750$.

Checking Basic Concepts Sections 6.5 and 6.6

1. (a) $\dfrac{3 - \dfrac{1}{x^2}}{3 + \dfrac{1}{x^2}} = \dfrac{3 - \dfrac{1}{x^2}}{3 + \dfrac{1}{x^2}} \cdot \dfrac{x^2}{x^2} = \dfrac{3x^2 - 1}{3x^2 + 1}$

    (b) $\dfrac{\dfrac{2}{x-1} - \dfrac{2}{x+1}}{\dfrac{4}{x^2-1}} = \dfrac{\dfrac{2}{x-1} - \dfrac{2}{x+1}}{\dfrac{4}{x^2-1}} \cdot \dfrac{(x-1)(x+1)}{(x-1)(x+1)} = \dfrac{2(x+1) - 2(x-1)}{4} = \dfrac{4}{4} = 1$

2. (a) $y = kx \Rightarrow 6 = 8k \Rightarrow k = 0.75$

    (b) $y = 0.75x \Rightarrow y = 0.75(11) = 8.25$

3. (a) Direct. The ratios $\dfrac{y}{x}$ always equal 0.4. The equation is $y = 0.4x$.

    (b) Inverse. The products xy always equal 120. The equation is $y = \dfrac{120}{x}$.

Section 6.7 Division of Polynomials

1. term

3. $5 \cdot 4 + 1$

5. No, the divisor is not of the form $x - k$.

7. $\dfrac{4x - 6}{2} = \dfrac{4x}{2} - \dfrac{6}{2} = 2x - 3$

   Table $Y_1 = (4X - 6)/2$ and $Y_2 = 2X - 3$ with TblStart = 0 and ΔTbl = 1. See Figure 7.

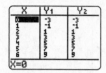 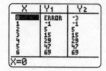 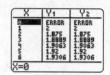

       Figure 7   Figure 9   Figure 11

9. $\dfrac{6x^3 - 9x}{3x} = \dfrac{6x^3}{3x} - \dfrac{9x}{3x} = 2x^2 - 3$

   Table $Y_1 = (6X^3 - 9X)/(3X)$ and $Y_2 = 2X^2 - 3$ with TblStart = 0 and ΔTbl = 1. See Figure 9.

11. $\left(4x^2 - x + 1\right) \div 2x^2 = \dfrac{4x^2}{2x^2} - \dfrac{x}{2x^2} + \dfrac{1}{2x^2} = 2 - \dfrac{1}{2x} + \dfrac{1}{2x^2}$

   Table $Y_1 = \left(4X^2 - X + 1\right)/\left(2X^2\right)$ and $Y_2 = 2 - 1/(2X) + 1/\left(2X^2\right)$ with TblStart = 0 and ΔTbl = 1.

   See Figure 11.

13. $\dfrac{9x^2 - 12x - 3}{3} = \dfrac{9x^2}{3} - \dfrac{12x}{3} - \dfrac{3}{3} = 3x^2 - 4x - 1$

15. $\dfrac{12a^3 - 18a}{6a} = \dfrac{12a^3}{6a} - \dfrac{18a}{6a} = 2a^2 - 3$

17. $\left(16x^3 - 24x\right) \div (12x) = \dfrac{16x^3}{12x} - \dfrac{24x}{12x} = \dfrac{4}{3}x^2 - 2$

19. $\left(a^2b^2 - 4ab + ab^2\right) \div ab = \dfrac{a^2b^2}{ab} - \dfrac{4ab}{ab} + \dfrac{ab^2}{ab} = ab - 4 + b$

21. $\dfrac{6m^4n^4 + 3m^2n^2 - 12}{3m^2n^2} = \dfrac{6m^4n^4}{3m^2n^2} + \dfrac{3m^2n^2}{3m^2n^2} - \dfrac{12}{3m^2n^2} = 2m^2n^2 + 1 - \dfrac{4}{m^2n^2}$

23. $3x - 7$ The solution is: $3x - 7$

$$x - 3 \overline{) 3x^2 - 16x + 21}$$

          $\underline{3x^2 - 9x}$

               $-7x + 21$

               $\underline{-7x + 21}$

                  0

25.

$$\begin{array}{r} x^2 - 1 \\ 2x+3 \overline{)\,2x^3 + 3x^2 - 2x - 2} \\ \underline{2x^3 + 3x^2} \\ -2x - 2 \\ \underline{-2x - 3} \\ 1 \end{array}$$

The solution is: $x^2 - 1 + \dfrac{1}{2x+3}$

27.

$$\begin{array}{r} 10x + 10 \\ x-1 \overline{)\,10x^2 + 0x - 5} \\ \underline{10x^2 - 10x} \\ 10x - 5 \\ \underline{10x - 10} \\ 5 \end{array}$$

The solution is: $10x + 10 + \dfrac{5}{x-1}$

29.

$$\begin{array}{r} 4x^2 - 1 \\ x+2 \overline{)\,4x^3 + 8x^2 - x - 2} \\ \underline{4x^3 + 8x^2} \\ 0 - x - 2 \\ \underline{-x - 2} \\ 0 \end{array}$$

The solution is: $4x^2 - 1$

31.

$$\begin{array}{r} x^2 - 4x + 19 \\ x+4 \overline{)\,x^3 + 0x^2 + 3x - 4} \\ \underline{x^3 + 4x^2} \\ -4x^2 + 3x \\ \underline{-4x^2 - 16x} \\ 19x - 4 \\ \underline{19x + 76} \\ -80 \end{array}$$

The solution is: $x^2 - 4x + 19 - \dfrac{80}{x+4}$

33.

$$\begin{array}{r} x^2+3x-6 \\ 3x-1\overline{)3x^3+8x^2-21x+7} \\ \underline{3x^3-x^2} \\ 9x^2-21x \\ \underline{9x^2-\ 3x} \\ -18x+7 \\ \underline{-18x+6} \\ 1 \end{array}$$

The solution is: $x^2+3x-6+\dfrac{1}{3x-1}$

35.

$$\begin{array}{r} a^3-a \\ 2a+5\overline{)2a^4+5a^3-2a^2-5a} \\ \underline{2a^4+5a^3} \\ 0-2a^2-5a \\ \underline{-2a^2-5a} \\ 0 \end{array}$$

The solution is: a^3-a

37.

$$\begin{array}{r} 3x+4 \\ x^2+0x-4\overline{)3x^3+4x^2-12x-16} \\ \underline{3x^3+0x^2-12x} \\ 4x^2+0x-16 \\ \underline{4x^2+0x-16} \\ 0 \end{array}$$

The solution is: $3x+4$

39.

$$\begin{array}{r} x+1 \\ x^2+0x-1\overline{)x^3+x^2-x+0} \\ \underline{x^3+0x^2-x} \\ x^2+0x+0 \\ \underline{x^2+0x-1} \\ 1 \end{array}$$

The solution is: $x+1+\dfrac{1}{x^2-1}$

41.

$$\begin{array}{r} 2a^2-a+3 \\ a^2-a+5\overline{)2a^4-3a^3+14a^2-8a+10} \\ \underline{2a^4-2a^3+10a^2} \\ -a^3+4a^2-8a+10 \\ \underline{-a^3+a^2-5a} \\ 3a^2-3a+10 \\ \underline{3a^2-3a+15} \\ -5 \end{array}$$

The solution is: $2a^2-a+3-\dfrac{5}{a^2-a+5}$

43. $\left(a^2+ab+b^2\right)(a-b)+1 = a^3+a^2b+ab^2-a^2b-ab^2-b^3+1 = a^3-b^3+1$

45. $\begin{array}{r} 1\,\underline{|} \quad\ 1\ \ \ 3\ \ -1 \\ \phantom{1\underline{|}}\ \ 1\ \ \ \ 4 \\ \hline \phantom{1\underline{|}}\ 1\ \ \ 4\ \ \ \ 3 \end{array}$ The solution is: $x+4+\dfrac{3}{x-1}$

47. $\begin{array}{r} 7\,\underline{|} \quad\ 3\ \ -22\ \ \ 7 \\ \phantom{7\underline{|}}\ \ 21\ \ -7 \\ \hline \phantom{7\underline{|}}\ 3\ \ \ -1\ \ \ \ 0 \end{array}$ The solution is: $3x-1$

49. $\begin{array}{r} -4\,\underline{|} \quad\ 1\ \ \ 7\ \ \ 14\ \ \ 8 \\ \phantom{-4\underline{|}}\ -4\ -12\ -8 \\ \hline \phantom{-4\underline{|}}\ 1\ \ \ 3\ \ \ \ 2\ \ \ \ 0 \end{array}$ The solution is: x^2+3x+2 .

51. $\begin{array}{r} 2\,\underline{|} \quad\ 2\ \ \ 1\ \ \ 0\ \ -1 \\ \phantom{2\underline{|}}\ \ 4\ \ 10\ \ 20 \\ \hline \phantom{2\underline{|}}\ 2\ \ \ 5\ \ 10\ \ 19 \end{array}$ The solution is: $2x^2+5x+10+\dfrac{19}{x-2}$.

53. $\begin{array}{r} -2\,\underline{|} \quad\ 2\ \ \ 0\ \ \ 3\ \ \ 0\ \ -4 \\ \phantom{-2\underline{|}}\ -4\ \ \ 8\ -22\ \ 44 \\ \hline \phantom{-2\underline{|}}\ 2\ -4\ \ 11\ -22\ \ 40 \end{array}$ The solution is: $2x^3-4x^2+11x-22+\dfrac{40}{x+2}$.

55. $\begin{array}{r} 1\,\underline{|} \quad\ 1\ \ \ 0\ \ \ 0\ \ \ 0\ \ -1 \\ \phantom{1\underline{|}}\ \ 1\ \ \ 1\ \ \ 1\ \ \ 1 \\ \hline \phantom{1\underline{|}}\ 1\ \ \ 1\ \ \ 1\ \ \ 1\ \ \ \ 0 \end{array}$ The solution is: b^3+b^2+b+1 .

57. Use polynomial division to find $\left(6x^2+7x-5\right)\div(2x-1)=3x+5$. That is, $L=3x+5$.

     When $x=8, L=3(8)+5=29$ feet.

59. The volume of a box is $V=lwh$, so $h=\dfrac{V}{lw}$. Divide the volume by the length times the width.

     Use polynomial division to find $\left(x^3+7x^2+14x+8\right)\div\left(x^2+5x+4\right)=x+2$. That is, $h=x+2$.

61. (a) By synthetic division the remainder for $\left(x^2-3x+2\right)\div(x-1)$ is 0. $p(1)=(1)^2-3(1)+2=0$

     (b) By synthetic division the remainder for $\left(3x^2+5x-2\right)\div(x+2)$ is 0.

          $p(-2)=3(-2)^2+5(-2)-2=0$

     (c) By synthetic division the remainder for $\left(3x^3-4x^2-5x+3\right)\div(x-2)$ is 1.

          $p(2)=3(2)^3-4(2)^2-5(2)+3=1$

     (d) By synthetic division the remainder for $\left(x^3-2x^2-x+2\right)\div(x+1)$ is 0.

          $p(-1)=(-1)^3-2(-1)^2-(-1)+2=0$

     (e) By synthetic division the remainder for $\left(x^4-5x^2-1\right)\div(x-3)$ is 35.

          $p(3)=(3)^4-5(3)^2-1=35$

63. (a) $p(2) = (2)^2 + (2) - 6 = 0$

Yes, it is a factor because by synthetic division $(x^2 + x - 6) \div (x - 2) = x + 3$.

(b) $p(1) = (1)^2 + 4(1) - 5 = 0$

Yes, it is a factor because by synthetic division $(x^2 + 4x - 5) \div (x - 1) = x + 5$.

(c) $p(-2) = (-2)^2 + 8(-2) + 11 = -1$ No, it is not a factor because by synthetic division

$(x^2 + 8x + 11) \div (x + 2) = x + 6$ Remainder -1.

(d) $p(-1) = (-1)^3 + (-1)^2 + (-1) + 1 = 0$

Yes, it is a factor because by synthetic division $(x^3 + x^2 + x + 1) \div (x + 1) = x^2 + 1$.

(e) $p(2) = (2)^3 - 3(2)^2 - (2) - 3 = -9$

No, it is not a factor because $(x^3 - 3x^2 - x - 3) \div (x - 2) = x^2 - x - 3$ Remainder -9.

Checking Basic Concepts Section 6.7

1. (a) $\dfrac{2x - x^2}{x^2} = \dfrac{2x}{x^2} - \dfrac{x^2}{x^2} = \dfrac{2}{x} - 1$

(b) $\dfrac{6a^2 - 9a + 15}{3a} = \dfrac{6a^2}{3a} - \dfrac{9a}{3a} + \dfrac{15}{3a} = 2a - 3 + \dfrac{5}{a}$

2.
$$\begin{array}{r}
5x - 3 \\
2x+1 \overline{)10x^2 - x + 4} \\
\underline{10x^2 + 5x} \\
-6x + 4 \\
\underline{-6x - 3} \\
7
\end{array}$$

The solution is: $5x - 3 + \dfrac{7}{2x+1}$

3. (a)
$$\begin{array}{r}
1 \rvert \quad 2 \quad -5 \quad 0 \quad -1 \\
 2 \quad -3 \quad -3 \\
\hline
2 \quad -3 \quad -3 \quad -4
\end{array}$$
The solution is: $2x^2 - 3x - 3 - \dfrac{4}{x-1}$.

(b)
$$\begin{array}{r}
-2 \rvert \quad 2 \quad -5 \quad 0 \quad -1 \\
 -4 \quad 18 \quad -36 \\
\hline
2 \quad -9 \quad 18 \quad -37
\end{array}$$
The solution is: $2x^2 - 9x + 18 - \dfrac{37}{x+2}$

Chapter 6 Review

1. Symbolic: $f(x) = \dfrac{1}{x-1}$

2. Symbolic: $f(x) = \dfrac{x-3}{x}$

3. See Figure 3. Because the denominator cannot be equal to 0, the domain is $\{x \mid x \neq -2\}$.

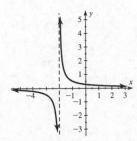

Figure 3

4. $f(3) = \dfrac{1}{(3)^2 - 1} = \dfrac{1}{9-1} = \dfrac{1}{8}$; The values $x = \pm 1$ are not in the domain of f.

5. $f(-3) = \dfrac{3}{(-3)+2} = \dfrac{3}{-1} = -3$ and $f(2) = \dfrac{3}{(2)+2} = \dfrac{3}{4} = 0.75$

6. $f(-2) = \dfrac{2(-2)}{(-2)^2 - 4} = \dfrac{-4}{0}$ is undefined and $f(3) = \dfrac{2(3)}{(3)^2 - 4} = \dfrac{6}{5} = 1.2$

7. $f(0) = 3$ and $f(2) = 1$; The equation of the vertical asymptote is $x = 1$.

8. $f(-3) = 2$ and $f(-2)$ is undefined; The equation of the vertical asymptote is $x = -2$.

9. The LCD is x. The first step is to multiply both sides of the equation by the LCD.

$$x \cdot \left(\dfrac{3-x}{x}\right) = 2 \cdot x \Rightarrow 3 - x = 2x \Rightarrow 3 = 3x \Rightarrow x = 1$$

10. The LCD is $5(x+3)$. The first step is to multiply both sides of the equation by the LCD.

$$5(x+3) \cdot \left(\dfrac{1}{x+3}\right) = \dfrac{1}{5} \cdot 5(x+3) \Rightarrow 5 = x + 3 \Rightarrow x = 2$$

11. The LCD is $x-2$. The first step is to multiply both sides of the equation by the LCD.

$$(x-2) \cdot \left(\dfrac{x}{x-2}\right) = (x-2) \cdot \dfrac{2x-2}{x-2} \Rightarrow x = 2x - 2 \Rightarrow x = 2$$

Some expressions in the original equation are not defined when $x = 2$. No solutions.

12. The LCD is $2 - 3x$. The first step is to multiply both sides of the equation by the LCD.

$$(2-3x) \cdot \left(\dfrac{4}{2-3x}\right) = -1 \cdot (2-3x) \Rightarrow 4 = 3x - 2 \Rightarrow 3x = 6 \Rightarrow x = 2$$

13. (a) $(f+g)(3) = \left(2(3)^2 - 3(3)\right) + (2(3)-3) = (2(9)-9) + (6-3) = (18-9) + 3 = 9 + 3 = 12$

   (b) $(fg)(3) = \left(2(3)^2 - 3(3)\right)(2(3)-3) = (2(9)-9)(6-3) = (18-9)(3) = 9(3) = 27$

14. (a) $(f-g)(x) = \left(x^2-1\right) - (x-1) = x^2 - x$

   (b) $(f/g)(x) = \dfrac{x^2-1}{x-1} = \dfrac{(x-1)(x+1)}{x-1} = x+1$

15. $\dfrac{4a}{6a^4} = \dfrac{2}{3a^3}$

16. $\dfrac{(x-3)(x+2)}{(x+1)(x-3)} = \dfrac{x+2}{x+1}$

17. $\dfrac{x^2-4}{x-2} = \dfrac{(x-2)(x+2)}{x-2} = x+2$

18. $\dfrac{x^2-6x-7}{2x^2-x-3} = \dfrac{(x-7)(x+1)}{(2x-3)(x+1)} = \dfrac{x-7}{2x-3}$

19. $\dfrac{8-x}{x-8} = \dfrac{-1(x-8)}{x-8} = -1$

20. $-\dfrac{3-2x}{2x-3} = \dfrac{-1(3-2x)}{2x-3} = \dfrac{2x-3}{2x-3} = 1$

21. $\dfrac{1}{2y} \cdot \dfrac{4y^2}{8} = \dfrac{1}{2} \cdot \dfrac{y}{2} = \dfrac{y}{4}$

22. $\dfrac{x+2}{x-5} \cdot \dfrac{x-5}{x+1} = \dfrac{x+2}{1} \cdot \dfrac{1}{x+1} = \dfrac{x+2}{x+1}$

23. $\dfrac{x^2+1}{x^2-1} \cdot \dfrac{x-1}{x+1} = \dfrac{x^2+1}{(x-1)(x+1)} \cdot \dfrac{x-1}{x+1} = \dfrac{x^2+1}{x+1} \cdot \dfrac{1}{x+1} = \dfrac{x^2+1}{(x+1)^2}$

24. $\dfrac{x^2+2x+1}{x^2-9} \cdot \dfrac{x+3}{x+1} = \dfrac{(x+1)(x+1)}{(x-3)(x+3)} \cdot \dfrac{x+3}{x+1} = \dfrac{x+1}{x-3} \cdot \dfrac{1}{1} = \dfrac{x+1}{x-3}$

25. $\dfrac{1}{3y} \div \dfrac{1}{9y^4} = \dfrac{1}{3y} \cdot \dfrac{9y^4}{1} = \dfrac{1}{1} \cdot \dfrac{3y^3}{1} = 3y^3$

26. $\dfrac{2x+2}{3x-3} \div \dfrac{x+1}{x-1} = \dfrac{2(x+1)}{3(x-1)} \cdot \dfrac{x-1}{x+1} = \dfrac{2}{3} \cdot \dfrac{1}{1} = \dfrac{2}{3}$

27. $\dfrac{x^2+2x}{x^2-25} \div \dfrac{x+2}{x+5} = \dfrac{x(x+2)}{(x-5)(x+5)} \cdot \dfrac{x+5}{x+2} = \dfrac{x}{x-5} \cdot \dfrac{1}{1} = \dfrac{x}{x-5}$

28. $\dfrac{x^2+2x-15}{x^2+4x+3} \div \dfrac{x-3}{x+1} = \dfrac{(x+5)(x-3)}{(x+3)(x+1)} \cdot \dfrac{x+1}{x-3} = \dfrac{x+5}{x+3} \cdot \dfrac{1}{1} = \dfrac{x+5}{x+3}$

29. $\dfrac{x^2+x-2}{2x^2-7x+3}\cdot\dfrac{x^2-3x}{x^2+2x-3}=\dfrac{(x+2)(x-1)}{(2x-1)(x-3)}\cdot\dfrac{x(x-3)}{(x+3)(x-1)}=\dfrac{x(x+2)}{(2x-1)(x+3)}$

30. $\dfrac{x^2+3x+2}{x^2+7x+12}\div\dfrac{x^2+4x+4}{x^2+4x+3}=\dfrac{(x+2)(x+1)}{(x+3)(x+4)}\cdot\dfrac{(x+3)(x+1)}{(x+2)(x+2)}=\dfrac{(x+1)^2}{(x+2)(x+4)}$

31. $36=2\cdot2\cdot3\cdot3$ and $24=2\cdot2\cdot2\cdot3$ and $16=2\cdot2\cdot2\cdot2$. The LCM is $2\cdot2\cdot2\cdot2\cdot3\cdot3=144$.

32. $4ab=2\cdot2\cdot a\cdot b$ and $a^2b=a\cdot a\cdot b$. The LCM is $2\cdot2\cdot a\cdot a\cdot b=4a^2b$.

33. $9x^2y=3\cdot3\cdot x\cdot x\cdot y$ and $6xy^3=2\cdot3\cdot x\cdot y\cdot y\cdot y$. The LCM is $2\cdot3\cdot3\cdot x\cdot x\cdot y\cdot y\cdot y=18x^2y^3$.

34. $x-1=(x-1)$ and $x+2=(x+2)$. The LCM is $(x-1)(x+2)$.

35. $x^2-9=(x-3)(x+3)$ and $x(x+3)=x(x+3)$. The LCM is $x(x-3)(x+3)$.

36. $x^2=x\cdot x$ and $x-3=(x-3)$ and $x^2-6x+9=(x-3)^2$. The LCM is $x^2(x-3)^2$.

37. $\dfrac{1}{x+4}+\dfrac{3}{x+4}=\dfrac{1+3}{x+4}=\dfrac{4}{x+4}$

38. $\dfrac{2}{x}+\dfrac{x-3}{x}=\dfrac{2+x-3}{x}=\dfrac{x-1}{x}$

39. $\dfrac{1}{x+1}-\dfrac{x}{x+1}=\dfrac{1-x}{x+1}$

40. $\dfrac{2x}{x-2}-\dfrac{2}{x-2}=\dfrac{2x-2}{x-2}$

41. $\dfrac{4}{1-t}+\dfrac{t}{t-1}=\dfrac{4}{-1(t-1)}+\dfrac{t}{t-1}=\dfrac{-4}{t-1}+\dfrac{t}{t-1}=\dfrac{t-4}{t-1}$

42. $\dfrac{2}{y-2}-\dfrac{2}{y+2}=\dfrac{2(y+2)}{(y-2)(y+2)}-\dfrac{2(y-2)}{(y-2)(y+2)}=\dfrac{2y+4-2y+4}{(y-2)(y+2)}=\dfrac{8}{(y-2)(y+2)}$

43. $\dfrac{4b}{a^2c}-\dfrac{3a}{b^2c}=\dfrac{4b^3}{a^2b^2c}-\dfrac{3a^3}{a^2b^2c}=\dfrac{4b^3-3a^3}{a^2b^2c}$

44. $\dfrac{r}{5t^2}+\dfrac{t}{5r^2}=\dfrac{r^3}{5r^2t^2}+\dfrac{t^3}{5r^2t^2}=\dfrac{r^3+t^3}{5r^2t^2}$

45. $\dfrac{4}{a^2-b^2}-\dfrac{2}{a+b}=\dfrac{4}{(a-b)(a+b)}-\dfrac{2(a-b)}{(a-b)(a+b)}=\dfrac{4-2a+2b}{(a-b)(a+b)}=-\dfrac{2(a-b-2)}{(a-b)(a+b)}$

46. $\dfrac{a}{a-b}+\dfrac{b}{a+b}=\dfrac{a(a+b)}{(a-b)(a+b)}+\dfrac{b(a-b)}{(a-b)(a+b)}=\dfrac{a^2+2ab-b^2}{(a-b)(a+b)}$

47. $\dfrac{1}{x^2-3x+2}+\dfrac{1}{x^2+x-2}=\dfrac{1}{(x-2)(x-1)}+\dfrac{1}{(x+2)(x-1)}=$

$\dfrac{x+2}{(x-2)(x-1)(x+2)}+\dfrac{x-2}{(x-2)(x-1)(x+2)}=\dfrac{x+2+x-2}{(x-2)(x-1)(x+2)}=\dfrac{2x}{(x-2)(x-1)(x+2)}$

48. $\dfrac{1}{x^2-5x+6}-\dfrac{1}{x^2+x-6}=\dfrac{1}{(x-2)(x-3)}-\dfrac{1}{(x+3)(x-2)}=$

$\dfrac{x+3}{(x-3)(x-2)(x+3)}-\dfrac{x-3}{(x-3)(x-2)(x+3)}=\dfrac{x+3-x+3}{(x-3)(x-2)(x+3)}=\dfrac{6}{(x-3)(x-2)(x+3)}$

49. $\dfrac{1}{x-2}+\dfrac{2}{x+2}-\dfrac{x}{x-2}=\dfrac{x+2}{(x-2)(x+2)}+\dfrac{2(x-2)}{(x-2)(x+2)}-\dfrac{x(x+2)}{(x-2)(x+2)}=$

$\dfrac{x+2+2x-4-x^2-2x}{(x-2)(x+2)}=\dfrac{-x^2+x-2}{(x-2)(x+2)}$

50. $\dfrac{2x}{2x-1}-\dfrac{3}{2x+1}-\dfrac{1}{2x-1}=\dfrac{2x(2x+1)}{(2x-1)(2x+1)}-\dfrac{3(2x-1)}{(2x-1)(2x+1)}-\dfrac{2x+1}{(2x-1)(2x+1)}=$

$\dfrac{4x^2+2x-6x+3-2x-1}{(2x-1)(2x+1)}=\dfrac{4x^2-6x+2}{(2x-1)(2x+1)}=\dfrac{2(x-1)(2x-1)}{(2x-1)(2x+1)}=\dfrac{2x-2}{2x+1}$

51. The LCD is $2x$. The first step is to multiply both sides of the equation by the LCD.

$\dfrac{4}{x}-\dfrac{5}{2x}=\dfrac{1}{2}\Rightarrow 2x\cdot\dfrac{4}{x}-2x\cdot\dfrac{5}{2x}=\dfrac{1}{2}\cdot 2x\Rightarrow 8-5=x\Rightarrow x=3$

52. Cross multiply to obtain the following.

$\dfrac{1}{x-4}=\dfrac{3}{2x-1}\Rightarrow 3(x-4)=1(2x-1)\Rightarrow 3x-12=2x-1\Rightarrow x=11$

53. The LCD is x^2. The first step is to multiply both sides of the equation by the LCD.

$x^2\cdot\left(\dfrac{2}{x^2}-\dfrac{1}{x}\right)=1\cdot x^2\Rightarrow 2-x=x^2\Rightarrow x^2+x-2=0\Rightarrow (x+2)(x-1)=0\Rightarrow x=-2 \text{ or } x=1$

54. The LCD is $x(x+4)$. The first step is to multiply both sides of the equation by the LCD.

$x(x+4)\cdot\left(\dfrac{1}{x+4}-\dfrac{1}{x}\right)=1\cdot x(x+4)\Rightarrow x-(x+4)=x^2+4x\Rightarrow x^2+4x+4=0\Rightarrow$

$(x+2)(x+2)=0\Rightarrow x=-2$

55. The LCD is $3(x-1)(x+1)$. The first step is to multiply both sides of the equation by the LCD.

$3(x-1)(x+1)\cdot\left(\dfrac{1}{x^2-1}-\dfrac{1}{x-1}\right)=\dfrac{2}{3}\cdot 3(x-1)(x+1)\Rightarrow 3-3(x+1)=2(x-1)(x+1)\Rightarrow$

$3-3x-3=2x^2-2\Rightarrow 2x^2+3x-2=0\Rightarrow (x+2)(2x-1)=0\Rightarrow x=-2 \text{ or } x=\dfrac{1}{2}$

56. The LCD is $(x-3)(x+3)$. The first step is to multiply both sides of the equation by the LCD.

$(x^2-9)\cdot\left(\dfrac{1}{x-3}+\dfrac{1}{x+3}\right)=\dfrac{-5}{x^2-9}\cdot(x^2-9)\Rightarrow x+3+x-3=-5\Rightarrow 2x=-5\Rightarrow x=-\dfrac{5}{2}$

57. The LCD is $(x+2)(x-2)^2$. The first step is to multiply both sides of the equation by the LCD.

$$(x+2)(x-2)^2 \cdot \frac{1}{(x-2)^2} - (x+2)(x-2)^2 \cdot \frac{1}{x^2-4} = \frac{2}{(x-2)^2} \cdot (x+2)(x-2)^2 \Rightarrow$$

$$x+2-(x-2)=2(x+2) \Rightarrow 4 = 2x+4 \Rightarrow 2x = 0 \Rightarrow x = 0$$

58. The LCD is $2(x+1)(x+2)$. The first step is to multiply both sides of the equation by the LCD.

$$2(x+1)(x+2) \cdot \frac{1}{x^2+3x+2} + 2(x+1)(x+2) \cdot \frac{x}{x+2} = \frac{1}{2} \cdot 2(x+1)(x+2) \Rightarrow$$

$$2+2x(x+1)=(x+1)(x+2) \Rightarrow 2x^2+2x+2 = x^2+3x+2 \Rightarrow x^2-x=0 \Rightarrow$$

$$x(x-1)=0 \Rightarrow x=0 \text{ or } x=1$$

59. $\dfrac{-3x-3}{x-3} + \dfrac{x^2+x}{x-3} = 7x+7 \Rightarrow \dfrac{-3x-3+x^2+x}{x-3} = 7x+7$

$\Rightarrow \dfrac{x^2-2x-3}{x-3} = 7x+7 \Rightarrow \dfrac{(x-3)(x+1)}{x-3} = 7x+7 \Rightarrow x+1 = 7x+7 \Rightarrow -6x = 6 \Rightarrow x = -1$

60. $\dfrac{1}{x-4} + \dfrac{x}{x+4} = \dfrac{8}{x^2-16} \Rightarrow (x-4)(x+4) \cdot \dfrac{1}{x-4} + (x-4)(x+4) \cdot \dfrac{x}{x+4}$

$= \dfrac{8}{(x-4)(x+4)} \cdot (x-4)(x+4) \Rightarrow x+4 + x(x-4) = 8$

$\Rightarrow x+4+x^2-4x = 8 \Rightarrow x^2-3x-4=0 \Rightarrow (x-4)(x+1)=0 \Rightarrow x-4=0 \text{ or } x+1=0 \Rightarrow x=-1$

($x=4$ is not a solution because this value makes expressions in the original equation undefined)

61. $y_1 = y_2$ when $x = 2$. By substitution of $x = 2$, this answer checks.

62. $y_1 = y_2$ when $x = -2$ and $x = 3$. By substitution of $x = -2$ and $x = 3$, these answers check.

63. $m = \dfrac{y_2-y_1}{x_2-x_1} \Rightarrow m(x_2-x_1) = y_2-y_1 \Rightarrow m(x_2-x_1)+y_1 = y_2 \Rightarrow y_2 = m(x_2-x_1)+y_1$

64. $T = \dfrac{a}{b+2} \Rightarrow T(b+2) = a \Rightarrow a = T(b+2)$

65. $\dfrac{1}{f} = \dfrac{1}{p} + \dfrac{1}{q} \Rightarrow pq = fq+fp \Rightarrow pq-fp = fq \Rightarrow p(q-f) = fq \Rightarrow p = \dfrac{fq}{q-f}$

66. $I = \dfrac{2a+3b}{ab} \Rightarrow abI = 2a+3b \Rightarrow abI-3b = 2a \Rightarrow b(aI-3) = 2a \Rightarrow b = \dfrac{2a}{aI-3}$

67. $\dfrac{\frac{3}{5}}{\frac{10}{13}} = \dfrac{3}{5} \cdot \dfrac{13}{10} = \dfrac{39}{50}$

68. $\dfrac{\dfrac{4}{ab}}{\dfrac{2}{bc}} = \dfrac{4}{ab} \cdot \dfrac{bc}{2} = \dfrac{2c}{a}$

69. $\dfrac{\dfrac{2n-1}{n}}{\dfrac{3n}{2n+1}} = \dfrac{2n-1}{n} \cdot \dfrac{2n+1}{3n} = \dfrac{4n^2-1}{3n^2}$

70. $\dfrac{\dfrac{1}{x-y}}{\dfrac{1}{x^2-y^2}} = \dfrac{1}{x-y} \cdot \dfrac{x^2-y^2}{1} = \dfrac{(x-y)(x+y)}{x-y} = x+y$

71. $\dfrac{2+\dfrac{3}{x}}{2-\dfrac{3}{x}} = \dfrac{2+\dfrac{3}{x}}{2-\dfrac{3}{x}} \cdot \dfrac{x}{x} = \dfrac{2x+3}{2x-3}$

72. $\dfrac{\dfrac{1}{x}+\dfrac{1}{2}}{\dfrac{1}{4}-\dfrac{2}{x}} = \dfrac{\dfrac{1}{x}+\dfrac{1}{2}}{\dfrac{1}{4}-\dfrac{2}{x}} \cdot \dfrac{4x}{4x} = \dfrac{4+2x}{x-8} = \dfrac{2(2+x)}{x-8}$

73. $\dfrac{\dfrac{4}{x}+\dfrac{1}{x-1}}{\dfrac{1}{x}-\dfrac{2}{x-1}} = \dfrac{\dfrac{4}{x}+\dfrac{1}{x-1}}{\dfrac{1}{x}-\dfrac{2}{x-1}} \cdot \dfrac{x(x-1)}{x(x-1)} = \dfrac{4(x-1)+x}{x-1-2x} = \dfrac{5x-4}{-x-1} = \dfrac{5x-4}{-1(x+1)} = \dfrac{4-5x}{x+1}$

74. $\dfrac{\dfrac{1}{x+3}-\dfrac{1}{x-3}}{\dfrac{4}{x+3}-\dfrac{2}{x-3}} = \dfrac{\dfrac{1}{x+3}-\dfrac{1}{x-3}}{\dfrac{4}{x+3}-\dfrac{2}{x-3}} \cdot \dfrac{(x+3)(x-3)}{(x+3)(x-3)} = \dfrac{x-3-(x+3)}{4(x-3)-2(x+3)} = \dfrac{-6}{2x-18} = \dfrac{-3}{x-9}$

75. $\dfrac{x}{6} = \dfrac{6}{20} \Rightarrow 20x = 36 \Rightarrow x = \dfrac{36}{20} \Rightarrow x = \dfrac{9}{5}$

76. $\dfrac{11}{x} = \dfrac{5}{7} \Rightarrow 5x = 77 \Rightarrow x = \dfrac{77}{5}$

77. $\dfrac{x+1}{5} = \dfrac{x}{3} \Rightarrow 5x = 3x+3 \Rightarrow 2x = 3 \Rightarrow x = \dfrac{3}{2}$

78. $\dfrac{3}{7} = \dfrac{4}{x-1} \Rightarrow 3x-3 = 28 \Rightarrow 3x = 31 \Rightarrow x = \dfrac{31}{3}$

79. $\dfrac{x+1}{5} = \dfrac{10}{15} \Rightarrow 15x+15 = 50 \Rightarrow 15x = 35 \Rightarrow x = \dfrac{35}{15} = \dfrac{7}{3}$

80. $\dfrac{7}{8} = \dfrac{11}{x} \Rightarrow 7x = 88 \Rightarrow x = \dfrac{88}{7} \approx 12.6$

81. $y = kx \Rightarrow 8 = 2k \Rightarrow k = 4 \Rightarrow y = 4x$, so when $x = 7$, $y = 4(7) = 28$

82. $y = \dfrac{k}{x} \Rightarrow 5 = \dfrac{k}{10} \Rightarrow k = 50 \Rightarrow y = \dfrac{50}{x}$, so when $x = 25$, $y = \dfrac{50}{25} = 2$

83. $z = kxy \Rightarrow 483 = k(23)(7) \Rightarrow 483 = 161k \Rightarrow k = \dfrac{483}{161} \Rightarrow k = 3$

84. $z = kxy^2 \Rightarrow 891 = k(22)(3)^2 \Rightarrow 891 = 198k \Rightarrow k = \dfrac{891}{198} \Rightarrow k = \dfrac{9}{2} \Rightarrow z = \dfrac{9}{2}xy^2$

   So when $x = 10$ and $y = 4$, $z = \dfrac{9}{2}(10)(4)^2 = 720$

85. Inverse. The products xy always equal 200. $y = \dfrac{k}{x} \Rightarrow 100 = \dfrac{k}{2} \Rightarrow k = 200;$

   The equation is $y = \dfrac{200}{x}$.

86. Direct. The ratios $\dfrac{y}{x}$ always equal 3. $y = kx \Rightarrow 9 = 3k \Rightarrow k = 3;$ The equation is $y = 3x.$

87. Direct. The ratios $\dfrac{y}{x}$ always equal $\dfrac{1}{2}$. The constant of variation is $k = \dfrac{1}{2}.$

88. Inverse. The products xy always equal 6. The constant of variation is $k = 6.$

89. $\dfrac{10x+15}{5} = \dfrac{10x}{5} + \dfrac{15}{5} = 2x + 3$

90. $\dfrac{2x^2 + x}{x} = \dfrac{2x^2}{x} + \dfrac{x}{x} = 2x + 1$

91. $\left(4x^3 - x^2 + 2x\right) \div 2x^2 = \dfrac{4x^3 - x^2 + 2x}{2x^2} = \dfrac{4x^3}{2x^2} - \dfrac{x^2}{2x^2} + \dfrac{2x}{2x^2} = 2x - \dfrac{1}{2} + \dfrac{1}{x}$

92. $\left(4a^3 b - 6ab^2\right) \div \left(2a^2 b^2\right) = \dfrac{4a^3 b - 6ab^2}{2a^2 b^2} = \dfrac{4a^3 b}{2a^2 b^2} - \dfrac{6ab^2}{2a^2 b^2} = \dfrac{2a}{b} - \dfrac{3}{a}$

93.
$$
\begin{array}{r}
2x - 3 \\
x+1\overline{\smash{\big)}\,2x^2 - x - 2} \\
\underline{2x^2 + 2x} \\
-3x - 2 \\
\underline{-3x - 3} \\
1
\end{array}
$$

The solution is: $2x - 3 + \dfrac{1}{x+1}$

94.
$$\begin{array}{r} x^2 - x + 2 \\ x+1\overline{\smash{\big)}\,x^3 + 0x^2 + x - 1} \\ \underline{x^3 + x^2} \\ -x^2 + x \\ \underline{-x^2 - x} \\ 2x - 1 \\ \underline{2x + 2} \\ -3 \end{array}$$

The solution is: $x^2 - x + 2 - \dfrac{3}{x+1}$

95.
$$\begin{array}{r} 2x^2 - x + 1 \\ 3x-2\overline{\smash{\big)}\,6x^3 - 7x^2 + 5x - 1} \\ \underline{6x^3 - 4x^2} \\ -3x^2 + 5x \\ \underline{-3x^2 + 2x} \\ 3x - 1 \\ \underline{3x - 2} \\ 1 \end{array}$$

The solution is: $2x^2 - x + 1 + \dfrac{1}{3x-2}$

96.
$$\begin{array}{r} 2x - 9 \\ x^2+0x-2\overline{\smash{\big)}\,2x^3 - 9x^2 + 21x - 21} \\ \underline{2x^3 + 0x^2 - 4x} \\ -9x^2 + 25x - 21 \\ \underline{-9x^2 + 0x + 18} \\ 25x - 39 \end{array}$$

The solution is: $2x - 9 + \dfrac{25x - 39}{x^2 - 2}$

97.
$$\begin{array}{r|rrr} 3 & 2 & -11 & 13 \\ & & 6 & -15 \\ \hline & 2 & -5 & -2 \end{array}$$

The solution is: $2x - 5 - \dfrac{2}{x-3}$

98.
$$\begin{array}{r|rrrr} -4 & 3 & 10 & -4 & 19 \\ & & -12 & 8 & -16 \\ \hline & 3 & -2 & 4 & 3 \end{array}$$

The solution is: $3x^2 - 2x + 4 + \dfrac{3}{x+4}$

99. (a) Table $Y_1 = 1/(10 - X)$ with TblStart = 9 and ΔTbl = 0.1. See Figure 100.

   (b) As x approaches 10 the waiting time increases. As vehicles arrive at a faster rate, the wait in line increases.

100. (a) $D(0) = \dfrac{1600}{9.6 - 30(0)} = \dfrac{1600}{9.6} = 166.\overline{6}$; $D(0.1) = \dfrac{1600}{9.6 - 30(0.1)} = \dfrac{1600}{6.6} = 242.\overline{42}$

   The downhill grade adds $242.\overline{42} - 166.\overline{6} \approx 75.8$ feet to the braking distance.

   (b) $200 = \dfrac{1600}{9.6 - 30x} \Rightarrow 200(9.6 - 30x) = 1600 \Rightarrow 9.6 - 30x = 8 \Rightarrow -30x = -1.6 \Rightarrow x = 0.05\overline{3}$

101. (a) $N(14) = \dfrac{14^2}{225 - 15(14)} = \dfrac{196}{15} \approx 13$ cars

   (b) As x increases the number of cars in line increases. This agrees with intuition.

102. (a) $P(0) = \dfrac{5}{(0)^2 + 1} = \dfrac{5}{1} = 5$; Initially there were 5000 fish in the lake.

   (b) Graph $Y_1 = 5/(X^2 + 1)$ in [0, 6, 1] by [0, 6, 1]. See Figure 102.

   (c) Over this 6-year period, the fish population decreased.

   (d) $1 = \dfrac{5}{x^2 + 1} \Rightarrow x^2 + 1 = 5 \Rightarrow x^2 = 4 \Rightarrow x = 2$ (the value $x = -2$ has no physical meaning)

   The population was 1 thousand after 2 years.

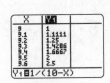

$[0, 6, 1]$ by $[0, 6, 1]$

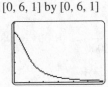

$[0, 2, 1]$ by $[0, 2, 1]$

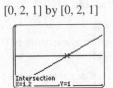

Figure 100   Figure 102   Figure 103

103. (a) $\dfrac{x}{2} + \dfrac{x}{3} = 1$

   (b) $6 \cdot \left(\dfrac{x}{2} + \dfrac{x}{3} \right) = 1 \cdot 6 \Rightarrow 3x + 2x = 6 \Rightarrow 5x = 6 \Rightarrow x = \dfrac{6}{5} = 1.2$ hours

   (c) Graph $Y_1 = X/2 + X/3$ and $Y_2 = 1$ in [0, 2, 1] by [0, 2, 1]. See Figure 103. $x = 1.2$ hours

104. Let x represent the speed of the current. Then $12 - x$ represents the boat's upstream speed and

   $12 + x$ represents the boat's downstream speed. Since $t = \dfrac{d}{r}$, the equation is $\dfrac{48}{12 - x} + \dfrac{48}{12 + x} = 9$.

   The LCD is $(12 - x)(12 + x)$. The first step is to multiply both sides of the equation by the LCD.

   $(12 - x)(12 + x) \cdot \left(\dfrac{48}{12 - x} + \dfrac{48}{12 + x} \right) = 9 \cdot (12 - x)(12 + x) \Rightarrow 576 + 48x + 576 - 48x = 1296 - 9x^2$

   $\Rightarrow 9x^2 - 144 = 0 \Rightarrow 9x^2 = 144 \Rightarrow x^2 = 16 \Rightarrow x = \pm\sqrt{16} \Rightarrow x = 4$ mph ($x = -4$ has no meaning)

105. $\dfrac{650}{74} = \dfrac{387}{x} \Rightarrow 650x = 28{,}638 \Rightarrow x = \dfrac{28{,}638}{650} \approx 44$ minutes

106. $\dfrac{5}{3} = \dfrac{x}{26} \Rightarrow 3x = 130 \Rightarrow x = \dfrac{130}{3} \approx 43.3$ feet

107. (a) Plot the data in [0, 300, 50] by [0, 250, 50]. See Figure 107. The data represent inverse variation.

   (b) $R = \dfrac{k}{x} \Rightarrow 242 = \dfrac{k}{50} \Rightarrow k = 12{,}100$; Equation: $R = \dfrac{12{,}100}{W}$

(c) $R = \dfrac{12,100}{55} = 220$ ohms

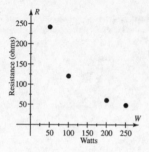

Figure 107

108. (a) $D = kW \Rightarrow 0.3 = 5k \Rightarrow k = 0.06$; Equation: $D = 0.06W$

(b) $D = 0.06(7) = 0.42$ inches

109. The change in temperature is 18°F so $\dfrac{18}{4000} = \dfrac{x}{6000} \Rightarrow 4000x = 108,000 \Rightarrow x = 27°$ change at 6000

feet. The temperature at 6000 feet is $80 - 27 = 53°$F.

110. (a)   Direct. The ratios $\dfrac{y}{x}$ always equal $\dfrac{7}{2}$.

(b) $y = kx \Rightarrow 3.5 = 1k \Rightarrow k = 3.5$,   Equation: $y = 3.5x$

(c) $y = 3.5(2.3) = 8.05\%$

Chapter 6 Test

1. $f(x) = \dfrac{x}{x+2}$

2. (a) $f(-2) = \dfrac{1}{4(-2)^2 - 1} = \dfrac{1}{16 - 1} = \dfrac{1}{15}$

(b)   Here $4x^2 - 1 \neq 0 \Rightarrow x \neq -\dfrac{1}{2}$ and $x \neq \dfrac{1}{2}$. The domain is $\left\{ x \mid x \neq \dfrac{1}{2}, x \neq -\dfrac{1}{2} \right\}$.

3. See Figure 3.

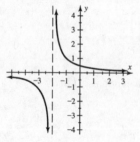

Figure 3

4. $\dfrac{2a^3}{4a^2} = \dfrac{a}{2}$

5. $\dfrac{1-2t}{2t-1} = \dfrac{-1(2t-1)}{2t-1} = -1$

6. $\dfrac{x^2-2x-15}{2x^2-x-21} = \dfrac{(x-5)(x+3)}{(2x-7)(x+3)} = \dfrac{x-5}{2x-7}$

7. $\dfrac{x^2+4}{x^2-4} \cdot \dfrac{x-2}{x+2} = \dfrac{x^2+4}{(x-2)(x+2)} \cdot \dfrac{x-2}{x+2} = \dfrac{x^2+4}{x+2} \cdot \dfrac{1}{x+2} = \dfrac{x^2+4}{(x+2)^2}$

8. $\dfrac{1}{4y^2} \div \dfrac{1}{8y^4} = \dfrac{1}{4y^2} \cdot \dfrac{8y^4}{1} = \dfrac{1}{1} \cdot \dfrac{2y^2}{1} = 2y^2$

9. $\dfrac{x}{x+5} + \dfrac{1-x}{x+5} = \dfrac{x+1-x}{x+5} = \dfrac{1}{x+5}$

10. $\dfrac{2x}{x-2} - \dfrac{1}{x+2} = \dfrac{2x(x+2)}{(x-2)(x+2)} - \dfrac{x-2}{(x-2)(x+2)} = \dfrac{2x(x+2)-(x-2)}{(x-2)(x+2)}$

$= \dfrac{2x^2+4x-x+2}{(x-2)(x+2)} = \dfrac{2x^2+3x+2}{(x-2)(x+2)}$

11. $\dfrac{a^2}{3b} - \dfrac{2b^3}{5a} = \dfrac{a^2}{3b} \cdot \dfrac{5a}{5a} - \dfrac{2b^3}{5a} \cdot \dfrac{3b}{3b} = \dfrac{5a^3}{15ab} - \dfrac{6b^4}{15ab} = \dfrac{5a^3-6b^4}{15ab}$

12. $\dfrac{x}{x-2} - \dfrac{4}{x^2} - \dfrac{2}{x} = \dfrac{x}{(x-2)} \cdot \dfrac{x^2}{x^2} - \dfrac{4}{x^2} \cdot \dfrac{x-2}{x-2} - \dfrac{2}{x} \cdot \dfrac{x(x-2)}{x(x-2)} = \dfrac{x^3}{x^2(x-2)} - \dfrac{4x-8}{x^2(x-2)} - \dfrac{2x^2-4x}{x^2(x-2)}$

$= \dfrac{x^3-4x+8-2x^2+4x}{x^2(x-2)} = \dfrac{x^3-2x^2+8}{x^2(x-2)}$

13. $\dfrac{3+\dfrac{3}{x}}{3-\dfrac{3}{x}} = \dfrac{3+\dfrac{3}{x}}{3-\dfrac{3}{x}} \cdot \dfrac{x}{x} = \dfrac{3x+3}{3x-3} = \dfrac{3(x+1)}{3(x-1)} = \dfrac{x+1}{x-1}$

14. $\dfrac{1}{z+4} - \dfrac{z}{(z+4)^2} = \dfrac{z+4}{(z+4)(z+4)} - \dfrac{z}{(z+4)(z+4)} = \dfrac{z+4-z}{(z+4)(z+4)} = \dfrac{4}{(z+4)^2}$

15. $\dfrac{\dfrac{1}{x-2}+\dfrac{x}{x-2}}{\dfrac{1}{3}-\dfrac{5}{x-2}} = \dfrac{\dfrac{1}{x-2}+\dfrac{x}{x-2}}{\dfrac{1}{3}-\dfrac{5}{x-2}} \cdot \dfrac{3(x-2)}{3(x-2)} = \dfrac{3+3x}{x-2-15} = \dfrac{3x+3}{x-17} = \dfrac{3(x+1)}{x-17}$

16. The LCD is $7(5t+1)$. The first step is to multiply both sides of the equation by the LCD.

$7(5t+1) \cdot \dfrac{t}{5t+1} = \dfrac{2}{7} \cdot 7(5t+1) \Rightarrow 7t = 10t+2 \Rightarrow -3t = 2 \Rightarrow t = -\dfrac{2}{3}$

17. The LCD is $(2x-1)(x+4)$. The first step is to multiply both sides of the equation by the LCD.

$$(2x-1)(x+4)\cdot\frac{x}{2x-1}=\frac{x+2}{x+4}\cdot(2x-1)(x+4)$$

$$\Rightarrow (x+4)x=(x+2)(2x-1)\Rightarrow x^2+4x=2x^2+3x-2$$

$$\Rightarrow 0=x^2-x-2\Rightarrow 0=(x+1)(x-2)\Rightarrow \text{Either } x=-1 \text{ or } x=2$$

18. The LCD is $2-x$. The first step is to multiply both sides of the equation by the LCD.

$$(2-x)\cdot\left(\frac{x+4}{2-x}-\frac{2x-1}{2-x}\right)=0\cdot(2-x)\Rightarrow x+4-(2x-1)=0\Rightarrow x+4-2x+1=0\Rightarrow x=5$$

19. The LCD is $x+1$. The first step is to multiply both sides of the equation by the LCD.

$$(x+1)\cdot\left(\frac{5}{x+1}\right)=4\cdot(x+1)\Rightarrow 5=4x+4\Rightarrow 1=4x\Rightarrow \frac{1}{4}=x$$

20. The LCD is $(x+2)(x-2)$ The first step is to multiply both sides of the equation by the LCD.

$$(x+2)(x-2)\cdot\left(\frac{1}{x+2}+\frac{1}{x-2}\right)=\frac{4}{(x+2)(x-2)}\cdot(x+2)(x-2)\Rightarrow$$

$(x-2)+(x+2)=4\Rightarrow 2x=4\Rightarrow x=2$ ($x=2$ is not a solution because this value makes expressions in

the original equation undefined) Therefore, there are no solutions to this equation.

21. The LCD is $(x-2)(x+2)$. The first step is to multiply both sides of the equation by the LCD.

$$(x-2)(x+2)\cdot\left(\frac{1}{x^2-4}-\frac{1}{x-2}\right)=\frac{1}{x+2}\cdot(x-2)(x+2)\Rightarrow 1-(x+2)=x-2\Rightarrow$$

$$-x-1=x-2\Rightarrow -2x=-1\Rightarrow x=\frac{1}{2}$$

22. $F=\dfrac{Gm}{r^2}\Rightarrow Fr^2=Gm\Rightarrow m=\dfrac{Fr^2}{G}$

23. Graph $Y_1=\sqrt{12}\Big/\left(X^2+2\right)-(1.7X)/(X-1.2)$ and $Y_2=\sqrt{3}$ in $[-5, 5, 1]$ by $[-5, 5, 1]$.

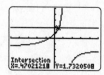

Figure 23

The solution is approximately 0.47.

24. $(f-g)(2)=(2-1)-\left(2^2-2(2)+1\right)=1-(4-4+1)=1-1=0;$

$$(f/g)(x)=\frac{x-1}{x^2-2x+1}=\frac{x-1}{(x-1)(x-1)}=\frac{1}{x-1}$$

25. $\dfrac{7}{12}=\dfrac{x}{20}\Rightarrow 12x=140\Rightarrow x=\dfrac{140}{12}=\dfrac{35}{3}$

26. $y = kx \Rightarrow 8 = 23k \Rightarrow k = \dfrac{8}{23} \Rightarrow y = \dfrac{8}{23}x$, so when $x = 10$, $y = \dfrac{8}{23}(10) = \dfrac{80}{23}$

27. Inverse. The products xy always equal 100. $y = \dfrac{k}{x} \Rightarrow 50 = \dfrac{k}{2} \Rightarrow k = 100$; The equation is $y = \dfrac{100}{x}$.

28. Direct. The ratios $\dfrac{y}{x}$ always equal 1.5. $y = kx \Rightarrow 3 = 2k \Rightarrow k = 1.5$; The equation is $y = 1.5x$.

29. $\dfrac{4a^3 + 10a}{2a} = \dfrac{4a^3}{2a} + \dfrac{10a}{2a} = 2a^2 + 5$

30. Using synthetic division:
$$\begin{array}{r|rrrr} -2 & 3 & 5 & 0 & -2 \\ & & -6 & 2 & -4 \\ \hline & 3 & -1 & 2 & -6 \end{array}$$
The solution is: $3x^2 - x + 2 - \dfrac{6}{x+2}$

31. First convert 15 feet to $15 \cdot 12 = 180$ inches.

$\dfrac{73}{50} = \dfrac{x}{180} \Rightarrow 50x = 13{,}140 \Rightarrow x = \dfrac{13{,}140}{50} = 262.8$ inches, or $\dfrac{262.8}{12} = 21.9$ feet

32. (a) Graph $Y_1 = 1/(25 - X)$ in [0, 25, 5] by [0, 2, 0.5]. Vertical asymptote: $x = 25$. See Figure 32.

    (b) $1 = \dfrac{1}{25 - x} \Rightarrow 25 - x = 1 \Rightarrow -x = -24 \Rightarrow x = 24$ vehicles per minute

    [0, 25, 5] by [0, 2, 0.5]

    Figure 32

33. (a) $\dfrac{x}{24} + \dfrac{x}{30} = 1$

    (b) The LCD is 120. The first step is to multiply both sides of the equation by the LCD.

    $120 \cdot \left(\dfrac{x}{24} + \dfrac{x}{30} \right) = 1 \cdot 120 \Rightarrow 5x + 4x = 120 \Rightarrow 9x = 120 \Rightarrow x = \dfrac{120}{9} = \dfrac{40}{3} \approx 13.3$ hours

34. The change in dew point is 11°F so $\dfrac{11}{10{,}000} = \dfrac{x}{7500} \Rightarrow 10{,}000x = 82{,}500 \Rightarrow x = 8.25°$ change at

    7500 feet. So the dew point at 7500 feet is $50°\text{F} - 8.25°\text{F} = 41.75°\text{F}$.

Chapter 6 Extended and Discovery Exercises

1. It quadruples.

2. It doubles.

3. (a) $1.125 = k(2.1)^{1.25} \Rightarrow k = \dfrac{1.125}{2.1^{1.25}} \approx 0.445$

(b) $0.8 = 0.445x^{1.25} \Rightarrow x^{1.25} = \dfrac{0.8}{0.445} \Rightarrow \left(x^{1.25}\right)^{0.8} = \left(\dfrac{0.8}{0.445}\right)^{0.8} \Rightarrow x \approx 1.60$ grams

4. $V = kr^2 \Rightarrow 150 = k(5)^2 \Rightarrow 150 = 25k \Rightarrow k = \dfrac{150}{25} = 6 \Rightarrow V = 6r^2$ so when $r = 7, V = 6(7)^2 = 294\,\text{in}^3$

5. It is reduced to $\dfrac{1}{4}$ of its value.

6. It is reduced to $\dfrac{1}{8}$ of its value.

7. (a) $200 = \dfrac{k}{4000^2} \Rightarrow k = 200\left(4000^2\right) = 3.2 \times 10^9$

    (b)   Graph $Y_1 = \left(3.2 \times 10^{\wedge}9\right)/X^2$ in [0, 10,000, 1000] by [0, 200, 50]. See Figure 7.

          Fifty pounds is $\tfrac{1}{4}$ of 200 pounds and so the person should be twice as far or 8000 miles from

          Earth's center.

    (c)   One percent is $\tfrac{1}{100}$ of the weight of the object on Earth's surface, thus the object must be 10

          times as far or 40,000 miles from Earth's center.

[0, 10,000, 1000] by [0, 200, 50]   [0, 5, 1] by [0, 30, 5]

                        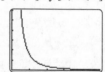

       Figure 7   Figure 8

8. (a)   Since $k = I \cdot d^2, k = 31.68(0.5)^2 = 7.92$.

    (b)   Graph $Y_1 = 7.92/X^2$ in [0, 5, 1] by [0, 30, 5]. See Figure 8.

          As the distance from the light bulb increases the intensity decreases.

    (c)   When the distance is doubled, the intensity is reduced to $\tfrac{1}{4}$ of its value.

    (d) $1 = \dfrac{7.92}{d^2} \Rightarrow d^2 = 7.92 \Rightarrow d = \sqrt{7.92} \approx 2.81$ meters

Chapters 1-6 Cumulative Review Exercises

1. $y = \sqrt{74 + (7)} = \sqrt{81} = 9$

2. $r = 16 - (-4)^2 = 16 - 16 = 0$

3. Natural: 1; Whole: 0, 1; Integer: $-\dfrac{12}{4}, 0, 1$; Rational: $-\dfrac{12}{4}, 0, 1, 2.\overline{11}, \dfrac{13}{2}$; Irrational: $\sqrt{3}$

4. By substitution of values, the formula that best fits the data is (a).

5. $\dfrac{18x^{-2}y^3}{3x^2y^{-3}} = 6x^{-2-2}y^{3-(-3)} = 6x^{-4}y^6 = \dfrac{6y^6}{x^4}$

6. $\left(\dfrac{2c^2}{3d^3}\right)^{-2} = \left(\dfrac{3d^3}{2c^2}\right)^2 = \dfrac{3^2\left(d^3\right)^2}{2^2\left(c^2\right)^2} = \dfrac{9d^6}{4c^4}$

7. Move the decimal point 10 places to the left: $67,300,000,000 = 6.73 \times 10^{10}$.

8. The domain corresponds to the x-coordinates and the range corresponds to the y-coordinates.

   $D = \{-3, -1, 0, 4\}$; $R = \{-2, 0, 1, 5\}$

9. When $x = -2$, $y = 3(-2) - 1 = -7$. The other ordered pairs can be found similarly.

   The ordered pairs are $(-2, -7), (-1, -4), (0, -1), (1, 2), (2, 5)$. See Figure 9.

10. When $x = -2$, $y = \dfrac{4 - (-2)^2}{2} = \dfrac{4 - 4}{2} = \dfrac{0}{2} = 0$. The other ordered pairs can be found similarly.

    The ordered pairs are $(-2, 0), \left(-1, \dfrac{3}{2}\right), (0, 2), \left(1, \dfrac{3}{2}\right), (2, 0)$. See Figure 10.

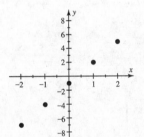

Figure 9

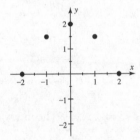

Figure 10

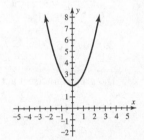

Figure 11

11. See Figure 11.

12. The function is defined for all values of the variable except 1. The domain is $\{x \mid x \neq 1\}$.

13. The slope is $m = \dfrac{3 - (-5)}{2 - (-2)} = \dfrac{8}{4} = 2$. Since $f(x) = -1$ when $x = 0$ the y-intercept is -1.

    Here $f(x) = 2x - 1$.

14. See Figure 14.

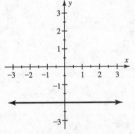

Figure 14

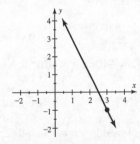

Figure 17

15. The equation is in the form $f(x) = mx + b$. The slope is -3 and the y-intercept is 2.

16. $m = \dfrac{9-(-3)}{2-6} = \dfrac{12}{-4} = -3$

17. See Figure 17.

18. The graph falls 1 unit for each 2 units of run. The slope is $-\dfrac{1}{2}$. The y-intercept is 2.

 So $y = -\dfrac{1}{2}x + 2$.

19. Since the line is parallel to $y = 2x + 3$, the slope is $m = 2$. Using the point-slope form gives

 $y = 2(x-1) + 4 \Rightarrow y = 2x - 2 + 4 \Rightarrow y = 2x + 2$

20. Since the line is perpendicular to $y = \dfrac{2}{3}x - 2$, the slope is $m = -\dfrac{3}{2}$. Using the point-slope form

 gives $y = -\dfrac{3}{2}(x-2) + 1 \Rightarrow y = -\dfrac{3}{2}x + 3 + 1 \Rightarrow y = -\dfrac{3}{2}x + 4$

21. $\dfrac{2}{5}(x+1) - 6 = -4 \Rightarrow \dfrac{2}{5}(x+1) = 2 \Rightarrow x + 1 = 5 \Rightarrow x = 4$

22. $\dfrac{1}{4}\left(\dfrac{t-5}{3}\right) - 6 = \dfrac{2}{3}t - (3t+7) \Rightarrow t - 5 - 72 = 8t - 12(3t+7) \Rightarrow t - 77 = -28t - 84 \Rightarrow$

 $29t = -7 \Rightarrow t = -\dfrac{7}{29}$

23. The graphs intersect at the point $(3, -2)$, so $y_1 = y_2$ when $x = 3$.

24. $\dfrac{2x+3}{4} \le \dfrac{1}{3} \Rightarrow 6x + 9 \le 4 \Rightarrow 6x \le -5 \Rightarrow x \le -\dfrac{5}{6}$. The interval is $\left(-\infty, -\dfrac{5}{6}\right]$.

25. $\dfrac{1}{2}z - 5 > \dfrac{3}{4}z - (2z+5) \Rightarrow 2z - 20 > 3z - 4(2z+5) \Rightarrow 7z > 0 \Rightarrow z > 0$. The interval is $(0, \infty)$.

26. $-6 \le -\dfrac{2}{3}x - 4 < -2 \Rightarrow -2 \le -\dfrac{2}{3}x < 2 \Rightarrow 3 \ge x > -3 \Rightarrow -3 < x \le 3$.

 The interval is $(-3, 3]$. See Figure 26.

27. $3x - 1 \le 5$ or $2x + 5 > 13 \Rightarrow 3x \le 6$ or $2x > 8 \Rightarrow x \le 2$ or $x > 4$.

 The interval is $(-\infty, 2] \cup (4, \infty)$. See Figure 27.

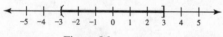

Figure 26

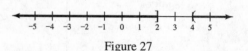

Figure 27

28. $\left|\dfrac{1}{3}x + 6\right| = 4 \Rightarrow \dfrac{1}{3}x + 6 = -4 \Rightarrow \dfrac{1}{3}x = -10 \Rightarrow x = -30$ or $\dfrac{1}{3}x + 6 = 4 \Rightarrow \dfrac{1}{3}x = -2 \Rightarrow x = -6$

29. The solutions to $|2x - 3| < 11$ satisfy $c < x < d$ where c and d are the solutions to $|2x - 3| = 11$.

 $|2x - 3| = 11$ is equivalent to $2x - 3 = -11 \Rightarrow x = -4$ and $2x - 3 = 11 \Rightarrow x = 7$.

 The interval is $(-4, 7)$.

30. First divide each side of $-3|t-5| \leq -18$ by -3 to obtain $|t-5| \geq 6$.

   The solutions to $|t-5| \geq 6$ satisfy $t \leq c$ or $t \geq d$ where c and d are the solutions to $|t-5| = 6$.

   $|t-5| = 6$ is equivalent to $t-5 = -6 \Rightarrow t = -1$ and $t-5 = 6 \Rightarrow t = 11$.

   The interval is $(-\infty, -1] \cup [11, \infty)$.

31. By substitution, $(-8, -1)$ is a solution to the given system of equations.

32. See Figure 32.

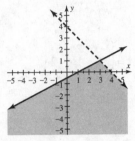

   Figure 32

33. Multiplying the second equation by 4 and adding the two equations will eliminate the variable y.

$$\begin{array}{l} 2x-8y=5 \\ 16x+8y=4 \\ \hline 18x=9 \end{array}$$ Thus, $x = \dfrac{1}{2}$. And so $2\left(\dfrac{1}{2}\right)-8y=5 \Rightarrow y=-\dfrac{1}{2}$. The solution is $\left(\dfrac{1}{2}, -\dfrac{1}{2}\right)$.

34. Multiplying the second equation by 2 and adding the two equations will eliminate the variable x.

$$\begin{array}{l} 2x-3y=\ 12 \\ -2x+4y=-12 \\ \hline y=0 \end{array}$$ Thus, $y = 0$. And so $2x-3(0)=12 \Rightarrow x=6$. The solution is $(6, 0)$.

35. From the graph of the region of feasible solutions (not shown), the vertices are (0, 0), (0, 3), (2, 2),

   and (3, 0). The maximum value of R occurs at one of the vertices. For $(0, 0)$, $R = 2(0)+3(0)=0$.

   For $(0, 3)$, $R = 2(0)+3(3)=9$. For $(2, 2)$, $R = 2(2)+3(2)=10$. For $(3, 0)$, $R = 2(3)+3(0)=6$.

   The maximum value is $R = 10$.

36. Multiply the second equation by 3 and add the first and second equations to eliminate the variable y.

$$\begin{array}{l} 2x+3y\ -z=\ 3 \\ 9x-3y+12z=30 \\ \hline 11x+11z=33 \end{array} \text{ or } x+z=3$$

   Add the second and third equations together to eliminate the variable y. $\begin{array}{l} 3x-y+4z=10 \\ 2x+y-2z=-1 \\ \hline 5x+2z=9 \end{array}$

   Multiply the first *new* equation by -2 and add the first *new* equation and second *new* equation to

   eliminate z. $\begin{array}{l} -2x-2z=-6 \\ 5x+2z=\ 9 \\ \hline 3x=3 \end{array}$ And so $x = 1$. Substitute $x = 1$ into the first *new* equation:

   $(1)+z=3 \Rightarrow z=2$

   Substitute $x = 1$ and $z = 2$ into the *original* first equation: $2(1)+3y-(2)=3 \Rightarrow y=1$

   The solution is (1, 1, 2).

37. $\begin{bmatrix} 1 & 1 & -1 & | & 4 \\ -1 & -1 & -1 & | & 0 \\ 1 & -2 & 1 & | & -9 \end{bmatrix} \begin{matrix} \\ R_2 + R_1 \to \\ R_3 - R_1 \to \end{matrix} \begin{bmatrix} 1 & 1 & -1 & | & 4 \\ 0 & 0 & -2 & | & 4 \\ 0 & -3 & 2 & | & -13 \end{bmatrix} \begin{matrix} \\ Exchange \\ R_2 \leftrightarrow R_3 \end{matrix} \begin{bmatrix} 1 & 1 & -1 & | & 4 \\ 0 & -3 & 2 & | & -13 \\ 0 & 0 & -2 & | & 4 \end{bmatrix}$

$\begin{matrix} \\ \\ (-1/2)R_3 \to \end{matrix} \begin{bmatrix} 1 & 1 & -1 & | & 4 \\ 0 & -3 & 2 & | & -13 \\ 0 & 0 & 1 & | & -2 \end{bmatrix} \begin{matrix} R_1 + R_3 \to \\ R_2 - 2R_3 \to \\ \\ \end{matrix} \begin{bmatrix} 1 & 1 & 0 & | & 2 \\ 0 & -3 & 0 & | & -9 \\ 0 & 0 & 1 & | & -2 \end{bmatrix} \begin{matrix} R_1 + (1/3)R_2 \to \\ (-1/3)R_2 \to \\ \\ \end{matrix} \begin{bmatrix} 1 & 0 & 0 & | & -1 \\ 0 & 1 & 0 & | & 3 \\ 0 & 0 & 1 & | & -2 \end{bmatrix}$

The solution is $(-1, 3, -2)$.

38. $\det A = 4(3) - 1(-2) = 12 + 2 = 14$

39. The triangle has vertices $(-3, 2)$, $(2, 1)$ and $(1, -2)$. The matrix needed is $A = \begin{bmatrix} -3 & 2 & 1 \\ 2 & 1 & -2 \\ 1 & 1 & 1 \end{bmatrix}$.

    The area is $D = \left| \dfrac{1}{2} \det([A]) \right| = 8 \text{ in}^2$.

40. $E = \det \begin{bmatrix} 8 & 7 \\ 18 & 5 \end{bmatrix} = 40 - 126 = -86$; $F = \det \begin{bmatrix} 6 & 8 \\ -8 & 18 \end{bmatrix} = 108 + 64 = 172$

    $D = \det \begin{bmatrix} 6 & 7 \\ -8 & 5 \end{bmatrix} = 30 + 56 = 86$; The solution is $x = \dfrac{E}{D} = \dfrac{-86}{86} = -1$ and $y = \dfrac{F}{D} = \dfrac{172}{86} = 2$.

41. $3x^2 \left(x^3 + 5x - 2 \right) = 3x^5 + 15x^3 - 6x^2$

42. $(2x - 5)(x + 3) = 2x^2 + 6x - 5x - 15 = 2x^2 + x - 15$

43. $6x + 3x^2 = 0 \Rightarrow 3x(2 + x) = 0 \Rightarrow$ Either $3x = 0 \Rightarrow x = 0$ or $2 + x = 0 \Rightarrow x = -2$

44. $3a^3 - a^2 + 15a - 5 = a^2(3a - 1) + 5(3a - 1) = (3a - 1)\left(a^2 + 5\right)$

45. $4x^2 + 5x - 6 = (4x - 3)(x + 2)$

46. $6x^3 - 9x^2 - 6x = 3x\left(2x^2 - 3x - 2\right) = 3x(x - 2)(2x + 1)$

47. $9a^2 - 4b^2 = (3a - 2b)(3a + 2b)$

48. $64t^3 + 27 = (4t + 3)\left(16t^2 - 12t + 9\right)$

49. $3x^2 - 11x = 4 \Rightarrow 3x^2 - 11x - 4 = 0 \Rightarrow (3x + 1)(x - 4) = 0 \Rightarrow x = -\dfrac{1}{3}$ or $x = 4$

50. $x^4 - x^3 - 30x^2 = 0 \Rightarrow x^2\left(x^2 - x - 30\right) = 0 \Rightarrow x^2(x + 5)(x - 6) = 0 \Rightarrow x = 0$ or $x = -5$ or $x = 6$

51. $\dfrac{x^2 + 3x - 10}{x^2 - 4} \cdot \dfrac{x - 2}{x + 5} = \dfrac{(x + 5)(x - 2)(x - 2)}{(x - 2)(x + 2)(x + 5)} = \dfrac{x - 2}{x + 2}$

52. $\dfrac{x^2 + 2x - 24}{x^2 + 3x - 18} \div \dfrac{x + 3}{x^2 - 9} = \dfrac{x^2 + 2x - 24}{x^2 + 3x - 18} \cdot \dfrac{x^2 - 9}{x + 3} = \dfrac{(x + 6)(x - 4)(x - 3)(x + 3)}{(x + 6)(x - 3)(x + 3)} = x - 4$

53. $\dfrac{2}{t+2} - \dfrac{t}{t^2-4} = \dfrac{2(t-2)}{(t-2)(t+2)} - \dfrac{t}{(t-2)(t+2)} = \dfrac{2t-4-t}{(t-2)(t+2)} = \dfrac{t-4}{t^2-4}$

54. $\dfrac{4a}{3ab^2} + \dfrac{b}{a^2c} = \dfrac{4a^2c}{3a^2b^2c} + \dfrac{3b^3}{3a^2b^2c} = \dfrac{4a^2c+3b^3}{3a^2b^2c}$

55. $\dfrac{1}{x+4} + \dfrac{1}{x-4} = \dfrac{-7}{x^2-16} \Rightarrow 1(x-4)+1(x+4)=-7 \Rightarrow 2x=-7 \Rightarrow x=-\dfrac{7}{2}$

56. $\dfrac{1}{y^2+3y-4} - \dfrac{y}{y-1} = -1 \Rightarrow 1-y(y+4)=-1(y+4)(y-1) \Rightarrow 1-y^2-4y=-y^2-3y+4 \Rightarrow$

   $1-4y=4-3y \Rightarrow y=-3$

57. $J = \dfrac{y+z}{z} \Rightarrow Jz = y+z \Rightarrow Jz-z=y \Rightarrow z(J-1)=y \Rightarrow z=\dfrac{y}{J-1}$

58. $\dfrac{\dfrac{4}{x^2}+\dfrac{1}{x}}{\dfrac{4}{x^2}-\dfrac{1}{x}} = \dfrac{\dfrac{4}{x^2}+\dfrac{1}{x}}{\dfrac{4}{x^2}-\dfrac{1}{x}} \cdot \dfrac{x^2}{x^2} = \dfrac{4+x}{4-x}$

59. $y=\dfrac{k}{x} \Rightarrow 2=\dfrac{k}{4} \Rightarrow k=8 \Rightarrow y=\dfrac{8}{x}$, so when $x=16$, $y=\dfrac{8}{16}=\dfrac{1}{2}$

60.
$$
\begin{array}{r}
x^2-2x+6 \\
x+2\overline{\smash{)}\,x^3+0x^2+2x+11} \\
\underline{x^3+2x^2} \\
-2x^2+2x \\
\underline{-2x^2-4x} \\
6x+11 \\
\underline{6x+12} \\
-1
\end{array}
$$
The solution is: $x^2-2x+6-\dfrac{1}{x+2}$

61. $(f+g)(3) = \left(3^2-3(3)+2\right)+(3-2) = (9-9+2)+1 = 2+1 = 3$

62. $(f/g)(x) = \dfrac{x^2-3x+2}{x-2} = \dfrac{(x-2)(x-1)}{x-2} = x-1$

63. (a) See Figure 63. These data are linear.

   (b) $m = \dfrac{100-0}{212-32} = \dfrac{100}{180} = \dfrac{5}{9}$. Letting $h=32$ and $k=0$, the function is $f(x)=\dfrac{5}{9}(x-32)$.

   (c) $f(104) = \dfrac{5}{9}(104-32) = \dfrac{5}{9}(72) = 40°C$

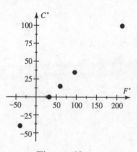

Figure 63

64. Let x represent the length of each of the shorter sides. Then $\frac{1}{2}x+7$ represents the length of the

longer side. The perimeter is 22 so $x+x+\frac{1}{2}x+7=22 \Rightarrow \frac{5}{2}x=15 \Rightarrow x=6$.

The sides are 6, 6, and 10 inches.

65. Let x and y represent the time spent on the stair climber and the bicycle respectively. First convert 90

minutes to $\frac{90}{60}$ hrs $=1.5$ hrs Then the system needed is $x+y=1.5$ and $690x+540y=885$.

Multiplying the first equation by -540 and adding the two equations will eliminate the variable y.

$$\begin{array}{r} -540x-540y=-810 \\ 690x+540y=885 \\ \hline 150x=75 \end{array}$$ Thus, $x=0.5$. And so $(0.5)+y=1.5 \Rightarrow y=1$. The athlete spent 0.5 hour

on the stair climber and 1 hour on the stationary bicycle.

66. Let x and y represent the number of \$2 and \$5 tickets respectively. Then the system needed is

$x+y=30$ and $2x+5y=78$. Multiplying the first equation by -2 and adding the two equations

will eliminate the x. $$\begin{array}{r} -2x-2y=-60 \\ 2x+5y=78 \\ \hline 3y=18 \end{array}$$ Thus, $y=6$. And so $x+(6)=30 \Rightarrow x=24$. There were 24

children and 6 adults.

67. Let x represent the width of the rectangle. Then $x+4$ represents the length.

$$x(x+4)=165 \Rightarrow x^2+4x-165=0 \Rightarrow (x-11)(x+15)=0 \Rightarrow x=11 \text{ or } x=-15$$

The solution $x=-15$ has no physical meaning. The rectangle is 11 feet by 15 feet.

68. (a) $-16t^2+44t=0 \Rightarrow 4t^2-11t=0 \Rightarrow t(4t-11)=0 \Rightarrow t=0 \text{ or } 4t-11=0 \Rightarrow t=0 \text{ or } t=\frac{11}{4}$

        The ball strikes the ground after $\frac{11}{4}=2.75$ seconds.

(b) $-16t^2+44t=18 \Rightarrow -16t^2+44t-18=0 \Rightarrow 8t^2-22t+9=0 \Rightarrow (2t-1)(4t-9) \Rightarrow$

        $t=\frac{1}{2} \text{ or } t=\frac{9}{4}$. The ball is 18 feet high after $\frac{1}{2}=0.5$ second and $\frac{9}{4}=2.25$ seconds.

69. (a) $\dfrac{x}{15} + \dfrac{x}{10} = 1$

    (b) $\dfrac{x}{15} + \dfrac{x}{10} = 1 \Rightarrow 2x + 3x = 30 \Rightarrow 5x = 30 \Rightarrow x = 6 \text{ hours}$

70. $\dfrac{6}{4} = \dfrac{x}{32} \Rightarrow 4x = 192 \Rightarrow x = 48 \text{ feet}$

Chapter 7: Radical Expressions and Functions

7.1: Radical Expressions and Functions

1. ± 3

3. 2

5. b

7. undefined

9. d

11. See Figure 11.

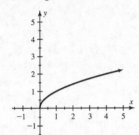

Figure 11

13. The domain is $\{x \mid x \geq 0\}$. See Figure 11 above.

15. $\sqrt{9} = 3$

17. $\sqrt{0.36} = 0.6$

19. $\sqrt{\dfrac{16}{25}} = \dfrac{4}{5}$

21. $\sqrt{x^2} = x$ since $x > 0$

23. $\sqrt[3]{27} = 3$

25. $\sqrt[3]{-64} = -4$

27. $\sqrt[3]{\dfrac{8}{27}} = \dfrac{2}{3}$

29. $-\sqrt[3]{x^9} = -\sqrt[3]{\left(x^3\right)^3} = -x^3$

31. $\sqrt[3]{(2x)^6} = \sqrt[3]{\left((2x)^2\right)^3} = (2x)^2 = 4x^2$

33. $\sqrt[4]{81} = 3$

35. $\sqrt[5]{-243} = \sqrt[5]{(-3)^5} = -3$

37. $\sqrt[4]{-16}$ is not possible to evaluate over the real numbers since the index is even and the radicand is negative.

39. $-\sqrt{5} \approx -2.24$

41. $\sqrt[3]{5} \approx 1.71$

43. $\sqrt[5]{-7} \approx -1.48$

45. $\sqrt{(-4)^2} = \sqrt{16} = 4$

47. $\sqrt{y^2} = |y|$

49. $\sqrt{(x-5)^2} = |x-5|$

51. $\sqrt{x^2 - 2x + 1} = \sqrt{(x-1)^2} = |x-1|$

53. $\sqrt[4]{y^4} = |y|$

55. $\sqrt[4]{x^{12}} = \sqrt[4]{(x^3)^4} = |x^3|$

57. $\sqrt[5]{x^5} = x$

59. $f(10) = \sqrt{10-1} = \sqrt{9} = 3;\ f(0)$ is not possible because $\sqrt{0-1} = \sqrt{-1}$

61. $f(-1) = \sqrt{3-3(-1)} = \sqrt{3+3} = \sqrt{6};\ f(5)$ is not possible because $\sqrt{3-3(5)} = \sqrt{3-15} = \sqrt{-12}$

63. $f(-4) = \sqrt{(-4)^2 - (-4)} = \sqrt{16+4} = \sqrt{20} = \sqrt{4 \cdot 5} = 2\sqrt{5};\ f(3) = \sqrt{3^2 - 3} = \sqrt{9-3} = \sqrt{6}$

65. $f(-3) = \sqrt[3]{(-3)^2 - 8} = \sqrt[3]{9-8} = \sqrt[3]{1} = 1;\ f(4) = \sqrt[3]{4^2 - 8} = \sqrt[3]{16-8} = \sqrt[3]{8} = 2$

67. $f(1) = \sqrt[3]{1-9} = \sqrt[3]{-8} = -2;\ f(10) = \sqrt[3]{10-9} = \sqrt[3]{1} = 1$

69. $f(-2) = \sqrt[3]{3-(-2)^2} = \sqrt[3]{3-4} = \sqrt[3]{-1} = -1;\ f(3) = \sqrt[3]{3-3^2} = \sqrt[3]{3-9} = \sqrt[3]{-6}$ or $-\sqrt[3]{6}$

71. $T(64) = \frac{1}{2}\sqrt{64} = \frac{1}{2}(8) = 4$

73. $f(4) = \sqrt{4+5} + \sqrt{4} = \sqrt{9} + \sqrt{4} = 3+2 = 5$

75. $x+2 \geq 0 \Rightarrow x \geq -2 \Rightarrow$ Domain: $[-2, \infty)$

77. $x-2 \geq 0 \Rightarrow x \geq 2 \Rightarrow$ Domain: $[2, \infty)$

79. $2x-4 \geq 0 \Rightarrow 2x \geq 4 \Rightarrow x \geq 2 \Rightarrow$ Domain: $[2, \infty)$

81. $1-x \geq 0 \Rightarrow -x \geq -1 \Rightarrow x \leq 1 \Rightarrow$ Domain: $(-\infty, 1]$

83. $8-5x \geq 0 \Rightarrow -5x \geq -8 \Rightarrow x \leq \frac{8}{5} \Rightarrow$ Domain: $\left(-\infty, \frac{5}{8}\right]$

85. $3x^2 + 4 \geq 0 \Rightarrow 3x^2 \geq -4 \Rightarrow x^2 \geq -\frac{4}{3} \Rightarrow$ Domain: $(-\infty, \infty)$

87. $2x+1 > 0 \Rightarrow 2x > -1 \Rightarrow x > -\frac{1}{2} \Rightarrow$ Domain: $\left(-\frac{1}{2}, \infty\right)$

89. See Figure 89. This graph is shifted 2 units left.

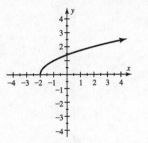

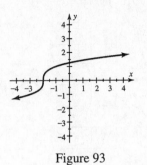

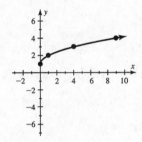

| Figure 89 | Figure 91 | Figure 93 |

91. See Figure 91. This graph is shifted 2 units upward.

93. See Figure 93. This graph is shifted 2 units left.

95. $f(x) = \sqrt{x} + 1$

x	$\sqrt{x}+1$
-1	—
0	1
1	2
4	3
9	4

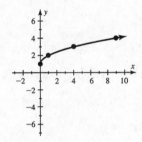

97. $f(x) = \sqrt{3x}$

x	$\sqrt{3x}$
-1	—
0	0
$\frac{1}{3}$	1
$\frac{4}{3}$	2
3	3

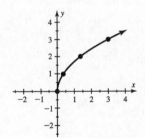

99. $f(x) = 2\sqrt[3]{x}$

x	$2\sqrt[3]{x}$
-8	-4
-1	-2
0	0
1	2
8	4

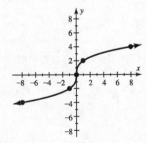

101. $f(x) = \sqrt[3]{x} - 1$

x	$\sqrt[3]{x} - 1$
−7	−2
0	−1
1	0
2	1
9	2

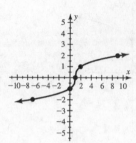

103. $s = \dfrac{1}{2}(3+4+5) = 6 \implies A = \sqrt{6(6-3)(6-4)(6-5)} = \sqrt{6(3)(2)(1)} = \sqrt{36} = 6$

105. $T(4) = \dfrac{\sqrt{4}}{2} = \dfrac{2}{2} = 1$ second

107. $d = 1.22\sqrt{10,000} \Rightarrow d = 1.22(100) \Rightarrow d = 122$ miles

109. $R(16) - R(15) = 108\sqrt{16} - 108\sqrt{15} \approx 13.72$. The result is about \$14 thousand. Since the revenue is lower than the salary of the 16th employee it is not a good decision to hire the additional employee.

111. (a) $P(25,000) = 400\sqrt{25,000} + 8000 \approx 71245.56$, If \$25,000 is spent on equipment per worker, each worker will produce about \$71,246 worth of goods.

(b)   See Figure 111.

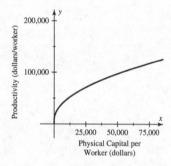

Figure 111

(c) $P(50,000) - P(25,000) = \left(400\sqrt{50,000} + 8000\right) - \left(400\sqrt{25,000} + 8000\right) \approx 26,197$

$P(75,000) - P(50,000) = \left(400\sqrt{75,000} + 8000\right) - \left(400\sqrt{50,000} + 8000\right) \approx 20,102$

An additional \$25,000 spent on equipment per worker. There is a point where the business starts to lose money.

Section 7.2 Rational Exponents

1. $4^{1/2} = \sqrt{4} = 2$

3. $4^{-1/2} = \dfrac{1}{\sqrt{4}} = \dfrac{1}{2}$

5. $\sqrt{x} = x^{1/2}$

7. $\sqrt[n]{a}$

9. c, $\sqrt{x^3} = x^{3/2}$

11. g, $25^{-1/2} = \dfrac{1}{\sqrt{25}} = \dfrac{1}{5}$

13. d, $x^{1/5} = \sqrt[5]{x}$

15. a, $\sqrt{x} \cdot \sqrt[3]{x} = x^{1/2} \cdot x^{1/3} = x^{(1/2)+(1/3)} = x^{5/6} = \sqrt[6]{x^5}$

17. $7^{1/2} = \sqrt{7}$

19. $a^{1/3} = \sqrt[3]{a}$

21. $x^{5/6} = \sqrt[6]{x^5}$

23. $(x+5)^{1/2} = \sqrt{x+5}$

25. $b^{-2/3} = \dfrac{1}{b^{2/3}} = \dfrac{1}{\sqrt[3]{b^2}}$

27. $\sqrt{t} = t^{1/2}$

29. $\sqrt[3]{(x+1)} = (x+1)^{1/3}$

31. $\dfrac{1}{\sqrt{x+1}} = \dfrac{1}{(x+1)^{1/2}} = (x+1)^{-1/2}$

33. $\sqrt{a^2 - b^2} = \left(a^2 - b^2\right)^{1/2}$

35. $\dfrac{1}{\sqrt[3]{x^7}} = \dfrac{1}{(x)^{7/3}} = (x)^{-7/3}$

37. $16^{1/5} \approx 1.74$

39. $5^{1/3} \approx 1.71$

41. $9^{3/5} \approx 3.74$

43. $4^{-3/7} \approx 0.55$

45. $9^{1/2} = \sqrt{9} = 3$

47. $8^{1/3} = \sqrt[3]{8} = 2$

49. $\left(\dfrac{4}{9}\right)^{1/2} = \sqrt{\dfrac{4}{9}} = \dfrac{2}{3}$

51. $(-8)^{2/3} = \sqrt[3]{(-8)^2} = \sqrt[3]{64} = 4$

53. $\left(\dfrac{1}{8}\right)^{-1/3} = 8^{1/3} = \sqrt[3]{8} = 2$

55. $16^{-3/4} = \dfrac{1}{\sqrt[4]{16^3}} = \dfrac{1}{\left(\sqrt[4]{16}\right)^3} = \dfrac{1}{(2)^3} = \dfrac{1}{8}$

57. $\left(4^{1/2}\right)^{-3} = \dfrac{1}{\left(\sqrt{4}\right)^3} = \dfrac{1}{2^3} = \dfrac{1}{8}$

59. $z^{1/4} = \sqrt[4]{z}$

61. $y^{-2/5} = \dfrac{1}{\sqrt[5]{y^2}}$

63. $(3x)^{1/3} = \sqrt[3]{3x}$

65. $\sqrt{y} = y^{1/2}$

67. $\sqrt{x} \cdot \sqrt{x} = \left(x^{1/2}\right)^2 = x$

69. $\sqrt[3]{8x^2} = \sqrt[3]{8} \cdot \sqrt[3]{x^2} = 2x^{2/3}$

71. $\dfrac{\sqrt{49x}}{\sqrt[3]{x^2}} = \dfrac{\sqrt{49} \cdot \sqrt{x}}{x^{2/3}} = 7x^{1/2-2/3} = 7x^{-1/6} = \dfrac{7}{x^{1/6}}$

73. $\left(b^{1/a}\, b^{2/a}\right)^a = \left(b^{1/a+2/a}\right)^a = \left(b^{3/a}\right)^a = b^{\frac{3}{a}\cdot a} = b^3$

75. $\left(x^2\right)^{3/2} = x^{2\cdot 3/2} = x^3$

77. $\sqrt[3]{x^3 y^6} = \left(x^3 y^6\right)^{1/3} = x^{3\cdot 1/3} \cdot y^{6\cdot 1/3} = xy^2$

79. $\sqrt{y^3} \cdot \sqrt[3]{y^2} = (y^3)^{1/2} \cdot (y^2)^{1/3} = y^{3\cdot 1/2} \cdot y^{2\cdot 1/3} = y^{3/2} \cdot y^{2/3} = y^{3/2+2/3} = y^{13/6}$

81. $\left(\dfrac{x^6}{27}\right)^{2/3} = \dfrac{x^{6\cdot 2/3}}{27^{2/3}} = \dfrac{x^4}{\left(\sqrt[3]{27}\right)^2} = \dfrac{x^4}{3^2} = \dfrac{x^4}{9}$

83. $\left(\dfrac{x^2}{y^6}\right)^{-1/2} = \left(\dfrac{y^6}{x^2}\right)^{1/2} = \dfrac{y^{6\cdot 1/2}}{x^{2\cdot 1/2}} = \dfrac{y^3}{x}$

85. $\sqrt{\sqrt{y}} = \left(y^{1/2}\right)^{1/2} = y^{1/2\cdot 1/2} = y^{1/4}$

87. $\left(a^{-1/2}\right)^{4/3} = a^{-1/2\cdot 4/3} = a^{-2/3} = \dfrac{1}{a^{2/3}}$

89. $\dfrac{\left(k^{1/2}\right)^{-3}}{\left(k^2\right)^{1/4}} = \dfrac{k^{-3/2}}{k^{1/2}} = k^{-3/2-1/2} = k^{-4/2} = k^{-2} = \dfrac{1}{k^2}$

91. $\sqrt{b} \cdot \sqrt[4]{b} = b^{1/2} \cdot b^{1/4} = b^{1/2+1/4} = b^{3/4}$

93. $p^{1/2}\left(p^{3/2} + p^{1/2}\right) = p^{1/2+3/2} + p^{1/2+1/2} = p^2 + p$

95. $\sqrt[3]{x}\left(\sqrt{x} - \sqrt[3]{x^2}\right) = x^{1/3}\left(x^{1/2} - x^{2/3}\right) = x^{1/3+1/2} - x^{1/3+2/3} = x^{5/6} - x$

97. $\dfrac{\sqrt[3]{27x}}{\sqrt{x}} = \dfrac{\sqrt[3]{27} \cdot \sqrt[3]{x}}{\sqrt{x}} = \dfrac{3x^{1/3}}{x^{1/2}} = 3x^{(1/3)-(1/2)} = 3x^{-1/6} = \dfrac{3}{x^{1/6}}$

99. (a)

x	0	20	40	60	80
$A(x)$	0	27%	37%	44%	50%

(b) The abandonment rates level off. The longer a person watches a video, the more likely they are to continue watching.

101. $N(h) = 1.6h^{-1/2} = 1.6(2.5)^{-1/2} = \dfrac{1.6}{\sqrt{2.5}} \approx 1.01$; The stepping frequency is about 1 step per second.

103. (a) $A = 100\sqrt[3]{8^2} = 100\sqrt[3]{64} = 100 \cdot 4 = 400$ square inches

(b) $A = 100\sqrt[3]{W^2} \Rightarrow A = 100W^{2/3}$

105. (a) $H(12) = 35.2(12)^{3/40} \approx 42.4$ cm, $H(24) = 35.2(24)^{3/40} \approx 44.7$ cm

(b) According to our results above an infant's head size increases the most in the first year.

107. See exercise 101 for an example. *Answers may vary.*

Checking Basic Concepts Sections 7.1 and 7.2

1. (a) ± 7

   (b) 7

2. (a) $\sqrt[3]{-8} = -2$

   (b) $-\sqrt[4]{81} = -3$

3. (a) $x^{3/2} = \sqrt{x^3}$ or $\left(\sqrt{x}\right)^3$

   (b) $x^{2/3} = \sqrt[3]{x^2}$ or $\left(\sqrt[3]{x}\right)^2$

   (c) $x^{-2/5} = \dfrac{1}{\sqrt[5]{x^2}}$ or $\dfrac{1}{\left(\sqrt[5]{x}\right)^2}$

4. $\sqrt{(x-1)^2} = |x-1|$

5. (a) $f(9) = \sqrt{9} = \sqrt{3^2} = 3$

   (b) $g(125) = \sqrt[3]{125} = \sqrt[3]{5^3} = 5$

(c) $h\left(\dfrac{12}{7}\right)=\left(\dfrac{12}{7}\right)^{7/12}\approx 1.37$

Section 7.3 Simplifying Radical Expressions

1. Yes

3. $\sqrt[3]{ab}$

5. $\dfrac{a}{b}$

7. No; $\sqrt{50}=\sqrt{25\cdot 2}=\sqrt{25}\cdot\sqrt{2}=5\sqrt{2}$ and $\sqrt{25}+\sqrt{25}=5+5=2(5)$ or 10

9. No, because $1^3\neq 3$.

11. $\sqrt{3}\cdot\sqrt{3}=\sqrt{3\cdot 3}=\sqrt{9}=3$

13. $\sqrt{2}\cdot\sqrt{50}=\sqrt{2\cdot 50}=\sqrt{100}=10$

15. $\sqrt[3]{4}\cdot\sqrt[3]{16}=\sqrt[3]{4\cdot 16}=\sqrt[3]{64}=4$

17. $\sqrt{\dfrac{9}{25}}=\dfrac{\sqrt{9}}{\sqrt{25}}=\dfrac{3}{5}$

19. $\sqrt{\dfrac{1}{2}}\cdot\sqrt{\dfrac{1}{8}}=\sqrt{\dfrac{1\cdot 1}{2\cdot 8}}=\sqrt{\dfrac{1}{16}}=\dfrac{\sqrt{1}}{\sqrt{16}}=\dfrac{1}{4}$

21. $\sqrt[3]{\dfrac{2}{3}}\cdot\sqrt[3]{\dfrac{4}{3}}\cdot\sqrt[3]{\dfrac{1}{3}}=\sqrt[3]{\dfrac{2}{3}\cdot\dfrac{4}{3}\cdot\dfrac{1}{3}}=\sqrt[3]{\dfrac{8}{27}}=\dfrac{2}{3}$

23. $\sqrt{x^3}\cdot\sqrt{x^3}=\sqrt{x^3\cdot x^3}=\sqrt{x^6}=x^{6/2}=x^3$

25. $\sqrt[3]{\dfrac{7}{27}}=\dfrac{\sqrt[3]{7}}{\sqrt[3]{27}}=\dfrac{\sqrt[3]{7}}{3}$

27. $\sqrt[4]{\dfrac{x}{81}}=\dfrac{\sqrt[4]{x}}{\sqrt[4]{81}}=\dfrac{\sqrt[4]{x}}{3}$

29. $\sqrt{\dfrac{9}{z^2}}=\dfrac{\sqrt{9}}{\sqrt{z^2}}=\dfrac{3}{z}$

31. $\sqrt{\dfrac{x}{2}}\cdot\sqrt{\dfrac{x}{8}}=\sqrt{\dfrac{x\cdot x}{2\cdot 8}}=\sqrt{\dfrac{x^2}{16}}=\dfrac{\sqrt{x^2}}{\sqrt{16}}=\dfrac{x}{4}$

33. $\dfrac{\sqrt{45}}{\sqrt{5}}=\sqrt{\dfrac{45}{5}}=\sqrt{9}=3$

35. $\sqrt[3]{-4}\cdot\sqrt[3]{-16}=\sqrt[3]{-4\cdot(-16)}=\sqrt[3]{64}=4$

37. $\sqrt[4]{9}\cdot\sqrt[4]{9}=\sqrt[4]{9\cdot 9}=\sqrt[4]{81}=3$

39. $\dfrac{\sqrt[5]{64}}{\sqrt[5]{-2}} = \sqrt[5]{\dfrac{64}{-2}} = \sqrt[5]{-32} = -2$

41. $\dfrac{\sqrt{a^2 b}}{\sqrt{b}} = \sqrt{\dfrac{a^2 b}{b}} = \sqrt{a^2} = a$

43. $\dfrac{\sqrt[3]{54}}{\sqrt[3]{2}} = \sqrt[3]{\dfrac{54}{2}} = \sqrt[3]{27} = 3$

45. $\sqrt{4x^4} = \sqrt{4} \cdot \sqrt{\left(x^2\right)^2} = 2x^2$

47. $\sqrt[3]{-5a^6} = \sqrt[3]{-5} \cdot \sqrt[3]{\left(a^2\right)^3} = \sqrt[3]{-5} \cdot a^2 = -a^2 \sqrt[3]{5}$

49. $\sqrt[4]{16x^4 y} = \sqrt[4]{16} \cdot \sqrt[4]{x^4} \cdot \sqrt[4]{y} = 2x\sqrt[4]{y}$

51. $\sqrt{3x} \cdot \sqrt{12x} = \sqrt{3 \cdot 12 \cdot x \cdot x} = \sqrt{36x^2} = \sqrt{36} \cdot \sqrt{x^2} = 6x$

53. $\sqrt[3]{8x^6 y^3 z^9} = \sqrt[3]{8} \cdot \sqrt[3]{\left(x^2\right)^3} \cdot \sqrt[3]{y^3} \cdot \sqrt[3]{\left(z^3\right)^3} = 2x^2 yz^3$

55. $\sqrt[4]{\dfrac{3}{4}} \cdot \sqrt[4]{\dfrac{27}{4}} = \sqrt[4]{\dfrac{3}{4} \cdot \dfrac{27}{4}} = \sqrt[4]{\dfrac{81}{16}} = \dfrac{\sqrt[4]{81}}{\sqrt[4]{16}} = \dfrac{3}{2}$

57. $\sqrt[3]{12} \cdot \sqrt[3]{ab} = \sqrt[3]{12ab}$

59. $\sqrt[4]{25z} \cdot \sqrt[4]{25z} = \sqrt[4]{625z^2} = \sqrt[4]{625} \cdot \sqrt[4]{z^2} = 5z^{\frac{2}{4}} = 5z^{\frac{1}{2}} = 5\sqrt{z}$

61. $\sqrt[5]{\dfrac{7a}{b^2}} \cdot \sqrt[5]{\dfrac{b^2}{7a^6}} = \sqrt[5]{\dfrac{7ab^2}{7a^6 b^2}} = \sqrt[5]{\dfrac{1}{a^5}} = \dfrac{1}{a}$

63. $\sqrt{x+4} \cdot \sqrt{x-4} = \sqrt{(x+4)(x-4)} = \sqrt{x^2 - 16}$

65. $\sqrt[3]{a+1} \cdot \sqrt[3]{a^2 - a + 1} = \sqrt[3]{(a+1)\left(a^2 - a + 1\right)} = \sqrt[3]{a^3 + 1}$

67. $\dfrac{\sqrt{x^2 + 2x + 1}}{\sqrt{x+1}} = \sqrt{\dfrac{x^2 + 2x + 1}{x+1}} = \sqrt{\dfrac{(x+1)(x+1)}{x+1}} = \sqrt{x+1}$

69. $\sqrt{500} = \sqrt{100 \cdot 5} = \sqrt{100} \cdot \sqrt{5} = 10\sqrt{5}$; the answer is 10.

71. $\sqrt{8} = \sqrt{4 \cdot 2} = \sqrt{4} \cdot \sqrt{2} = 2\sqrt{2}$; the answer is 2.

73. $\sqrt{45} = \sqrt{9 \cdot 5} = \sqrt{9} \cdot \sqrt{5} = 3\sqrt{5}$; the answer is 3.

75. $\sqrt{200} = \sqrt{100 \cdot 2} = \sqrt{100} \cdot \sqrt{2} = 10\sqrt{2}$

77. $\sqrt[3]{81} = \sqrt[3]{27 \cdot 3} = \sqrt[3]{27} \cdot \sqrt[3]{3} = 3\sqrt[3]{3}$

79. $\sqrt[4]{64} = \sqrt[4]{16 \cdot 4} = \sqrt[4]{16} \cdot \sqrt[4]{4} = 2\sqrt[4]{4} = 2\sqrt[4]{2^2} = 2\sqrt{2}$

81. $\sqrt[5]{-64} = \sqrt[5]{-2^6} = \sqrt[5]{-2^5 \cdot 2} = \sqrt[5]{-2^5} \cdot \sqrt[5]{2} = -2\sqrt[5]{2}$

83. $\sqrt{b^5} = \sqrt{\left(b^2\right)^2 \cdot b} = \sqrt{\left(b^2\right)^2} \cdot \sqrt{b} = b^2\sqrt{b}$

85. $\sqrt{8n^3} = \sqrt{(2n)^2 \cdot 2n} = \sqrt{(2n)^2} \cdot \sqrt{2n} = 2n\sqrt{2n}$

87. $\sqrt{12a^2b^5} = \sqrt{\left(2ab^2\right)^2 \cdot 3b} = \sqrt{\left(2ab^2\right)^2} \cdot \sqrt{3b} = 2ab^2\sqrt{3b}$

89. $\sqrt[3]{-125x^4y^5} = \sqrt[3]{(-5xy)^3 \cdot xy^2} = \sqrt[3]{(-5xy)^3} \cdot \sqrt[3]{xy^2} = -5xy\sqrt[3]{xy^2}$

91. $\sqrt[3]{5t} \cdot \sqrt[3]{125t} = \sqrt[3]{625t^2} = \sqrt[3]{5^4t^2} = \sqrt[3]{5^3 \cdot 5t^2} = \sqrt[3]{5^3} \cdot \sqrt[3]{5t^2} = 5\sqrt[3]{5t^2}$

93. $\sqrt[4]{\dfrac{9t^5}{r^8}} \cdot \sqrt[4]{\dfrac{9r}{5t}} = \sqrt[4]{\dfrac{81rt^5}{5r^8t}} = \sqrt[4]{\dfrac{81t^4}{5r^7}} = \dfrac{\sqrt[4]{(3t)^4}}{\sqrt[4]{r^4 \cdot 5r^3}} = \dfrac{3t}{r\sqrt[4]{5r^3}}$

95. $\sqrt[3]{\dfrac{27x^2}{y^3}} = \dfrac{\sqrt[3]{27} \cdot \sqrt[3]{x^2}}{\sqrt[3]{y^3}} = \dfrac{3\sqrt[3]{x^2}}{y}$

97. $\sqrt{\dfrac{7a^2}{27}} \cdot \sqrt{\dfrac{7a}{3}} = \sqrt{\dfrac{49a^3}{81}} = \dfrac{\sqrt{49} \cdot \sqrt{a^3}}{\sqrt{81}} = \dfrac{7\sqrt{a^2} \cdot \sqrt{a}}{9} = \dfrac{7a\sqrt{a}}{9}$

99. $\left(\sqrt[mn]{a^mb^m}\right)^n = \left(\sqrt[m]{a^mb^m}\right)^{n/n} = \sqrt[m]{(ab)^m} = (ab)^{m/m} = ab$

101. $\sqrt{3} \cdot \sqrt[3]{3} = 3^{1/2} \cdot 3^{1/3} = 3^{1/2+1/3} = 3^{5/6} = \sqrt[6]{3^5}$

103. $\sqrt[4]{8} \cdot \sqrt[3]{4} = \sqrt[4]{2^3} \cdot \sqrt[3]{2^2} = 2^{3/4} \cdot 2^{2/3} = 2^{3/4+2/3} = 2^{17/12} = 2^{12/12+5/12} = 2 \cdot 2^{5/12} = 2\sqrt[12]{2^5}$

105. $\sqrt[4]{27} \cdot \sqrt[3]{9} \cdot \sqrt{3} = \sqrt[4]{3^3} \cdot \sqrt[3]{3^2} \cdot \sqrt{3} = 3^{3/4} \cdot 3^{2/3} \cdot 3^{1/2} = 3^{3/4+2/3+1/2} = 3^{23/12} = 3^{12/12} \cdot 3^{11/12} = 3\sqrt[12]{3^{11}}$

107. $\sqrt[4]{x^3} \cdot \sqrt[3]{x} = x^{3/4} \cdot x^{1/3} = x^{3/4+1/3} = x^{13/12} = x^{12/12} \cdot x^{1/12} = x\sqrt[12]{x}$

109. $\sqrt[4]{rt} \cdot \sqrt[3]{r^2t} = (rt)^{1/4} \cdot \left(r^2t\right)^{1/3} = r^{1/4}t^{1/4} \cdot r^{2/3}t^{1/3} = r^{1/4+2/3}t^{1/4+1/3} = r^{11/12}t^{7/12} = \sqrt[12]{r^{11}t^7}$

111. (a) $S = \sqrt{25M} = \sqrt{5^2}\sqrt{M} = 5\sqrt{M}$

   (b) $S = 5\sqrt{100} = 50$ mph

Section 7.4 Operations on Radical Expressions

1. $2\sqrt{a}$

3. like

5. No; 6 and $3\sqrt{5}$ are not like radicals.

7. $\sqrt{t} + 5$

9. Not possible, since $\sqrt{12} = 2\sqrt{3}$ and $\sqrt{24} = 2\sqrt{6}$.

11. Since $\sqrt{28} = \sqrt{4 \cdot 7} = \sqrt{4} \cdot \sqrt{7} = 2\sqrt{7}$ and $\sqrt{63} = \sqrt{9 \cdot 7} = \sqrt{9} \cdot \sqrt{7} = 3\sqrt{7},$ the like radicals are

$\sqrt{7}, 2\sqrt{7},$ and $3\sqrt{7}.$

13. Since $\sqrt[3]{16} = \sqrt[3]{8 \cdot 2} = \sqrt[3]{8} \cdot \sqrt[3]{2} = 2\sqrt[3]{2}$ and $\sqrt[3]{-54} = \sqrt[3]{-27 \cdot 2} = \sqrt[3]{-27} \cdot \sqrt[3]{2} = -3\sqrt[3]{2},$ the like radicals are

$2\sqrt[3]{2}$ and $-3\sqrt[3]{2}.$

15. Not possible, since $\sqrt{x^2 y} = x\sqrt{y}$ and $\sqrt{4y^2} = 2y.$

17. Since $\sqrt[3]{8xy} = \sqrt[3]{8} \cdot \sqrt[3]{xy} = 2\sqrt[3]{xy}$ and $\sqrt[3]{x^4 y^4} = \sqrt[3]{(xy)^3 \cdot xy} = \sqrt[3]{(xy)^3} \cdot \sqrt[3]{xy} = xy\sqrt[3]{xy},$ the like radicals

are $2\sqrt[3]{xy}$ and $xy\sqrt[3]{xy}.$

19. $2\sqrt{3} + 7\sqrt{3} = 9\sqrt{3}$

21. $4\sqrt[3]{5} + 2\sqrt[3]{5} = 6\sqrt[3]{5}$

23. Not possible, since 7 and $4\sqrt{7}$ are not like radicals

25. Not possible, since $2\sqrt{3}$ and $3\sqrt{2}$ are not like radicals

27. Not possible, since $\sqrt{3}$ and $\sqrt[3]{3}$ are not like radicals

29. $\sqrt[3]{16} + 3\sqrt[3]{2} = \sqrt[3]{8 \cdot 2} + 3\sqrt[3]{2} = \sqrt[3]{8} \cdot \sqrt[3]{2} + 3\sqrt[3]{2} = 2\sqrt[3]{2} + 3\sqrt[3]{2} = 5\sqrt[3]{2}$

31. $\sqrt{2} + \sqrt{18} + \sqrt{32} = \sqrt{2} + \sqrt{9 \cdot 2} + \sqrt{16 \cdot 2} = \sqrt{2} + \sqrt{9} \cdot \sqrt{2} + \sqrt{16} \cdot \sqrt{2} = \sqrt{2} + 3\sqrt{2} + 4\sqrt{2} = 8\sqrt{2}$

33. $11\sqrt{11} - 5\sqrt{11} = 6\sqrt{11}$

35. $\sqrt{x} + \sqrt{x} - \sqrt{y} = 2\sqrt{x} - \sqrt{y}$

37. $\sqrt[3]{z} + \sqrt[3]{z} = 2\sqrt[3]{z}$

39. $2\sqrt[3]{6} - 7\sqrt[3]{6} = -5\sqrt[3]{6}$

41. $\sqrt[3]{y^6} - \sqrt[3]{y^3} = \sqrt[3]{(y^2)^3} - \sqrt[3]{y^3} = y^2 - y$

43. $3\sqrt{28} + 3\sqrt{7} = 3\sqrt{4 \cdot 7} + 3\sqrt{7} = 3 \cdot 2\sqrt{7} + 3\sqrt{7} = 9\sqrt{7}$

45. $\sqrt[4]{48} + 4\sqrt[4]{3} = \sqrt[4]{16 \cdot 3} + 4\sqrt[4]{3} = \sqrt[4]{16} \cdot \sqrt[4]{3} + 4\sqrt[4]{3} = 2\sqrt[4]{3} + 4\sqrt[4]{3} = 6\sqrt[4]{3}$

47. $\sqrt{9x} + \sqrt{16x} = \sqrt{9} \cdot \sqrt{x} + \sqrt{16} \cdot \sqrt{x} = 3\sqrt{x} + 4\sqrt{x} = 7\sqrt{x}$

49. $3\sqrt{2k} + \sqrt{8k} + \sqrt{18k} = 3\sqrt{2k} + \sqrt{4 \cdot 2k} + \sqrt{9 \cdot 2k} = 3\sqrt{2k} + \sqrt{4} \cdot \sqrt{2k} + \sqrt{9} \cdot \sqrt{2k}$

$= 3\sqrt{2k} + 2\sqrt{2k} + 3\sqrt{2k} = 8\sqrt{2k}$

51. $\sqrt{44} - 4\sqrt{11} = \sqrt{4 \cdot 11} - 4\sqrt{11} = 2\sqrt{11} - 4\sqrt{11} = -2\sqrt{11}$

53. $2\sqrt[3]{16} + \sqrt[3]{2} - \sqrt{2} = 2\sqrt[3]{8 \cdot 2} + \sqrt[3]{2} - \sqrt{2} = 2 \cdot 2\sqrt[3]{2} + \sqrt[3]{2} - \sqrt{2} = 5\sqrt[3]{2} - \sqrt{2}$

55. $\sqrt[3]{xy} - 2\sqrt[3]{xy} = -\sqrt[3]{xy}$

57. $\sqrt{4x+8} + \sqrt{x+2} = \sqrt{4(x+2)} + \sqrt{x+2} = 2\sqrt{x+2} + \sqrt{x+2} = 3\sqrt{x+2}$

59. $\sqrt{9x+18}-\sqrt{4x+8}=\sqrt{9(x+2)}-\sqrt{4(x+2)}=3\sqrt{x+2}-2\sqrt{x+2}=\sqrt{x+2}$

61. $\sqrt{x^3+x^2}-\sqrt{x+1}=\sqrt{x^2(x+1)}-\sqrt{x+1}=x\sqrt{x+1}-1\sqrt{x+1}=(x-1)\sqrt{x+1}$

63. $\sqrt{25x^3}-\sqrt{x^3}=\sqrt{25}\cdot\sqrt{x^2}\cdot\sqrt{x}-\sqrt{x^2}\cdot\sqrt{x}=5x\sqrt{x}-x\sqrt{x}=4x\sqrt{x}$

65. $\sqrt[3]{\dfrac{7x}{8}}-\dfrac{\sqrt[3]{7x}}{3}=\dfrac{\sqrt[3]{7x}}{\sqrt[3]{8}}-\dfrac{\sqrt[3]{7x}}{3}=\dfrac{\sqrt[3]{7x}}{2}-\dfrac{\sqrt[3]{7x}}{3}=\dfrac{3\sqrt[3]{7x}}{6}-\dfrac{2\sqrt[3]{7x}}{6}=\dfrac{\sqrt[3]{7x}}{6}$

67. $\dfrac{4\sqrt{3}}{3}+\dfrac{\sqrt{3}}{6}=\dfrac{4\sqrt{3}}{3}\cdot\dfrac{2}{2}+\dfrac{\sqrt{3}}{6}=\dfrac{8\sqrt{3}}{6}+\dfrac{\sqrt{3}}{6}=\dfrac{8\sqrt{3}+\sqrt{3}}{6}=\dfrac{9\sqrt{3}}{6}=\dfrac{3\sqrt{3}}{2}$

69. $\dfrac{15\sqrt{8}}{4}-\dfrac{2\sqrt{2}}{5}=\dfrac{15\cdot2\sqrt{2}}{4}\cdot\dfrac{5}{5}-\dfrac{2\sqrt{2}}{5}\cdot\dfrac{4}{4}=\dfrac{150\sqrt{2}}{20}-\dfrac{8\sqrt{2}}{20}=\dfrac{150\sqrt{2}-8\sqrt{2}}{20}=\dfrac{142\sqrt{2}}{20}=\dfrac{71\sqrt{2}}{10}$

71. $2\sqrt[4]{64}-\sqrt[4]{324}+\sqrt[4]{4}=2\sqrt[4]{16\cdot4}-\sqrt[4]{81\cdot4}+\sqrt[4]{4}=4\sqrt[4]{4}-3\sqrt[4]{4}+\sqrt[4]{4}=2\sqrt[4]{4}=2\cdot2^{\frac{2}{4}}=2\sqrt{2}$

73. $5\sqrt[4]{x^5}-\sqrt[4]{x}=5\sqrt[4]{x^4\cdot x}-\sqrt[4]{x}=5x\sqrt[4]{x}-\sqrt[4]{x}=(5x-1)\sqrt[4]{x}$

75. $\sqrt{64x^3}-\sqrt{x}+3\sqrt{x}=\sqrt{(8x)^2\cdot x}-\sqrt{x}+3\sqrt{x}=8x\sqrt{x}-\sqrt{x}+3\sqrt{x}=2\sqrt{x}(4x+1)$

77. $\sqrt[4]{81a^5b^5}-\sqrt[4]{ab}=\sqrt[4]{(3ab)^4\cdot ab}-\sqrt[4]{ab}=3ab\sqrt[4]{ab}-\sqrt[4]{ab}=(3ab-1)\sqrt[4]{ab}$

79. $5\sqrt[3]{\dfrac{n^4}{125}}-2\sqrt[3]{n}=5\sqrt[3]{\dfrac{n^3}{125}\cdot n}-2\sqrt[3]{n}=5\cdot\dfrac{n}{5}\sqrt[3]{n}-2\sqrt[3]{n}=n\sqrt[3]{n}-2\sqrt[3]{n}=(n-2)\sqrt[3]{n}$

81. $(f+g)(x)=f(x)+g(x)=\left(5\sqrt{x}-2\right)+\left(-2\sqrt{x}+3\right)=3\sqrt{x}+1$

    $(f-g)(x)=f(x)-g(x)=\left(5\sqrt{x}-2\right)-\left(-2\sqrt{x}+3\right)=7\sqrt{x}-5$

83. $(f+g)(x)=f(x)+g(x)=\left(\sqrt[3]{8x}+1\right)+\left(2\sqrt[3]{x}-1\right)=2\sqrt[3]{x}+1+2\sqrt[3]{x}-1=4\sqrt[3]{x}$

    $(f-g)(x)=f(x)-g(x)=\left(\sqrt[3]{8x}+1\right)-\left(2\sqrt[3]{x}-1\right)=2\sqrt[3]{x}+1-\left(2\sqrt[3]{x}-1\right)=2$

85. $\left(\sqrt{x}-3\right)\left(\sqrt{x}+2\right)=\left(\sqrt{x}\right)^2+2\sqrt{x}-3\sqrt{x}-6=x-\sqrt{x}-6$

87. $\left(3+\sqrt{7}\right)\left(3-\sqrt{7}\right)=3^2-\left(\sqrt{7}\right)^2=9-7=2$

89. $\left(11-\sqrt{2}\right)\left(11+\sqrt{2}\right)=11^2-\left(\sqrt{2}\right)^2=121-2=119$

91. $\left(\sqrt{x}+8\right)\left(\sqrt{x}-8\right)=\left(\sqrt{x}\right)^2-8^2=x-64$

93. $\left(\sqrt{ab}-\sqrt{c}\right)\left(\sqrt{ab}+\sqrt{c}\right)=\left(\sqrt{ab}\right)^2-\left(\sqrt{c}\right)^2=ab-c$

95. $\left(\sqrt{x}-7\right)\left(\sqrt{x}+8\right)=\left(\sqrt{x}\right)^2+8\sqrt{x}-7\sqrt{x}-56=x+\sqrt{x}-56$

97. $\dfrac{1}{\sqrt{7}}=\dfrac{1}{\sqrt{7}}\cdot\dfrac{\sqrt{7}}{\sqrt{7}}=\dfrac{\sqrt{7}}{\left(\sqrt{7}\right)^2}=\dfrac{\sqrt{7}}{7}$

99. $\dfrac{4}{\sqrt{3}}=\dfrac{4}{\sqrt{3}}\cdot\dfrac{\sqrt{3}}{\sqrt{3}}=\dfrac{4\sqrt{3}}{\left(\sqrt{3}\right)^2}=\dfrac{4\sqrt{3}}{3}$

101. $\dfrac{5}{3\sqrt{5}}=\dfrac{5}{3\sqrt{5}}\cdot\dfrac{\sqrt{5}}{\sqrt{5}}=\dfrac{5\sqrt{5}}{3\left(\sqrt{5}\right)^2}=\dfrac{5\sqrt{5}}{3\cdot5}=\dfrac{\sqrt{5}}{3}$

103. $\sqrt{\dfrac{b}{12}}=\dfrac{\sqrt{b}}{\sqrt{12}}\cdot\dfrac{\sqrt{12}}{\sqrt{12}}=\dfrac{\sqrt{12b}}{\left(\sqrt{12}\right)^2}=\dfrac{\sqrt{4\cdot3b}}{12}=\dfrac{\sqrt{4}\cdot\sqrt{3b}}{12}=\dfrac{2\sqrt{3b}}{12}=\dfrac{\sqrt{3b}}{6}$

105. $\dfrac{rt}{2\sqrt{r^3}}=\dfrac{rt}{2\sqrt{r^2\cdot r}}\cdot\dfrac{\sqrt{r}}{\sqrt{r}}=\dfrac{rt\sqrt{r}}{2r\left(\sqrt{r}\right)^2}=\dfrac{t\sqrt{r}}{2r}$

107. $\dfrac{1}{3-\sqrt{2}}=\dfrac{1}{3-\sqrt{2}}\cdot\dfrac{3+\sqrt{2}}{3+\sqrt{2}}=\dfrac{3+\sqrt{2}}{9-2}=\dfrac{3+\sqrt{2}}{7}$

109. $\dfrac{\sqrt{2}}{\sqrt{5}+2}=\dfrac{\sqrt{2}}{\sqrt{5}+2}\cdot\dfrac{\sqrt{5}-2}{\sqrt{5}-2}=\dfrac{\sqrt{10}-2\sqrt{2}}{5-4}=\dfrac{\sqrt{10}-2\sqrt{2}}{1}=\sqrt{10}-2\sqrt{2}$

111. $\dfrac{\sqrt{7}-2}{\sqrt{7}+2}=\dfrac{\sqrt{7}-2}{\sqrt{7}+2}\cdot\dfrac{\sqrt{7}-2}{\sqrt{7}-2}=\dfrac{7-4\sqrt{7}+4}{7-4}=\dfrac{11-4\sqrt{7}}{3}$

113. $\dfrac{1}{\sqrt{7}-\sqrt{6}}=\dfrac{1}{\sqrt{7}-\sqrt{6}}\cdot\dfrac{\sqrt{7}+\sqrt{6}}{\sqrt{7}+\sqrt{6}}=\dfrac{\sqrt{7}+\sqrt{6}}{7-6}=\dfrac{\sqrt{7}+\sqrt{6}}{1}=\sqrt{7}+\sqrt{6}$

115. $\dfrac{\sqrt{z}}{\sqrt{z}-3}=\dfrac{\sqrt{z}}{\sqrt{z}-3}\cdot\dfrac{\sqrt{z}+3}{\sqrt{z}+3}=\dfrac{z+3\sqrt{z}}{z-9}$

117. $\dfrac{\sqrt{a}+\sqrt{b}}{\sqrt{a}-\sqrt{b}}=\dfrac{\sqrt{a}+\sqrt{b}}{\sqrt{a}-\sqrt{b}}\cdot\dfrac{\sqrt{a}+\sqrt{b}}{\sqrt{a}+\sqrt{b}}=\dfrac{a+2\sqrt{ab}+b}{a-b}$

119. $\dfrac{1}{\sqrt{x+1}-\sqrt{x}}=\dfrac{1}{\sqrt{x+1}-\sqrt{x}}\cdot\dfrac{\sqrt{x+1}+\sqrt{x}}{\sqrt{x+1}+\sqrt{x}}=\dfrac{\sqrt{x+1}+\sqrt{x}}{x+1-x}=\dfrac{\sqrt{x+1}+\sqrt{x}}{1}=\sqrt{x+1}+\sqrt{x}$

121. $\dfrac{3}{\sqrt[3]{x}}=\dfrac{3}{x^{1/3}}\cdot\dfrac{x^{2/3}}{x^{2/3}}=\dfrac{3x^{2/3}}{x}=\dfrac{3\sqrt[3]{x^2}}{x}$

123. $\dfrac{1}{\sqrt[3]{x^2}}=\dfrac{1}{x^{2/3}}\cdot\dfrac{x^{1/3}}{x^{1/3}}=\dfrac{x^{1/3}}{x}=\dfrac{\sqrt[3]{x}}{x}$

125. $\sqrt{27}+\sqrt{48}+\sqrt{75}=\sqrt{9\cdot3}+\sqrt{16\cdot3}+\sqrt{25\cdot3}=3\sqrt{3}+4\sqrt{3}+5\sqrt{3}=12\sqrt{3}\approx20.8$ cm

127. A square with a diagonal of length $\sqrt{3}$ has sides of length $\dfrac{\sqrt{3}}{\sqrt{2}}$. The perimeter of this square is

$4\cdot\dfrac{\sqrt{3}}{\sqrt{2}}=4\cdot\dfrac{\sqrt{3}}{\sqrt{2}}\cdot\dfrac{\sqrt{2}}{\sqrt{2}}=\dfrac{4\sqrt{6}}{2}=2\sqrt{6}$.

129. A square with a diagonal of length 60 feet has sides of length $\dfrac{60}{\sqrt{2}}$ feet. The perimeter of this square

is $4 \cdot \dfrac{60}{\sqrt{2}} = \dfrac{240}{\sqrt{2}} \cdot \dfrac{\sqrt{2}}{\sqrt{2}} = \dfrac{240\sqrt{2}}{2} = 120\sqrt{2}$ feet.

131. A square with an area of x square feet has sides of length $\sqrt{x}$ feet. Solve for c in the Pythagorean

Theorem as follows: $a^2 + b^2 = c^2 \Rightarrow \left(\sqrt{x}\right)^2 + \left(\sqrt{x}\right)^2 = c^2 \Rightarrow 2x = c^2 \Rightarrow \sqrt{2x} = c$. The length of the

diagonal is $\sqrt{2x}$ feet.

Checking Basic Concepts Sections 7.3 and 7.4

1. (a) $\left(64^{-3/2}\right)^{1/3} = 64^{-3/2 \cdot 1/3} = 64^{-1/2} = \dfrac{1}{64^{1/2}} = \dfrac{1}{\sqrt{64}} = \dfrac{1}{8}$

   (b) $\sqrt{5} \cdot \sqrt{20} = \sqrt{5 \cdot 20} = \sqrt{100} = 10$

   (c) $\sqrt[3]{-8x^4 y} = \sqrt[3]{(-2x)^3 \cdot xy} = -2x\sqrt[3]{xy}$

   (d) $\sqrt{\dfrac{4b}{5}} \cdot \sqrt{\dfrac{4b^3}{5}} = \sqrt{\dfrac{4b \cdot 4b^3}{5 \cdot 5}} = \dfrac{\sqrt{16b^4}}{\sqrt{25}} = \dfrac{4b^2}{5}$

2. $\sqrt[3]{7} \cdot \sqrt{7} = 7^{1/3} \cdot 7^{1/2} = 7^{1/3+1/2} = 7^{5/6} = \sqrt[6]{7^5}$

3. (a) $\sqrt{3} \cdot \sqrt{12} = \sqrt{3 \cdot 12} = \sqrt{36} = 6$

   (b) $\dfrac{\sqrt[3]{81}}{\sqrt[3]{3}} = \sqrt[3]{\dfrac{81}{3}} = \sqrt[3]{27} = 3$

   (c) $\sqrt{36x^6} = \sqrt{36} \cdot \sqrt{\left(x^3\right)^2} = 6x^3$

4. (a) $5\sqrt{6} + 2\sqrt{6} + \sqrt{7} = 7\sqrt{6} + \sqrt{7}$

   (b) $8\sqrt[3]{x} - 3\sqrt[3]{x} = 5\sqrt[3]{x}$

   (c) $\sqrt{9x} - \sqrt{4x} = \sqrt{9} \cdot \sqrt{x} - \sqrt{4} \cdot \sqrt{x} = 3\sqrt{x} - 2\sqrt{x} = \sqrt{x}$

5. (a) $\sqrt[3]{xy^4} - \sqrt[3]{x^4 y} = \sqrt[3]{y^3 \cdot xy} - \sqrt[3]{x^3 \cdot xy} = y\sqrt[3]{xy} - x\sqrt[3]{xy} = (y - x)\sqrt[3]{xy}$

   (b) $\left(4 - \sqrt{2}\right)\left(4 + \sqrt{2}\right) = 4^2 - \left(\sqrt{2}\right)^2 = 16 - 2 = 14$

6. $\dfrac{6}{2\sqrt{6}} = \dfrac{6}{2\sqrt{6}} \cdot \dfrac{\sqrt{6}}{\sqrt{6}} = \dfrac{6\sqrt{6}}{2 \cdot 6} = \dfrac{\sqrt{6}}{2}$

7. $\dfrac{2}{\sqrt{5}-1} = \dfrac{2}{\sqrt{5}-1} \cdot \dfrac{\sqrt{5}+1}{\sqrt{5}+1} = \dfrac{2\left(\sqrt{5}+1\right)}{5-1} = \dfrac{2\left(\sqrt{5}+1\right)}{4} = \dfrac{\sqrt{5}+1}{2}$

Section 7.5 More Radical Functions

1. See Figure 1.

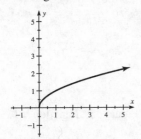

Figure 1

3. $\{x \mid x \geq 0\}$

5. $f(x) = x^p$, where p is rational

7. The variable cannot be negative. The domain is $\{x \mid x \geq 0\}$.

9. $f(-2) = \sqrt{(-2)^2 - 1} = \sqrt{4-1} = \sqrt{3} \approx 1.73$, $\quad f(-0) = \sqrt{(0)^2 - 1} = \sqrt{-1} \Rightarrow$ undefined

11. $f(-5) = \sqrt[4]{1-(-5)} = \sqrt[4]{6} \approx 1.57$, $\quad f(2) = \sqrt[4]{1-(2)} = \sqrt[4]{-1} \Rightarrow$ undefined

13. $f(-3) = \sqrt[5]{4 - 3(-3)} = \sqrt[5]{4+9} = \sqrt[5]{13} \approx 1.67$, $\quad f(1) = \sqrt[5]{4 - 3(1)} = \sqrt[5]{1} = 1$

15. $f(-5) = \sqrt[3]{1-(-5)} = \sqrt[3]{1+5} = \sqrt[3]{6} \approx 1.82$, $\quad f(2) = \sqrt[3]{1-(2)} = \sqrt[3]{-1} = -1$

17. $f(x) = x^{1/2} = \sqrt{x}$

19. $f(x) = x^{2/3} = \sqrt[3]{x^2}$

21. $f(x) = x^{-1/5} = \dfrac{1}{x^{1/5}} = \dfrac{1}{\sqrt[5]{x}}$

23. $f(4) = 4^{5/2} = \left(\sqrt{4}\right)^5 = 2^5 = 32$; $f(5) = 5^{5/2} \approx 55.90$

25. $f(-32) = (-32)^{-7/5} = \dfrac{1}{(-32)^{7/5}} = \dfrac{1}{\left(\sqrt[5]{-32}\right)^7} = \dfrac{1}{(-2)^7} = -\dfrac{1}{128} \approx -0.01$; $f(10) = 10^{-7/5} = \dfrac{1}{10^{7/5}} \approx 0.04$

27. $f(256) = 256^{1/4} = \sqrt[4]{256} = 4$; $f(-10) = (-10)^{1/4} = \sqrt[4]{-10} \Rightarrow$ Not possible

29. $f(32) = 32^{2/5} = \left(\sqrt[5]{32}\right)^2 = 2^2 = 4$; $f(-32) = (-32)^{2/5} = \left(\sqrt[5]{-32}\right)^2 = (-2)^2 = 4$

31. See Figure 31. Domain: $[0, \infty)$

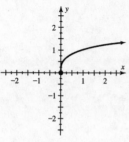

Figure 31

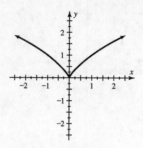

Figure 33

33. See Figure 33. Domain: $(-\infty, \infty)$

35. Graph $Y_1 = X^{\wedge}(1/5)$ and $Y_2 = X^{\wedge}(1/3)$ in [0, 6, 1] by [0, 6, 1]. See Figure 35. Function $g(x)$ increases faster.

[0, 6, 1] by [0, 6, 1]

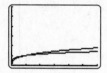

Figure 35

[0, 6, 1] by [0, 6, 1]

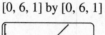

Figure 37

37. Graph $Y_1 = X^{\wedge}1.2$ and $Y_2 = X^{\wedge}0.45$ in [0, 6, 1] by [0, 6, 1]. See Figure 37. Function $f(x)$ increases faster.

39. $x^p > x^q$

41. (a) $(f+g)(2) = \sqrt{8 \cdot 2} + \sqrt{2 \cdot 2} = \sqrt{16} + \sqrt{4} = 4 + 2 = 6$

(b) $(f-g)(x) = \sqrt{8x} - \sqrt{2x} = \sqrt{4 \cdot 2x} - \sqrt{2x} = 2\sqrt{2x} - \sqrt{2x} = \sqrt{2x}$

(c) $(fg)(x) = \left(\sqrt{8x}\right)\left(\sqrt{2x}\right) = \sqrt{8x \cdot 2x} = \sqrt{16x^2} = 4|x|$

(d) $(f/g)(x) = \dfrac{\sqrt{8x}}{\sqrt{2x}} = \sqrt{\dfrac{8x}{2x}} = \sqrt{4} = 2$

43. b

45. c

47. d

49. b

51. $S(150) = 342(150)^{0.425} \approx 2877$ square inches

53. According to the graph the result is about 35%.

55. No, for $x \geq 10$ the accuracy is less than double.

57. (a) $T(0.8c) = 10\sqrt{1 - (0.8c/c)^2} = 10\sqrt{1 - 0.8^2} = 10\sqrt{1 - 0.64} = 10\sqrt{0.36} = 10 \cdot 0.6 = 6$ years

(b) The twin in the spaceship will be 4 years younger than the twin on Earth.

59. (a) See Figure 59a.

(b) $A(2) = k(2)^{2/3} = 0.254 \Rightarrow k = \dfrac{0.254}{2^{2/3}} \Rightarrow k \approx 0.16$

(c)   See Figure 59c. Yes the graph does pass through the data points.

(d) $A(2.5) = 0.16(2.5)^{2/3} \approx 0.295$ square meters

[0, 5, 1] by [0, 0.5, 0.1]   [0, 5, 1] by [0, 0.5, 0.1]

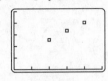   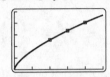

Figure 59a   Figure 59c.

61. (a) $L = kW^{1/3} \Rightarrow 0.422 = k \cdot 0.1^{1/3} \Rightarrow k = \dfrac{0.422}{0.1^{1/3}} \approx 0.91$

(b)   Plot the data and graph $Y_1 = 0.91X \wedge (1/3)$ in [0, 1.5, 0.1] by [0, 1, 0.1]. See Figure 61.

It increases.

(c) $L = 0.91(0.7)^{1/3} \approx 0.808$ meters

(d) $L = 0.91(0.65)^{1/3} \approx 0.788$;   A bird weighing 0.65 kg has a wing span of about 0.788 meters.

[0, 1600, 400] by [0, 220, 20]

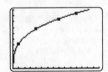

Figure 61

Section 7.6 Equations Involving Radical Expressions

1. Square each side.

3. Yes

5. The Pythagorean theorem is used to find an unknown side of a right triangle. *Answers may vary.*

7. $d = \sqrt{(x_2 - x_1)^2 + (y_2 - y_1)^2}$

9. $\sqrt{2} \cdot \sqrt{2} = \sqrt{2 \cdot 2} = \sqrt{4} = 2$

11. $\sqrt{x} \cdot \sqrt{x} = \sqrt{x \cdot x} = \sqrt{x^2} = x$

13. $\left(\sqrt{2x+1}\right)^2 = 2x + 1$

15. $\left(\sqrt[3]{5x^2}\right)^3 = 5x^2$

17. $\sqrt{x} = 8 \Rightarrow \left(\sqrt{x}\right)^2 = 8^2 \Rightarrow x = 64$

19. $\sqrt[4]{x} = 3 \Rightarrow \left(\sqrt[4]{x}\right)^4 = 3^4 \Rightarrow x = 81$

21. $\sqrt{2t+4} = 4 \Rightarrow \left(\sqrt{2t+4}\right)^2 = 4^2 \Rightarrow 2t+4 = 16 \Rightarrow 2t = 12 \Rightarrow t = 6$

23. $\sqrt{x+1} - 3 = 4 \Rightarrow \sqrt{x+1} = 7 \Rightarrow x+1 = 7^2 \Rightarrow x = 49 - 1 \Rightarrow x = 48$

25. $2\sqrt{x-2} + 1 = 5 \Rightarrow 2\sqrt{x-2} = 4 \Rightarrow \sqrt{x-2} = 2 \Rightarrow x - 2 = 2^2 \Rightarrow x = 4 + 2 \Rightarrow x = 6$

27. $\sqrt{x+6} = x \Rightarrow \left(\sqrt{x+6}\right)^2 = x^2 \Rightarrow x+6 = x^2 \Rightarrow x^2 - x - 6 = 0 \Rightarrow (x+2)(x-3) = 0 \Rightarrow$

  $x = -2$ or $x = 3$. The solution $x = -2$ does not check. The solution is $x = 3$.

29. $\sqrt[3]{x} = 3 \Rightarrow \left(\sqrt[3]{x}\right)^3 = 3^3 \Rightarrow x = 27$

31. $\sqrt[3]{2z-4} = -2 \Rightarrow 2z - 4 = (-2)^3 \Rightarrow 2z - 4 = -8 \Rightarrow 2z = -4 \Rightarrow z = -2$

33. $\sqrt[4]{t+1} = 2 \Rightarrow t+1 = 2^4 \Rightarrow t+1 = 16 \Rightarrow t = 15$

35. $\sqrt{5z-1} = \sqrt{z+1} \Rightarrow \left(\sqrt{5z-1}\right)^2 = \left(\sqrt{z+1}\right)^2 \Rightarrow 5z-1 = z+1 \Rightarrow 4z = 2 \Rightarrow z = \frac{1}{2}$

37. $\sqrt{1-x} = 1 - x \Rightarrow \left(\sqrt{1-x}\right)^2 = (1-x)^2 \Rightarrow 1 - x = 1 - 2x + x^2 \Rightarrow x^2 - x = 0 \Rightarrow$

  $x(x-1) = 0 \Rightarrow x = 0$ or $x = 1$

39. $\sqrt{b^2-4} = b - 2 \Rightarrow \left(\sqrt{b^2-4}\right)^2 = (b-2)^2 \Rightarrow b^2 - 4 = b^2 - 4b + 4 \Rightarrow 4b = 8 \Rightarrow b = 2$

41. $\sqrt{1-2x} = x + 7 \Rightarrow \left(\sqrt{1-2x}\right)^2 = (x+7)^2 \Rightarrow 1 - 2x = x^2 + 14x + 49 \Rightarrow x^2 + 16x + 48 = 0 \Rightarrow$

  $(x+12)(x+4) = 0 \Rightarrow x = -12$ or $x = -4$. The solution $x = -12$ does not check.

  The solution is $x = -4$.

43. $\sqrt{x} = \sqrt{x-5} + 1 \Rightarrow \left(\sqrt{x}\right)^2 = \left(\sqrt{x-5} + 1\right)^2 \Rightarrow x = (x-5) + 2\sqrt{x-5} + 1 \Rightarrow$

  $2\sqrt{x-5} = 4 \Rightarrow \left(2\sqrt{x-5}\right)^2 = 4^2 \Rightarrow 4(x-5) = 16 \Rightarrow 4x - 20 = 16 \Rightarrow 4x = 36 \Rightarrow x = 9$

45. $\sqrt{2t-2} + \sqrt{t} = 7 \Rightarrow \sqrt{2t-2} = 7 - \sqrt{t} \Rightarrow \left(\sqrt{2t-2}\right)^2 = \left(7 - \sqrt{t}\right)^2 \Rightarrow 2t - 2 = 49 - 14\sqrt{t} + t \Rightarrow$

  $14\sqrt{t} = 51 - t \Rightarrow \left(14\sqrt{t}\right)^2 = (51-t)^2 \Rightarrow 196t = 2601 - 102t + t^2 \Rightarrow t^2 - 298t + 2601 = 0 \Rightarrow$

  $(t-9)(t-289) = 0 \Rightarrow t = 9$ or $t = 289$. The solution $t = 289$ does not check.

  The solution is $t = 9$.

47. $x^2 = 49 \Rightarrow \sqrt{x^2} = \sqrt{49} \Rightarrow |x| = 7 \Rightarrow x = \pm 7$

49. $2z^2 = 200 \Rightarrow z^2 = 100 \Rightarrow \sqrt{z^2} = \sqrt{100} \Rightarrow |z| = 10 \Rightarrow z = \pm 10$

51. $(t+1)^2 = 16 \Rightarrow \sqrt{(t+1)^2} = \sqrt{16} \Rightarrow |t+1| = 4 \Rightarrow t+1 = \pm 4 \Rightarrow t = 3 \text{ or } t = -5$

53. $(4-2x)^2 = 100 \Rightarrow \sqrt{(4-2x)^2} = \sqrt{100} \Rightarrow |4-2x| = 10 \Rightarrow 4-2x = \pm 10$

$\Rightarrow 4-2x = 10 \text{ or } 4-2x = -10 \Rightarrow -2x = 6 \text{ or } -2x = -14 \Rightarrow x = -3 \text{ or } x = 7$

55. $b^3 = 64 \Rightarrow \sqrt[3]{b^3} = \sqrt[3]{64} \Rightarrow b = 4$

57. $2t^3 = -128 \Rightarrow t^3 = -64 \Rightarrow \sqrt[3]{t^3} = \sqrt[3]{-64} \Rightarrow t = -4$

59. $(x+1)^3 = 8 \Rightarrow \sqrt[3]{(x+1)^3} = \sqrt[3]{8} \Rightarrow x+1 = 2 \Rightarrow x = 1$

61. $(2-5z)^3 = -125 \Rightarrow \sqrt[3]{(2-5z)^3} = \sqrt[3]{-125} \Rightarrow 2-5z = -5 \Rightarrow -5z = -7 \Rightarrow z = \dfrac{7}{5}$

63. $x^4 = 16 \Rightarrow \sqrt[4]{x^4} = \sqrt[4]{16} \Rightarrow |x| = 2 \Rightarrow x = \pm 2$

65. $x^5 = 12 \Rightarrow \sqrt[5]{x^5} = \sqrt[5]{12} \Rightarrow x = \sqrt[5]{12}$

67. $2(x+2)^4 = 162 \Rightarrow (x+2)^4 = 81 \Rightarrow \sqrt[4]{(x+2)^4} = \sqrt[4]{81} \Rightarrow |x+2| = 3 \Rightarrow x+2 = \pm 3 \Rightarrow x = 1 \text{ or } x = -5$

69. Graphical: Graph $Y_1 = \sqrt[3]{(X+5)}$ and $Y_2 = 2$ in $[-7, 7, 1]$ by $[0, 4, 1]$. See Figure 69.

The solution is $x = 3$.

[–7, 7, 1] by [0, 4, 1]   [0, 3, 1] by [–2, 2, 1]

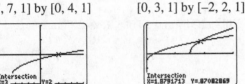

Figure 69   Figure 71

71. Graphical: Graph $Y_1 = \sqrt{(2X-3)}$ and $Y_2 = \sqrt{(X)} - (1/2)$ in $[0, 3, 1]$ by $[-2, 2, 1]$. See Figure 71.

The solution is $x \approx 1.88$.

73. Graphical: Graph $Y_1 = X^{\wedge}(5/3)$ and $Y_2 = 2 - 3X^2$ in $[-4, 4, 1]$ by $[-4, 4, 1]$. See Figures 73a & 73b.

The solutions are $x = -1$ or $x \approx 0.70$.

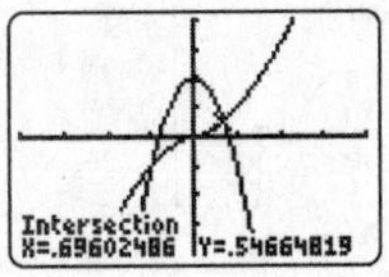

[–4, 4, 1] by [–4, 4, 1]   [–4, 4, 1] by [–4, 4, 1]   [–3, 3, 1] by [–3, 3, 1]

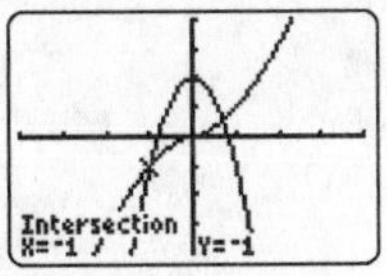

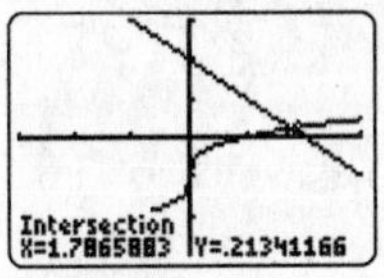

Figure 73a   Figure 73b   Figure 75

75. Graphical: Graph $Y_1 = X^{\wedge}(1/3) - 1$ and $Y_2 = 2 - X$ in $[-3, 3, 1]$ by $[-3, 3, 1]$. See Figure 75.

The solution is $z \approx 1.79$.

77. Graphical: Graph $Y_1 = \sqrt{(X+2)} + \sqrt{(3X+2)}$ and $Y_2 = 2$ in $[-2, 2, 1]$ by $[0, 5, 1]$. See Figure 77.

The solution is $y \approx -0.47$.

$[-2, 2, 1]$ by $[0, 5, 1]$ $[0, 30, 5]$ by $[0, 10, 1]$

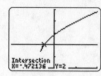   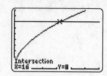   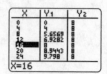

Figure 77   Figure 79b   Figure 79c

79. (a) $2\sqrt{x} = 8 \Rightarrow \left(2\sqrt{x}\right)^2 = 8^2 \Rightarrow 4x = 64 \Rightarrow x = 16$

    (b)   Graph $Y_1 = 2\sqrt{(X)}$ and $Y_2 = 8$ in $[0, 30, 5]$ by $[0, 10, 1]$. See Figure 79b.

           The solution is $x = 16$.

    (c)   Table $Y_1 = 2\sqrt{(X)}$ and $Y_2 = 8$ with TblStart = 0 and ΔTbl = 4. See Figure 79c.

           The solution is $x = 16$.

81. (a) $\sqrt{6z-2} = 8 \Rightarrow \left(\sqrt{6z-2}\right)^2 = 8^2 \Rightarrow 6z - 2 = 64 \Rightarrow 6z = 66 \Rightarrow z = 11$

    (b)   Graph $Y_1 = \sqrt{(6X-2)}$ and $Y_2 = 8$ in $[0, 20, 2]$ by $[0, 10, 1]$. See Figure 81b.

           The solution is $z = 11$.

    (c)   Table $Y_1 = \sqrt{(6X-2)}$ and $Y_2 = 8$ with TblStart = 7 and ΔTbl = 1. See Figure 81c.

           The solution is $z = 11$.

$[0, 20, 2]$ by $[0, 10, 1]$

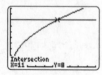   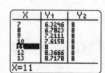

Figure 81b   Figure 81c

83. $T = 2\pi\sqrt{\dfrac{L}{32}} \Rightarrow \dfrac{T}{2\pi} = \sqrt{\dfrac{L}{32}} \Rightarrow \left(\dfrac{T}{2\pi}\right)^2 = \dfrac{L}{32} \Rightarrow \dfrac{T^2}{4\pi^2} = \dfrac{L}{32} \Rightarrow 32 \cdot \dfrac{T^2}{4\pi^2} = L \Rightarrow L = \dfrac{8T^2}{\pi^2}$

85. $r = \sqrt{\dfrac{A}{\pi}} \Rightarrow r^2 = \dfrac{A}{\pi} \Rightarrow \pi r^2 = A \Rightarrow A = \pi r^2$

87. Yes, since $6^2 + 8^2 = 10^2$. That is $36 + 64 = 100$.

89. Yes, since $\left(\sqrt{5}\right)^2 + \left(\sqrt{9}\right)^2 = \left(\sqrt{14}\right)^2$. That is $5 + 9 = 14$.

91. Yes, since $7^2 + 24^2 = 25^2$. That is $49 + 576 = 625$.

93. No, since $8^2 + 8^2 \neq 16^2$. That is $64 + 64 \neq 256$.

95. $4^2 + 4^2 = c^2 \Rightarrow c^2 = 16 + 16 \Rightarrow c^2 = 32 \Rightarrow c = \sqrt{32} = 4\sqrt{2}$

97. $24^2 + b^2 = 25^2 \Rightarrow 576 + b^2 = 625 \Rightarrow b^2 = 49 \Rightarrow b = \sqrt{49} = 7$

99. $3^2 + 4^2 = c^2 \Rightarrow c^2 = 9 + 16 \Rightarrow c^2 = 25 \Rightarrow c = \sqrt{25} = 5$

101. $\left(\sqrt{3}\right)^2 + b^2 = 8^2 \Rightarrow 3 + b^2 = 64 \Rightarrow b^2 = 61 \Rightarrow b = \sqrt{61}$

103. $a^2 + 48^2 = 50^2 \Rightarrow a^2 + 2304 = 2500 \Rightarrow a^2 = 196 \Rightarrow a = \sqrt{196} = 14$

105. From $(-2, 1)$ to $(2, 3)$, $d = \sqrt{\left(2-(-2)\right)^2 + \left(3-1\right)^2} = \sqrt{4^2 + 2^2} = \sqrt{16+4} = \sqrt{20} = 2\sqrt{5}$.

107. From $(10, 40)$ to $(30, -20)$,

$$d = \sqrt{\left(30-10\right)^2 + \left(-20-40\right)^2} = \sqrt{20^2 + \left(-60\right)^2} = \sqrt{400+3600} = \sqrt{4000} = 20\sqrt{10}.$$

109. $d = \sqrt{\left(4-(-1)\right)^2 + \left(10-2\right)^2} = \sqrt{5^2 + 8^2} = \sqrt{25+64} = \sqrt{89}$.

111. $d = \sqrt{\left(4-0\right)^2 + \left(0-(-3)\right)^2} = \sqrt{4^2 + 3^2} = \sqrt{16+9} = \sqrt{25} = 5$.

113. $\sqrt{\left(0-x\right)^2 + \left(6-3\right)^2} = 5 \Rightarrow \sqrt{\left(-x\right)^2 + 3^2} = 5 \Rightarrow \sqrt{x^2+9} = 5 \Rightarrow \left(\sqrt{x^2+9}\right)^2 = 5^2 \Rightarrow$

$x^2 + 9 = 25 \Rightarrow x^2 = 16 \Rightarrow x = \sqrt{16} = \pm 4$. Since x is positive, $x = 4$.

115. $\sqrt{\left(62-x\right)^2 + \left(6-(-5)\right)^2} = 61 \Rightarrow \sqrt{\left(62-x\right)^2 + 11^2} = 61 \Rightarrow \sqrt{\left(3844-124x+x^2\right)+121} = 61 \Rightarrow$

$\sqrt{x^2 - 124x + 3965} = 61 \Rightarrow \left(\sqrt{x^2 - 124x + 3965}\right)^2 = 61^2 \Rightarrow x^2 - 124x + 3965 = 3721 \Rightarrow$

$x^2 - 124x + 244 = 0 \Rightarrow (x-2)(x-122) = 0 \Rightarrow x = 2$ or $x = 122$

117. $400 = 100\sqrt[3]{W^2} \Rightarrow \dfrac{400}{100} = \sqrt[3]{W^2} \Rightarrow 4 = \sqrt[3]{W^2} \Rightarrow 4^3 = \left(\sqrt[3]{W^2}\right)^3 \Rightarrow 64 = W^2 \Rightarrow W = 8$ lb

119. $50 = 7.3\sqrt[16]{x^7} \Rightarrow \dfrac{50}{7.3} = \sqrt[16]{x^7} \Rightarrow \dfrac{50}{7.3} = x^{7/16} \Rightarrow \left(\dfrac{50}{7.3}\right)^{16/7} = x \Rightarrow 81.3 \approx x$, The result is about 81 sec.

121. $D(6) = 1.22\sqrt{6} \approx 2.988 \approx 3$ miles

123. $1.22\sqrt{h} = 20 \Rightarrow \sqrt{h} = \dfrac{20}{1.22} \Rightarrow \left(\sqrt{h}\right)^2 = \left(\dfrac{20}{1.22}\right)^2 \Rightarrow h \approx 268.745 \approx 269$ feet

125. $d^2 = 11.4^2 + 15.2^2 \Rightarrow d^2 = 129.96 + 231.04 \Rightarrow d^2 = 361 \Rightarrow d = \sqrt{361} = 19$ inches

127. The height can be found using proportions: $\dfrac{16}{9} = \dfrac{29}{x} \Rightarrow 16x = 261 \Rightarrow x = \dfrac{261}{16} \approx 16.3$ inches.

Then $d^2 = 29^2 + 16.3^2 \Rightarrow d^2 = 841 + 265.69 \Rightarrow d^2 = 1106.69 \Rightarrow d = \sqrt{1106.69} \approx 33.3$ inches.

129. (a) $\dfrac{60}{11}\sqrt{d} = 60 \Rightarrow \sqrt{d} = 60\left(\dfrac{11}{60}\right) \Rightarrow \sqrt{d} = 11 \Rightarrow \left(\sqrt{d}\right)^2 = 11^2 \Rightarrow d = 121$ feet

(b) $\dfrac{60}{11}\sqrt{d} = 100 \Rightarrow \sqrt{d} = 100\left(\dfrac{11}{60}\right) \Rightarrow \sqrt{d} = \dfrac{55}{3} \Rightarrow \left(\sqrt{d}\right)^2 = \left(\dfrac{55}{3}\right)^2 \Rightarrow d \approx 336$ feet

131. (a) $V = 30\sqrt{\dfrac{285}{178}} \approx 38$ mph. The accident vehicle was traveling about 38 mph.

(b) $45\sqrt{\dfrac{D}{255}} = 60 \Rightarrow \sqrt{\dfrac{D}{255}} = \dfrac{60}{45} \Rightarrow \left(\sqrt{\dfrac{D}{255}}\right)^2 = \left(\dfrac{4}{3}\right)^2 \Rightarrow \dfrac{D}{255} = \dfrac{16}{9} \Rightarrow D = 255\left(\dfrac{16}{9}\right) \approx 453$ feet

133. (a) $W(2v) = 3.8(2v)^3 = 3.8 \cdot 8 \cdot v^3 = 8 \cdot \left(3.8v^3\right)$; The wattage generated increases by a factor of 8.

(b) $W = 3.8v^3 \Rightarrow \dfrac{W}{3.8} = v^3 \Rightarrow \sqrt[3]{\dfrac{W}{3.8}} = \sqrt[3]{v^3} \Rightarrow v = \sqrt[3]{\dfrac{W}{3.8}}$

(c) $v = \sqrt[3]{\dfrac{30,400}{3.8}} = \sqrt[3]{8000} = 20$ mph

135. $c^2 = a^2 + a^2 \Rightarrow c^2 = 2a^2 \Rightarrow c = \sqrt{2a^2} = a\sqrt{2}$

Checking Basic Concepts Sections 7.5 and 7.6

1. (a)   See Figure 1a. $f(-1)$ is undefined

(b)   See Figure 1b. $f(-1) = -1$

(c)   See Figure 1c. $f(-1) = 1$

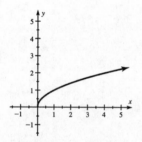

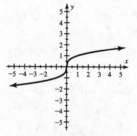

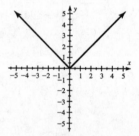

Figure 1a   Figure 1b   Figure 1c

2. $f(64) = 0.2(64)^{2/3} = 0.2\left(\sqrt[3]{64}\right)^2 = 0.2(4)^2 = 0.2 \cdot 16 = 3.2$

3. $x - 4 \geq 0 \Rightarrow x \geq 4 \Rightarrow$ Domain: $[4, \infty)$

4. (a) $\sqrt{2x-4} = 2 \Rightarrow \left(\sqrt{2x-4}\right)^2 = 2^2 \Rightarrow 2x - 4 = 4 \Rightarrow 2x = 8 \Rightarrow x = 4$

(b) $\sqrt[3]{x-1} = 3 \Rightarrow \left(\sqrt[3]{x-1}\right)^3 = 3^3 \Rightarrow x - 1 = 27 \Rightarrow x = 28$

(c) $\sqrt{3x} = 1 + \sqrt{x+1} \Rightarrow \left(\sqrt{3x}\right)^2 = \left(1 + \sqrt{x+1}\right)^2 \Rightarrow 3x = 1 + 2\sqrt{x+1} + x + 1 \Rightarrow$

$3x = x + 2 + 2\sqrt{x+1} \Rightarrow 2x - 2 = 2\sqrt{x+1} \Rightarrow x - 1 = \sqrt{x+1} \Rightarrow (x-1)^2 = \left(\sqrt{x+1}\right)^2 \Rightarrow$

$x^2 - 2x + 1 = x + 1 \Rightarrow x^2 - 3x = 0 \Rightarrow x(x-3) = 0 \Rightarrow x = 0$ or $x = 3$

The solution $x = 0$ does not check. The solution is $x = 3$.

5. $d = \sqrt{\left(2-(-3)\right)^2 + \left(-7-5\right)^2} = \sqrt{5^2 + \left(-12\right)^2} = \sqrt{25+144} = \sqrt{169} = 13$

6. $h^2 + 12.8^2 = 16^2 \Rightarrow h^2 + 163.84 = 256 \Rightarrow h^2 = 92.16 \Rightarrow h = \sqrt{92.16} = 9.6$ inches

7. $(x+1)^4 = 16 \Rightarrow \sqrt[4]{(x+1)^4} = \sqrt[4]{16} \Rightarrow |x+1| = 2 \Rightarrow x+1 = \pm 2 \Rightarrow x = -3$ or $x = 1$

Section 7.7 Complex Numbers

1. $2+3i$; *Answers may vary.*

3. i

5. $i\sqrt{a}$

7. $a+bi$

9. 4

11. 0

13. $\sqrt{-5} = i\sqrt{5}$

15. $\sqrt{-100} = i\sqrt{100} = i \cdot 10 = 10i$

17. $\sqrt{-144} = i\sqrt{144} = i \cdot 12 = 12i$

19. $\sqrt{-12} = i\sqrt{12} = i \cdot \sqrt{4 \cdot 3} = i \cdot \sqrt{4} \cdot \sqrt{3} = i \cdot 2 \cdot \sqrt{3} = 2i\sqrt{3}$

21. $\sqrt{-18} = i\sqrt{18} = i \cdot \sqrt{9 \cdot 2} = i \cdot \sqrt{9} \cdot \sqrt{2} = i \cdot 3 \cdot \sqrt{2} = 3i\sqrt{2}$

23. $(5+3i)+(-2-3i) = (5+(-2))+(3+(-3))i = 3+0i = 3$

25. $(2i)+(-8+5i) = (0+(-8))+(2+5)i = -8+7i$

27. $(2-7i)-(1+2i) = (2-1)+(-7-2)i = 1-9i$

29. $(5i)-(10-2i) = (0-10)+(5-(-2))i = -10+7i$

31. $(3+2i)(-1+5i) = -3+15i-2i+10i^2 = -3+13i+10(-1) = -3+13i-10 = -13+13i$

33. $4(5-3i) = 20-12i$

35. $(5+4i)(5-4i) = 25-16i^2 = 25-16(-1) = 25+16 = 41$

37. $(-4i)(5i) = -4 \cdot 5 \cdot i^2 = -20(-1) = 20$

39. $3i+(2-3i)-(1-5i) = 3i+2-3i-1+5i = 1+5i$

41. $(2+i)^2 = 4+4i+i^2 = 4+4i-1 = 3+4i$

43. $2i(-3+i) = -6i+2i^2 = -6i+2(-1) = -2-6i$

45. $i(1+i)^2 = i(1+2i+i^2) = i(1+2i-1) = i(2i) = 2i^2 = 2(-1) = -2$

47. $(a+3bi)(a-3bi) = a^2 - 9b^2i^2 = a^2 - 9b^2(-1) = a^2 + 9b^2$

49. When 11 is divided by 4, the result is 2 with remainder 3. Thus $i^{11} = i^3 = -i$.

51. When 21 is divided by 4, the result is 5 with remainder 1. Thus $i^{21} = i^1 = i$.

53. When 58 is divided by 4, the result is 14 with remainder 2. Thus $i^{58} = i^2 = -1$.

55. When 64 is divided by 4, the result is 16 with remainder 0. Thus $i^{64} = i^0 = 1$.

57. $3 - 4i$

59. Since $-6i = 0 - 6i,$ the complex conjugate is $0 + 6i = 6i$.

61. $5 + 4i$

63. Since $-1 = -1 + 0i,$ the complex conjugate is $-1 - 0i = -1$.

65. $\dfrac{2}{1+i} = \dfrac{2}{1+i} \cdot \dfrac{1-i}{1-i} = \dfrac{2(1-i)}{1^2 - i^2} = \dfrac{2-2i}{1+1} = \dfrac{2-2i}{2} = \dfrac{2}{2} - \dfrac{2}{2}i = 1-i$

67. $\dfrac{3i}{5-2i} = \dfrac{3i}{5-2i} \cdot \dfrac{5+2i}{5+2i} = \dfrac{3i(5+2i)}{5^2 - 4i^2} = \dfrac{15i+6i^2}{25+4} = \dfrac{15i-6}{29} = -\dfrac{6}{29} + \dfrac{15}{29}i$

69. $\dfrac{8+9i}{5+2i} = \dfrac{8+9i}{5+2i} \cdot \dfrac{5-2i}{5-2i} = \dfrac{40-16i+45i-18i^2}{5^2 - 4i^2} = \dfrac{40+29i-18(-1)}{25+4} = \dfrac{58+29i}{29} = 2+i$

71. $\dfrac{5+7i}{1-i} = \dfrac{5+7i}{1-i} \cdot \dfrac{1+i}{1+i} = \dfrac{5+5i+7i+7i^2}{1^2 - i^2} = \dfrac{5+12i+7(-1)}{1+1} = \dfrac{-2+12i}{2} = -1+6i$

73. $\dfrac{2-i}{i} = \dfrac{2-i}{i} \cdot \dfrac{-i}{-i} = \dfrac{-2i+i^2}{-i^2} = \dfrac{-2i+(-1)}{-(-1)} = \dfrac{-2i-1}{1} = -1-2i$

75. $\dfrac{1}{i} + \dfrac{1}{2i} = \dfrac{2}{2i} + \dfrac{1}{2i} = \dfrac{3}{2i} = \dfrac{3}{2i} \cdot \dfrac{-2i}{-2i} = \dfrac{-6i}{-4i^2} = \dfrac{-6i}{-4(-1)} = \dfrac{-6i}{4} = -\dfrac{3}{2}i$

77. $\dfrac{1}{-1+i} - \dfrac{2}{i} = \dfrac{i}{i(-1+i)} - \dfrac{2(-1+i)}{i(-1+i)} = \dfrac{i}{-1-i} - \dfrac{-2+2i}{-1-i} = \dfrac{i+2-2i}{-1-i} = \dfrac{2-i}{-1-i}$

$= \dfrac{2-i}{-1-i} \cdot \dfrac{-1+i}{-1+i} = \dfrac{-2+2i+i-i^2}{1-i^2} = \dfrac{-2+3i-(-1)}{1-(-1)} = \dfrac{-1+3i}{2} = -\dfrac{1}{2} + \dfrac{3}{2}i$

79. $Z = \dfrac{40+70i}{2+3i} = \dfrac{40+70i}{2+3i} \cdot \dfrac{2-3i}{2-3i} = \dfrac{80-120i+140i-210i^2}{2^2 - 9i^2} = \dfrac{290+20i}{13} = \dfrac{290}{13} + \dfrac{20}{13}i$

81. They are graphed using a real axis and an imaginary axis.

Checking Basic Concepts Section 7.7

1. (a) $\sqrt{-64} = i\sqrt{64} = i(8) = 8i$

   (b) $\sqrt{-17} = i\sqrt{17}$

2. (a) $(2-3i)+(1-i) = (2+1)+(-3+(-1))i = 3-4i$

(b) $4i-(2+i)=(0-2)+(4-1)i=-2+3i$

(c) $(3-2i)(1+i)=3+3i-2i-2i^2=3+i-2(-1)=3+i+2=5+i$

(d) $\dfrac{3}{2-2i}=\dfrac{3}{2-2i}\cdot\dfrac{2+2i}{2+2i}=\dfrac{3(2+2i)}{2^2-4i^2}=\dfrac{6+6i}{4+4}=\dfrac{6+6i}{8}=\dfrac{6}{8}+\dfrac{6}{8}i=\dfrac{3}{4}+\dfrac{3}{4}i$

Chapter 7 Review

1. $\sqrt{4}=2$

2. $\sqrt{36}=6$

3. $\sqrt{9x^2}=\sqrt{9}\cdot\sqrt{x^2}=3|x|$

4. $\sqrt{(x-1)^2}=|x-1|$

5. $\sqrt[3]{-64}=-4$

6. $\sqrt[3]{-125}=-5$

7. $\sqrt[3]{x^6}=\sqrt[3]{\left(x^2\right)^3}=x^2$

8. $\sqrt[3]{27x^3}=\sqrt[3]{27}\cdot\sqrt[3]{x^3}=3x$

9. $\sqrt[4]{16}=2$

10. $\sqrt[5]{-1}=-1$

11. $\sqrt[4]{x^8}=\sqrt[4]{\left(x^2\right)^4}=x^2$

12. $\sqrt[5]{(x+1)^5}=x+1$

13. $14^{1/2}=\sqrt{14}$

14. $(-5)^{1/3}=\sqrt[3]{-5}$

15. $\left(\dfrac{x}{y}\right)^{3/2}=\left(\sqrt{\dfrac{x}{y}}\right)^3$ or $\sqrt{\left(\dfrac{x}{y}\right)^3}$

16. $(xy)^{-2/3}=\dfrac{1}{(xy)^{2/3}}=\dfrac{1}{\sqrt[3]{(xy)^2}}$ or $\dfrac{1}{\left(\sqrt[3]{xy}\ \right)^2}$

17. $(-27)^{2/3}=\left(\sqrt[3]{-27}\ \right)^2=(-3)^2=9$

18. $16^{1/4}=\sqrt[4]{16}=2$

19. $16^{3/2}=\left(\sqrt{16}\ \right)^3=4^3=64$

20. $81^{3/4} = \left(\sqrt[4]{81}\right)^3 = 3^3 = 27$

21. $\left(z^3\right)^{2/3} = z^{3 \cdot 2/3} = z^2$

22. $\left(x^2 y^4\right)^{1/2} = x^{2 \cdot 1/2} \cdot y^{4 \cdot 1/2} = xy^2$

23. $\left(\dfrac{x^2}{y^6}\right)^{3/2} = \dfrac{x^{2 \cdot 3/2}}{y^{6 \cdot 3/2}} = \dfrac{x^3}{y^9}$

24. $\left(\dfrac{x^3}{y^6}\right)^{-1/3} = \dfrac{x^{3 \cdot (-1/3)}}{y^{6 \cdot (-1/3)}} = \dfrac{x^{-1}}{y^{-2}} = \dfrac{y^2}{x}$

25. $\sqrt{2} \cdot \sqrt{32} = \sqrt{64} = 8$

26. $\sqrt[3]{-4} \cdot \sqrt[3]{2} = \sqrt[3]{-8} = -2$

27. $\sqrt[3]{x^4} \cdot \sqrt[3]{x^2} = \sqrt[3]{x^6} = \sqrt[3]{\left(x^2\right)^3} = x^2$

28. $\dfrac{\sqrt{80}}{\sqrt{20}} = \sqrt{\dfrac{80}{20}} = \sqrt{4} = 2$

29. $\sqrt[3]{-\dfrac{x}{8}} = -\dfrac{\sqrt[3]{x}}{\sqrt[3]{8}} = -\dfrac{\sqrt[3]{x}}{2}$

30. $\sqrt{\dfrac{1}{3}} \cdot \sqrt{\dfrac{1}{3}} = \left(\sqrt{\dfrac{1}{3}}\right)^2 = \dfrac{1}{3}$

31. $\sqrt{48} = \sqrt{16 \cdot 3} = \sqrt{16} \cdot \sqrt{3} = 4\sqrt{3}$

32. $\sqrt{54} = \sqrt{9 \cdot 6} = \sqrt{9} \cdot \sqrt{6} = 3\sqrt{6}$

33. $\sqrt[3]{\dfrac{3}{x}} \cdot \sqrt[3]{\dfrac{9}{x^2}} = \sqrt[3]{\dfrac{3 \cdot 9}{x \cdot x^2}} = \sqrt[3]{\dfrac{27}{x^3}} = \dfrac{\sqrt[3]{27}}{\sqrt[3]{x^3}} = \dfrac{3}{x}$

34. $\sqrt{32a^3 b^2} = \sqrt{(4ab)^2 \cdot 2a} = \sqrt{(4ab)^2} \cdot \sqrt{2a} = 4ab\sqrt{2a}$

35. $\sqrt{3xy} \cdot \sqrt{27xy} = \sqrt{3 \cdot 27 \cdot xy \cdot xy} = \sqrt{81(xy)^2} = \sqrt{(9xy)^2} = 9xy$

36. $\sqrt[3]{-25z^2} \cdot \sqrt[3]{-5z^2} = \sqrt[3]{-25 \cdot (-5) \cdot z^2 \cdot z^2} = \sqrt[3]{125z^4} = \sqrt[3]{(5z)^3 \cdot z} = \sqrt{(5z)^3} \cdot \sqrt[3]{z} = 5z\sqrt[3]{z}$

37. $\sqrt{x^2 + 2x + 1} = \sqrt{(x+1)^2} = x + 1$

38. $\sqrt[4]{\dfrac{2a^2}{b}} \cdot \sqrt[4]{\dfrac{8a^3}{b^3}} = \sqrt[4]{\dfrac{16a^5}{b^4}} = \sqrt[4]{\left(\dfrac{2a}{b}\right)^4 \cdot a} = \sqrt[4]{\left(\dfrac{2a}{b}\right)^4} \cdot \sqrt[4]{a} = \dfrac{2a\sqrt[4]{a}}{b}$

39. $2\sqrt{x} \cdot \sqrt[3]{x} = 2x^{1/2} \cdot x^{1/3} = 2x^{1/2+1/3} = 2x^{5/6} = 2\sqrt[6]{x^5}$

40. $\sqrt[3]{rt} \cdot \sqrt[4]{r^2t^4} = (rt)^{1/3} \cdot \left(r^2t^4\right)^{1/4} = r^{1/3}t^{1/3} \cdot r^{1/2}t = r^{1/3+1/2}t^{1/3+1} = r^{5/6}t^{4/3} = r^{5/6}t^{8/6} = \sqrt[6]{r^5t^8}$ or $t\sqrt[6]{r^5t^2}$

41. $3\sqrt{3} + \sqrt{3} = 4\sqrt{3}$

42. $\sqrt[3]{x} + 2\sqrt[3]{x} = 3\sqrt[3]{x}$

43. $3\sqrt[3]{5} - 6\sqrt[3]{5} = -3\sqrt[3]{5}$

44. $\sqrt[4]{y} - 2\sqrt[4]{y} = -\sqrt[4]{y}$

45. $2\sqrt{12} + 7\sqrt{3} = 2\sqrt{4 \cdot 3} + 7\sqrt{3} = 2\sqrt{4} \cdot \sqrt{3} + 7\sqrt{3} = 4\sqrt{3} + 7\sqrt{3} = 11\sqrt{3}$

46. $3\sqrt{18} - 2\sqrt{2} = 3\sqrt{9 \cdot 2} - 2\sqrt{2} = 3\sqrt{9} \cdot \sqrt{2} - 2\sqrt{2} = 9\sqrt{2} - 2\sqrt{2} = 7\sqrt{2}$

47. $7\sqrt[3]{16} - \sqrt[3]{2} = 7\sqrt[3]{8 \cdot 2} - \sqrt[3]{2} = 7\sqrt[3]{8} \cdot \sqrt[3]{2} - \sqrt[3]{2} = 14\sqrt[3]{2} - \sqrt[3]{2} = 13\sqrt[3]{2}$

48. $\sqrt{4x+4} + \sqrt{x+1} = \sqrt{4(x+1)} + \sqrt{x+1} = \sqrt{4} \cdot \sqrt{x+1} + \sqrt{x+1} = 3\sqrt{x+1}$

49. $\sqrt{4x^3} - \sqrt{x} = \sqrt{(2x)^2 \cdot x} - \sqrt{x} = \sqrt{(2x)^2} \cdot \sqrt{x} - \sqrt{x} = 2x\sqrt{x} - \sqrt{x} = (2x-1)\sqrt{x}$

50. $\sqrt[3]{ab^4} + 2\sqrt[3]{a^4b} = \sqrt[3]{b^3 \cdot ab} + 2\sqrt[3]{a^3 \cdot ab} = \sqrt[3]{b^3} \cdot \sqrt[3]{ab} + 2\sqrt[3]{a^3} \cdot \sqrt[3]{ab} = (b+2a)\sqrt[3]{ab}$

51. $\left(1+\sqrt{2}\right)\left(3+\sqrt{2}\right) = 3 + \sqrt{2} + 3\sqrt{2} + \left(\sqrt{2}\right)^2 = 3 + 4\sqrt{2} + 2 = 5 + 4\sqrt{2}$

52. $\left(7-\sqrt{5}\right)\left(1+\sqrt{3}\right) = 7 + 7\sqrt{3} - \sqrt{5} - \sqrt{15}$

53. $\left(3+\sqrt{6}\right)\left(3-\sqrt{6}\right) = 3^2 - \left(\sqrt{6}\right)^2 = 9 - 6 = 3$

54. $\left(10-\sqrt{5}\right)\left(10+\sqrt{5}\right) = 10^2 - \left(\sqrt{5}\right)^2 = 100 - 5 = 95$

55. $\left(\sqrt{a}+\sqrt{2b}\right)\left(\sqrt{a}-\sqrt{2b}\right) = \left(\sqrt{a}\right)^2 - \left(\sqrt{2b}\right)^2 = a - 2b$

56. $\left(\sqrt{xy}-1\right)\left(\sqrt{xy}+2\right) = \left(\sqrt{xy}\right)^2 + 2\sqrt{xy} - \sqrt{xy} - 2 = xy + \sqrt{xy} - 2$

57. $\dfrac{4}{\sqrt{5}} = \dfrac{4}{\sqrt{5}} \cdot \dfrac{\sqrt{5}}{\sqrt{5}} = \dfrac{4\sqrt{5}}{5}$

58. $\dfrac{r}{2\sqrt{t}} = \dfrac{r}{2\sqrt{t}} \cdot \dfrac{\sqrt{t}}{\sqrt{t}} = \dfrac{r\sqrt{t}}{2t}$

59. $\dfrac{1}{\sqrt{2}+3} = \dfrac{1}{\sqrt{2}+3} \cdot \dfrac{\sqrt{2}-3}{\sqrt{2}-3} = \dfrac{\sqrt{2}-3}{2-9} = \dfrac{\sqrt{2}-3}{-7} = \dfrac{3-\sqrt{2}}{7}$

60. $\dfrac{2}{5-\sqrt{7}} = \dfrac{2}{5-\sqrt{7}} \cdot \dfrac{5+\sqrt{7}}{5+\sqrt{7}} = \dfrac{10+2\sqrt{7}}{25-7} = \dfrac{10+2\sqrt{7}}{18} = \dfrac{5+\sqrt{7}}{9}$

61. $\dfrac{1}{\sqrt{8}-\sqrt{7}} = \dfrac{1}{\sqrt{8}-\sqrt{7}} \cdot \dfrac{\sqrt{8}+\sqrt{7}}{\sqrt{8}+\sqrt{7}} = \dfrac{\sqrt{8}+\sqrt{7}}{8-7} = \dfrac{\sqrt{8}+\sqrt{7}}{1} = \sqrt{8}+\sqrt{7}$

62. $\dfrac{\sqrt{a}-\sqrt{b}}{\sqrt{a}+\sqrt{b}} = \dfrac{\sqrt{a}-\sqrt{b}}{\sqrt{a}+\sqrt{b}} \cdot \dfrac{\sqrt{a}-\sqrt{b}}{\sqrt{a}-\sqrt{b}} = \dfrac{a-2\sqrt{ab}+b}{a-b}$

63. See Figure 63.

64. See Figure 64.

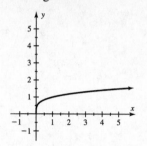

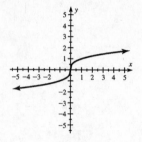

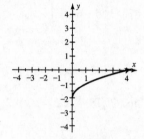

Figure 63   Figure 64   Figure 67

65. $f(x) = x^{1/2} = \sqrt{x}$; $f(4) = 4^{1/2} = \sqrt{4} = 2$

66. $f(x) = x^{2/7} = \sqrt[7]{x^2}$; $f(4) = 4^{2/7} = \sqrt[7]{4^2} = \sqrt[7]{16}$

67. See Figure 67. This graph is shifted 2 units downward.

68. See Figure 68. This graph is shifted 1 unit to the right.

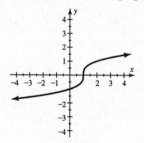

Figure 68

69. $x - 1 \geq 0 \Rightarrow x \geq 1 \Rightarrow \text{Domain}: [1, \infty)$

70. $6 - 2x \geq 0 \Rightarrow -2x \geq -6 \Rightarrow x \leq 3 \Rightarrow \text{Domain}: (-\infty, 3]$

71. $x^2 + 1 \geq 0 \Rightarrow x^2 \geq -1 \Rightarrow \text{Domain}: (-\infty, \infty)$

72. $x + 2 > 0 \Rightarrow x > -2 \Rightarrow \text{Domain}: (-2, \infty)$

73. $\sqrt{x+2} = x \Rightarrow \left(\sqrt{x+2}\right)^2 = x^2 \Rightarrow x + 2 = x^2 \Rightarrow x^2 - x - 2 = 0 \Rightarrow (x-2)(x+1) = 0 \Rightarrow x = 2 \text{ or } x = -1.$

The solution $x = -1$ does not check. The solution is $x = 2$.

74. $\sqrt{2x-1} = \sqrt{x+3} \Rightarrow \left(\sqrt{2x-1}\right)^2 = \left(\sqrt{x+3}\right)^2 \Rightarrow 2x - 1 = x + 3 \Rightarrow x = 4$

75. $\sqrt[3]{x-1} = 2 \Rightarrow \left(\sqrt[3]{x-1}\right)^3 = 2^3 \Rightarrow x - 1 = 8 \Rightarrow x = 9$

76. $\sqrt[3]{3x} = 3 \Rightarrow \left(\sqrt[3]{3x}\right)^3 = 3^3 \Rightarrow 3x = 27 \Rightarrow x = 9$

77. $\sqrt{2x} = x - 4 \Rightarrow \left(\sqrt{2x}\right)^2 = (x-4)^2 \Rightarrow 2x = x^2 - 8x + 16 \Rightarrow x^2 - 10x + 16 = 0 \Rightarrow$

$(x-2)(x-8) = 0 \Rightarrow x = 2$ or $x = 8$. The solution $x = 2$ does not check. The solution is $x = 8$.

78. $\sqrt{x} + 1 = \sqrt{x+2} \Rightarrow \left(\sqrt{x}+1\right)^2 = \left(\sqrt{x+2}\right)^2 \Rightarrow x + 2\sqrt{x} + 1 = x + 2 \Rightarrow 2\sqrt{x} = 1 \Rightarrow$

$\left(2\sqrt{x}\right)^2 = 1^2 \Rightarrow 4x = 1 \Rightarrow x = \dfrac{1}{4}$

79. Graph $Y_1 = (2X-1)\wedge(1/3)$ and $Y_2 = 2$ in $[-4, 6, 1]$ by $[-3, 3, 1]$. See Figure 79.

The solution is $x = 4.5$.

$[-4, 6, 1]$ by $[-3, 3, 1]$ $[-5, 5, 1]$ by $[-5, 5, 1]$

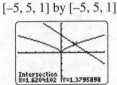

  Figure 79   Figure 80

80. Graph $Y_1 = X\wedge(2/3)$ and $Y_2 = 3 - X$ in $[-5, 5, 1]$ by $[-5, 5, 1]$. See Figure 80.

The solution is $x \approx 1.62$.

81. $c^2 = 4^2 + 7^2 \Rightarrow c^2 = 16 + 49 \Rightarrow c^2 = 65 \Rightarrow c = \sqrt{65}$

82. $5^2 + b^2 = 8^2 \Rightarrow 25 + b^2 = 64 \Rightarrow b^2 = 39 \Rightarrow b = \sqrt{39}$

83. $d = \sqrt{\left(2-(-2)\right)^2 + (-2-3)^2} = \sqrt{4^2 + (-5)^2} = \sqrt{16+25} = \sqrt{41}$

84. $d = \sqrt{(-4-2)^2 + \left(1-(-3)\right)^2} = \sqrt{(-6)^2 + 4^2} = \sqrt{36+16} = \sqrt{52} = 2\sqrt{13}$

85. $x^2 = 121 \Rightarrow x = \pm\sqrt{121} \Rightarrow x = \pm 11$

86. $2z^2 = 32 \Rightarrow z^2 = 16 \Rightarrow z = \pm\sqrt{16} \Rightarrow z = \pm 4$

87. $(x-1)^2 = 16 \Rightarrow x - 1 = \pm\sqrt{16} \Rightarrow x - 1 = \pm 4 \Rightarrow x = 1 \pm 4 \Rightarrow x = -3$ or $x = 5$

88. $x^3 = 64 \Rightarrow x = \sqrt[3]{64} \Rightarrow x = 4$

89. $(x-1)^3 = 8 \Rightarrow x - 1 = \sqrt[3]{8} \Rightarrow x - 1 = 2 \Rightarrow x = 3$

90. $(2x-1)^3 = 27 \Rightarrow 2x - 1 = \sqrt[3]{27} \Rightarrow 2x - 1 = 3 \Rightarrow 2x = 4 \Rightarrow x = 2$

91. $x^4 = 256 \Rightarrow \sqrt[4]{x^4} = \sqrt[4]{256} \Rightarrow |x| = 4 \Rightarrow x = \pm 4$

92. $x^5 = -1 \Rightarrow \sqrt[5]{x^5} = \sqrt[5]{-1} \Rightarrow x = -1$

93. $(x-3)^5 = -32 \Rightarrow \sqrt[5]{(x-3)^5} = \sqrt[5]{-32} \Rightarrow x - 3 = -2 \Rightarrow x = 1$

94. $3(x+1)^4 = 3 \Rightarrow (x+1)^4 = 1 \Rightarrow \sqrt[4]{(x+1)^4} = \sqrt[4]{1} \Rightarrow |x+1| = 1 \Rightarrow x + 1 = \pm 1 \Rightarrow x = -2$ or $x = 0$

95. $(1-2i) + (-3+2i) = (1+(-3)) + (-2+2)i = -2 + 0i = -2$

96. $(1+3i)-(3-i)=(1-3)+(3-(-1))i=-2+4i$

97. $(1-i)(2+3i)=2+3i-2i-3i^2=2+i-3(-1)=2+i+3=5+i$

98. $\dfrac{3+i}{1-i}=\dfrac{3+i}{1-i}\cdot\dfrac{1+i}{1+i}=\dfrac{3+3i+i+i^2}{1^2-i^2}=\dfrac{3+4i-1}{1+1}=\dfrac{2+4i}{2}=\dfrac{2}{2}+\dfrac{4}{2}i=1+2i$

99. $\dfrac{i(4+i)}{2-3i}=\dfrac{4i+i^2}{2-3i}=\dfrac{-1+4i}{2-3i}=\dfrac{-1+4i}{2-3i}\cdot\dfrac{2+3i}{2+3i}=\dfrac{-2-3i+8i+12i^2}{4-9i^2}=\dfrac{-2+5i+12(-1)}{4-9(-1)}$

      $=\dfrac{-14+5i}{13}=-\dfrac{14}{13}+\dfrac{5}{13}i$

100. $(1-i)^2(1+i)=(1-2i+i^2)(1+i)=(1-2i-1)(1+i)=-2i(1+i)=-2i-2i^2=2-2i$

101. $\dfrac{\sqrt{h}}{2}=4.6\Rightarrow\sqrt{h}=9.2\Rightarrow\left(\sqrt{h}\right)^2=9.2^2\Rightarrow h=84.64\approx85$ feet

102. $d^2=90^2+90^2\Rightarrow d^2=8100+8100\Rightarrow d^2=16,200\Rightarrow d=\sqrt{16,200}\approx127.3$ feet

103. $T=\dfrac{1}{4}\sqrt{10}\approx0.79$ seconds

104. (a) $A=s^2=\left(\sqrt{5}\right)^2=5$ square units

    (b) $V=s^3=\left(\sqrt{5}\right)^3=5\sqrt{5}$ cubic units

    (c) $d^2=\left(\sqrt{5}\right)^2+\left(\sqrt{5}\right)^2\Rightarrow d^2=5+5\Rightarrow d^2=10\Rightarrow d=\sqrt{10}$ units

    (d) $d^2=\left(\sqrt{5}\right)^2+\left(\sqrt{10}\right)^2\Rightarrow d^2=5+10\Rightarrow d^2=15\Rightarrow d=\sqrt{15}$ units

105. $2\pi\sqrt{\dfrac{L}{32.2}}=1\Rightarrow\sqrt{\dfrac{L}{32.2}}=\dfrac{1}{2\pi}\Rightarrow\left(\sqrt{\dfrac{L}{32.2}}\right)^2=\left(\dfrac{1}{2\pi}\right)^2\Rightarrow\dfrac{L}{32.2}=\dfrac{1}{4\pi^2}\Rightarrow L=\dfrac{32.2}{4\pi^2}\approx0.82$ feet

106. $2\pi\sqrt{\dfrac{L}{5.1}}=1\Rightarrow\sqrt{\dfrac{L}{5.1}}=\dfrac{1}{2\pi}\Rightarrow\left(\sqrt{\dfrac{L}{5.1}}\right)^2=\left(\dfrac{1}{2\pi}\right)^2\Rightarrow\dfrac{L}{5.1}=\dfrac{1}{4\pi^2}\Rightarrow L=\dfrac{5.1}{4\pi^2}\approx0.13$ feet. It is shorter.

107. $4(1+r)^{210}=281\Rightarrow(1+r)^{210}=\dfrac{281}{4}\Rightarrow\left((1+r)^{210}\right)^{1/210}=\left(\dfrac{281}{4}\right)^{1/210}\Rightarrow1+r=\left(\dfrac{281}{4}\right)^{1/210}\Rightarrow$

      $r=\left(\dfrac{281}{4}\right)^{1/210}-1\approx0.02$. From 1790 to 2000 the annual percentage growth rate was about 2%.

108. (a) $L=\sqrt{3.75(500)}=\sqrt{1875}\approx43$ mph

    (b) $L=1.5\sqrt{500}\approx34$ mph. A steeper bank allows for a higher speed limit.

      This agrees with intuition.

109. $x^2=7\Rightarrow x=\sqrt{7}\approx2.65$ feet

Chapter 7 Test

1. $\sqrt[3]{-27} = \sqrt[3]{(-3)^3} = -3$

2. $\sqrt{(z+1)^2} = |z+1|$

3. $\sqrt{25x^4} = \sqrt{25} \cdot \sqrt{(x^2)^2} = 5x^2$

4. $\sqrt[3]{8z^6} = \sqrt[3]{8} \cdot \sqrt[3]{(z^2)^3} = 2z^2$

5. $\sqrt[4]{16x^4y^5} = \sqrt[4]{(2xy)^4 \cdot y} = \sqrt[4]{(2xy)^4} \cdot \sqrt[4]{y} = 2xy\sqrt[4]{y}$

6. $\left(\sqrt{3} - \sqrt{2}\right)\left(\sqrt{3} + \sqrt{2}\right) = \left(\sqrt{3}\right)^2 - \left(\sqrt{2}\right)^2 = 3 - 2 = 1$

7. $7^{2/5} = \sqrt[5]{7^2}$ or $\left(\sqrt[5]{7}\right)^2$

8. $\left(\dfrac{x}{y}\right)^{-2/3} = \left(\dfrac{y}{x}\right)^{2/3} = \sqrt[3]{\left(\dfrac{y}{x}\right)^2}$ or $\left(\sqrt[3]{\dfrac{y}{x}}\right)^2$

9. $(-8)^{4/3} = \left(\sqrt[3]{-8}\right)^4 = (-2)^4 = 16$

10. $36^{-3/2} = \dfrac{1}{36^{3/2}} = \dfrac{1}{\left(\sqrt{36}\right)^3} = \dfrac{1}{6^3} = \dfrac{1}{216}$

11. $\sqrt[3]{x^4} = x^{4/3}$

12. $\sqrt{x} \cdot \sqrt[5]{x} = x^{1/2} \cdot x^{1/5} = x^{(1/2+1/5)} = x^{7/10}$

13. $4 - x \geq 0 \Rightarrow -x \geq -4 \Rightarrow x \leq 4 \Rightarrow \text{Domain} : (-\infty, 4]$

14. See Figure 14.

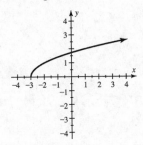

Figure 14

15. $\left(2z^{1/2}\right)^3 = 2^3 \cdot z^{1/2 \cdot 3} = 8z^{3/2}$

16. $\left(\dfrac{y^2}{z^3}\right)^{-1/3} = \left(\dfrac{z^3}{y^2}\right)^{1/3} = \dfrac{z^{3 \cdot 1/3}}{y^{2 \cdot 1/3}} = \dfrac{z}{y^{2/3}}$

17. $\sqrt{3} \cdot \sqrt{27} = \sqrt{81} = 9$

18. $\dfrac{\sqrt{y^3}}{\sqrt{4y}} = \sqrt{\dfrac{y^3}{4y}} = \sqrt{\dfrac{y^2}{4}} = \dfrac{\sqrt{y^2}}{\sqrt{4}} = \dfrac{y}{2}$

19. $7\sqrt{7} - 3\sqrt{7} + \sqrt{5} = 4\sqrt{7} + \sqrt{5}$

20. $7\sqrt[3]{x} - \sqrt[3]{x} = 6\sqrt[3]{x}$

21. $4\sqrt{18} + \sqrt{8} = 4(3)\sqrt{2} + 2\sqrt{2} = (12 + 2)\sqrt{2} = 14\sqrt{2}$

22. $\dfrac{\sqrt[3]{32}}{\sqrt[3]{4}} = \dfrac{\sqrt[3]{8 \cdot 4}}{\sqrt[3]{4}} = \dfrac{2\sqrt[3]{4}}{\sqrt[3]{4}} = 2$

23. (a) $\sqrt{x-2} = 5 \Rightarrow \left(\sqrt{x-2}\right)^2 = 5^2 \Rightarrow x - 2 = 25 \Rightarrow x = 27$

   (b) $\sqrt[3]{x+1} = 2 \Rightarrow \left(\sqrt[3]{x+1}\right)^3 = 2^3 \Rightarrow x + 1 = 8 \Rightarrow x = 7$

   (c) $(x-1)^3 = 8 \Rightarrow \sqrt[3]{(x-1)^3} = \sqrt[3]{8} \Rightarrow x - 1 = 2 \Rightarrow x = 3$

   (d) $\sqrt{2x+2} = x - 11 \Rightarrow \left(\sqrt{2x+2}\right)^2 = (x-11)^2 \Rightarrow 2x + 2 = x^2 - 22x + 121 \Rightarrow$

   $x^2 - 24x + 119 = 0 \Rightarrow (x-7)(x-17) = 0 \Rightarrow x = 7 \text{ or } x = 17.$

   The value $x = 7$ does not check. $x = 17$

24. (a) $\dfrac{2}{3\sqrt{7}} = \dfrac{2}{3\sqrt{7}} \cdot \dfrac{\sqrt{7}}{\sqrt{7}} = \dfrac{2\sqrt{7}}{21}$

   (b) $\dfrac{1}{1+\sqrt{5}} = \dfrac{1}{1+\sqrt{5}} \cdot \dfrac{1-\sqrt{5}}{1-\sqrt{5}} = \dfrac{1-\sqrt{5}}{1-5} = \dfrac{1-\sqrt{5}}{-4} = \dfrac{-1+\sqrt{5}}{4}$

25. Graph $Y_1 = \sqrt{3X} - X + 1$ and $Y_2 = (X-1)^{\wedge}(1/3)$. See Figure 25.

   The solution is approximately $x = 2.63$.

   $[-5, 5, 1]$ by $[-5, 5, 1]$

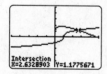

   Figure 25

26. $7^2 + b^2 = 13^2 \Rightarrow 49 + b^2 = 169 \Rightarrow b^2 = 120 \Rightarrow b = \sqrt{120} \approx 10.95$

27. $d = \sqrt{\left(-1-(-3)\right)^2 + (7-5)^2} = \sqrt{2^2 + 2^2} = \sqrt{4+4} = \sqrt{8} = 2\sqrt{2}$

28. $(-5+i) + (7-20i) = (-5+7) + (1+(-20))i = 2 - 19i$

29. $(3i) - (6-5i) = (0-6) + (3-(-5))i = -6 + 8i$

30. $\left(\dfrac{1}{2} - i\right)\left(\dfrac{1}{2} + i\right) = \dfrac{1}{4} + \dfrac{1}{2}i - \dfrac{1}{2}i - i^2 = \dfrac{1}{4} - (-1) = \dfrac{5}{4}$

31. $\dfrac{2i}{5+2i} = \dfrac{2i}{5+2i} \cdot \dfrac{5-2i}{5-2i} = \dfrac{2i(5-2i)}{5^2-4i^2} = \dfrac{10i-4i^2}{25+4} = \dfrac{4+10i}{29} = \dfrac{4}{29} + \dfrac{10}{29}i$

32. (a) $d = 1.22\sqrt{200} \approx 17.25$. The result is about 17.25 miles.

(b) $25 = 1.22\sqrt{x} \Rightarrow \dfrac{25}{1.22} = \sqrt{x} \Rightarrow \left(\dfrac{25}{1.22}\right)^2 = x \Rightarrow 419.9 \approx x$. The result is about 420 feet.

33. $27.4W^{1/3} = 30 \Rightarrow W^{1/3} = \dfrac{30}{27.4} \Rightarrow \left(W^{1/3}\right)^3 = \left(\dfrac{30}{27.4}\right)^3 \Rightarrow W = \left(\dfrac{30}{27.4}\right)^3 \approx 1.31$ lb

Chapter 7 Extended and Discovery Exercises

1. (a) $k(47)^{1.12}(11)^{1.98} = 11.4 \Rightarrow k = \dfrac{11.4}{(47)^{1.12}(11)^{1.98}} \approx 0.001325$

(b) $V = 0.001325(105)^{1.12}(20)^{1.98} \approx 91.6 \text{ ft}^3$

2. (a) $S = 15.7(154)^{0.425}(65)^{0.725} \approx 2753.963261 \approx 2754 \text{ in}^2$

(b) It increases by a factor of $2^{0.425} \approx 1.34$.

(c) It increases by a factor of $2^{0.725} \approx 1.65$.

3. (a) Since the segment AB is on land, the expression is $30x$.

(b) The legs of right triangle BCD have lengths $CB = 1000 - x$ and $CD = 500$. Let $d = BD$.

$$d^2 = (1000-x)^2 + 500^2 \Rightarrow d = \sqrt{(1000-x)^2 + 500^2}.$$

(c) Since the segment BD is underwater, the expression is $50\sqrt{(1000-x)^2 + 500^2}$.

(d) The expression is $30x + 50\sqrt{(1000-x)^2 + 500^2}$.

(e) Graph $Y_1 = 30X + 50\sqrt{\left((1000-X)^2 + 500^2\right)}$ in [0, 1000, 100] by [40,000, 60,000, 5000].

See Figure 3. The minimum cost is $50,000 when $x = 625$ feet.

[0, 1000, 100] by [40,000, 60,000, 5000]

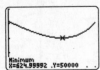

Figure 3

Chapters 1-7 Cumulative Review Exercises

1. $S = 4\pi(3)^2 = 4\pi(9) = 36\pi$

2. $D = \{-1, 0, 1\}; \ R = \{2, 4\}$

3. (a) $\left(\dfrac{ab^2}{b^{-1}}\right)^{-3} = \dfrac{a^{-3}b^{-6}}{b^3} = a^{-3}b^{-6-3} = a^{-3}b^{-9} = \dfrac{1}{a^3 b^9}$

   (b) $\dfrac{\left(x^2 y\right)^3}{x^2 \left(y^2\right)^{-3}} = \dfrac{x^6 y^3}{x^2 y^{-6}} = x^{6-2} y^{3-(-6)} = x^4 y^9$

   (c) $(rt)^2 \left(r^2 t\right)^3 = r^2 t^2 r^6 t^3 = r^{2+6} t^{2+3} = r^8 t^5$

4. $0.00043 = 4.3 \times 10^{-4}$

5. $f(3) = \dfrac{3}{3-2} = \dfrac{3}{1} = 3$; the denominator cannot equal 0, so $x \neq 2$.

6. All real numbers

7. See Figure 7.

8. See Figure 8.

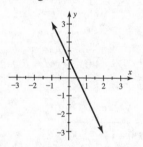 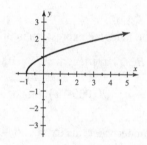 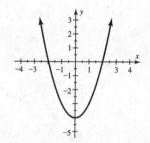

Figure 7   Figure 8   Figure 9

9. See Figure 9.

   (a) D: All real numbers; $R: y \geq -4$

   (b) $f(-2) = (-2)^2 - 4 = 4 - 4 = 0$

   (c) The x-intercepts are -2 and 2.

   (d) $x^2 - 4 = 0 \Rightarrow (x-2)(x+2) = 0 \Rightarrow x = 2$ or $x = -2$

10. The line perpendicular to $y = -2x$ has slope $m = \dfrac{1}{2}$.

$$y - 2 = \dfrac{1}{2}\left(x - (-1)\right) \Rightarrow y = \dfrac{1}{2}x + \dfrac{1}{2} + 2 \Rightarrow y = \dfrac{1}{2}x + \dfrac{5}{2}$$

11. $m = \dfrac{-5-7}{2-(-2)} = \dfrac{-12}{4} = -3; \ y - (-5) = -3(x-2) \Rightarrow y = -3x + 6 - 5 \Rightarrow y = -3x + 1; \ f(x) = 1 - 3x$

12. See Figure 12.

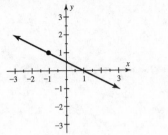   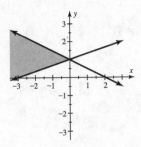

　　　　Figure 12   Figure 18

13. $5x-(3-x)=\dfrac{1}{2}x \Rightarrow 5x-3+x-\dfrac{1}{2}x=0 \Rightarrow \dfrac{11}{2}x-3=0 \Rightarrow \dfrac{11}{2}x=3 \Rightarrow x=\dfrac{6}{11}$

14. $2x-5\le 4-x \Rightarrow 3x\le 9 \Rightarrow x\le 3; \left(-\infty,\, 3\right]$

15. $\left|x-2\right|\le 3 \Rightarrow x-2\le 3$ and $x-2\ge -3 \Rightarrow x\le 5$ and $x\ge -1;\ \left[-1,\, 5\right]$

16. $-1\le 1-2x\le 6 \Rightarrow -2\le -2x\le 5 \Rightarrow 1\ge x\ge -\dfrac{5}{2} \Rightarrow -\dfrac{5}{2}\le x\le 1;\ \left[-\dfrac{5}{2},\, 1\right]$

17. (a)   Add the equations. $\begin{array}{r}2x-y=4\\ x+y=8\\ \hline 3x\quad\ =12\end{array} \Rightarrow x=4$ $x+y=8 \Rightarrow 4+y=8 \Rightarrow y=4.$

    The solution is $\left(4,\, 4\right)$.

(b)   Multiply the second equation by -3 and add to the first equation. $\begin{array}{r}3x-4y=2\\ -3x+4y=-3\\ \hline 0=-1\end{array}$

    The result is a contradiction; there are no solutions

18. Note that $x+2y\le 2 \Rightarrow y\le -\dfrac{1}{2}x+1$ and $-x+3y\ge 3 \Rightarrow y\ge \dfrac{1}{3}x+1.$ See Figure 18.

19. Add the first two equations to eliminate the variable z. $\begin{array}{r}x+2y-z=6\\ x-3y+z=-2\\ \hline 2x-y\quad\ =4\end{array}$ Add the first and third

equations to eliminate the variable z. $\begin{array}{r}x+2y-z=6\\ x+y+z=6\\ \hline 2x+3y\quad\ =12\end{array}$ Multiply this new equation by -1 and add it

to the result of the first sum. $\begin{array}{r}-2x-3y=-12\\ 2x-y=\ \ 4\\ \hline -4y=-8\end{array} \Rightarrow y=2;\ 2x-2=4 \Rightarrow 2x=6 \Rightarrow x=3$

Substitute $x=3$ and $y=2$ into the third equation to solve for z:

$3+2+z=6 \Rightarrow 5+z=6 \Rightarrow z=1.$ The solution is $\left(3,\, 2,\, 1\right)$.

20. $4(1 \cdot 1 - (-2) \cdot 0) - 2(2 \cdot 1 - 0 \cdot 0) + (-1)(2(-2) - 0 \cdot 1)$

    $= 4(1 - 0) - 2(2 - 0) - 1(-4 - 0) = 4(1) - 2(2) - 1(-4) = 4 - 4 + 4 = 4$

21. $4x(4 - x^3) = 16x - 4x^4 = -4x^4 + 16x$

22. $(x - 4)(x + 4) = x^2 - 16$

23. $(5x + 3)(x - 2) = 5x^2 - 10x + 3x - 6 = 5x^2 - 7x - 6$

24. $(4x + 9)^2 = (4x)^2 + 2(4x)(9) + (9)^2 = 16x^2 + 72x + 81$

25. $9x^2 - 16 = (3x - 4)(3x + 4)$

26. $x^2 - 4x + 4 = (x - 2)(x - 2) = (x - 2)^2$

27. $15x^3 - 9x^2 = 3x^2(5x - 3)$

28. $12x^2 - 5x - 3 = (4x - 3)(3x + 1)$

29. $r^3 - 1 = (r - 1)(r^2 + r + 1)$

30. $x^3 - 3x^2 + 5x - 15 = x^2(x - 3) + 5(x - 3) = (x - 3)(x^2 + 5)$

31. $x^2 - 3x + 2 = 0 \Rightarrow (x - 2)(x - 1) = 0 \Rightarrow x = 2 \text{ or } x = 1$

32. $x^3 = 4x \Rightarrow x^3 - 4x = 0 \Rightarrow x(x^2 - 4) = 0 \Rightarrow x(x - 2)(x + 2) = 0 \Rightarrow x = 0 \text{ or } x = 2 \text{ or } x = -2$

33. $\dfrac{x^2 + 3x + 2}{x - 3} \div \dfrac{x + 1}{2x - 6} = \dfrac{(x + 2)(x + 1)}{x - 3} \cdot \dfrac{2(x - 3)}{x + 1} = 2(x + 2)$

34. $\dfrac{2}{x - 1} + \dfrac{5}{x} = \dfrac{2}{x - 1} \cdot \dfrac{x}{x} + \dfrac{5}{x} \cdot \dfrac{x - 1}{x - 1} = \dfrac{2x + 5(x - 1)}{x(x - 1)} = \dfrac{2x + 5x - 5}{x(x - 1)} = \dfrac{7x - 5}{x(x - 1)}$

35. $\sqrt{36x^2} = \sqrt{36} \cdot \sqrt{x^2} = 6x$

36. $\sqrt[3]{64} = 4$

37. $16^{-3/2} = \left(\sqrt{16}\right)^{-3} = 4^{-3} = \dfrac{1}{4^3} = \dfrac{1}{64}$

38. $\sqrt[4]{625} = \sqrt[4]{(5)^4} = 5$

39. $\sqrt{2x} \cdot \sqrt{8x} = \sqrt{(2x)(8x)} = \sqrt{16x^2} = 4x$

40. $\sqrt{x} \cdot \sqrt[4]{x} = x^{1/2} \cdot x^{1/4} = x^{3/4} \text{ or } \sqrt[4]{x^3}$

41. $\dfrac{\sqrt[3]{16x^4}}{\sqrt[3]{2x}} = \sqrt[3]{\dfrac{16x^4}{2x}} = \sqrt[3]{8x^3} = 2x$

42. $4\sqrt{12x} - 2\sqrt{3x} = 4\sqrt{4 \cdot 3x} - 2\sqrt{3x} = 4\sqrt{4} \cdot \sqrt{3x} - 2\sqrt{3x} = 4(2)\sqrt{3x} - 2\sqrt{3x} = 8\sqrt{3x} - 2\sqrt{3x} = 6\sqrt{3x}$

43. $\left(2x+\sqrt{3}\right)\left(x-\sqrt{3}\right)=2x^2-2x\sqrt{3}+x\sqrt{3}-\left(\sqrt{3}\right)^2=2x^2-x\sqrt{3}-3$

44. The radicand must be greater than or equal to $0 \Rightarrow 1-x \geq 0 \Rightarrow x \leq 1;\ (-\infty,\ 1]$

45. See Figure 45.

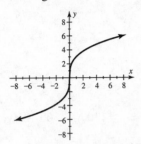

Figure 45

46. $d=\sqrt{\left(x_2-x_1\right)^2+\left(y_2-y_1\right)^2}=\sqrt{\left(1-(-2)\right)^2+\left(2-3\right)^2}=\sqrt{3^2+(-1)^2}=\sqrt{9+1}=\sqrt{10}$

47. $(1-i)(2+3i)=2+3i-2i-3i^2=2+i-3(-1)=2+3+i=5+i$

48. Use the Pythagorean Theorem: $a^2+b^2=c^2 \Rightarrow 5^2+12^2=c^2 \Rightarrow 25+144=c^2 \Rightarrow 169=c^2 \Rightarrow c=13$

49. $2\sqrt{x+3}=x \Rightarrow \sqrt{x+3}=\dfrac{x}{2} \Rightarrow x+3=\left(\dfrac{x}{2}\right)^2 \Rightarrow x+3=\dfrac{x^2}{4} \Rightarrow 4(x+3)=4\left(\dfrac{x^2}{4}\right) \Rightarrow$

$4x+12=x^2 \Rightarrow x^2-4x-12=0 \Rightarrow (x-6)(x+2)=0 \Rightarrow x=6\ \left(x=-2\ \text{does not check}\right).$

50. $\sqrt[3]{x-1}=3 \Rightarrow x-1=3^3 \Rightarrow x-1=27 \Rightarrow x=28$

51. $\sqrt{x}+4=2\sqrt{x+5} \Rightarrow \left(\sqrt{x}+4\right)^2=4(x+5) \Rightarrow x+8\sqrt{x}+16=4x+20 \Rightarrow$

$8\sqrt{x}=3x+4 \Rightarrow 64x=(3x+4)^2 \Rightarrow 64x=9x^2+24x+16 \Rightarrow 9x^2-40x+16=0$

$\Rightarrow (9x-4)(x-4)=0 \Rightarrow x=\dfrac{4}{9}\ \text{or}\ x=4$

52. $\dfrac{1}{3}x^4=27 \Rightarrow x^4=81 \Rightarrow x=\sqrt[4]{81} \Rightarrow x=\pm3$

53. Graph $Y_1=\left(X^2-2\right)^{(1/3)}+X$ and $Y_2=\sqrt{(X)}$. See Figure 53. The solution is $x \approx 1.41$.

[−4.7, 4.7, 1] by [−3.1, 3.1, 1]   [−6, 6, 1] by [−4, 4, 1].

     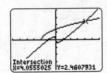

Figure 53   Figure 54

54. Graph $Y_1=X-X^{\left(1/3\right)}$ and $Y_2=\sqrt{(X+2)}$. See Figure 54. The solution is $x \approx 4.06$.

55. The tank initially contains 300 gallons of water. Water is leaving the tank at 15 gallons per minute.

56. See Figure 56.

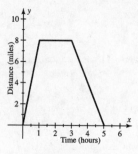

Figure 56

57. Let x represent the amount invested at 5%. Then

$0.05x + 0.04(2000 - x) = 93 \Rightarrow 0.05x + 80 - 0.04x = 93 \Rightarrow 0.01x = 13 \Rightarrow x = 1300;$ $1300 is invested

at 5% and $700 is invested at 4%.

58. Let x represent the length of a side of the base. Then $x^2(x-4) = 256 \Rightarrow x^3 - 4x^2 - 256 = 0$

Table $Y_1 = X^3 - 4X^2 - 256$ with TblStart $= 1$ and ΔTbl $= 1$ (not shown). $x = 8$, so the

dimensions are 8 inches by 8 inches by 4 inches.

59. 5 feet 3 inches $= 5.25$ feet; let h represent the height of the building; $\dfrac{5.25}{7.5} = \dfrac{h}{3.2}$

$\Rightarrow (5.25)(32) = 7.5h \Rightarrow 168 = 7.5h \Rightarrow h = 22.4$ feet

60. Let $x, y, z,$ represent the measures of the angles of the triangle from smallest to largest, respectively.

$$x + y + z = 180$$

The system needed is $z = x + y - 20$. Substitute $z = x + y - 20$ into the third equation:

$$y + z = x + 90$$

$y + x + y - 20 = x + 90 \Rightarrow 2y = 110 \Rightarrow y = 55$. Substitute $y = 55$ and $z = x + y - 20$ into the first

equation: $x + 55 + x + 55 - 20 = 180 \Rightarrow 2x + 90 = 180 \Rightarrow 2x = 90 \Rightarrow x = 45$

Then $z = x + y - 20 \Rightarrow z = 45 + 55 - 20 \Rightarrow z = 80.$

The angle measures are 45°, 55°, and 80°.

Chapter 8: Quadratic Functions and Equations

8.1: Quadratic Functions and Their Graphs

1. parabola

3. axis of symmetry

5. See Figure 5. *Answers may vary.*

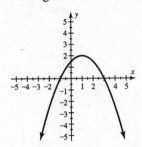

Figure 5

7. narrower

9. $ax^2 + bx + c$ with $a \neq 0$

11. True

13. False, The maximum y-value on the parabola is b.

15. $x = -\dfrac{b}{2a} = -\dfrac{(-4)}{2(1)} = 2$, $f(2) = (2)^2 - 4(2) - 2 = 4 - 8 - 2 = -6$; The vertex is $(2, -6)$.

17. $x = -\dfrac{b}{2a} = -\dfrac{(-2)}{2(-\frac{1}{3})} = -3$, $f(-3) = -\dfrac{1}{3}(-3)^2 - 2(-3) + 1 = -3 + 6 + 1 = 4$; The vertex is $(-3, 4)$.

19. $x = -\dfrac{b}{2a} = -\dfrac{(0)}{2(-2)} = 0$, $f(0) = 3 - 2(0)^2 = 3 - 0 = 3$; The vertex is $(0, 3)$.

21. $x = -\dfrac{b}{2a} = -\dfrac{(0.6)}{2(-0.3)} = 1$, $f(1) = -0.3(1)^2 + 0.6(1) + 1.1 = -0.3 + 0.6 + 1.1 = 1.4$; The vertex is $(1, 1.4)$.

23. $x = -\dfrac{b}{2a} = -\dfrac{(6)}{2(-1)} = \dfrac{6}{2} = 3$, $f(3) = 6(3) - (3)^2 = 18 - 9 = 9$; The vertex is $(3, 9)$.

25. $f(-2) = 0$ and $f(0) = -4$

27. $f(-3) = -2$ and $f(1) = -2$

29. The vertex is $(1, -2)$. The axis of symmetry is $x = 1$. The parabola opens upward.

31. The vertex is $(-2, 3)$. The axis of symmetry is $x = -2$. The parabola opens downward.

33. (a) See Figure 33.

    (b) The vertex is $(0, 0)$. The axis of symmetry is $x = 0$.

    (c) $f(-2) = \frac{1}{2}(-2)^2 = \frac{1}{2}(4) = 2$ and $f(3) = \frac{1}{2}(3)^2 = \frac{1}{2}(9) = 4.5$

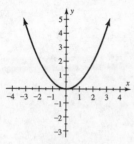

Figure 33

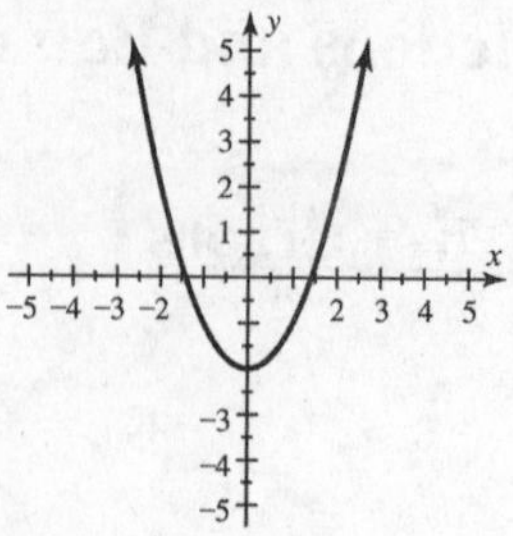

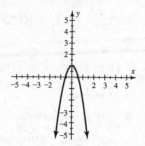

Figure 35

Figure 37

35. (a)   See Figure 35.

    (b)   The vertex is $(0, -2)$. The axis of symmetry is $x = 0$.

    (c) $f(-2) = (-2)^2 - 2 = 4 - 2 = 2$ and $f(3) = (3)^2 - 2 = 9 - 2 = 7$

37. (a)   See Figure 37.

    (b)   The vertex is $(0, 1)$. The axis of symmetry is $x = 0$.

    (c) $f(-2) = -3(-2)^2 + 1 = -12 + 1 = -11$ and $f(3) = -3(3)^2 + 1 = -27 + 1 = -26$

39. (a)   See Figure 39.

    (b)   The vertex is $(1, 0)$. The axis of symmetry is $x = 1$.

    (c) $f(-2) = ((-2) - 1)^2 = (-3)^2 = 9$ and $f(3) = ((3) - 1)^2 = (2)^2 = 4$

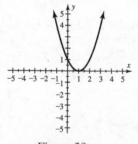

Figure 39

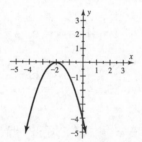

Figure 41

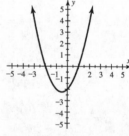

Figure 43

41. (a)   See Figure 41.

    (b)   The vertex is $(-2, 0)$. The axis of symmetry is $x = -2$.

    (c) $f(-2) = -(-2 + 2)^2 = -(0)^2 = 0$ and $f(3) = -(3 + 2)^2 = -(5)^2 = -25$

43. (a)   See Figure 43.

    (b)   The vertex is $(-0.5, -2.25)$. The axis of symmetry is $x = -0.5$.

    (c) $f(-2) = (-2)^2 + (-2) - 2 = 4 - 2 - 2 = 0$ and $f(3) = (3)^2 + (3) - 2 = 9 + 3 - 2 = 10$

45. (a)   See Figure 45.

    (b)   The vertex is $(0, -3)$. The axis of symmetry is $x = 0$.

    (c) $f(-2) = 2(-2)^2 - 3 = 8 - 3 = 5$ and $f(3) = 2(3)^2 - 3 = 18 - 3 = 15$

47. (a)   See Figure 47.

    (b)   The vertex is $(1, 1)$. The axis of symmetry is $x = 1$.

    (c) $f(-2) = 2(-2) - (-2)^2 = -4 - 4 = -8$ and $f(3) = 2(3) - (3)^2 = 6 - 9 = -3$

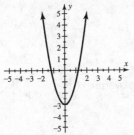

Figure 45

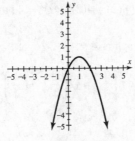

Figure 47

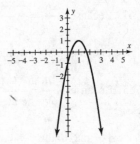

Figure 49

49. (a)   See Figure 49.

    (b)   The vertex is $(1, 1)$. The axis of symmetry is $x = 1$.

    (c) $f(-2) = -2(-2)^2 + 4(-2) - 1 = -8 - 8 - 1 = -17$ and $f(3) = -2(3)^2 + 4(3) - 1 = -18 + 12 - 1 = -7$

51. (a)   See Figure 51.

    (b)   The vertex is $(2, 4)$. The axis of symmetry is $x = 2$.

    (c) $f(-2) = \frac{1}{4}(-2)^2 - (-2) + 5 = 1 + 2 + 5 = 8$ and $f(3) = \frac{1}{4}(3)^2 - (3) + 5 = 2.25 - 3 + 5 = 4.25$

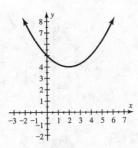

Figure 51

53. Because $-\dfrac{b}{2a} = -\dfrac{2}{2(1)} = -1$ and $f(-1) = (-1)^2 + 2(-1) - 1 = -2$, the vertex is $(-1, -2)$.

    The minimum y-value on the graph is -2.

55. Because $-\dfrac{b}{2a} = -\dfrac{-5}{2(1)} = \dfrac{5}{2}$ and $f\left(\dfrac{5}{2}\right) = \left(\dfrac{5}{2}\right)^2 - 5\left(\dfrac{5}{2}\right) = -\dfrac{25}{4}$, the vertex is $\left(\dfrac{5}{2}, -\dfrac{25}{4}\right)$.

    The minimum y-value on the graph is $-\dfrac{25}{4}$.

57. Because $-\dfrac{b}{2a} = -\dfrac{2}{2(2)} = -\dfrac{1}{2}$ and $f\left(-\dfrac{1}{2}\right) = 2\left(-\dfrac{1}{2}\right)^2 + 2\left(-\dfrac{1}{2}\right) - 3 = -\dfrac{7}{2}$, the vertex is $\left(-\dfrac{1}{2}, -\dfrac{7}{2}\right)$.

    The minimum y-value on the graph is $-\dfrac{7}{2}$.

59. Because $-\dfrac{b}{2a} = -\dfrac{2}{2(-1)} = 1$ and $f(1) = -(1)^2 + 2(1) + 5 = 6$, the vertex is $(1,\ 6)$.

The maximum y-value on the graph is 6.

61. Because $-\dfrac{b}{2a} = -\dfrac{4}{2(-1)} = 2$ and $f(2) = 4(2) - (2)^2 = 4$, the vertex is $(2,\ 4)$.

The maximum y-value on the graph is 4.

63. Because $-\dfrac{b}{2a} = -\dfrac{1}{2(-2)} = \dfrac{1}{4}$ and $f\left(\dfrac{1}{4}\right) = -2\left(\dfrac{1}{4}\right)^2 + \left(\dfrac{1}{4}\right) - 5 = -\dfrac{39}{8}$, the vertex is $\left(\dfrac{1}{4},\ -\dfrac{39}{8}\right)$.

The maximum y-value on the graph is $-\dfrac{39}{8}$.

65. Let x be one of the numbers, then $20 - x$ is the other number. $x(20 - x) = 20x - x^2$. The graph of

$y = -x^2 + 20x$ is a parabola that opens downward, so its vertex is the maximum point on the graph.

The x-value of the vertex is on the axis of symmetry, so $x = -\dfrac{b}{2a} = -\dfrac{20}{2(-1)} = -\dfrac{20}{-2} = 10$. So the

numbers are 10 and 10.

67. See Figure 67. Compared to $y = x^2$, the graph is reflected across the x-axis.

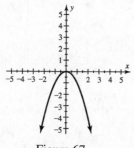
Figure 67

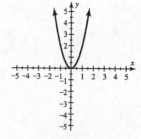
Figure 69

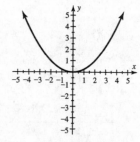
Figure 71

69. See Figure 69. Compared to $y = x^2$, the graph is narrower.

71. See Figure 71. Compared to $y = x^2$, the graph is wider.

73. See Figure 73. Compared to $y = x^2$, the graph is reflected across the x-axis and is wider.

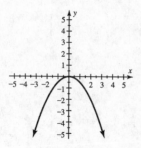
Figure 73

75. (a) $a = \dfrac{1}{2}$, so $a > 0$ and the graph opens upward. Because $0 < |a| < 1$, the graph is wider than the

graph of $y = x^2$.

(b)   The axis of symmetry is $x = -\dfrac{b}{2a} = -\dfrac{1}{2(\frac{1}{2})} = -\dfrac{1}{1} = -1$.

$f(-1) = \dfrac{1}{2}(-1)^2 + (-1) - \dfrac{3}{2} = \dfrac{1}{2}(1) - \dfrac{2}{2} - \dfrac{3}{2} = \dfrac{1}{2} - \dfrac{2}{2} - \dfrac{3}{2} = -\dfrac{4}{2} = -2$ so the vertex is $(-1, -2)$.

(c)   The y-intercept is $c = -\dfrac{3}{2}$. To find the x-intercepts, let $y = 0$ and solve for x:

$\dfrac{1}{2}x^2 + x - \dfrac{3}{2} = 0 \Rightarrow 2\left(\dfrac{1}{2}x^2 + x - \dfrac{3}{2}\right) = 2(0) \Rightarrow x^2 + 2x - 3 = 0 \Rightarrow$

$(x+3)(x-1) = 0 \Rightarrow x = -3$ or $x = 1$

(d)   See Figure 75.

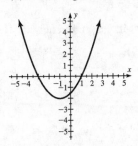

Figure 75   Figure 77   Figure 79

77. (a) $a = -1$, so $a < 0$ and the graph opens downward. Because $|a| = 1$, the graph has the same width

as the graph of $y = x^2$.

(b)   The axis of symmetry is $x = -\dfrac{b}{2a} = -\dfrac{2}{2(-1)} = -\dfrac{2}{-2} = 1$. $f(1) = 2(1) - 1^2 = 2 - 1 = 1$, so the

vertex is $(1, 1)$.

(c)   The y-intercept is $c = 0$. To find the x-intercepts, let $y = 0$ and solve for x:

$2x - x^2 = 0 \Rightarrow x(2 - x) = 0 \Rightarrow x = 0$ and $x = 2$.

(d)   See Figure 77.

79. (a) $a = 2$, so $a > 0$ and the graph opens upward. Because $|a| > 1$, the graph is narrower than the

graph of $y = x^2$.

(b)   The axis of symmetry is $x = -\dfrac{b}{2a} = -\dfrac{2}{2(2)} = -\dfrac{2}{4} = -\dfrac{1}{2}$.

$f\left(-\dfrac{1}{2}\right) = 2\left(-\dfrac{1}{2}\right)^2 + 2\left(-\dfrac{1}{2}\right) - 4 = 2\left(\dfrac{1}{4}\right) - 1 - 4 = \dfrac{1}{2} - \dfrac{2}{2} - \dfrac{8}{2} = -\dfrac{9}{2}$ so the vertex is $\left(-\dfrac{1}{2}, -\dfrac{9}{2}\right)$.

(c)   The y-intercept is $c = -4$. To find the x-intercepts, let $y = 0$ and solve for x:

$$2x^2 + 2x - 4 = 0 \Rightarrow 2(x^2 + x - 2) = 0 \Rightarrow 2(x + 2)(x - 1) = 0 \Rightarrow x = -2 \text{ or } x = 1.$$

(d)   See Figure 79.

81. Because $-\dfrac{b}{2a} = -\dfrac{\frac{200}{3}}{2\left(-\frac{100}{9}\right)} = 3$ and $f(3) = -\dfrac{100}{9}(3)^2 + \dfrac{200}{3}(3) = 100$, the vertex is $(3,\ 100)$.

The number of pieces that yields maximum satisfaction is 3.

83. d.   The stone's distance from the ground would increase and then decrease.

85. a.   The temperature would first decrease but after the repair, it would increase.

87. (a)   When the ball is hit, $t = 0$. Then $h(0) = -16(0)^2 + 64(0) + 2 = 2$ feet.

(b)   Find the x-coordinate of the vertex. $-\dfrac{b}{2a} = -\dfrac{64}{2(-16)} = 2$ seconds

(c)   Find the y-coordinate of the vertex. $h(2) = -16(2)^2 + 64(2) + 2 = 66$ feet.

89. The x-coordinate of the vertex represents the time when the ball reaches its maximum height. Here

we have $x = \dfrac{-b}{2a} = \dfrac{-66}{2(-16)} = \dfrac{66}{32} \approx 2$ seconds. The maximum height is

$h(2) = -16(2)^2 + 66(2) + 6 \approx 74$ feet.

91. (a)   The revenue is increasing when $x \le 50$, and it is decreasing when $x \ge 50$.

(b)   From the graph, the maximum revenue is $2500 when 50 tickets are sold.

(c)   If x represents the number of tickets sold, then $100 - x$ represents the price of one ticket.
The total revenue is given by $f(x) = x(100 - x)$.

(d)   Since $f(x) = -x^2 + 100x$, the number of tickets that should be sold to maximize revenue is

$-\dfrac{b}{2a} = -\dfrac{100}{2(-1)} = 50$. The maximum revenue is $f(50) = 50(100 - 50) = 50(50) = \$2500$.

93. (a) $V(1) = 10.75(1)^2 - 24(1) + 35 = 10.75 - 24 + 35 = 21.75$

$V(2) = 10.75(2)^2 - 24(2) + 35 = 43 - 48 + 35 = 30$

$V(3) = 10.75(3)^2 - 24(3) + 35 = 96.75 - 72 + 35 = 59.75$

$V(4) = 10.75(4)^2 - 24(4) + 35 = 172 - 96 + 35 = 111$

In 2007 there were 21.75 million unique Facebook visitors in one month. The other values can
be interpreted similarly.

(b)   The increases between consecutive years are 8.25 million, 29.75 million, and 51.25 million. A
linear function does not model the data because these three values are not equal.

95. Because there are 1200 feet of fence and the width of the enclosure is x, the length is given by $1200 - 2x$. The area of the enclosure is $A(x) = x(1200 - 2x)$ or $A(x) = -2x^2 + 1200x$. The value of x that will maximize the area is the x-coordinate of the vertex, $-\dfrac{b}{2a} = -\dfrac{1200}{2(-2)} = 300$. Thus the width is 300 feet and the length is $1200 - 2(300) = 1200 - 600 = 600$ feet. The enclosure measures 300 feet by 600 feet.

97. $S(8) = -0.227(8)^2 + 8.155(8) - 8.8 \approx 42$ inches

99. (a) $C(1990) = \dfrac{1}{300}(1990)^2 - \dfrac{199}{15}(1990) + \dfrac{39,619}{3} = 6$ In 1990, emissions were 6 billion metric tons.

   (b) $C(2020) = \dfrac{1}{300}(2020)^2 - \dfrac{199}{15}(2020) + \dfrac{39,619}{3} = 9$, $C(2020) - C(1990) = 9 - 6 = 3$

   There is an expected increase of 3 billion metric tons.

[20, 40, 5] by [0, 30, 5]   [20, 40, 5] by [0, 30, 5]

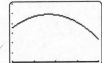

Figure 99a   Figure 99b

Section 8.2 Parabolas and Modeling

1. $x^2 + 2$

3. (1, 2)

5. $f(x) = ax^2 + bx + c$ or $f(x) = a(x - h)^2 + k$

7. downward

9. a

11. The graph is shifted 3 units downward.

x	-2	-1	0	1	2
$y = x^2$	4	1	0	1	4
$y = x^2 - 3$	1	-2	-3	-2	1

13. The graph is shifted 3 units to the right.

x	-2	-1	0	1	2
$y = x^2$	4	1	0	1	4

x	1	2	3	4	5
$y = (x - 3)^2$	4	1	0	1	4

15. (a)   The vertex form of the function is $f(x) = (x - 0)^2 + (-4)$. See Figure 15.

   (b)   The vertex is $(0, -4)$.

   (c)   Compared to the graph of $y = x^2$, the graph is shifted down 4 units.

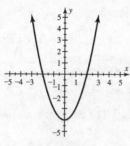

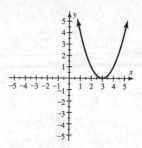

Figure 15   Figure 17   Figure 19

17. (a)   The vertex form of the function is $f(x) = 2(x-0)^2 + 1$. See Figure 17.

(b)   The vertex is $(0,\ 1)$.

(c)   Compared to the graph of $y = x^2$, the graph is narrower and is shifted up 1 unit.

19. (a)   The vertex form of the function is $f(x) = (x-3)^2 + 0$. See Figure 19.

(b)   The vertex is $(3,\ 0)$.

(c)   Compared to the graph of $y = x^2$, the graph is shifted right 3 units.

21. (a)   The vertex form of the function is $f(x) = -(x-0)^2 + 0$. See Figure 21.

(b)   The vertex is $(0,\ 0)$.

(c)   Compared to the graph of $y = x^2$, the graph is reflected across the x-axis.

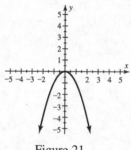

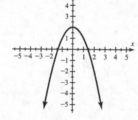

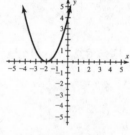

Figure 21   Figure 23   Figure 25

23. (a)   The vertex form of the function is $f(x) = -(x-0)^2 + 2$. See Figure 23.

(b)   The vertex is $(0,\ 2)$.

(c)   Compared to the graph of $y = x^2$, the graph is reflected across the x-axis and shifted up 2

units.

25. (a)   The vertex form of the function is $f(x) = (x-(-2))^2 + 0$. See Figure 25.

(b)   The vertex is $(-2,\ 0)$.

(c)   Compared to the graph of $y = x^2$, the graph is shifted left 2 units

27. (a)   The vertex form of the function is $f(x) = (x-(-1))^2 + (-2)$. See Figure 27.

(b)   The vertex is $(-1,\ -2)$.

(c)   Compared to the graph of $y = x^2$, the graph is shifted left 1 unit and down 2 units.

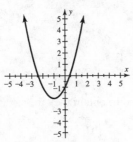

Figure 27

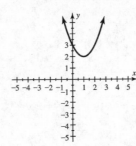

Figure 29

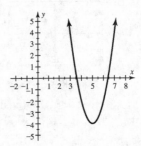

Figure 31

29. (a)   The vertex form of the function is $f(x) = (x-1)^2 + 2$. See Figure 29.

(b)   The vertex is $(1,\ 2)$.

(c)   Compared to the graph of $y = x^2$, the graph is shifted right 1 unit and up 2 units.

31. (a)   The vertex form of the function is $f(x) = 2(x-5)^2 + (-4)$. See Figure 31.

(b)   The vertex is $(5,\ -4)$.

(c)   Compared to the graph of $y = x^2$, the graph is narrower, shifted right 5 units and down 4 units.

33. (a)   The vertex form of the function is $f(x) = -\dfrac{1}{2}(x-(-3))^2 + 1$. See Figure 33.

(b)   The vertex is $(-3,\ 1)$.

(c)   Compared to the graph of $y = x^2$, the graph is wider, reflected across the x-axis, shifted left 3 units and up 1 unit.

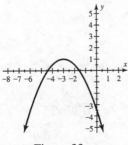

Figure 33

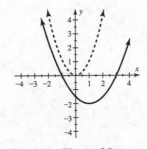

Figure 35

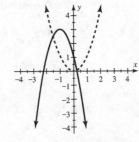

Figure 37

35. The graph of $f(x)$ is the graph of $y = x^2$ translated 1 unit right and 2 units downward. The graph of $f(x)$ is wider. See Figure 35.

37. The graph of $f(x)$ is the graph of $y = x^2$ translated 1 unit left and 3 units upward. The graph of $f(x)$ opens downward and is narrower. See Figure 37.

39. See Figure 39.

[–20, 20, 2] by [–20, 20, 2]

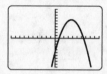

Figure 39

41. Since $h = 3$, $k = 4$ and $a = 3$, the equation is $y = 3(x-3)^2 + 4$. Expand to obtain the other form:

$$y = 3(x^2 - 6x + 9) + 4 \Rightarrow y = 3x^2 - 18x + 27 + 4 \Rightarrow y = 3x^2 - 18x + 31$$

43. Since $h = 5$, $k = -2$ and $a = -\dfrac{1}{2}$, the equation is $y = -\dfrac{1}{2}(x-5)^2 - 2$. Expand to obtain the other

form: $y = -\dfrac{1}{2}(x^2 - 10x + 25) - 2 \Rightarrow y = -\dfrac{1}{2}x^2 + 5x - \dfrac{25}{2} - 2 \Rightarrow y = -\dfrac{1}{2}x^2 + 5x - \dfrac{29}{2}$

45. Since $h = 1$, $k = 2$ and $a = 1$, the equation is $y = (x-1)^2 + 2$.

47. Since $h = 0$, $k = -3$ and $a = -1$, the equation is $y = -(x-0)^2 - 3$ or $y = -x^2 - 3$.

49. Since $h = 0$, $k = -3$ and $a = 1$, the equation is $y = (x-0)^2 - 3$ or $y = x^2 - 3$.

51. Since $h = -1$, $k = 2$ and $a = -1$, the equation is $y = -(x+1)^2 + 2$.

53. (a) We can use the vertex formula $x = -\dfrac{b}{2a}$, to find the x-coordinate of the vertex with $a = 4$

and $b = -8$. $x = -\dfrac{(-8)}{2(4)} = \dfrac{8}{8} = 1$. To find the y-coordinate, let $x = 1$ in $y = 4x^2 - 8x + 5$,

$y = 4(1)^2 - 8(1) + 5 = 1$ The vertex is (1, 1).

    (b) We can now find the vertex form with $a = 4$, $h = 1$, and $k = 1$.

$$y = a(x-h)^2 + k \Rightarrow y = 4(x-1)^2 + 1$$

55. (a) We can use the vertex formula $x = -\dfrac{b}{2a}$, to find the x-coordinate of the vertex with $a = -1$

and $b = -2$. $x = -\dfrac{(-2)}{2(-1)} = \dfrac{2}{-2} = -1$. To find the y-coordinate, let $x = -1$ in $y = -x^2 - 2x - 3$,

$y = -(-1)^2 - 2(-1) - 3 = -2$ The vertex is $(-1, -2)$.

    (b) We can now find the vertex form with $a = -1$, $h = -1$, and $k = -2$.

$$y = a(x-h)^2 + k \Rightarrow y = -(x-(-1))^2 + (-2) \Rightarrow y = -(x+1)^2 - 2$$

57. (a) We can use the vertex formula $x = -\dfrac{b}{2a}$, to find the x-coordinate of the vertex with $a = -2$

and $b = -4$. $x = -\dfrac{(-4)}{2(-2)} = \dfrac{4}{-4} = -1$. To find the y-coordinate, let $x = -1$ in

$y = -2x^2 - 4x + 1$, $y = -2(-1)^2 - 4(-1) + 1 = 3$ The vertex is $(-1, 3)$.

    (b)   We can now find the vertex form with $a = -2$, $h = -1$, and $k = 3$.

$$y = a(x-h)^2 + k \Rightarrow y = -2(x-(-1))^2 + 3 \Rightarrow y = -2(x+1)^2 + 3$$

59. The area of the rectangle in question is $1 \cdot 1 = 1$. Thus, to find the complete area of the square (complete the square) we must add the computed area of 1.

61. $y = x^2 + 2x \Rightarrow y = (x^2 + 2x + 1) - 1 \Rightarrow y = (x+1)^2 - 1$. The vertex is $(-1, -1)$.

63. $y = x^2 - 4x \Rightarrow y = (x^2 - 4x + 4) - 4 \Rightarrow y = (x-2)^2 - 4$. The vertex is $(2, -4)$.

65. $y = x^2 + 2x - 3 \Rightarrow y = (x^2 + 2x + 1) - 3 - 1 \Rightarrow y = (x+1)^2 - 4$. The vertex is $(-1, -4)$.

67. $y = x^2 - 4x + 5 \Rightarrow y = (x^2 - 4x + 4) + 5 - 4 \Rightarrow y = (x-2)^2 + 1$. The vertex is $(2, 1)$.

69. $y = x^2 + 3x - 2 \Rightarrow y = \left(x^2 + 3x + \dfrac{9}{4}\right) - 2 - \dfrac{9}{4} \Rightarrow y = \left(x + \dfrac{3}{2}\right)^2 - \dfrac{17}{4}$. The vertex is $\left(-\dfrac{3}{2}, -\dfrac{17}{4}\right)$.

71. $y = x^2 - 7x + 1 \Rightarrow y = \left(x^2 - 7x + \dfrac{49}{4}\right) + 1 - \dfrac{49}{4} \Rightarrow y = \left(x - \dfrac{7}{2}\right)^2 - \dfrac{45}{4}$. The vertex is $\left(\dfrac{7}{2}, -\dfrac{45}{4}\right)$.

73. $y = 3x^2 + 6x - 1 \Rightarrow y = 3(x^2 + 2x + 1) - 1 - 3 \Rightarrow y = 3(x+1)^2 - 4$. The vertex is $(-1, -4)$.

75. $y = 2x^2 - 3x \Rightarrow y = 2\left(x^2 - \dfrac{3}{2}x + \dfrac{9}{16}\right) - \dfrac{9}{8} \Rightarrow y = 2\left(x - \dfrac{3}{4}\right)^2 - \dfrac{9}{8}$. The vertex is $\left(\dfrac{3}{4}, -\dfrac{9}{8}\right)$.

77. $y = -2x^2 - 8x + 5 \Rightarrow y = -2(x^2 + 4x + 4) + 5 + 8 \Rightarrow y = -2(x+2)^2 + 13$. The vertex is $(-2, 13)$.

79. Since $y = 2$ when $x = 1$, $2 = a(1)^2 \Rightarrow 2 = 1a \Rightarrow a = 2$.

81. Since $y = 1.2$ when $x = 2$, $1.2 = a(2)^2 \Rightarrow 1.2 = 4a \Rightarrow a = 0.3$.

83. A scatterplot of the data (not shown) indicates that the data point $(1, -3)$ is the vertex of the parabola. The function has the form $f(x) = a(x-1)^2 - 3$. $f(2) = a(2-1)^2 - 3 = -1 \Rightarrow a - 3 = -1 \Rightarrow a = 2$.. Thus $f(x) = 2(x-1)^2 - 3$.

85. A scatterplot of the data (not shown) indicates that the data point $(1980, 6)$ is the vertex of the parabola. The function has the form $f(x) = a(x-1980)^2 + 6$.

    $f(1990) = a(1990-1980)^2 + 6 = 55 \Rightarrow 100a + 6 = 55 \Rightarrow a = 0.5$. Thus $f(x) = 0.5(x-1980)^2 + 6$.

87. (a)   See Figure 87.

    (b)   Since $D = 12$ when $x = 12$, we have $12 = a(12)^2 \Rightarrow 12 = 144a \Rightarrow a = \dfrac{12}{144} \Rightarrow a = \dfrac{1}{12}$.

    The function is $D(x) = \dfrac{1}{12}x^2$.

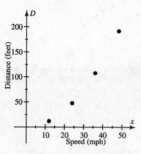

Figure 87

89. (a)   The GDP decreases and then increases.

  (b)   No, because the data decreases and then increases.

  (c)   Quadratic; This type of function can model data that decreases and then increases.

  (d)   (1995, 600). This point has the minimum y-value and the data is symmetric about this point.

  (e)   The scatterplot of the data indicates that the data point (1995, 600) is the vertex of the

    parabola. The function has the form $C(x) = a(x-1995)^2 + 600$.

    $C(1980) = a(1980-1995)^2 + 600 = 1000 \Rightarrow 224a = 400 \Rightarrow a \approx 1.8$.

    So $C(x) = 1.8(x-1995)^2 + 600$.

  (f)   See Figure 89.

  [1970, 2030, 10] by [0, 2500, 500]

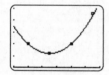

    Figure 89

91. (a)   Graph the data (not shown) to see that the data point (1982, 1) is near the vertex of a parabola.

    Graph $Y_1 = (X-1982)\wedge 2+1$ to see if it is a reasonable model. A better model is

    $f(x) = 2(x-1982)^2 + 1$. *Answers may vary.*

  (b)   Using the model, $f(1992) = 2(1992-1982)^2 + 1 = 2(10)^2 + 1 = 2(100) + 1 = 201$. The model

    predicts 201 thousand deaths in 1992, which is close to the actual value of 202

    thousand. *Answers may vary.*

93. As we read the graph from left to right we can see that the function is increasing when $x > 2$ and

  decreasing when $x < 2$.

95. As we read the graph from left to right we can see that the function is increasing when $x < 1$ and

  decreasing when $x > 1$.

97. We can use the vertex formula $x = -\dfrac{b}{2a}$, to find the x-coordinate of the vertex with $a = 3$ and

$b = -4$. $x = -\dfrac{(-4)}{2(3)} = \dfrac{4}{6} = \dfrac{2}{3}$. Since the value of a is positive, the graph is opening upward and

the function is increasing when $x > \dfrac{2}{3}$ and decreasing when $x < \dfrac{2}{3}$.

99. We can use the vertex formula $x = -\dfrac{b}{2a}$, to find the x-coordinate of the vertex with $a = -1$ and

$b = -3$. $x = -\dfrac{(-3)}{2(-1)} = -\dfrac{3}{2}$. Since the value of a is negative, the graph is opening downward and

the function is increasing when $x < -\dfrac{3}{2}$ and decreasing when $x > -\dfrac{3}{2}$.

101. We can use the vertex formula $x = -\dfrac{b}{2a}$, to find the x-coordinate of the vertex with $a = -4$ and

$b = -1$. $x = -\dfrac{(-1)}{2(-4)} = -\dfrac{1}{8}$. Since the value of a is negative, the graph is opening downward and

the function is increasing when $x < -\dfrac{1}{8}$ and decreasing when $x > -\dfrac{1}{8}$.

Checking Basic Concepts Sections 8.1 and 8.2

1. (a)   See Figure 1a. The vertex is $(0, -2)$, and the axis of symmetry is $x = 0$.

   (b)   See Figure 1b. The vertex is $(1, -3)$, and the axis of symmetry is $x = 1$.

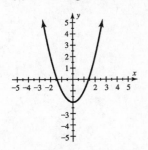 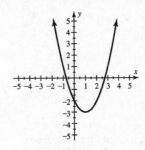

Figure 1a   Figure 1b

2. The graph of y_1 opens upward whereas y_2 opens downward. Also, y_1 is narrower than y_2.

3. Because $-\dfrac{b}{2a} = -\dfrac{12}{2(-3)} = 2$ and $f(2) = -3(2)^2 + 12(2) - 5 = 7$, the vertex is $(2, 7)$.

   The maximum y-value on the graph is 7.

4. (a)   See Figure 4a. Compared to the graph of $y = x^2$, the graph is shifted right 1 unit and

   up 2 units.

   (b)   See Figure 4b. Compared to the graph of $y = x^2$, the graph is reflected across the x-axis and

   shifted left 3 units.

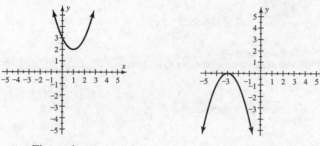

Figure 4a   Figure 4b

5. (a) $y = x^2 + 14x - 7 \Rightarrow y = (x^2 + 14x + 49) - 7 - 49 \Rightarrow y = (x + 7)^2 - 56$.

  (b) $y = 4x^2 + 8x - 2 \Rightarrow y = 4(x^2 + 2x + 1) - 2 - 4 \Rightarrow y = 4(x + 1)^2 - 6$.

Section 8.3 Quadratic Equations

1. $x^2 + 3x - 2 = 0$; *Answers may vary*. A quadratic equation can have 0, 1 or 2 solutions.

3. Factoring, square root property, completing the square

5. See Figure 5. *Answers may vary*

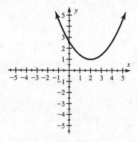

Figure 5

7. $x = \pm 8$; the square root property.

9. Yes

11. No, there is no x^2 term.

13. Yes

15. No, the term $\sqrt{x}$ is not allowed in a quadratic equation.

17. (a) $1 \pm \sqrt{7} \approx 1 \pm 2.65 = 3.65$ or -1.65

  (b) $-2 \pm \sqrt{11} \approx -2 \pm 3.32 = 1.32$ or -5.32

19. (a) $\dfrac{3 \pm \sqrt{13}}{5} \approx \dfrac{3 \pm 3.61}{5} = \dfrac{6.61}{5}$ or $-\dfrac{0.61}{5} \approx 1.32$ or -0.12

  (b) $\dfrac{-5 \pm \sqrt{6}}{9} \approx \dfrac{-5 \pm 2.45}{9} = \dfrac{-2.55}{9}$ or $\dfrac{-7.45}{9} \approx -0.28$ or -0.83

21. $-2, 1$

23. No real solutions

25. $-2, 3$

27. -0.5

29. (a) $x^2 - 4x - 5 = 0 \Rightarrow (x+1)(x-5) = 0 \Rightarrow x+1 = 0 \text{ or } x-5 = 0 \Rightarrow x = -1 \text{ or } 5$

   (b) A graph of $y = x^2 - 4x - 5$ (not shown) intersects the x-axis at -1 and 5.

   (c) A table of $y = x^2 - 4x - 5$ (not shown) yields y-values of zero when $x = -1$ and 5.

31. (a) $x^2 + 2x = 3 \Rightarrow x^2 + 2x - 3 = 0 \Rightarrow (x+3)(x-1) = 0 \Rightarrow x+3 = 0 \text{ or } x-1 = 0 \Rightarrow x = -3 \text{ or } 1$

   (b) A graph of $y = x^2 + 2x - 3$ (not shown) intersects the x-axis at -3 and 1.

   (c) A table of $y = x^2 + 2x - 3$ (not shown) yields y-values of zero when $x = -3$ and 1.

33. (a) $x^2 = 9 \Rightarrow x^2 - 9 = 0 \Rightarrow (x+3)(x-3) = 0 \Rightarrow x+3 = 0 \text{ or } x-3 = 0 \Rightarrow x = -3 \text{ or } 3$

   (b) A graph of $y = x^2 - 9$ (not shown) intersects the x-axis at -3 and 3.

   (c) A table of $y = x^2 - 9$ (not shown) yields y-values of zero when $x = -3$ and 3.

35. (a) $4x^2 - 4x - 3 = 0 \Rightarrow (2x+1)(2x-3) = 0 \Rightarrow 2x+1 = 0 \text{ or } 2x-3 = 0 \Rightarrow x = -\dfrac{1}{2} \text{ or } \dfrac{3}{2}$

   (b) A graph of $y = 4x^2 - 4x - 3$ (not shown) intersects the x-axis at $-\dfrac{1}{2}$ and $\dfrac{3}{2}$.

   (c) A table of $y = 4x^2 - 4x - 3$ (not shown) yields y-values of zero when $x = -\dfrac{1}{2}$ and $\dfrac{3}{2}$.

37. (a) $x^2 + 2x = -1 \Rightarrow x^2 + 2x + 1 = 0 \Rightarrow (x+1)(x+1) = 0 \Rightarrow x+1 = 0 \Rightarrow x = -1$.

   (b) A graph of $y = x^2 + 2x + 1$ (not shown) intersects the x-axis at -1.

   (c) A table of $y = x^2 + 2x + 1$ (not shown) yields y-values of zero when $x = -1$.

39. (a) $x^2 + 2 = 0 \Rightarrow x^2 = -2 \Rightarrow$ no real solutions.

   (b) A graph of $y = x^2 + 2$ (not shown) does not intersect the x-axis.

   (c) A table of $y = x^2 + 2$ (not shown) does not yield y-values of zero.

41. $x^2 + 2x - 35 = 0 \Rightarrow (x+7)(x-5) = 0 \Rightarrow \text{Either } x+7 = 0 \Rightarrow x = -7 \text{ or } x-5 = 0 \Rightarrow x = 5$

43. $6x^2 - x - 1 = 0 \Rightarrow (3x+1)(2x-1) = 0 \Rightarrow \text{Either } 3x+1 = 0 \Rightarrow x = -\dfrac{1}{3} \text{ or } 2x-1 = 0 \Rightarrow x = \dfrac{1}{2}$

45. $4x^2 + 13x + 9 = x \Rightarrow 4x^2 + 12x + 9 = 0 \Rightarrow (2x+3)(2x+3) = 0 \Rightarrow 2x+3 = 0 \Rightarrow x = -\dfrac{3}{2}$

47. $25x^2 - 350 = 125x \Rightarrow 25x^2 - 125x - 350 = 0 \Rightarrow 25(x^2 - 5x - 14) = 0 \Rightarrow 25(x-7)(x+2) = 0 \Rightarrow$

   $\text{Either } x-7 = 0 \Rightarrow x = 7 \text{ or } x+2 = 0 \Rightarrow x = -2$

49. $10x^2 - 27x + 18 = 0 \Rightarrow (5x-6)(2x-3) = 0 \Rightarrow \text{Either } 5x-6 = 0 \Rightarrow x = \dfrac{6}{5} \text{ or } 2x-3 = 0 \Rightarrow x = \dfrac{3}{2}$

51. $x^2 = 144 \Rightarrow x = \pm\sqrt{144} \Rightarrow x = \pm 12$

53. $5x^2 - 64 = 0 \Rightarrow 5x^2 = 64 \Rightarrow x^2 = \dfrac{64}{5} \Rightarrow x = \pm\sqrt{\dfrac{64}{5}} \Rightarrow x = \pm\dfrac{8}{\sqrt{5}}$ or $\pm\dfrac{8\sqrt{5}}{5}$

55. $(x+1)^2 = 25 \Rightarrow x+1 = \pm\sqrt{25} \Rightarrow x+1 = \pm 5 \Rightarrow x = -1 \pm 5 \Rightarrow x = -6$ or 4

57. $(x-1)^2 = 64 \Rightarrow x-1 = \pm\sqrt{64} \Rightarrow x-1 = \pm 8 \Rightarrow x = 1 \pm 8 \Rightarrow x = -7$ or 9

59. $(2x-1)^2 = 5 \Rightarrow 2x-1 = \pm\sqrt{5} \Rightarrow 2x = 1 \pm\sqrt{5} \Rightarrow x = \dfrac{1 \pm\sqrt{5}}{2}$

61. $10(x-5)^2 = 50 \Rightarrow (x-5)^2 = 5 \Rightarrow x-5 = \pm\sqrt{5} \Rightarrow x = 5 \pm\sqrt{5}$

63. $\left(\dfrac{4}{2}\right)^2 = 2^2 = 4$

65. $\left(\dfrac{-5}{2}\right)^2 = \dfrac{25}{4}$

67. The term needed to complete the square is $\left(\dfrac{-8}{2}\right)^2 = (-4)^2 = 16$. The resulting perfect square is

$(x-4)^2$.

69. The term needed to complete the square is $\left(\dfrac{9}{2}\right)^2 = \dfrac{81}{4}$. The resulting perfect square is $\left(x + \dfrac{9}{2}\right)^2$.

71. $x^2 - 2x = 24 \Rightarrow x^2 - 2x + 1 = 24 + 1 \Rightarrow (x-1)^2 = 25 \Rightarrow x-1 = \pm\sqrt{25} \Rightarrow x-1 = \pm 5 \Rightarrow$

    $x = 1 \pm 5 \Rightarrow x = -4$ or 6

73. $x^2 + 6x - 2 = 0 \Rightarrow x^2 + 6x + 9 = 2 + 9 \Rightarrow (x+3)^2 = 11 \Rightarrow x+3 = \pm\sqrt{11} \Rightarrow x = -3 \pm\sqrt{11}$

75. $x^2 - 3x = 5 \Rightarrow x^2 - 3x + \dfrac{9}{4} = 5 + \dfrac{9}{4} \Rightarrow \left(x - \dfrac{3}{2}\right)^2 = \dfrac{29}{4} \Rightarrow x - \dfrac{3}{2} = \pm\sqrt{\dfrac{29}{4}} \Rightarrow x = \dfrac{3}{2} \pm\dfrac{\sqrt{29}}{2} = \dfrac{3 \pm\sqrt{29}}{2}$

77. $x^2 - 5x + 1 = 0 \Rightarrow x^2 - 5x = -1 \Rightarrow x^2 - 5x + \dfrac{25}{4} = -1 + \dfrac{25}{4} \Rightarrow \left(x - \dfrac{5}{2}\right)^2 = \dfrac{21}{4} \Rightarrow$

    $x - \dfrac{5}{2} = \pm\sqrt{\dfrac{21}{4}} \Rightarrow x = \dfrac{5}{2} \pm\dfrac{\sqrt{21}}{2} = \dfrac{5 \pm\sqrt{21}}{2}$

79. $x^2 - 4 = 2x \Rightarrow x^2 - 2x = 4 \Rightarrow x^2 - 2x + 1 = 4 + 1 \Rightarrow (x-1)^2 = 5 \Rightarrow x-1 = \pm\sqrt{5} \Rightarrow x = 1 \pm\sqrt{5}$

81. $2x^2 - 3x = 4 \Rightarrow x^2 - \dfrac{3}{2}x = 2 \Rightarrow x^2 - \dfrac{3}{2}x + \dfrac{9}{16} = 2 + \dfrac{9}{16} \Rightarrow \left(x - \dfrac{3}{4}\right)^2 = \dfrac{41}{16} \Rightarrow x - \dfrac{3}{4} = \pm\sqrt{\dfrac{41}{16}} \Rightarrow$

    $x = \dfrac{3}{4} \pm\dfrac{\sqrt{41}}{4} = \dfrac{3 \pm\sqrt{41}}{4}$

83. $4x^2 - 8x - 7 = 0 \Rightarrow x^2 - 2x - \dfrac{7}{4} = 0 \Rightarrow x^2 - 2x + 1 = \dfrac{7}{4} + 1 \Rightarrow (x-1)^2 = \dfrac{11}{4} \Rightarrow$

$x - 1 = \pm\sqrt{\dfrac{11}{4}} \Rightarrow x = 1 \pm \dfrac{\sqrt{11}}{2} \Rightarrow x = \dfrac{2 \pm \sqrt{11}}{2}$

85. $36x^2 + 18x + 1 = 0 \Rightarrow 36\left(x^2 + \dfrac{1}{2}x + \dfrac{1}{16}\right) = -1 + \dfrac{36}{16} \Rightarrow 36\left(x + \dfrac{1}{4}\right)^2 = \dfrac{20}{16} \Rightarrow$

$\left(x + \dfrac{1}{4}\right)^2 = \dfrac{20}{576} \Rightarrow x + \dfrac{1}{4} = \pm\sqrt{\dfrac{20}{576}} \Rightarrow x = -\dfrac{1}{4} \pm \dfrac{\sqrt{20}}{24} = \dfrac{-6 \pm 2\sqrt{5}}{24} = \dfrac{-3 \pm \sqrt{5}}{12}$

87. $3x^2 + 12x = 36 \Rightarrow 3x^2 + 12x - 36 = 0 \Rightarrow 3(x^2 + 4x - 12) = 0 \Rightarrow 3(x+6)(x-2) = 0 \Rightarrow$

Either $x + 6 = 0 \Rightarrow x = -6$ or $x - 2 = 0 \Rightarrow x = 2$

89. $x^2 + 4x = -2 \Rightarrow x^2 + 4x + 4 = -2 + 4 \Rightarrow (x+2)^2 = 2 \Rightarrow x + 2 = \pm\sqrt{2} \Rightarrow x = -2 \pm \sqrt{2}$

91. $3x^2 - 4 = 2 \Rightarrow 3x^2 = 6 \Rightarrow x^2 = 2 \Rightarrow x = \pm\sqrt{2}$

93. $-6x^2 + 70 = 16x \Rightarrow -6x^2 - 16x + 70 = 0 \Rightarrow -2(3x^2 + 8x - 35) = 0 \Rightarrow -2(3x-7)(x+5) = 0 \Rightarrow$

Either $3x - 7 = 0 \Rightarrow x = \dfrac{7}{3}$ or $x + 5 = 0 \Rightarrow x = -5$

95. $-3x(x-8) = 6 \Rightarrow -3x^2 + 24x = 6 \Rightarrow x^2 - 8x = -2 \Rightarrow x^2 - 8x + 16 = -2 + 16 \Rightarrow (x-4)^2 = 14$

$\Rightarrow x - 4 = \pm\sqrt{14} \Rightarrow x = 4 \pm \sqrt{14}$

97. $ax^2 - c = 0 \Rightarrow ax^2 = c \Rightarrow x^2 = \dfrac{c}{a} \Rightarrow x = \pm\sqrt{\dfrac{c}{a}}$

99. (a) $x^2 - 3x - 18 = 0 \Rightarrow (x+3)(x-6) = 0 \Rightarrow$ Either $x + 3 = 0 \Rightarrow x = -3$ or $x - 6 = 0 \Rightarrow x = 6$

(b) Graph $Y_1 = X^2 - 3X - 18$ in $[-5, 8, 1]$ by $[-25, 5, 5]$. See Figures 99a & 99b.

The solutions are the x-intercepts, $x = -3$ or $x = 6$.

(c) Table $Y_1 = X^2 - 3X - 18$ with TblStart $= -6$ and ΔTbl $= 3$. See Figure 99c.

Since $Y_1 = 0$ when $x = -3$ or when $x = 6$, the solutions are $x = -3$ or $x = 6$.

$[-5, 8, 1]$ by $[-25, 5, 5]$ $[-5, 8, 1]$ by $[-25, 5, 5]$

          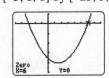   

Figure 99a   Figure 99b   Figure 99c

101. (a) $x^2 - 8x + 15 = 0 \Rightarrow (x-3)(x-5) = 0 \Rightarrow$ Either $x - 3 = 0 \Rightarrow x = 3$ or $x - 5 = 0 \Rightarrow x = 5$

(b) Graph $Y_1 = X^2 - 8X + 15$ in $[0, 10, 1]$ by $[-5, 10, 1]$. See Figures 101a & 101b.

The solutions are the x-intercepts, $x = 3$ or $x = 5$.

(c) Table $Y_1 = X^2 - 8X + 15$ with TblStart = 1 and ΔTbl = 1. See Figure 101c.

Since $Y_1 = 0$ when $x = 3$ or when $x = 5$, the solutions are $x = 3$ or $x = 5$.

[0, 10, 1] by [–5, 10, 1]   [0, 10, 1] by [–5, 10, 1]

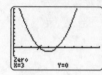   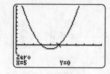   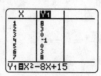

Figure 101a   Figure 101b   Figure 101c

103. (a) $4(x^2 + 35) = 48x \Rightarrow x^2 + 35 = 12x \Rightarrow x^2 - 12x + 35 = 0 \Rightarrow (x-5)(x-7) = 0 \Rightarrow$

Either $x - 5 = 0 \Rightarrow x = 5$ or $x - 7 = 0 \Rightarrow x = 7$

(b) Graph $Y_1 = 4(X^2 + 35) - 48X$ in [–2, 10, 1] by [–6, 6, 1]. See Figures 103a & 103b.

The solutions are the x-intercepts, $x = 5$ or $x = 7$.

(c) Table $Y_1 = 4(X^2 + 35) - 48X$ with TblStart = 3 and ΔTbl = 1. See Figure 103c.

Since $Y_1 = 0$ when $x = 5$ or when $x = 7$, the solutions are $x = 5$ or $x = 7$.

[–2, 10, 1] by [–6, 6, 1]   [–2, 10, 1] by [–6, 6, 1]

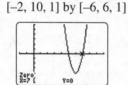   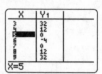

Figure 103a   Figure 103b   Figure 103c

105. $x = y^2 - 1 \Rightarrow x + 1 = y^2 \Rightarrow \pm\sqrt{x+1} = y \Rightarrow y = \pm\sqrt{x+1}$

107. $K = \dfrac{1}{2}mv^2 \Rightarrow 2K = mv^2 \Rightarrow \dfrac{2K}{m} = v^2 \Rightarrow \pm\sqrt{\dfrac{2K}{m}} = v \Rightarrow v = \pm\sqrt{\dfrac{2K}{m}}$

109. $E = \dfrac{k}{r^2} \Rightarrow Er^2 = k \Rightarrow r^2 = \dfrac{k}{E} \Rightarrow r = \pm\sqrt{\dfrac{k}{E}}$

111. $LC = \dfrac{1}{(2\pi f)^2} \Rightarrow 4\pi^2 f^2 LC = 1 \Rightarrow f^2 = \dfrac{1}{4\pi^2 LC} \Rightarrow f = \pm\sqrt{\dfrac{1}{4\pi^2 LC}} \Rightarrow f = \pm\dfrac{1}{2\pi\sqrt{LC}}$

113. (a) $\dfrac{1}{2}x^2 = 450 \Rightarrow x^2 = 900 \Rightarrow x = \pm\sqrt{900} \Rightarrow x = 30$ mph ($x = -30$ has no physical meaning)

(b) $\dfrac{1}{2}x^2 = 800 \Rightarrow x^2 = 1600 \Rightarrow x = \pm\sqrt{1600} \Rightarrow x = 40$ mph ($x = -40$ has no physical meaning)

115. $60 - 16t^2 = 0 \Rightarrow -16t^2 = -60 \Rightarrow t^2 = \dfrac{60}{16} \Rightarrow t = \pm\sqrt{\dfrac{60}{16}} \Rightarrow t = \dfrac{\sqrt{60}}{4} \approx 1.9$ seconds. The value $t \approx -1.9$

seconds has no physical meaning. The toy takes about 1.9 seconds to hit the ground. This is not

twice the time it takes to fall from a height of 30 feet.

37. (a)   Since the parabola opens downward, $a < 0$.

    (b)   The solution is the x-intercept, $x = 2$

    (c)   Since there is one real solution, the discriminant is zero.

39. (a) $(1)^2 - 4(3)(-2) = 25$

    (b)   Since the discriminant is positive, there are two real solutions.

    (c)   Graph $Y_1 = 3X^2 + X - 2$ in [–3, 3, 1] by [–3, 3, 1]. See Figure 39. There are two x-intercepts.

    [–3, 3, 1] by [–3, 3, 1]   [0, 4, 1] by [–3, 3, 1]

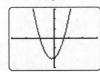

        Figure 39   Figure 41

41. (a) $(-4)^2 - 4(1)(4) = 0$

    (b)   Since the discriminant is zero, there is only one real solution.

    (c)   Graph $Y_1 = X^2 - 4X + 4$ in [0, 4, 1] by [–3, 3, 1]. See Figure 41. There is one x-intercept.

43. (a) $\left(\dfrac{3}{2}\right)^2 - 4\left(\dfrac{1}{2}\right)(2) = -\dfrac{7}{4}$

    (b)   Since the discriminant is negative, there are no real solutions.

    (c)   Graph $Y_1 = (1/2)X^2 + (3/2)X + 2$ in [–10, 10, 1] by [0, 10, 1]. See Figure 43. There are no

        x-intercepts.

    [–10, 10, 1] by [0, 10, 1]   [–5, 5, 1] by [–8, 5, 1]

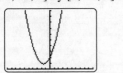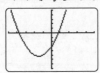

        Figure 43   Figure 45

45. (a) $(3)^2 - 4(1)(-3) = 21$

    (b)   Since the discriminant is positive, there are two real solutions.

    (c)   Graph $Y_1 = X^2 + 3X - 3$ in [–5, 5, 1] by [–8, 5, 1]. See Figure 45. There are two x-intercepts.

47. $x = \dfrac{-(-2) \pm \sqrt{(-2)^2 - 4(1)(-1)}}{2(1)} = \dfrac{2 \pm \sqrt{8}}{2} = \dfrac{2 \pm 2\sqrt{2}}{2} = \dfrac{2(1 \pm \sqrt{2})}{2} = 1 \pm \sqrt{2}$

49. $x = \dfrac{-(-1) \pm \sqrt{(-1)^2 - 4(-2)(3)}}{2(-2)} = \dfrac{1 \pm \sqrt{25}}{-4} = \dfrac{1 \pm 5}{-4} = -\dfrac{3}{2}$ or 1

51. $x = \dfrac{-1 \pm \sqrt{(1)^2 - 4(1)(5)}}{2(1)} = \dfrac{-1 \pm \sqrt{-19}}{2} \Rightarrow$ No real solutions. No x-intercepts.

53. $x = \dfrac{-(0) \pm \sqrt{(0)^2 - 4(1)(9)}}{2(1)} = \dfrac{0 \pm \sqrt{-36}}{2} \Rightarrow$ No real solutions. No x-intercepts.

93. $2x^2 - 4x + 6 = 0 \Rightarrow x^2 - 2x + 3 = 0 \Rightarrow x^2 - 2x + 1 = -3 + 1 \Rightarrow (x-1)^2 = -2 \Rightarrow$

$x - 1 = \pm\sqrt{-2} \Rightarrow x = 1 \pm \sqrt{-2} \Rightarrow x = 1 \pm i\sqrt{2}$

95. $x^2 - 3x + 2 = 0 \Rightarrow (x-1)(x-2) = 0 \Rightarrow x = 1$ or $x = 2$

97. Multiply both sides of the equation by 4 to obtain the following result.

$0.5x^2 - 1.75x - 1 = 0 \Rightarrow 2x^2 - 7x - 4 = 0 \Rightarrow (2x+1)(x-4) = 0 \Rightarrow x = -\dfrac{1}{2}$ or $x = 4$

99. $x = \dfrac{-(-5) \pm \sqrt{(-5)^2 - 4(1)(2)}}{2(1)} = \dfrac{5 \pm \sqrt{17}}{2}$

101. $x = \dfrac{-1 \pm \sqrt{(1)^2 - 4(2)(8)}}{2(2)} = \dfrac{-1 \pm \sqrt{-63}}{4} = \dfrac{-1 \pm 3i\sqrt{7}}{4} = -\dfrac{1}{4} \pm \dfrac{3}{4}i\sqrt{7}$

103. $4x^2 - 1 = 0 \Rightarrow (2x-1)(2x+1) = 0 \Rightarrow$ Either $2x - 1 = 0 \Rightarrow x = \dfrac{1}{2}$ or $2x + 1 = 0 \Rightarrow x = -\dfrac{1}{2}$

105. $3x^2 + 6 = 0 \Rightarrow 3(x^2 + 2) = 0 \Rightarrow x^2 + 2 = 0 \Rightarrow x^2 = -2 \Rightarrow x = \pm\sqrt{-2} \Rightarrow x = \pm i\sqrt{2}$

107. $9x^2 + 1 = 6x \Rightarrow 9x^2 - 6x + 1 = 0 \Rightarrow (3x-1)(3x-1) = 0 \Rightarrow 3x - 1 = 0 \Rightarrow x = \dfrac{1}{3}$

109. $\dfrac{1}{9}x^2 + \dfrac{11}{3}x - 42 = 0 \Rightarrow x^2 + 33x - 378 = 0 \Rightarrow x = (x-9)(x+42) = 0 \Rightarrow x = 9$ mph

The value $x = -42$ has no physical meaning.

111. $\dfrac{1}{9}x^2 + \dfrac{11}{3}x - 390 = 0 \Rightarrow x^2 + 33x - 3510 = 0 \Rightarrow x = (x-45)(x+78) = 0 \Rightarrow x = 45$ mph

The value $x = -78$ has no physical meaning.

113. $-0.095x^2 + 1.85x + 6 = 12 \Rightarrow -0.095x^2 + 1.85x - 6 = 0 \Rightarrow$

$x = \dfrac{-(1.85) \pm \sqrt{(1.85)^2 - 4(-0.095)(-6)}}{2(-0.095)} \Rightarrow x \approx 4.11$ or $x \approx 15.36$

Since x represents the number of years after 1989 the result is about 1993 and 2004.

115. (a) $G(0) = 0.4(0)^2 + 1.8(0) + 6 = 6$, In October 2010 Groupon's value was \$6 billion.

(b) $0.4x^2 + 1.8x + 6 = 15 \Rightarrow 0.4x^2 + 1.8x - 9 = 0 \Rightarrow$

$x = \dfrac{-(1.8) \pm \sqrt{(1.8)^2 - 4(0.4)(-9)}}{2(0.4)} \Rightarrow x = 3$ or $x = -7.5.$

Since x represents the number of months since October 2010 the result is January 2011. The value $x = -7.5$ has no physical meaning.

117. $2.39x^2 + 5.04x + 5.1 = 200 \Rightarrow 2.39x^2 + 5.04x - 194.9 = 0$

$x = \dfrac{-5.04 \pm \sqrt{(5.04)^2 - 4(2.39)(-194.9)}}{2(2.39)} = \dfrac{-5.04 \pm \sqrt{1888.6456}}{4.78} \Rightarrow x \approx 8.04$; about 1992

The value $x \approx -10.15$ has no meaning in this problem. Our answer agrees with the graph.

119. Let x represent the speed of the airplane without wind. The first trip takes $\dfrac{500}{x-20}$ hours. The second

trip takes $\dfrac{500}{x+10}$ hours. The total is $\dfrac{500}{x-20}+\dfrac{500}{x+10}=4$. Solve this equation for x.

$$\frac{500}{x-20}+\frac{500}{x+10}=4 \Rightarrow \frac{125}{x-20}+\frac{125}{x+10}=1 \Rightarrow 125(x+10)+125(x-20)=(x-20)(x+10) \Rightarrow$$

$$250x-1250=x^2-10x-200 \Rightarrow x^2-260x+1050=0 \Rightarrow$$

$$x=\frac{-(-260)\pm\sqrt{(-260)^2-4(1)(1050)}}{2(1)}=\frac{260\pm\sqrt{63,400}}{2}=130\pm5\sqrt{634}\approx 256 \text{ mph}$$

121. Let x represent the height of the screen. Then $x+3$ represents the width. The equation is
$x(x+3)=154$.

(a)   Graph $Y_1 = X^2 + 3X - 154$ in [–20, 20, 2] by [–200, 100, 50]. See Figure 121a.

$x=11,\ x+3=14$ The screen is 11 inches by 14 inches.

(b)   Table $Y_1 = X^2 + 3X - 154$ with TblStart = 7 and ΔTbl = 1. See Figure 121b.

$x=11,\ x+3=14$ The screen is 11 inches by 14 inches.

(c) $x^2+3x-154=0 \Rightarrow (x-11)(x+14)=0 \Rightarrow x=11$; The value $x=-14$ has no physical

meaning. The screen is 11 inches by 14 inches.

[–20, 20, 2] by [–200, 100, 50]

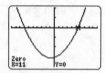   

Figure 121a   Figure 121b

123. (a)   The rate of change is not constant.

(b)   From the table, the height was 7 centimeters after about 75 seconds. *Answers may vary.*

(c) $x=\dfrac{-(-0.15)\pm\sqrt{(0.15)^2-4(0.0004)(9)}}{2(0.0004)}=\dfrac{0.15\pm\sqrt{0.0081}}{0.0008}=\dfrac{0.15\pm0.09}{0.0008}\Rightarrow x=75$ seconds

The value $x=300$ is not possible for this problem.

Checking Basic Concepts Sections 8.3 and 8.4

1. Symbolic: $2x^2-7x+3=0 \Rightarrow (2x-1)(x-3)=0 \Rightarrow$ Either $2x-1=0 \Rightarrow x=\dfrac{1}{2}$ or $x-3=0 \Rightarrow x=3$

Graphical: Graph $Y_1 = 2X^2 - 7X + 3$ in [–5, 5, 1] by [–5, 5, 1].

See Figures 1a & 1b. $x=\dfrac{1}{2}$ or $x=3$

[–5, 5, 1] by [–5, 5, 1]

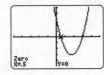   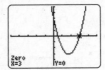

Figure 1a   Figure 1b

2. $x^2 = 5 \Rightarrow x = \pm\sqrt{5}$

3. $x^2 - 4x + 1 = 0 \Rightarrow x^2 - 4x + 4 = -1 + 4 \Rightarrow (x-2)^2 = 3 \Rightarrow x - 2 = \pm\sqrt{3} \Rightarrow x = 2 \pm \sqrt{3}$

4. $x^2 + y^2 = 1 \Rightarrow y^2 = 1 - x^2 \Rightarrow y = \pm\sqrt{1 - x^2}$

5. (a) $2x^2 = 3x + 1 \Rightarrow 2x^2 - 3x - 1 = 0 \Rightarrow x = \dfrac{-(-3) \pm \sqrt{(-3)^2 - 4(2)(-1)}}{2(2)} = \dfrac{3 \pm \sqrt{17}}{4}$

    (b) $9x^2 - 24x + 16 = 0 \Rightarrow x = \dfrac{-(-24) \pm \sqrt{(-24)^2 - 4(9)(16)}}{2(9)} = \dfrac{24 \pm \sqrt{0}}{18} = \dfrac{24}{18} = \dfrac{4}{3}$

    (c) $x = \dfrac{-1 \pm \sqrt{1^2 - 4(1)(2)}}{2(1)} = \dfrac{-1 \pm \sqrt{-7}}{2} = \dfrac{-1 \pm i\sqrt{7}}{2} = -\dfrac{1}{2} \pm i\dfrac{\sqrt{7}}{2}$

6. (a) $(-5)^2 - 4(1)(5) = 5$; Since the discriminant is positive, there are two real solutions.

    (b) $(-5)^2 - 4(2)(4) = -7$; Since the discriminant is negative, there are no real solutions.

    (c) $(-56)^2 - 4(49)(16) = 0$; Since the discriminant is zero, there is one real solution

Section 8.5 Quadratic Inequalities

1. An inequality has an inequality sign rather than an equals sign.

3. No, since $9 \not< 7$.

5. $-2 < x < 4$

7. Yes

9. Yes

11. No, this inequality is linear.

13. Yes, since $2(3)^2 + (3) - 1 = 20$ and $20 > 0$.

15. No, since $(0)^2 + 2 = 2$ and $2 \not\le 0$.

17. No, since $(-3)^2 - 3(-3) = 18$ and $18 \not\le 1$.

19. (a) The solutions are the x-intercepts, $x = -3$ or $x = 2$.

    (b) $-3 < x < 2$

    (c) $x < -3$ or $x > 2$

21. (a) The solutions are the x-intercepts, $x = -2$ or $x = 2$.

    (b) $-2 < x < 2$

(c) $x < -2$ or $x > 2$

23. (a)   The solutions are the x-intercepts, $x = -10$ or $x = 5$.

    (b) $x < -10$ or $x > 5$

    (c) $-10 < x < 5$

25. Graph $Y_1 = \pi X \wedge 2 - \sqrt{(3)} X$ and $Y_2 = 3/11$ in $[-4.7, 4.7, 1]$ by $[-3.1, 3.1, 1]$.

    See Figures 25a and 25b. The solution is $[-0.128, 0.679]$.

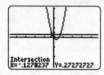

     Figure 25a   Figure 25b

27. (a)   The solutions are $x = -2$ or $x = 2$.

    (b) $-2 < x < 2$

    (c) $x < -2$ or $x > 2$

29. (a)   The solutions are $x = -4$ or $x = 0$.

    (b) $-4 < x < 0$

    (c) $x < -4$ or $x > 0$

31. $x^2 + 10x + 21 = 0 \Rightarrow (x + 7)(x + 3) = 0 \Rightarrow x = -7$ or $x = -3$

    Since the parabola opens upward, the solution is $[-7, -3]$.

33. $3x^2 - 9x + 6 = 0 \Rightarrow 3(x - 1)(x - 2) = 0 \Rightarrow x = 1$ or $x = 2$

    Since the parabola opens upward, the solution is $(-\infty, 1) \cup (2, \infty)$.

35. $x^2 = 10 \Rightarrow x = -\sqrt{10}$ or $x = \sqrt{10}$ Since the parabola opens upward, the solution is $(-\sqrt{10}, \sqrt{10})$.

37. $x(6 - x) = 0 \Rightarrow x = 0$ or $x = 6$ Since the parabola opens upward, the solution is $(-\infty, 0) \cup (6, \infty)$.

39. $x(4 - x) = 2 \Rightarrow -x^2 + 4x - 2 = 0 \Rightarrow x = \dfrac{-4 \pm \sqrt{(4)^2 - 4(-1)(-2)}}{2(-1)} = 2 \pm \sqrt{2}$

    Since the parabola opens downward, the solution is $(-\infty, 2 - \sqrt{2}] \cup [2 + \sqrt{2}, \infty)$.

41. (a) $x^2 - 4 = 0 \Rightarrow x^2 = 4 \Rightarrow x = -2$ or $x = 2$

    (b)   Since the parabola opens upward, the solution is $-2 < x < 2$.

    (c)   Since the parabola opens upward, the solution is $x < -2$ or $x > 2$.

43. (a) $x^2 + x - 1 = 0 \Rightarrow x = \dfrac{-1 \pm \sqrt{(1)^2 - 4(1)(-1)}}{2(1)} \Rightarrow x = \dfrac{-1 \pm \sqrt{5}}{2}$

    (b)   Since the parabola opens upward, the solution is $\dfrac{-1 - \sqrt{5}}{2} < x < \dfrac{-1 + \sqrt{5}}{2}$.

    (c)   Since the parabola opens upward, the solution is $x < \dfrac{-1 - \sqrt{5}}{2}$ or $x > \dfrac{-1 + \sqrt{5}}{2}$.

45. First replace the inequality symbol with an equals sign and solve the resulting equation.

    $x^2 + 4x + 3 = 0 \Rightarrow (x+3)(x+1) = 0 \Rightarrow x = -3$ or -1. The parabola given by $y = x^2 + 4x + 3$ lies

    below the x-axis when $-3 < x < -1$. The interval is $(-3, \ -1)$.

47. First replace the inequality symbol with an equals sign and solve the resulting equation.

    $2x^2 - x - 15 = 0 \Rightarrow (2x+5)(x-3) = 0 \Rightarrow x = -2.5$ or 3. The parabola given by $y = 2x^2 - x - 15$ lies

    above the x-axis when $x \le -2.5$ or $x \ge 3$. The interval is $(-\infty, \ -2.5] \cup [3, \ \infty)$.

49. First replace the inequality symbol with an equals sign and solve the resulting equation.

    $2x^2 = 8 \Rightarrow x^2 = 4 \Rightarrow x^2 - 4 = 0 \Rightarrow (x+2)(x-2) = 0 \Rightarrow x = -2$ or 2. The parabola given by

    $y = 2x^2 - 8$ lies below the x-axis when $-2 \le x \le 2$. The interval is $[-2, \ 2]$.

51. The value of x^2 is greater than -5 for all values of x. The parabola given by $y = x^2 + 5$ lies

    entirely above the x-axis. The interval is $(-\infty, \ \infty)$.

53. First replace the inequality symbol with an equals sign and solve the resulting equation.

    $-x^2 + 3x = 0 \Rightarrow -x(x-3) = 0 \Rightarrow x = 0$ or 3. The parabola given by $y = -x^2 + 3x$ lies above the

    x-axis when $0 < x < 3$. The interval is $(0, \ 3)$.

55. $x^2 + 2 \le 0$ has no solutions; the left side of the inequality is always greater than or equal to 2.

57. $(x-2)^2 \le 0 \Rightarrow (x-2)^2 = 0$ because $(x-2)^2$ can never be less than 0.

    $(x-2)^2 = 0 \Rightarrow x - 2 = 0 \Rightarrow x = 2$.

59. First replace the inequality symbol with an equals sign and solve the resulting equation.

    $(x+1)^2 = 0 \Rightarrow x + 1 = 0 \Rightarrow x = -1$. The parabola given by $y = (x+1)^2$ lies above the x-axis when

    $x < -1$ or $x > -1$. The interval is $(-\infty, -1) \cup (-1, \infty)$.

61. $x(1-x) \ge -2 \Rightarrow x(1-x) + 2 \ge 0$. Replace the inequality symbol with an equals sign and solve the

    resulting equation. $x(1-x) + 2 = 0 \Rightarrow x - x^2 + 2 = 0 \Rightarrow -x^2 + x + 2 = 0 \Rightarrow (-x+2)(x+1) = 0 \Rightarrow x = 2$

    or $x = -1$. The parabola given by $y = -x^2 + x + 2$ lies above or on the x-axis when $-1 \le x \le 2$.

    The interval is $[-1, 2]$.

63. (a)   Graph $Y_1 = 0.0000375X^2 - 0.175X + 1000$ and $Y_2 = 850$ in [0, 4000, 1000] by

    [500, 1200, 100]. See Figures 63a & 63b. The elevation is 850 feet or less from 1131 feet to

    3535 feet (approximately).

    (b)   The elevation is 850 feet or more before 1131 feet or after 3535 feet (approximately).

[0, 4000, 1000] by [500, 1200, 100]   [0, 4000, 1000] by [500, 1200, 100]

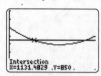   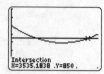

Figure 63a   Figure 63b

65. (a)   From the formula, $f(1985) = -0.05107(1985)^2 + 194.74(1985) - 184,949 \approx 383$.

From the graph $f(1985) \approx 383$.

(b)   The death rate was 500 or less about 1969 and after.

(c) $-0.05107x^2 + 194.74x - 184,949 = 500 \Rightarrow -0.05107x^2 + 194.74x - 185,449 = 0 \Rightarrow$

$$x = \frac{-194.74 \pm \sqrt{(194.74)^2 - 4(-0.05107)(-185,449)}}{2(-0.05107)} \approx 1845 \text{ or } 1969.$$

The only valid solution is 1969. The death rate was 500 or less about 1969 or after.

67. Let w represent the width of the pen. Then $w+5$ represents the length and the area is given by

$w(w+5) = w^2 + 5w$. We want to find w so that $w^2 + 5w \geq 176$ and $w^2 + 5w \leq 500$.

$w^2 + 5w - 176 = 0 \Rightarrow (w+16)(w-11) = 0 \Rightarrow w = -16$ or $w = 11$, so $w \geq 11$. The values $w \leq -16$ have

no meaning. $w^2 + 5w - 500 = 0 \Rightarrow (w+25)(w-20) = 0 \Rightarrow w = -25$ or $w = 20$, so $w \leq 20$. The values

$w \leq -25$ have no meaning. The width must be from 11 feet to 20 feet.

Section 8.6 Equations in Quadratic Form

1. $u^2 - 7u + 6 = 0 \Rightarrow (u-1)(u-6) = 0 \Rightarrow u = 1$ or 6

When $u = 1$, $x^2 = 1 \Rightarrow x = \pm 1$. When $u = 6$, $x^2 = 6 \Rightarrow x = \pm\sqrt{6}$.

3. $3u^2 + u - 10 = 0 \Rightarrow (u+2)(3u-5) = 0 \Rightarrow u = -2$ or $\dfrac{5}{3}$

When $u = -2$, $z^3 = -2 \Rightarrow z = -\sqrt[3]{2}$. When $u = \dfrac{5}{3}$, $z^3 = \dfrac{5}{3} \Rightarrow z = \sqrt[3]{\dfrac{5}{3}}$.

5. $4u^2 + 17u + 15 = 0 \Rightarrow (u+3)(4u+5) = 0 \Rightarrow u = -3$ or $-\dfrac{5}{4}$

When $u = -3$, $n^{-1} = -3 \Rightarrow n = -\dfrac{1}{3}$. When $u = -\dfrac{5}{4}$, $n^{-1} = -\dfrac{5}{4} \Rightarrow n = -\dfrac{4}{5}$.

7. Let $u = x^2$. Then $u^2 = 8u + 9 \Rightarrow u^2 - 8u - 9 = 0 \Rightarrow (u+1)(u-9) = 0 \Rightarrow u = -1$ or 9.

When $u = -1$, $x^2 = -1 \Rightarrow$ no solutions. When $u = 9$, $x^2 = 9 \Rightarrow x = \pm\sqrt{9} = -3$ or 3.

9. Let $u = x^3$. Then $3u^2 - 5u - 2 = 0 \Rightarrow (3u+1)(u-2) = 0 \Rightarrow u = -\dfrac{1}{3}$ or 2.

When $u = -\dfrac{1}{3}$, $x^3 = -\dfrac{1}{3} \Rightarrow x = -\sqrt[3]{\dfrac{1}{3}}$. When $u = 2$, $x^3 = 2 \Rightarrow x = \sqrt[3]{2}$.

11. Let $u = z^{-1}$. Then $2u^2 + 11u = 40 \Rightarrow 2u^2 + 11u - 40 = 0 \Rightarrow (u+8)(2u-5) = 0 \Rightarrow u = -8$ or $\dfrac{5}{2}$.

When $u = -8$, $z^{-1} = -8 \Rightarrow z = -\dfrac{1}{8}$. When $u = \dfrac{5}{2}$, $z^{-1} = \dfrac{5}{2} \Rightarrow z = \dfrac{2}{5}$.

13. Let $u = x^{1/3}$. Then $u^2 - 2u + 1 = 0 \Rightarrow (u-1)^2 = 0 \Rightarrow u = 1$. When $u = 1$, $x^{1/3} = 1 \Rightarrow x = 1^3 = 1$.

15. Let $u = x^{1/5}$. Then $u^2 - 33u + 32 = 0 \Rightarrow (u-1)(u-32) = 0 \Rightarrow u = 1$ or 32.

When $u = 1$, $x^{1/5} = 1 \Rightarrow x = 1^5 = 1$. When $u = 32$, $x^{1/5} = 32 \Rightarrow x = 32^5 = 33,554,432$.

17. Let $u = x^{1/2}$. Then $u^2 - 13u + 36 = 0 \Rightarrow (u-4)(u-9) = 0 \Rightarrow u = 4$ or 9.

When $u = 4$, $x^{1/2} = 4 \Rightarrow x = 4^2 = 16$. When $u = 9$, $x^{1/2} = 9 \Rightarrow x = 9^2 = 81$.

19. Let $u = z^{1/4}$. Then $u^2 - 2u + 1 = 0 \Rightarrow (u-1)^2 = 0 \Rightarrow u = 1$. When $u = 1$, $z^{1/4} = 1 \Rightarrow x = 1^4 = 1$.

21. Let $u = x + 1$. Then $u^2 - 5u - 14 = 0 \Rightarrow (u+2)(u-7) = 0 \Rightarrow u = -2$ or 7.

When $u = -2$, $x + 1 = -2 \Rightarrow x = -3$. When $u = 7$, $x + 1 = 7 \Rightarrow x = 6$.

23. Let $u = x^2 - 1$. Then $u^2 - 4 = 0 \Rightarrow (u+2)(u-2) = 0 \Rightarrow u = -2$ or 2.

When $u = -2$, $x^2 - 1 = -2 \Rightarrow x^2 = -1 \Rightarrow$ no solutions.

When $u = 2$, $x^2 - 1 = 2 \Rightarrow x^2 = 3 \Rightarrow x = -\sqrt{3}$ or $\sqrt{3}$.

25. $x^4 - 16 = 0 \Rightarrow (x^2 - 4)(x^2 + 4) = 0 \Rightarrow (x-2)(x+2)(x^2+4) = 0 \Rightarrow x = 2$ or $x = -2$ or

$x^2 = -4 \Rightarrow x = \pm\sqrt{-4} \Rightarrow x = \pm 2i$

27. $x^3 + x = 0 \Rightarrow x(x^2 + 1) = 0 \Rightarrow x = 0$ or $x^2 = -1 \Rightarrow x = \pm\sqrt{-1} \Rightarrow x = \pm i$

29. $x^4 - 2 = x^2 \Rightarrow x^4 - x^2 - 2 = 0 \Rightarrow (x^2 - 2)(x^2 + 1) = 0 \Rightarrow x^2 = 2$ or $x^2 = -1 \Rightarrow x = \pm\sqrt{2}$ or $x = \pm i$

31. $\dfrac{1}{x} + \dfrac{1}{x^2} = -\dfrac{1}{2} \Rightarrow 2x^2\left(\dfrac{1}{x} + \dfrac{1}{x^2}\right) = 2x^2\left(-\dfrac{1}{2}\right) \Rightarrow 2x + 2 = -x^2 \Rightarrow x^2 + 2x + 2 = 0 \Rightarrow x^2 + 2x = -2$

$\Rightarrow x^2 + 2x + 1 = -2 + 1 \Rightarrow (x+1)^2 = -1 \Rightarrow x + 1 = \pm\sqrt{-1} \Rightarrow x = -1 \pm i$

33. $\dfrac{2}{x-2} - \dfrac{1}{x} = -\dfrac{1}{2} \Rightarrow 2x(x-2)\left(\dfrac{2}{x-2} - \dfrac{1}{x}\right) = 2x(x-2)\left(-\dfrac{1}{2}\right) \Rightarrow 4x - 2(x-2) = -x(x-2)$

$\Rightarrow 4x - 2x + 4 = -x^2 + 2x \Rightarrow x^2 + 4 = 0 \Rightarrow x^2 = -4 \Rightarrow x = \pm\sqrt{-4} \Rightarrow x = \pm 2i$

Checking Basic Concepts Sections 8.5 and 8.6

1. $x^2 - x - 6 = 0 \Rightarrow (x+2)(x-3) = 0 \Rightarrow x = -2$ or $x = 3$

Since the parabola opens upward, the solution is $(-\infty, -2) \cup (3, \infty)$

2. $3x^2 + 5x + 2 = 0 \Rightarrow (x+1)(3x+2) = 0 \Rightarrow x = -1$ or $x = -\dfrac{2}{3}$

   Since the parabola opens upward, the solution is $\left[-1,\ -\dfrac{2}{3}\right]$.

3. Let $u = x^3$. Then $u^2 + 6u - 16 = 0 \Rightarrow (u+8)(u-2) = 0 \Rightarrow u = -8$ or 2.

   When $u = -8$, $x^3 = -8 \Rightarrow x = \sqrt[3]{-8} = -2$. When $u = 2$, $x^3 = 2 \Rightarrow x = \sqrt[3]{2}$.

4. Let $u = x^{1/3}$. Then $u^2 - 7u - 8 = 0 \Rightarrow (u+1)(u-8) = 0 \Rightarrow u = -1$ or 8.

   When $u = -1$, $x^{1/3} = -1 \Rightarrow x = (-1)^3 = -1$. When $u = 8$, $x^{1/3} = 8 \Rightarrow x = 8^3 = 512$.

5. $x^4 + 2x^2 + 1 = 0 \Rightarrow (x^2 + 1)^2 = 0 \Rightarrow x^2 + 1 = 0 \Rightarrow x^2 = -1 \Rightarrow x = \pm\sqrt{-1} \Rightarrow x = \pm i$

Chapter 8 Review

1. Vertex: $(-3, 4)$; Axis of symmetry: $x = -3$; Opens downward;

2. Vertex: $(1,\ 0)$; Axis of symmetry: $x = 1$; Opens upward;

3. (a) See Figure 3.

   (b) The vertex is $(0, -2)$. The axis of symmetry is $x = 0$.

   (c) $f(-1) = (-1)^2 - 2 = 1 - 2 = -1$

4. (a) See Figure 4.

   (b) The vertex is $(2, 1)$. The axis of symmetry is $x = 2$.

   (c) $f(3) = -(3)^2 + 4(3) - 3 = -9 + 12 - 3 = 0$

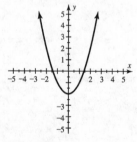   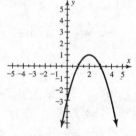   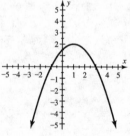

Figure 3   Figure 4   Figure 5

5. (a) See Figure 5.

   (b) The vertex is $(1, 2)$. The axis of symmetry is $x = 1$.

   (c) $f(-2) = -\dfrac{1}{2}(-2)^2 + (-2) + \dfrac{3}{2} = -2 - 2 + \dfrac{3}{2} = -2.5$

6. (a) See Figure 6.

   (b) The vertex is $(-2, -3)$. The axis of symmetry is $x = -2$.

   (c) $f(-3) = 2(-3)^2 + 8(-3) + 5 = 18 - 24 + 5 = -1$

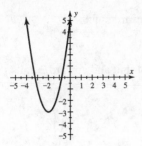

Figure 6

7. Because $-\dfrac{b}{2a} = -\dfrac{-6}{2(2)} = \dfrac{3}{2}$ and $f\left(\dfrac{3}{2}\right) = 2\left(\dfrac{3}{2}\right)^2 - 6\left(\dfrac{3}{2}\right) + 1 = -\dfrac{7}{2}$, the vertex is $\left(\dfrac{3}{2},\ -\dfrac{7}{2}\right)$.

The minimum y-value on the graph is $-\dfrac{7}{2}$.

8. Because $-\dfrac{b}{2a} = -\dfrac{2}{2(-3)} = \dfrac{1}{3}$ and $f\left(\dfrac{1}{3}\right) = -3\left(\dfrac{1}{3}\right)^2 + 2\left(\dfrac{1}{3}\right) - 5 = -\dfrac{14}{3}$, the vertex is $\left(\dfrac{1}{3},\ -\dfrac{14}{3}\right)$.

The maximum y-value on the graph is $-\dfrac{14}{3}$.

9. $x = -\dfrac{b}{2a} = -\dfrac{(-4)}{2(1)} = 2$, $f(2) = (2)^2 - 4(2) - 2 = -6$; The vertex is $(2, -6)$.

10. $x = -\dfrac{b}{2a} = -\dfrac{(0)}{2(-1)} = 0$, $f(0) = 5 - (0)^2 = 5$; The vertex is $(0, 5)$.

11. $x = -\dfrac{b}{2a} = -\dfrac{(1)}{2(-\frac{1}{4})} = 2$, $f(2) = -\dfrac{1}{4}(2)^2 + (2) + 1 = 2$; The vertex is $(2, 2)$.

12. $x = -\dfrac{b}{2a} = -\dfrac{(2)}{2(1)} = -1$, $f(-1) = 2 + 2(-1) + (-1)^2 = 1$; The vertex is $(-1, 1)$.

13. (a) See Figure 13.

  (b) This graph is a shift of the graph of $y = x^2$ upward 2 units.

14. (a) See Figure 14.

  (b) This graph is more narrow than the graph of $y = x^2$.

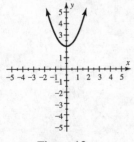

Figure 13

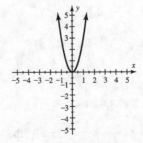

Figure 14

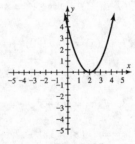

Figure 15

15. (a) See Figure 15.

  (b) This graph is a shift of the graph of $y = x^2$ right 2 units.

16. (a) See Figure 16.

(b)   This graph is a shift of the graph of $y = x^2$ left 1 unit and downward 3 units.

17. (a)   See Figure 17.

(b)   This graph is wider than the graph of $y = x^2$ and is shifted left 1 unit and upward 2 units.

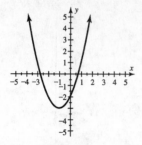

Figure 16   Figure 17   Figure 18

18. (a)   See Figure 18.

(b)   This graph is more narrow than the graph of $y = x^2$ and is shifted right 1 unit and downward 3 units.

19. $y = a(x-h)^2 + k$ where (h, k) is the vertex , gives $y = -4(x-2)^2 - 5$.

20. Opening downward means $a = -1$. $y = a(x-h)^2 + k$, gives $y = -1(x+4)^2 + 6$.

21. $y = x^2 + 4x - 7 \Rightarrow y = (x^2 + 4x + 4) - 7 - 4 \Rightarrow y = (x+2)^2 - 11$. The vertex is $(-2, \ -11)$.

22. $y = x^2 - 7x + 1 \Rightarrow y = \left(x^2 - 7x + \dfrac{49}{4}\right) + 1 - \dfrac{49}{4} \Rightarrow y = \left(x - \dfrac{7}{2}\right)^2 - \dfrac{45}{4}$. The vertex is $\left(\dfrac{7}{2}, \ -\dfrac{45}{4}\right)$.

23. $y = 2x^2 - 3x - 8 \Rightarrow y = 2\left(x^2 - \dfrac{3}{2}x + \dfrac{9}{16}\right) - 8 - \dfrac{9}{8} \Rightarrow y = 2\left(x - \dfrac{3}{4}\right)^2 - \dfrac{73}{8}$.

The vertex is $\left(\dfrac{3}{4}, \ -\dfrac{73}{8}\right)$.

24. $y = 3x^2 + 6x - 2 \Rightarrow y = 3(x^2 + 2x + 1) - 2 - 3 \Rightarrow y = 3(x+1)^2 - 5$. The vertex is $(-1, -5)$.

25. $a(1)^2 - 1 = 2 \Rightarrow a = 3$

26. $a(-1)^2 - 1 = -\dfrac{3}{4} \Rightarrow a = \dfrac{1}{4}$

27. $f(x) = -5(x-3)^2 + 4 \Rightarrow f(x) = -5(x^2 - 6x + 9) + 4$

$\Rightarrow f(x) = -5x^2 + 30x - 45 + 4 \Rightarrow f(x) = -5x^2 + 30x - 41$; the y-intercept is $c = -41$.

28. $f(x) = 3(x+2)^2 - 4 \Rightarrow f(x) = 3(x^2 + 4x + 4) - 4$

$\Rightarrow f(x) = 3x^2 + 12x + 12 - 4 \Rightarrow f(x) = 3x^2 + 12x + 8$; the y-intercept is $c = 8$.

29. $-2, 3$

30. -1

31. No real solutions.

32. –4, 6

33. –10, 5

34. –0.5, 0.25

35. (a)   Graph $Y_1 = X^2 - 5X - 50$ in [–10, 20, 5] by [–100, 20, 10]. See Figures 35a & 35b.

The solutions are the x-intercepts, $x = -5$ and $x = 10$.

(b)   Table $Y_1 = X^2 - 5X - 50$ with TblStart = –10 and ΔTbl = 5. See Figure 35c.

Since $Y_1 = 0$ when $x = -5$ and when $x = 10$, the solutions are $x = -5$ or $x = 10$.

[–10, 20, 5] by [–100, 20, 10]   [–10, 20, 5] by [–100, 20, 10]

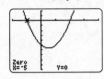   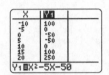

Figure 35a   Figure 35b   Figure 35c

36. (a)   Graph $Y_1 = (1/2)X^2 + X - (3/2)$ in [–5, 5, 1] by [–5, 5, 1]. See Figures 36a & 36b.

The solutions are the x-intercepts, $x = -3$ or $x = 1$.

(b)   Table $Y_1 = (1/2)X^2 + X - (3/2)$ with TblStart = –4 and ΔTbl = 1. See Figure 36c.

Since $Y_1 = 0$ when $x = -3$ and when $x = 1$, the solutions are $x = -3$ and $x = 1$.

[–5, 5, 1] by [–5, 5, 1]   [–5, 5, 1] by [–5, 5, 1]

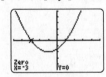   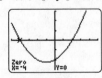

Figure 36a   Figure 36b   Figure 36c

37. (a)   Graph $Y_1 = (1/4)X^2 + (1/2)X - 2$ in [–5, 5, 1] by [–3, 3, 1]. See Figures 37a & 37b.

The solutions are the x-intercepts, $x = -4$ and $x = 2$.

(b)   Table $Y_1 = (1/4)X^2 + (1/2)X - 2$ with TblStart = –8 and ΔTbl = 2. See Figure 37c.

Since $Y_1 = 0$ when $x = -4$ and when $x = 2$, the solutions are $x = -4$, and $x = 2$.

[–5, 5, 1] by [–3, 3, 1]   [–5, 5, 1] by [–3, 3, 1]

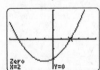   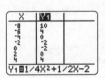

Figure 37a   Figure 37b   Figure 37c

38. (a)   Graph $Y_1 = (1/4)X^2 - (1/2)X - (3/4)$ in [–5, 5, 1] by [–3, 3, 1]. See Figures 38a & 38b.

The solutions are the x-intercepts, $x = -1$ and $x = 3$.

(b)   Table $Y_1 = (1/4)X^2 - (1/2)X - (3/4)$ with TblStart = –2 and ΔTbl = 1. See Figure 38c.

Since $Y_1 = 0$ when $x = -1$ and when $x = 3$, the solutions are $x = -1$ and $x = 3$.

[–5, 5, 1] by [–3, 3, 1]   [–5, 5, 1] by [–3, 3, 1]

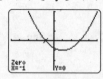

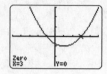

Figure 38a   Figure 38b   Figure 38c

39. $x^2 + x - 20 = 0 \Rightarrow (x+5)(x-4) = 0 \Rightarrow x = -5 \text{ or } x = 4$

40. $x^2 + 11x + 24 = 0 \Rightarrow (x+8)(x+3) = 0 \Rightarrow x = -8 \text{ or } x = -3$

41. $15x^2 - 4x - 4 = 0 \Rightarrow (5x+2)(3x-2) = 0 \Rightarrow x = -\dfrac{2}{5} \text{ or } x = \dfrac{2}{3}$

42. $7x^2 - 25x + 12 = 0 \Rightarrow (7x-4)(x-3) = 0 \Rightarrow x = \dfrac{4}{7} \text{ or } x = 3$

43. $x^2 = 100 \Rightarrow x = \pm\sqrt{100} \Rightarrow x = \pm 10$

44. $3x^2 = \dfrac{1}{3} \Rightarrow x^2 = \dfrac{1}{9} \Rightarrow x = \pm\sqrt{\dfrac{1}{9}} \Rightarrow x = \pm\dfrac{1}{3}$

45. $4x^2 - 6 = 0 \Rightarrow x^2 = \dfrac{6}{4} \Rightarrow x = \pm\sqrt{\dfrac{6}{4}} \Rightarrow x = \pm\dfrac{\sqrt{6}}{2}$

46. $5x^2 = x^2 - 4 \Rightarrow 4x^2 = -4 \Rightarrow x^2 = -1 \Rightarrow x = \pm\sqrt{-1} \Rightarrow$ No real solutions.

47. $x^2 + 6x = -2 \Rightarrow x^2 + 6x + 9 = -2 + 9 \Rightarrow (x+3)^2 = 7 \Rightarrow x + 3 = \pm\sqrt{7} \Rightarrow x = -3 \pm\sqrt{7}$

48. $x^2 - 4x = 6 \Rightarrow x^2 - 4x + 4 = 6 + 4 \Rightarrow (x-2)^2 = 10 \Rightarrow x - 2 = \pm\sqrt{10} \Rightarrow x = 2 \pm\sqrt{10}$

49. $x^2 - 2x - 5 = 0 \Rightarrow x^2 - 2x + 1 = 5 + 1 \Rightarrow (x-1)^2 = 6 \Rightarrow x - 1 = \pm\sqrt{6} \Rightarrow x = 1 \pm\sqrt{6}$

50. $2x^2 + 6x - 1 = 0 \Rightarrow x^2 + 3x + \dfrac{9}{4} = \dfrac{1}{2} + \dfrac{9}{4} \Rightarrow \left(x + \dfrac{3}{2}\right)^2 = \dfrac{11}{4} \Rightarrow x + \dfrac{3}{2} = \pm\sqrt{\dfrac{11}{4}} \Rightarrow x = \dfrac{-3 \pm\sqrt{11}}{2}$

51. $F = \dfrac{k}{(R+r)^2} \Rightarrow F(R+r)^2 = k \Rightarrow (R+r)^2 = \dfrac{k}{F} \Rightarrow R + r = \pm\sqrt{\dfrac{k}{F}} \Rightarrow R = -r \pm\sqrt{\dfrac{k}{F}}$

52. $2x^2 + 3y^2 = 12 \Rightarrow 3y^2 = 12 - 2x^2 \Rightarrow y^2 = \dfrac{12 - 2x^2}{3} \Rightarrow y = \pm\sqrt{\dfrac{12 - 2x^2}{3}}$

53. $x = \dfrac{-(-9) \pm \sqrt{(-9)^2 - 4(1)(18)}}{2(1)} = \dfrac{9 \pm \sqrt{9}}{2} = \dfrac{9 \pm 3}{2} \Rightarrow x = 3 \text{ or } x = 6$

54. $x = \dfrac{-(-24) \pm \sqrt{(-24)^2 - 4(1)(143)}}{2(1)} = \dfrac{24 \pm \sqrt{4}}{2} = \dfrac{24 \pm 2}{2} \Rightarrow x = 11 \text{ or } x = 13$

55. $x = \dfrac{-1 \pm \sqrt{(1)^2 - 4(6)(-1)}}{2(6)} = \dfrac{-1 \pm \sqrt{25}}{12} = \dfrac{-1 \pm 5}{12} \Rightarrow x = -\dfrac{1}{2} \text{ or } x = \dfrac{1}{3}$

56. $x = \dfrac{-(-5) \pm \sqrt{(-5)^2 - 4(5)(1)}}{2(5)} = \dfrac{5 \pm \sqrt{5}}{10}$

57. $x = \dfrac{-(-8) \pm \sqrt{(-8)^2 - 4(1)(-5)}}{2(1)} = \dfrac{8 \pm \sqrt{84}}{2} = \dfrac{8 \pm 2\sqrt{21}}{2} = \dfrac{2(4 \pm \sqrt{21})}{2} = 4 \pm \sqrt{21}$

58. $x = \dfrac{-(-6) \pm \sqrt{(-6)^2 - 4(2)(3)}}{2(2)} = \dfrac{6 \pm \sqrt{12}}{4} = \dfrac{6 \pm 2\sqrt{3}}{4} = \dfrac{2(3 \pm \sqrt{3})}{2(2)} = \dfrac{3 \pm \sqrt{3}}{2}$

59. $x^2 - 4 = 0 \Rightarrow (x - 2)(x + 2) = 0 \Rightarrow x = 2$ or $x = -2$

60. $4x^2 - 1 = 0 \Rightarrow (2x - 1)(2x + 1) = 0 \Rightarrow x = \dfrac{1}{2}$ or $x = -\dfrac{1}{2}$

61. $2x^2 + 15 = 11x \Rightarrow 2x^2 - 11x + 15 = 0 \Rightarrow (2x - 5)(x - 3) = 0 \Rightarrow x = \dfrac{5}{2}$ or $x = 3$

62. $2x^2 + 15 = 13x \Rightarrow 2x^2 - 13x + 15 = 0 \Rightarrow (2x - 3)(x - 5) = 0 \Rightarrow x = \dfrac{3}{2}$ or $x = 5$

63. $x(5 - x) = 2x + 1 \Rightarrow 5x - x^2 - 2x - 1 = 0 \Rightarrow -x^2 + 3x - 1 = 0;\ a = -1, b = 3, c = -1$

$x = \dfrac{-b \pm \sqrt{b^2 - 4(ac)}}{2a} \Rightarrow x = \dfrac{-3 \pm \sqrt{3^2 - 4(-1)(-1)}}{2(-1)} \Rightarrow x = \dfrac{-3 \pm \sqrt{5}}{-2} \Rightarrow x = \dfrac{3 \pm \sqrt{5}}{2}$

64. $-2x(x - 1) = x - \dfrac{1}{2} \Rightarrow 2(-2x)(x - 1) = 2\left(x - \dfrac{1}{2}\right) \Rightarrow -4x^2 + 4x = 2x - 1 \Rightarrow -4x^2 + 2x + 1 = 0;$

$a = -4, b = 2, c = 1 \quad x = \dfrac{-b \pm \sqrt{b^2 - 4ac}}{2a} \Rightarrow x = \dfrac{-2 \pm \sqrt{2^2 - 4(-4)(1)}}{2(-4)}$

$\Rightarrow x = \dfrac{-2 \pm \sqrt{20}}{-8} \Rightarrow x = \dfrac{-2 \pm 2\sqrt{5}}{-8} \Rightarrow x = \dfrac{1 \pm \sqrt{5}}{4}$

65. (a)   Since the parabola opens upward, $a > 0$.

    (b)   The solutions are the x-intercepts, $x = -2$ or $x = 3$.

    (c)   Since there are two unique solutions, the discriminant is positive.

66. (a)   Since the parabola opens upward, $a > 0$.

    (b)   The solution is the x-intercept, $x = 2$.

    (c)   Since there is one solution, the discriminant is zero.

67. (a)   Since the parabola opens downward, $a < 0$.

    (b)   There are no x-intercepts. No real solutions.

    (c)   Since there are no real solutions, the discriminant is negative.

68. (a)   Since the parabola opens downward, $a < 0$.

    (b)   The solutions are the x-intercepts, $x = -4$ or $x = 2$.

    (c)   Since there are two unique solutions, the discriminant is positive.

69. (a) $(-3)^2 - 4(2)(1) = 1$

    (b)   Since the discriminant is positive, there are two real solutions.

(c)   Graph $Y_1 = 2X^2 - 3X + 1$ in [0, 2, 1] by [–1, 1, 1]. See Figure 69. There are two x-intercepts.

70. (a) $(2)^2 - 4(7)(-5) = 144$

     (b)   Since the discriminant is positive, there are two real solutions.

     (c)   Graph $Y_1 = 7X^2 + 2X - 5$ in [–2, 2, 1] by [–10, 5, 1]. See Figure 70. There are two x-intercepts.

[0, 2, 1] by [–1, 1, 1]   [–2, 2, 1] by [–10, 5, 1]

     Figure 69   Figure 70

71. (a) $(1)^2 - 4(3)(2) = -23$

     (b)   Since the discriminant is negative, there are no real solutions.

     (c)   Graph $Y_1 = 3X^2 + X + 2$ in [–3, 3, 1] by [–5, 10, 1]. See Figure 71. There are no x-intercepts.

[–3, 3, 1] by [–5, 10, 1]   [0, 3, 1] by [–5, 10, 1]

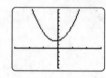   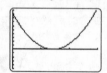

     Figure 71   Figure 72

72. (a) $(-12.6)^2 - 4(4.41)(9) = 0$

     (b)   Since the discriminant is zero, there is one real solution.

     (c)   Graph $4.41X^2 - 12.6X + 9$ in [0, 3, 1] by [–5, 10, 1]. See Figure 72.

        There is one x-intercept.

73. $x = \dfrac{-1 \pm \sqrt{(1)^2 - 4(1)(5)}}{2(1)} = \dfrac{-1 \pm \sqrt{-19}}{2} = -\dfrac{1}{2} \pm i\dfrac{\sqrt{19}}{2}$

74. $2x^2 + 8 = 0 \Rightarrow 2x^2 = -8 \Rightarrow x^2 = -4 \Rightarrow x = \pm\sqrt{-4} \Rightarrow x = \pm 2i$

75. $x = \dfrac{-(-1) \pm \sqrt{(-1)^2 - 4(2)(1)}}{2(2)} = \dfrac{1 \pm \sqrt{-7}}{4} = \dfrac{1}{4} \pm i\dfrac{\sqrt{7}}{4}$

76. $x = \dfrac{-(-2) \pm \sqrt{(-2)^2 - 4(7)(5)}}{2(7)} = \dfrac{2 \pm \sqrt{-136}}{14} = \dfrac{2 \pm 2\sqrt{-34}}{14} = \dfrac{2(1 \pm \sqrt{-34})}{14} = \dfrac{1}{7} \pm i\dfrac{\sqrt{34}}{7}$

77. (a)   The solutions are the x-intercepts, $x = -2$ or $x = 6$.

     (b) $-2 < x < 6$

     (c) $x < -2$ or $x > 6$

78. (a)   The solutions are the x-intercepts, $x = -2$ or $x = 0$.

     (b) $x < -2$ or $x > 0$

(c) $-2 < x < 0$

79. (a) The solutions are $x = -4$ or $x = 4$.

   (b) $-4 < x < 4$

   (c) $x < -4$ or $x > 4$

80. (a) The solutions are $x = -2$ or $x = 1$.

   (b) $-2 < x < 1$

   (c) $x < -2$ or $x > 1$

81. (a) $x^2 - 2x - 3 = 0 \Rightarrow (x+1)(x-3) = 0 \Rightarrow x = -1$ or $x = 3$

   (b) Since the parabola opens upward, the solution is $-1 < x < 3$.

   (c) Since the parabola opens upward, the solution is $x < -1$ or $x > 3$.

82. (a) $2x^2 - 7x - 15 = 0 \Rightarrow (2x+3)(x-5) = 0 \Rightarrow x = -\dfrac{3}{2}$ or $x = 5$

   (b) Since the parabola opens upward, the solution is $-\dfrac{3}{2} \le x \le 5$.

   (c) Since the parabola opens upward, the solution is $x \le -\dfrac{3}{2}$ or $x \ge 5$.

83. $x^2 + 4x + 3 = 0 \Rightarrow (x+3)(x+1) = 0 \Rightarrow x = -3$ or $x = -1$

   Since the parabola opens upward, the solution is $[-3,\ -1]$.

84. $5x^2 - 16x + 3 = 0 \Rightarrow (5x-1)(x-3) = 0 \Rightarrow x = \dfrac{1}{5}$ or $x = 3$

   Since the parabola opens upward, the solution is $\left(\dfrac{1}{5},\ 3\right)$.

85. $6x^2 - 13x + 2 = 0 \Rightarrow (6x-1)(x-2) = 0 \Rightarrow x = \dfrac{1}{6}$ or $x = 2$

   Since the parabola opens upward, the solution is $\left(-\infty,\ \dfrac{1}{6}\right) \cup (2,\ \infty)$.

86. $x^2 = 5 \Rightarrow \Rightarrow x = -\sqrt{5}$ or $x = \sqrt{5}$

   Since the parabola opens upward, the solution is $(-\infty,\ \sqrt{5}\,] \cup [\sqrt{5},\ \infty)$.

87. The graph of the parabola $y = (x-1)^2$ lies on or above the x-axis for all real numbers x. The interval is $(-\infty, \infty)$.

88. $x^2 + 3 < 2 \Rightarrow x^2 < -1$, which is not true for any real number x. The graph of the parabola $y = x^2 + 1$ lies above the x-axis for all real numbers, so there are no solutions.

89. Let $u = x^2$. Then $u^2 - 14u + 45 = 0 \Rightarrow (u-5)(u-9) = 0 \Rightarrow u = 5$ or 9.

   When $u = 5$, $x^2 = 5 \Rightarrow x = \pm\sqrt{5}$. When $u = 9$, $x^2 = 9 \Rightarrow x = \pm\sqrt{9} = \pm 3$.

90. Let $u = z^{-1}$. Then $2u^2 + u - 28 = 0 \Rightarrow (u+4)(2u-7) = 0 \Rightarrow u = -4$ or $\dfrac{7}{2}$.

When $u = -4$, $z^{-1} = -4 \Rightarrow z = -\dfrac{1}{4}$. When $u = \dfrac{7}{2}$, $z^{-1} = \dfrac{7}{2} \Rightarrow z = \dfrac{2}{7}$.

91. Let $u = x^{1/3}$. Then $u^2 - 9u + 8 = 0 \Rightarrow (u-1)(u-8) = 0 \Rightarrow u = 1$ or 8.

When $u = 1$, $x^{1/3} = 1 \Rightarrow x = 1^3 = 1$. When $u = 8$, $x^{1/3} = 8 \Rightarrow x = 8^3 = 512$.

92. Let $u = x - 1$. Then $u^2 + 2u + 1 = 0 \Rightarrow (u+1)^2 = 0 \Rightarrow u = -1$. When $u = -1$, $x - 1 = -1 \Rightarrow x = 0$.

93. $4x^4 + 4x^2 + 1 = 0 \Rightarrow (2x^2 + 1)^2 = 0 \Rightarrow 2x^2 + 1 = 0 \Rightarrow x^2 = -\dfrac{1}{2} \Rightarrow x = \pm\sqrt{-\dfrac{1}{2}} \Rightarrow x = \pm i \dfrac{\sqrt{1}}{\sqrt{2}}$

$\Rightarrow x = \pm i \dfrac{1}{\sqrt{2}} \Rightarrow x = \pm i \dfrac{\sqrt{2}}{2}$

94. $\dfrac{1}{x-2} - \dfrac{3}{x} = -1 \Rightarrow x(x-2)\left(\dfrac{1}{x-2} - \dfrac{3}{x}\right) = x(x-2)(-1) \Rightarrow x - 3(x-2) = -x^2 + 2x \Rightarrow x^2 - 4x + 6 = 0$

$\Rightarrow x^2 - 4x = -6 \Rightarrow x^2 - 4x + 4 = -6 + 4 \Rightarrow (x-2)^2 = -2 \Rightarrow x - 2 = \pm\sqrt{-2} \Rightarrow x = 2 \pm i\sqrt{2}$

95. (a) $f(x) = x(12 - 2x)$

(b)   Note that $f(x) = x(12 - 2x) \Rightarrow f(x) = -2x^2 + 12x + 0$. Then $-\dfrac{b}{2a} = -\dfrac{12}{2(-2)} = 3$.

The maximum area occurs when $x = 3$. The dimensions should be 6 inches by 3 inches.

96. (a) $-16t^2 + 44t + 4 = 32 \Rightarrow -16t^2 + 44t - 28 = 0 \Rightarrow 4t^2 - 11t + 7 = 0 \Rightarrow (t-1)(4t-7) = 0 \Rightarrow$

$t = 1$ or $t = \dfrac{7}{4}$. The height of the stone is 32 feet after 1 second and 1.75 seconds.

(b)   Find the vertex. $-\dfrac{b}{2a} = -\dfrac{44}{2(-16)} = 1.375$; $f(1.375) = -16(1.375)^2 + 44(1.375) + 4 = 34.25$

After 1.375 seconds, the stone is at a height of 34.25 feet.

97. (a) $f(x) = x(90 - 3x)$

(b)   Graph $Y_1 = X(90 - 3X)$ in $[0, 30, 5]$ by $[0, 800, 100]$. See Figure 97a.

(c)   Graph $Y_1 = X(90 - 3X)$ and $Y_2 = 600$ in $[0, 30, 5]$ by $[0, 800, 100]$.

See Figures 97b &97c. 10 or 20 rooms.

(d)   Graph $Y_1 = X(90 - 3X)$ in $[0, 30, 5]$ by $[0, 800, 100]$. See Figure 97d. Rent 15 rooms.

$[0, 30, 5]$ by $[0, 800, 100]$ $[0, 30, 5]$ by $[0, 800, 100]$

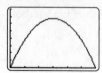

Figure 97a   Figure 97b

[0, 30, 5] by [0, 800, 100]   [0, 30, 5] by [0, 800, 100]

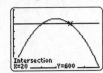   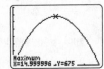

Figure 97c   Figure 97d

98. (a) $x(x+2) = 143$ or $x^2 + 2x - 143 = 0$

(b) $x^2 + 2x - 143 = 0 \Rightarrow (x+13)(x-11) = 0 \Rightarrow x = -13$ or 11. There are two possible number

pairs. Either $x = -13$, and the other number is -11, or $x = 11$, and the other number is 13.

99. (a) $\dfrac{x^2}{12} = 144 \Rightarrow x^2 = 1728 \Rightarrow x = \sqrt{1728} \approx 41.6$ mph $(-\sqrt{1728}$ mph has no meaning).

(b) $\dfrac{x^2}{12} = 300 \Rightarrow x^2 = 3600 \Rightarrow x = \sqrt{3600} = 60$ mph $(-60$ mph has no meaning).

100. (a) The vertex is (1950, 220). In 1950 the per-capita energy consumption was at a low of 220

million Btu.

(b) Graph $Y_1 = (1/4)(X - 1950)^2 + 220$ in [1950, 1970, 5] by [200, 350, 25]. See Figure 100. It

increased.

(c) $f(2010) = \dfrac{1}{4}(2010 - 1950)^2 + 220 = 1120$. The function is not a good model for 2010 because

the trend represented by the model did not continue after 1970.

[1950, 1970, 5] by [200, 350, 25].

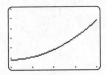

Figure 100

101. $\sqrt{123} \approx 11.1$; the screen is about 11.1 inches by 11.1 inches.

102. $x^2 + (x+70)^2 = 130^2 \Rightarrow x^2 + x^2 + 140x + 4900 = 16,900 \Rightarrow 2x^2 + 140x - 12,000 = 0 \Rightarrow$

$2(x-50)(x+120) = 0 \Rightarrow x = 50$ or $x = -120$. The solution is $x = 50$ feet

($x = -120$ has no meaning).

103. $(30+2x)(50+2x) - 30(50) = 250 \Rightarrow 1500 + 160x + 4x^2 - 1500 = 250 \Rightarrow 4x^2 + 160x - 250 = 0 \Rightarrow$

$x = \dfrac{-160 \pm \sqrt{(160)^2 - 4(4)(-250)}}{2(4)} = \dfrac{-160 \pm \sqrt{29,600}}{8} \Rightarrow x \approx -41.5$ or $x \approx 1.5$

The width of the strip of grass is about 1.5 feet. The value $x \approx -41.5$ has no physical meaning.

104. We must find r so that $750 \le \dfrac{1}{3}\pi r^2 (20) \le 1700$.

$$2250 \le \pi r^2 (20) \le 5100 \Rightarrow \frac{2250}{20\pi} \le r^2 \le \frac{5100}{20\pi} \Rightarrow \sqrt{\frac{2250}{20\pi}} \le r \le \sqrt{\frac{5100}{20\pi}} \Rightarrow 5.98 \le r \le 9.01$$

The values or r can range from about 6 inches to about 9 inches.

Chapter 8 Test

1. $x = -\dfrac{b}{2a} = -\dfrac{1}{2(-\frac{1}{2})} = \dfrac{1}{1} = 1$, $f(1) = -\dfrac{1}{2}(1)^2 + (1) + 1 = \dfrac{3}{2}$; Vertex: $\left(1, \ \dfrac{3}{2}\right)$; Axis of symmetry:

   $x = 1$. $f(-2) = -\frac{1}{2}(-2)^2 + (-2) + 1 = -\frac{1}{2}(4) + (-2) + 1 = -3$

2. Because $-\dfrac{b}{2a} = -\dfrac{3}{2(1)} = -\dfrac{3}{2}$ and $f\left(-\dfrac{3}{2}\right) = \left(-\dfrac{3}{2}\right)^2 + 3\left(-\dfrac{3}{2}\right) - 5 = -\dfrac{29}{4}$, the vertex is $\left(-\dfrac{3}{2}, \ -\dfrac{29}{4}\right)$.

   The minimum y-value on the graph is $-\dfrac{29}{4}$.

3. Since $f(x) = 0$ when $x = -2$, $a(-2)^2 + 2 = 0 \Rightarrow 4a = -2 \Rightarrow a = -\dfrac{1}{2}$.

4. (a) Same as $y = x^2$ except shifted 1 unit right. See Figure 4a.

   (b) Same as $y = x^2$ except shifted 2 units downward. See Figure 4b.   (c)   Same as

   $y = x^2$ except shifted 3 units right, 2 units upward, and wider. See Figure 4c.

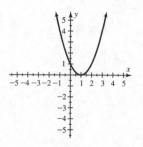   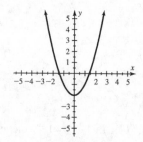   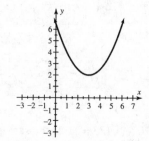

   Figure 4a   Figure 4b   Figure 4c

5. $y = x^2 - 6x + 2 \Rightarrow y = (x^2 - 6x + 9) + 2 - 9 \Rightarrow y = (x-3)^2 - 7$. The vertex is $(3, \ -7)$. The axis of

   symmetry is $x = 3$.

6. The solutions are the x-intercepts, $x = -1$ or $x = 2$. $f(1) = 2$.

7. $3x^2 + 11x - 4 = 0 \Rightarrow (x+4)(3x-1) = 0 \Rightarrow x = -4$ or $x = \dfrac{1}{3}$

8. $2x^2 = 2 - 6x^2 \Rightarrow 8x^2 = 2 \Rightarrow x^2 = \dfrac{1}{4} \Rightarrow x = \pm\sqrt{\dfrac{1}{4}} \Rightarrow x = -\dfrac{1}{2}$ or $x = \dfrac{1}{2}$

9. $x^2 - 8x = 1 \Rightarrow x^2 - 8x + 16 = 1 + 16 \Rightarrow (x-4)^2 = 17 \Rightarrow x - 4 = \pm\sqrt{17} \Rightarrow x = 4 \pm \sqrt{17}$

10. $x = \dfrac{-3 \pm \sqrt{(3)^2 - 4(-2)(1)}}{2(-2)} = \dfrac{-3 \pm \sqrt{17}}{-4} = \dfrac{3 \pm \sqrt{17}}{4}$

11. $9x^2 - 16 = 0 \Rightarrow (3x + 4)(3x - 4) = 0 \Rightarrow x = \pm\dfrac{4}{3}$

12. $F = \dfrac{Gm^2}{r^2} \Rightarrow Fr^2 = Gm^2 \Rightarrow m^2 = \dfrac{Fr^2}{G} \Rightarrow m = \pm\sqrt{\dfrac{Fr^2}{G}}$

13. (a)   Since the parabola opens downward, $a < 0$.

    (b)   The solutions are the x-intercepts, $x = -3$ or $x = 1$.

    (c)   Since there are two real solutions, the discriminant is positive.

14. (a) $(4)^2 - 4(-3)(-5) = -44$

    (b)   Since the discriminant is negative, there are no real solutions.

    (c)   Graph $Y_1 = -3X^2 + 4X - 5$ in [–5, 5, 1] by [–20, 10, 5]. See Figure 14. It does not intersect

          the x-axis.

    [–5, 5, 1] by [–20, 10, 5]

    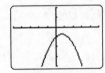

    Figure 14

15. (a)   The solutions are the x-intercepts, $x = -1$ or $x = 1$.

    (b) $-1 < x < 1$

    (c) $x < -1$ or $x > 1$

16. (a)   The solutions are the x-intercepts, $x = -10$ or $x = 20$.

    (b) $x < -10$ or $x > 20$

    (c) $-10 < x < 20$

17. (a) $8x^2 - 2x - 3 = 0 \Rightarrow (2x + 1)(4x - 3) = 0 \Rightarrow x = -\dfrac{1}{2}$ or $x = \dfrac{3}{4}$

    (b)   Since the parabola opens upward, the solution is $\left[-\dfrac{1}{2}, \dfrac{3}{4}\right]$.

    (c)   Since the parabola opens upward, the solution is $\left(-\infty, -\dfrac{1}{2}\right] \cup \left[\dfrac{3}{4}, \infty\right)$.

18. $x^2 + 2x = 0 \Rightarrow x(x + 2) = 0 \Rightarrow x = -2$ or $x = 0$. Solutions to $x^2 + 2x \le 0$ lie between and include

    these two values. The solution set is [–2, 0].

19. Let $u = x^3$. Then $u^2 - 3u + 2 = 0 \Rightarrow (u - 1)(u - 2) = 0 \Rightarrow u = 1$ or 2.

    When $u = 1$, $x^3 = 1 \Rightarrow x = \sqrt[3]{1} = 1$. When $u = 2$, $x^3 = 2 \Rightarrow x = \sqrt[3]{2}$.

20. $2x^2 + 4x + 3 = 0 \Rightarrow x = \dfrac{-4 \pm \sqrt{(4)^2 - 4(2)(3)}}{2(2)} = \dfrac{-4 \pm \sqrt{-8}}{4} = -1 \pm i\dfrac{\sqrt{2}}{2}$

21. Graph $Y_1 = \sqrt{2} - \pi X^2$ and $Y_2 = 2.12X - 0.5\pi$ in $[-5, 5, 1]$ by $[-7, 3, 1]$. See Figures 21a and 21b.

   The solutions are approximately $x = -1.37$ and $x = 0.69$.

   [−5, 5, 1] by [−7, 3, 1].   [−5, 5, 1] by [−7, 3, 1]   [0, 6, 1] by [0, 150, 50]

      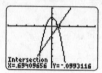   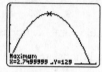

   Figure 21a   Figure 21b   Figure 24

22. $\dfrac{x^2}{9} = 250 \Rightarrow x^2 = 2250 \Rightarrow x = \sqrt{2250} \approx 47.4$ mph (the value $x \approx -47.4$ mph has no

   physical meaning).

23. (a) $2z + 2x + 20 = 200 \Rightarrow 2x + 2z = 180 \Rightarrow z = 90 - x$, so the formula is $f(x) = (x + 20)(90 - x)$.

   (b) Note that $f(x) = (x + 20)(90 - x) \Rightarrow f(x) = -x^2 + 70x + 1800$. The area is greatest for the

        value of x at the vertex of the parabola, $x = -\dfrac{b}{2a} = -\dfrac{70}{2(-1)} = 35$. The enclosed area is greatest

        when $x = 35$.

24. (a) Graph $Y_1 = -16X^2 + 88X + 8$ in $[0, 6, 1]$ by $[0, 150, 50]$. See Figure 24.

   (b) $t = \dfrac{-88 \pm \sqrt{(88)^2 - 4(-16)(8)}}{2(-16)} = \dfrac{-88 \pm \sqrt{8256}}{-32} \approx 5.6$ seconds (the value $t \approx -0.089$ has

        no meaning).

   (c) The stone reaches a maximum height of 129 feet after 2.75 seconds. See Figure 24.

Chapter 8 Extended and Discovery Exercises

1. (a) For the first 3 years of life, the likelihood of survival increases with age. After 3 years of life,
        it decreases with age.

   (b) Plot the data in $[0, 10, 1]$ by $[0, 75, 5]$. See Figure 1b. A quadratic function could model these
        data since the data points form a parabola.

   (c) Graph $Y_1 = -3.57X + 71.1$ in $[0, 10, 1]$ by $[0, 75, 5]$. See Figure 1c.

        Graph $Y_1 = -2.07X^2 + 17.1X + 33$ in $[0, 10, 1]$ by $[0, 75, 5]$. See Figure 1d.

        The function f_2 models the data better.

   (d) Evaluate f at $x = 6.5$ to find the likelihood of a 5.5-year-old sparrowhawk living 1 more year.

        $f_2(6.5) = -2.07(6.5)^2 + 17.1(6.5) + 33 \approx 56.7 \%$

[0, 10, 1] by [0, 75, 5]   [0, 10, 1] by [0, 75, 5]   [0, 10, 1] by [0, 75, 5]

    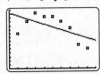

Figure 1b   Figure 1c   Figure 1d

2. (a)   Plot the data in [–5, 35, 5] by [0, 100, 10]. See Figure 2.

   (b)   A quadratic function could model these data since the data points form a parabola.

   (c)   Using the data point (12, 95) as the vertex, the function has the form $f(x) = a(x-12)^2 + 95$.

         Using the data point (17, 89), $a(17-12)^2 + 95 = 89 \Rightarrow 25a = -6 \Rightarrow a = -0.24$ The function is

         given by $f(x) = -0.24(x-12)^2 + 95$. *Answers may vary.*

   (d)   The x-coordinate of the vertex, 12°C, represents the the temperature that photosynthesis is most

         efficient.

[–5, 35, 5] by [0, 100, 10]   [–4, 4, 1] by [0, 6, 1]   [–4, 4, 1] by [0, 6, 1]

        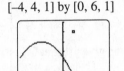

Figure 2   Figure 3a   Figure 3b

3. (a)   Plot (1, 5) and graph $Y_1 = -0.4X^2 + 4$ in [–4, 4, 1] by [0, 6, 1]. See Figure 3a.

   (b)   After 10 seconds the plane would have moved 2 kilometers. We will shift the parabola 2 units

         left. Plot (1, 5) and graph $Y_1 = -0.4(X+2)^2 + 4$ in [–4, 4, 1] by [0, 6, 1]. See Figure 3b.

4. We would need to translate the mountain both to the right and downward so that the airplane would

   appear to move to the left and upward. One example could be $f(x) = -0.4(x-2)^2 + 4 - 2$.

5. The discriminant is $(-1)^2 - 4(10)(-3) = 121 = 11^2$. The trinomial factors as $(2x+1)(5x-3)$.

6. The discriminant is $(-3)^2 - 4(4)(-6) = 105$. The trinomial will not factor.

7. The discriminant is $(2)^2 - 4(3)(-2) = 28$. The trinomial will not factor.

8. The discriminant is $(1)^2 - 4(2)(3) = -23$. The trinomial will not factor.

9. $x^3 - x^2 - 6x = 0 \Rightarrow x(x^2 - x - 6) = 0 \Rightarrow x(x-3)(x+2) = 0 \Rightarrow x = 0, 3,$ or -2

   In the interval $(-\infty, -2)$, we test $x = -3$. $f(-3) = (-3)^3 - (-3)^2 - 6(-3) = -18 < 0$

   In the interval $(-2, 0)$, we test $x = -1$. $f(-1) = (-1)^3 - (-1)^2 - 6(-1) = 4 > 0$

   In the interval $(0, 3)$, we test $x = 1$. $f(1) = (1)^3 - (1)^2 - 6(1) = -6 < 0$

   In the interval $(3, \infty)$, we test $x = 4$. $f(4) = (4)^3 - (4)^2 - 6(4) = 24 > 0$

   The solution to the given inequality is $(-2, 0) \cup (3, \infty)$.

10. $x^3 - 3x^2 + 2x = 0 \Rightarrow x(x^2 - 3x + 2) = 0 \Rightarrow x(x-1)(x-2) = 0 \Rightarrow x = 0, 1,$ or 2

In the interval $(-\infty,\ 0)$, we test $x = -1$. $f(-1) = (-1)^3 - 3(-1)^2 + 2(-1) = -6 < 0$

In the interval $(0,\ 1)$, we test $x = 0.5$. $f(0.5) = (0.5)^3 - 3(0.5)^2 + 2(0.5) = 0.375 > 0$

In the interval $(1,\ 2)$, we test $x = 1.5$. $f(1.5) = (1.5)^3 - 3(1.5)^2 + 2(1.5) = -0.375 < 0$

In the interval $(2,\ \infty)$, we test $x = 3$. $f(3) = (3)^3 - 3(3)^2 + 2(3) = 6 > 0$

The solution to the given inequality is $(-\infty,\ 0) \cup (1,\ 2)$.

11. $x^3 - 7x^2 + 14x - 8 = 0 \Rightarrow (x^3 - 8) - 7x^2 + 14x = 0 \Rightarrow (x-2)(x^2 + 2x + 4) - 7x(x-2) = 0 \Rightarrow$

   $(x-2)((x^2 + 2x + 4) - 7x) = 0 \Rightarrow (x-2)(x^2 - 5x + 4) = 0 \Rightarrow (x-2)(x-1)(x-4) = 0 \Rightarrow x = 1,\ 2,\ \text{or } 4$

In the interval $(-\infty,\ 1)$, we test $x = 0$. $f(0) = (0)^3 - 7(0)^2 + 14(0) - 8 = -8 \le 0$

In the interval $(1,\ 2)$, we test $x = 1.5$. $f(1.5) = (1.5)^3 - 7(1.5)^2 + 14(1.5) - 8 = 0.625 \ge 0$

In the interval $(2,\ 4)$, we test $x = 3$. $f(3) = (3)^3 - 7(3)^2 + 14(3) - 8 = -2 \le 0$

In the interval $(4,\ \infty)$, we test $x = 5$. $f(5) = (5)^3 - 7(5)^2 + 14(5) - 8 = 12 \ge 0$

The solution to the given inequality is $(-\infty,\ 1] \cup [2,\ 4]$.

12. $9x - x^3 = 0 \Rightarrow x(9 - x^2) = 0 \Rightarrow x(3 - x)(3 + x) = 0 \Rightarrow x = 0,\ 3,\ \text{or } -3$

In the interval $(-\infty,\ -3)$, we test $x = -4$. $f(-4) = 9(-4) - (-4)^3 = 28 \ge 0$

In the interval $(-3,\ 0)$, we test $x = -1$. $f(-1) = 9(-1) - (-1)^3 = -8 \le 0$

In the interval $(0,\ 3)$, we test $x = 1$. $f(1) = 9(1) - (1)^3 = 8 \ge 0$

In the interval $(3,\ \infty)$, we test $x = 4$. $f(4) = 9(4) - (4)^3 = -28 \le 0$

The solution to the given inequality is $(-\infty,\ -3] \cup [0,\ 3]$.

13. $x^4 - 5x^2 + 4 = 0 \Rightarrow (x^2 - 1)(x^2 - 4) = 0 \Rightarrow (x-1)(x+1)(x-2)(x+2) = 0 \Rightarrow x = -2,\ -1,\ 1,\ \text{or } 2$

In the interval $(-\infty,\ -2)$, we test $x = -3$. $f(-3) = (-3)^4 - 5(-3)^2 + 4 = 40 > 0$

In the interval $(-2,\ -1)$, we test $x = -1.5$. $f(-1.5) = (-1.5)^4 - 5(-1.5)^2 + 4 = -2.1875 < 0$

In the interval $(-1,\ 1)$, we test $x = 0$. $f(0) = (0)^4 - 5(0)^2 + 4 = 4 > 0$

In the interval $(1,\ 2)$, we test $x = 1.5$. $f(1.5) = (1.5)^4 - 5(1.5)^2 + 4 = -2.1875 < 0$

In the interval $(2,\ \infty)$, we test $x = 3$. $f(3) = (3)^4 - 5(3)^2 + 4 = 40 > 0$

The solution to the given inequality is $(-\infty,\ -2) \cup (-1,\ 1) \cup (2,\ \infty)$.

14. $1 - x^4 = 0 \Rightarrow (1 - x^2)(1 + x^2) = 0 \Rightarrow (1 - x)(1 + x)(1 + x^2) = 0 \Rightarrow x = -1 \text{ or } 1$

In the interval $(-\infty,\ -1)$, we test $x = -2$. $f(-2) = 1 - (-2)^4 = -15 < 0$

In the interval $(-1,\ 1)$, we test $x = 0$. $f(0) = 1 - (0)^4 = 1 > 0$

In the interval $(1,\ \infty)$, we test $x = 2$. $f(2) = 1 - (2)^4 = -15 < 0$

The solution to the given inequality is $(-\infty, -1) \cup (1, \infty)$.

15. $\dfrac{3-x}{3x} = 0 \Rightarrow 3-x = 0 \Rightarrow x = 3$. The expression is undefined when $3x = 0 \Rightarrow x = 0$.

In the interval $(-\infty, 0)$, we test $x = -1$. $f(-1) = \dfrac{3-(-1)}{3(-1)} = -\dfrac{4}{3} \le 0$

In the interval $(0, 3)$, we test $x = 1$. $f(1) = \dfrac{3-(1)}{3(1)} = \dfrac{2}{3} \ge 0$

In the interval $(3, \infty)$, we test $x = 4$. $f(4) = \dfrac{3-(4)}{3(4)} = -\dfrac{1}{12} \le 0$

The solution to the given inequality is $(0, 3]$.

16. $\dfrac{x-2}{x+2} = 0 \Rightarrow x-2 = 0 \Rightarrow x = 2$. The expression is undefined when $x+2 = 0 \Rightarrow x = -2$.

In the interval $(-\infty, -2)$, we test $x = -3$. $f(-3) = \dfrac{(-3)-2}{(-3)+2} = 5 > 0$

In the interval $(-2, 2)$, we test $x = 0$. $f(0) = \dfrac{(0)-2}{(0)+2} = -1 < 0$

In the interval $(2, \infty)$, we test $x = 3$. $f(3) = \dfrac{(3)-2}{(3)+2} = \dfrac{1}{5} > 0$

The solution to the given inequality is $(-\infty, -2) \cup (2, \infty)$.

17. $\dfrac{3-2x}{1+x} = 3 \Rightarrow 3-2x = 3+3x \Rightarrow 5x = 0 \Rightarrow x = 0$.

The expression is undefined when $1+x = 0 \Rightarrow x = -1$.

In the interval $(-\infty, -1)$, we test $x = -2$. $f(-2) = \dfrac{3-2(-2)}{1+(-2)} - 3 = -10 < 0$

In the interval $(-1, 0)$, we test $x = -0.5$. $f(-0.5) = \dfrac{3-2(-0.5)}{1+(-0.5)} - 3 = 5 > 0$

In the interval $(0, \infty)$, we test $x = 1$. $f(1) = \dfrac{3-2(1)}{1+(1)} - 3 = -\dfrac{5}{2} < 0$

The solution to the given inequality is $(-\infty, -1) \cup (0, \infty)$.

18. $\dfrac{x+1}{4-2x} = 1 \Rightarrow x+1 = 4-2x \Rightarrow 3x = 3 \Rightarrow x = 1$.

The expression is undefined when $4-2x = 0 \Rightarrow 2x = 4 \Rightarrow x = 2$.

In the interval $(-\infty, 1)$, we test $x = 0$. $f(0) = \dfrac{(0)+1}{4-2(0)} - 1 = -\dfrac{3}{4} \le 0$

In the interval $(1, 2)$, we test $x = 1.5$. $f(1.5) = \dfrac{(1.5)+1}{4-2(1.5)} - 1 = \dfrac{3}{2} \ge 0$

In the interval $(2, \infty)$, we test $x = 3$. $f(3) = \dfrac{(3)+1}{4-2(3)} - 1 = -3 \le 0$

The solution to the given inequality is $[1, 2)$.

19. The expression is undefined when $x^2 - 4 = 0 \Rightarrow (x+2)(x-2) = 0 \Rightarrow x = -2$ or 2.

In the interval $(-\infty, -2)$, we test $x = -3$. $f(-3) = \dfrac{5}{(-3)^2 - 4} = 1 > 0$

In the interval $(-2, 2)$, we test $x = 0$. $f(0) = \dfrac{5}{(0)^2 - 4} = -\dfrac{5}{4} < 0$

In the interval $(2, \infty)$, we test $x = 3$. $f(3) = \dfrac{5}{(3)^2 - 4} = 1 > 0$

The solution to the given inequality is $(-2, 2)$.

20. $\dfrac{x}{x^2 - 1} = 0 \Rightarrow x = 0$.   The expression is undefined when $x^2 - 1 = 0 \Rightarrow (x+1)(x-1) = 0 \Rightarrow x = -1$ or 1.

In the interval $(-\infty, -1)$, we test $x = -2$. $f(-2) = \dfrac{(-2)}{(-2)^2 - 1} = -\dfrac{2}{3} \le 0$

In the interval $(-1, 0)$, we test $x = -0.5$. $f(-0.5) = \dfrac{(-0.5)}{(-0.5)^2 - 1} = \dfrac{2}{3} \ge 0$

In the interval $(0, 1)$, we test $x = 0.5$. $f(0.5) = \dfrac{(0.5)}{(0.5)^2 - 1} = -\dfrac{2}{3} \le 0$

In the interval $(1, \infty)$, we test $x = 2$. $f(2) = \dfrac{(2)}{(2)^2 - 1} = \dfrac{2}{3} \ge 0$

The solution to the given inequality is $(-1, 0] \cup (1, \infty)$.

Chapters 1-8 Cumulative Review Exercises

1. $F = \dfrac{5}{(-2)^2 + 1} = \dfrac{5}{4+1} = \dfrac{5}{5} = 1$

2. Natural number: $\sqrt[3]{8} = 2$; whole number: $0, \sqrt[3]{8}$; integer: $0, -5, \sqrt[3]{8}$;

   rational number: $0.\overline{4}, 0, -5, \sqrt[3]{8}, -\dfrac{4}{3}$; irrational number: $\sqrt{7}$

3. (a) $\left(\dfrac{x^2 y^6}{x^{-3}}\right)^2 = \dfrac{x^4 y^{12}}{x^{-6}} = x^{4-(-6)} y^{12} = x^{10} y^{12}$

   (b) $\dfrac{(xy^{-3})^2}{x(y^{-2})^{-1}} = \dfrac{x^2 y^{-6}}{xy^2} = x^{2-1} y^{-6-2} = x^1 y^{-8} = \dfrac{x}{y^8}$

(c) $(a^2b)^2(ab^3)^{-4} = a^4b^2a^{-4}b^{-12} = a^{4-4}b^{2-12} = a^0b^{-10} = \dfrac{1}{b^{10}}$

4. $9{,}290{,}000 = 9.29 \times 10^6$

5. $f(-2) = \sqrt{2-(-2)} = \sqrt{4} = 2$; the radicand must be greater than or equal to zero, so

$2-x \geq 0 \Rightarrow x \leq 2$.

6. $(2, 5)$ lies on the graph of f.

7. See Figure 7.

8. See Figure 8.

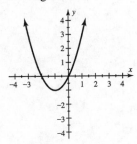

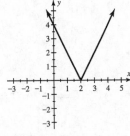

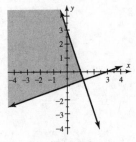

Figure 7   Figure 8   Figure 16

9. Parallel lines have the same slope, so $m = \dfrac{4-1}{-2-0} = \dfrac{3}{-2} = -\dfrac{3}{2}$.

$$y-(-1) = -\dfrac{3}{2}(x-4) \Rightarrow y = -\dfrac{3}{2}x+6-1 \Rightarrow y = -\dfrac{3}{2}x+5$$

10. $x = -3$

11. $2x-3(x+2) = 6 \Rightarrow 2x-3x-6 = 6 \Rightarrow -x = 12 \Rightarrow x = -12$

12. $7-x > 3x \Rightarrow 7 > 4x \Rightarrow \dfrac{7}{4} > x$; $\left(-\infty, \dfrac{7}{4}\right)$

13. $|3x-2| \leq 1 \Rightarrow 3x-2 \leq 1$ and $3x-2 \geq -1 \Rightarrow 3x \leq 3$ and $3x \geq 1 \Rightarrow x \leq 1$ and $x \geq \dfrac{1}{3}$; $\left[\dfrac{1}{3}, 1\right]$

14. $-4 \leq 1-x < 2 \Rightarrow -5 \leq -x < 1 \Rightarrow 5 \geq x > -1 \Rightarrow -1 < x \leq 5$; $(-1, 5]$

15. Multiply the first equation by 5 and add to the second equation. $\begin{array}{r} -5x-20y = -15 \\ 5x+y = -4 \\ \hline -19y = -19 \end{array} \Rightarrow y = 1$

Substitute $y = 1$ into the first equation: $-x-4(1) = -3 \Rightarrow -x = -3+4 \Rightarrow -x = 1 \Rightarrow x = -1$

The solution is $(-1, 1)$.

16. Note that $3x+y \leq 3 \Rightarrow y \leq -3x+3$ and $x-3y \leq 3 \Rightarrow -3y \leq -x+3 \Rightarrow y \geq \dfrac{1}{3}x-1$. See Figure 16.

17. Add the first two equations. $\begin{array}{r} x+y-z = 3 \\ x-y+z = 1 \\ \hline 2x = 4 \end{array} \Rightarrow x = 2$

Add the second and third equations.

$$x - y + z = 1$$
$$2x - y - z = 1$$
$$\underline{3x - 2y \quad = 2} \Rightarrow 3(2) - 2y = 2 \Rightarrow 6 - 2y = 2 \Rightarrow -2y = -4 \Rightarrow y = 2$$

Substitute $x = 2$ and $y = 2$ into the first equation. $2 + 2 - z = 3 \Rightarrow 4 - z = 3 \Rightarrow -z = -1 \Rightarrow z = 1$

The solution is $(2, 2, 1)$.

18. $(3x - 2)(2x + 7) = 6x^2 + 21x - 4x - 14 = 6x^2 + 17x - 14$

19. $3xy(x^2 + y^2) = 3x^3y + 3xy^3$

20. $(\sqrt{x} + 3)(\sqrt{x} - 3) = (\sqrt{x})^2 - (3)^2 = x - 9$

21. $x^3 - x^2 - 2x = x(x^2 - x - 2) = x(x - 2)(x + 1)$

22. $4x^2 - 25 = (2x - 5)(2x + 5)$

23. $x^2 - 3 = 0 \Rightarrow x^2 = 3 \Rightarrow x = \pm\sqrt{3}$

24. $x^2 + 1 = 2x \Rightarrow x^2 - 2x + 1 = 0 \Rightarrow (x - 1)^2 = 0 \Rightarrow x - 1 = 0 \Rightarrow x = 1$

25. $\dfrac{(x+3)^2}{x+2} \cdot \dfrac{x+2}{2x+6} = \dfrac{(x+3)(x+3)(x+2)}{(x+2)(2)(x+3)} = \dfrac{x+3}{2}$ or $\dfrac{1}{2}(x+3)$

26. $\dfrac{1}{x+2} - \dfrac{1}{x} = \dfrac{1}{x+2} \cdot \dfrac{x}{x} - \dfrac{1}{x} \cdot \dfrac{x+2}{x+2} = \dfrac{x}{x(x+2)} - \dfrac{x+2}{x(x+2)} = \dfrac{x-x-2}{x(x+2)} = \dfrac{-2}{x(x+2)}$

27. $\sqrt{16x^6} = \sqrt{16} \cdot \sqrt{x^6} = 4x^3$

28. $16^{-3/2} = (16^{1/2})^{-3} = 4^{-3} = \dfrac{1}{4^3} = \dfrac{1}{64}$

29. $\dfrac{\sqrt[3]{81x}}{\sqrt[3]{3x}} = \sqrt[3]{\dfrac{81x}{3x}} = \sqrt[3]{27} = 3$

30. $\sqrt{8x} + \sqrt{2x} = \sqrt{4 \cdot 2x} + \sqrt{2x} = 2\sqrt{2x} + \sqrt{2x} = 3\sqrt{2x}$

31. See Figure 31.

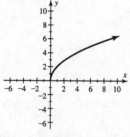   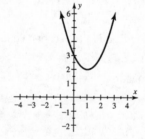

Figure 31   Figure 36

32. $d = \sqrt{(x_2 - x_1)^2 + (y_2 - y_1)^2} = \sqrt{(4 - (-1))^2 + (3 - 2)^2} = \sqrt{(5)^2 + (1)^2} = \sqrt{25 + 1} = \sqrt{26}$

33. $\dfrac{3-i}{2+i} = \dfrac{3-i}{2+i} \cdot \dfrac{2-i}{2-i} = \dfrac{6 - 3i - 2i + i^2}{4 - i^2} = \dfrac{6 - 5i - 1}{4 - (-1)} = \dfrac{5 - 5i}{5} = 1 - i$

34. $3\sqrt{x+1} = 2x \Rightarrow 9(x+1) = 4x^2 \Rightarrow 9x + 9 = 4x^2 \Rightarrow 4x^2 - 9x - 9 = 0 \Rightarrow (x - 3)(4x + 3) = 0 \Rightarrow x = 3$

$$\left(x=-\frac{3}{4}\text{ does not check}\right)$$

35. Graph $Y_1 = 2X$ and $Y_2 = \sqrt{\ }((2.1-X)+(0.1X)\wedge(1/3)$ in $[-4.7, 4.7, 1]$ by $[-3.1, 3.1, 1]$.

   See Figure 35. The solution is $x \approx 0.79$.

   $[-4.7, 4.7, 1]$ by $[-3.1, 3.1, 1]$

   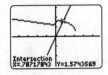

   Figure 35

36. See Figure 36.

   (a) $(1, 2)$

   (b) $f(-1) = (-1)^2 - 2(-1) + 3 = 1 + 2 + 3 = 6$

   (c) $x = 1$

   (d) f is increasing when $x \geq 1$

37. $f(x) = 2x^2 - 4x - 1 \Rightarrow f(x) = 2(x^2 - 2x) - 1 \Rightarrow f(x) = 2(x^2 - 2x + 1) - 1 - 2 \Rightarrow f(x) = 2(x-1)^2 - 3$

38. The graph is shifted 1 unit left and 2 units downward. The graph of $f(x)$ is narrower.

39. $x^2 + 6x = 2 \Rightarrow x^2 + 6x + 9 = 2 + 9 \Rightarrow (x+3)^2 = 11 \Rightarrow x + 3 = \pm\sqrt{11} \Rightarrow x = -3 \pm \sqrt{11}$

40. $2x^2 - 3x = 1 \Rightarrow 2x^2 - 3x - 1 = 0$; $a = 2, b = -3, c = -1$

   $$x = \frac{-b \pm \sqrt{b^2 - 4ac}}{2a} = \frac{-(-3) \pm \sqrt{(-3)^2 - 4(2)(-1)}}{2(2)} = \frac{3 \pm \sqrt{17}}{4}$$

41. $x(4-x) = 3 \Rightarrow 4x - x^2 = 3 \Rightarrow x^2 - 4x + 3 = 0 \Rightarrow (x-3)(x-1) = 0 \Rightarrow x = 3$ or $x = 1$

42. (a) $x = -2$ or $x = 1$

   (b) $-2 \leq x \leq 1$

43. $x^2 - 3x + 2 = 0 \Rightarrow (x-2)(x-1) = 0 \Rightarrow x = 1$ or $x = 2$. Since the parabola opens up, the solution is

   $x < 1$ or $x > 2$.

44. $x^4 - 256 = 0 \Rightarrow (x^2 - 16)(x^2 + 16) = 0 \Rightarrow (x-4)(x+4)(x^2 + 16) = 0 \Rightarrow x = 4$ or $x = -4$ or

   $x^2 = -16 \Rightarrow x = \pm\sqrt{-16} \Rightarrow x = \pm 4i$

45. c; $y = ax - b$ is a linear equation and $a > 0$.

46. f; $y = b$ is a horizontal line.

47. g; $y = -ax^2 + c$ is a quadratic equation whose graph is a parabola opening downward (since $a > 0$).

48. e; $y = \dfrac{a}{x}$ is a rational equation, undefined when $x = 0$ and $y = 0$.

49. d; $y = ax^3$ is a cubic equation.

50. a; $y = |ax + b|$ has only nonnegative values.

51. b; $y = a\sqrt{x}$ is a radical function defined only for $x \geq 0$.

52. h; $y = a\sqrt[3]{x}$ is a cube root equation.

53. (a) $G(0) = 300$ gallons ; initially, the tank holds 300 gallons.

    (b)   The t-intercept is 6; after 6 minutes the tank is empty.

    (c)   (0, 300) and (6, 0) are on the graph so $m = \dfrac{0 - 300}{6 - 0} = \dfrac{-300}{6} = -50$; water is pumped out at 50

          gallons per minute.

    (d)   Since (0, 300) is on the graph and $m = -50$, $G(t) = -50t + 300$ or $G(t) = 300 - 50t$

54. Let x, y, and z be the amounts invested at 4% , 5%, and 6% interest, respectively. The system needed

$$\begin{array}{lll} & x + y + z = 4000 & \qquad x + y + \ z = 4000 \\ \text{is} & z = y + 1000 & \Rightarrow \qquad y - \ z = -1000 \\ & 0.04x + 0.05y + 0.06z = 216 & \qquad 4x + 5y + 6z = 21{,}600 \end{array}$$

Multiply the first equation by -4 and add to the third equation

$$\begin{array}{r} -4x - 4y - 4z = -16{,}000 \\ \underline{4x + 5y + 6z = \ \ \ 21{,}600} \\ y + 2z = \ \ \ \ \ 5600 \end{array}$$

Multiply the second equation by 2 and add to this new equation.

$$\begin{array}{r} 2y - 2z = -2000 \\ \underline{y + 2z = \ \ 5600} \\ 3y \ \ \ \ \ \ \ = \ \ 3600 \qquad \Rightarrow y = 1200 \end{array}$$

Substitute $y = 1200$ into the second equation.

$1200 - z = -1000 \Rightarrow -z = -2200 \Rightarrow z = 2200$

Substitute $y = 1200$ and $z = 2200$ into the first equation.

$x + 1200 + 2200 = 4000 \Rightarrow x = 600$

$600 is invested at 4%, $1200 is invested at 5%, and $2200 is invested at 6%.

55. Let x represent the length of one of the two equal sides of the garden without the gate. Then the other

sides of the garden have lengths $\dfrac{1}{2}(490 - 2x + 10) = \dfrac{1}{2}(500 - 2x) = 250 - x$. $(250 - x)x = 250x - x^2$;

The graph of $y = -x^2 + 250x$ is a parabola opening downward; its vertex is its maximum point. The

x-value of the vertex is $x = -\dfrac{b}{2a} = -\dfrac{250}{2(-1)} = 125$. The dimensions of the garden of largest area are

125 feet by 125 feet.

56. Let h represent the height of the tree. $\dfrac{6}{10} = \dfrac{h}{55} \quad \Rightarrow \quad 10h = 6(55) \Rightarrow 10h = 330 \Rightarrow h = 33$ feet

Chapter 9: Exponential and Logarithmic Functions

9.1: Composite and Inverse Functions

1. $g(f(7))$

3. No

5. No

7. adding 10

9. 8; 6

11. one-to-one

13. (a) $f(-2)=(-2)^2=4$, then $(g \circ f)(-2)=g(f(-2))=g(4)=4+3=7$

   (b) $g(4)=4+3=7$, then $(f \circ g)(4)=f(g(4))=f(7)=7^2=49$

   (c) $(g \circ f)(x)=g(f(x))=g(x^2)=x^2+3$

   (d) $(f \circ g)(x)=f(g(x))=f(x+3)=(x+3)^2$

15. (a) $f(-2)=2(-2)=-4$, then $(g \circ f)(-2)=g(f(-2))=g(-4)=(-4)^3-1=-65$

   (b) $g(4)=4^3-1=63$, then $(f \circ g)(4)=f(g(4))=f(63)=2(63)=126$

   (c) $(g \circ f)(x)=g(f(x))=g(2x)=(2x)^3-1=8x^3-1$

   (d) $(f \circ g)(x)=f(g(x))=f(x^3-1)=2(x^3-1)=2x^3-2$

17. (a) $f(-2)=\dfrac{1}{2}(-2)=-1$, then $(g \circ f)(-2)=g(f(-2))=g(-1)=|(-1)-2|=3$

   (b) $g(4)=|4-2|=2$, then $(f \circ g)(4)=f(g(4))=f(2)=\dfrac{1}{2}(2)=1$

   (c) $(g \circ f)(x)=g(f(x))=g\left(\dfrac{1}{2}x\right)=\left|\dfrac{1}{2}x-2\right|$

   (d) $(f \circ g)(x)=f(g(x))=f(|x-2|)=\dfrac{1}{2}|x-2|$

19. (a) $f(-2)=\dfrac{1}{-2}=-\dfrac{1}{2}$, then $(g \circ f)(-2)=g(f(-2))=g\left(-\dfrac{1}{2}\right)=3-5\left(-\dfrac{1}{2}\right)=\dfrac{11}{2}$

   (b) $g(4)=3-5(4)=-17$, then $(f \circ g)(4)=f(g(4))=f(-17)=\dfrac{1}{-17}=-\dfrac{1}{17}$

   (c) $(g \circ f)(x)=g(f(x))=g\left(\dfrac{1}{x}\right)=3-5\left(\dfrac{1}{x}\right)=3-\dfrac{5}{x}$

   (d) $(f \circ g)(x)=f(g(x))=f(3-5x)=\dfrac{1}{3-5x}$

21. (a) $f(-2)=2(-2)=-4$, then $(g\circ f)(-2)=g(f(-2))=g(-4)=4(-4)^2-2(-4)+5=77$

    (b) $g(4)=4(4)^2-2(4)+5=61$, then $(f\circ g)(4)=f(g(4))=f(61)=2(61)=122$

    (c) $(g\circ f)(x)=g(f(x))=g(2x)=4(2x)^2-2(2x)+5=16x^2-4x+5$

    (d) $(f\circ g)(x)=f(g(x))=f(4x^2-2x+5)=2(4x^2-2x+5)=8x^2-4x+10$

23. (a) $(f\circ g)(0)=f(g(0))=f(-1)=1$

    (b) $(g\circ f)(-1)=g(f(-1))=g(1)=2$

25. (a) $(f\circ f)(-1)=f(f(-1))=f(1)=-1$

    (b) $(g\circ g)(0)=g(g(0))=g(-1)=1$

27. (a) $\left(f^{-1}\circ g\right)(-2)=f^{-1}(g(-2))=f^{-1}(0)=0$

    (b) $\left(g^{-1}\circ f\right)(2)=g^{-1}(f(2))=g^{-1}(-2)=2$

29. (a) $(f\circ g)(0)=f(g(0))=f(-1)=2$

    (b) $(g\circ f)(1)=g(f(1))=g(2)=-3$

    (c) $(f\circ f)(-1)=f(f(-1))=f(2)=-1$

31. $f(1)=f(-1)=5$; *Answers may vary.*

33. $f(1)=f(-1)=101$; *Answers may vary.*

35. $f(2)=f(-2)=4$; *Answers may vary.*

37. This graph passes the horizontal line test. The function is one-to-one.

39. This graph does not pass the horizontal line test. The function is not one-to-one.

41. This graph passes the horizontal line test. The function is one-to-one.

43. Divide x by 7. $f(x)=7x$; $g(x)=\dfrac{x}{7}$

45. Multiply x by 2 and then subtract 5. $f(x)=\dfrac{x+5}{2}$; $g(x)=2x-5$

47. Add 3 to x and then multiply the result by 2. $f(x)=\dfrac{1}{2}x-3$; $g(x)=2(x+3)$

49. Take the cube root of x and then subtract 5. $f(x)=(x+5)^3$; $g(x)=\sqrt[3]{x}-5$

51. $\left(f\circ f^{-1}\right)(x)=f\left(f^{-1}(x)\right)=f\left(\dfrac{x}{4}\right)=4\left(\dfrac{x}{4}\right)=x$ and $\left(f^{-1}\circ f\right)(x)=f^{-1}(f(x))=f^{-1}(4x)=\dfrac{4x}{4}=x$

53. $\left(f\circ f^{-1}\right)(x)=f\left(f^{-1}(x)\right)=f\left(\dfrac{x-5}{3}\right)=3\left(\dfrac{x-5}{3}\right)+5=x-5+5=x$

$$\left(f^{-1} \circ f\right)(x) = f^{-1}\left(f(x)\right) = f^{-1}(3x+5) = \frac{(3x+5)-5}{3} = \frac{3x}{3} = x$$

55. $\left(f \circ f^{-1}\right)(x) = f\left(f^{-1}(x)\right) = f\left(\sqrt[3]{x}\right) = \left(\sqrt[3]{x}\right)^3 = x$

$\left(f^{-1} \circ f\right)(x) = f^{-1}\left(f(x)\right) = f^{-1}\left(x^3\right) = \sqrt[3]{x^3} = x$

57. $\left(f \circ f^{-1}\right)(x) = f\left(f^{-1}(x)\right) = f\left(\dfrac{1}{x}\right) = \dfrac{1}{\frac{1}{x}} = x$ and $\left(f^{-1} \circ f\right)(x) = f^{-1}\left(f(x)\right) = f^{-1}\left(\dfrac{1}{x}\right) = \dfrac{1}{\frac{1}{x}} = x$

59. $f(x) = 12x \Rightarrow y = 12x$, interchange x and y and solve for y. $x = 12y \Rightarrow y = \dfrac{x}{12} \Rightarrow f^{-1}(x) = \dfrac{x}{12}$

61. $f(x) = x+8 \Rightarrow y = x+8$, interchange x and y and solve for y.

$x = y+8 \Rightarrow y = x-8 \Rightarrow f^{-1}(x) = x-8$

63. $f(x) = 5x-2 \Rightarrow y = 5x-2$, interchange x and y and solve for y.

$x = 5y-2 \Rightarrow 5y = x+2 \Rightarrow y = \dfrac{x+2}{5} \Rightarrow f^{-1}(x) = \dfrac{x+2}{5}$

65. $f(x) = -\dfrac{1}{2}x+1 \Rightarrow y = -\dfrac{1}{2}x+1$, interchange x and y and solve for y.

$x = -\dfrac{1}{2}y+1 \Rightarrow -\dfrac{1}{2}y = x-1 \Rightarrow y = -2(x-1) \Rightarrow f^{-1}(x) = -2(x-1)$

67. $f(x) = 8-x \Rightarrow y = 8-x$, interchange x and y and solve for y.

$x = 8-y \Rightarrow y = 8-x \Rightarrow f^{-1}(x) = 8-x$

69. $f(x) = \dfrac{x+1}{2} \Rightarrow y = \dfrac{x+1}{2}$, interchange x and y and solve for y.

$x = \dfrac{y+1}{2} \Rightarrow y+1 = 2x \Rightarrow y = 2x-1 \Rightarrow f^{-1}(x) = 2x-1$

71. $f(x) = \sqrt[3]{2x} \Rightarrow y = \sqrt[3]{2x}$, interchange x and y and solve for y.

$x = \sqrt[3]{2y} \Rightarrow 2y = x^3 \Rightarrow y = \dfrac{x^3}{2} \Rightarrow f^{-1}(x) = \dfrac{x^3}{2}$

73. $f(x) = x^3 - 8 \Rightarrow y = x^3 - 8$, interchange x and y and solve for y.

$x = y^3 - 8 \Rightarrow y^3 = x+8 \Rightarrow y = \sqrt[3]{x+8} \Rightarrow f^{-1}(x) = \sqrt[3]{x+8}$

75. See Figure 75. The domain of f = the range of $f^{-1} = \{0, 1, 2, 3, 4\}$.

The range of f = the domain of $f^{-1} = \{0, 5, 10, 15, 20\}$.

x	0	5	10	15	20
$f^{-1}(x)$	0	1	2	3	4

Figure 75

x	4	2	0	-2	-4
$f^{-1}(x)$	-5	0	5	10	15

Figure 77

77. See Figure 77. The domain of f = the range of $f^{-1} = \{-5, 0, 5, 10, 15\}$.

  The range of f = the domain of $f^{-1} = \{-4, -2, 0, 2, 4\}$.

79. See Figure 79.

81. See Figure 81.

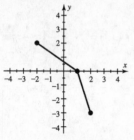

Figure 79

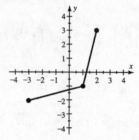

Figure 81

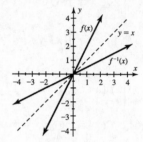

Figure 83

83. The graph of f^{-1} is a reflection of the graph of f across the line $y = x$. See Figure 83.

85. The graph of f^{-1} is a reflection of the graph of f across the line $y = x$. See Figure 85.

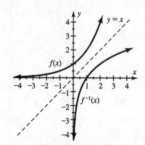

Figure 85

87. (a) $(fg)(2) = (2^2 - 2)(2^2 + 2) = (4 - 2)(4 + 2) = (2)(6) = 12$

  (b) $(f - g)(x) = (x^2 - 2) - (x^2 + 2) = x^2 - x^2 - 2 - 2 = -4$

  (c) $(f \circ g)(x) = f(g(x)) = f(x^2 + 2) = (x^2 + 2)^2 - 2 = x^4 + 4x^2 + 4 - 2 = x^4 + 4x^2 + 2$

89. (a) $(fg)(2) = \left(\dfrac{1}{2}\right)\left(\dfrac{2}{2}\right) = \dfrac{2}{4} = \dfrac{1}{2}$

  (b) $(f - g)(x) = \dfrac{1}{x} - \dfrac{2}{x} = -\dfrac{1}{x}$

  (c) $(f \circ g)(x) = f(g(x)) = f\left(\dfrac{2}{x}\right) = \dfrac{1}{\dfrac{2}{x}} = \dfrac{x}{2}$

91. (a) $(C \circ r)(5) = C(r(5)) = C(2 \cdot 5) = C(10) = 2\pi \cdot 10 = 20\pi$

  After 5 seconds, the wave has a circumference of $20\pi \approx 62.8$ feet.

  (b) $(C \circ r)(t) = C(r(t)) = C(2t) = 2\pi \cdot 2t = 4\pi t$

93. (a) $P(1980) = 16$; In 1980, 16% of people 25 or older completed 4 or more years of college.

(b) See Figure 93.

(c) $P^{-1}(16) = 1980$

x	8	16	27	29
$P^{-1}(x)$	1960	1980	2000	2010

Figure 93

95. (a) $T(1) = 75°$ and $M(75) = 150$

(b) $(M \circ T)(1) = M(T(1)) = M(75) = 150$

One hour after midnight there are 150 mosquitoes per 100 square feet.

(c) $(M \circ T)(h)$ calculates the number of mosquitoes per 100 square feet, h hours after midnight.

(d) For $T(h)$, $m = \dfrac{50-80}{6-0} = \dfrac{-30}{6} = -5$. Since the y-intercept is 80, the equation is

$T(h) = -5h + 80$. For $M(T)$, $m = \dfrac{150-100}{75-50} = \dfrac{50}{25} = 2$. Since the line passes through

(50, 100), the equation is $M(T) = 2(T-50) + 100 = 2T - 100 + 100 = 2T$.

(e) $(M \circ T)(h) = M(T(h)) = M(-5h+80) = 2(-5h+80) = -10h + 160$

97. (a) Yes, this is a one-to-one function because different inputs result in different outputs.

(b) $f(x) = \dfrac{5}{9}x + 32 \Rightarrow y = \dfrac{5}{9}x + 32$, interchange x and y and solve for y.

$x = \dfrac{5}{9}y + 32 \Rightarrow \dfrac{5}{9}y = x - 32 \Rightarrow y = \dfrac{9}{5}(x-32) \Rightarrow f^{-1}(x) = \dfrac{9}{5}(x-32)$

This formula converts x degrees Fahrenheit to an equivalent temperature in degrees Celsius.

99. Since there are 4 quarts in 1 gallon, the function $f(x) = 4x$ converts x gallons to quarts.

$f(x) = 4x \Rightarrow y = 4x$, interchange x and y and solve for y. $x = 4y \Rightarrow y = \dfrac{x}{4} \Rightarrow f^{-1}(x) = \dfrac{x}{4}$

This formula converts x quarts to gallons.

101. (a) There are 16 cups in a gallon. The function is $C(x) = 16x$.

(b) There are 48 teaspoons in a cup. The function is $T(x) = 48x$.

(c) $(T \circ C)(x) = T(C(x)) = T(16x) = 48(16x) = 768x$

(d) $(T \circ C)(3) = T(C(3)) = T(48) = 48(48) = 2304$, There are 2304 teaspoons in 3 gallons.

Section 9.2 Exponential Functions

1. $f(x) = Ca^x$

3. Domain: All real numbers; Range: All positive real numbers

5. $e \approx 2.718$

7. factor

9. $\dfrac{B-A}{A} \times 100$

11. $f(-2) = 3^{-2} = \dfrac{1}{3^2} = \dfrac{1}{9}$ and $f(2) = 3^2 = 9$

13. $f(0) = 5\left(2^0\right) = 5(1) = 5$ and $f(5) = 5\left(2^5\right) = 5(32) = 160$

15. $f(-2) = \left(\dfrac{1}{2}\right)^{-2} = 2^2 = 4$ and $f(3) = \left(\dfrac{1}{2}\right)^3 = \dfrac{1}{2^3} = \dfrac{1}{8}$

17. $f(-1) = 5(3)^{-(-1)} = 5(3)^1 = 15$ and $f(2) = 5(3)^{-2} = 5\left(\dfrac{1}{3^2}\right) = \dfrac{5}{9}$

19. $f(-3) = 1.8^{-3} \approx 0.17$ and $f(1.5) = 1.8^{1.5} \approx 2.41$

21. $f(-1) = 3(0.6)^{-1} = 5$ and $f(2) = 3(0.6)^2 = 1.08$

23. $f(0) = a^0 = 1$; $f(-1) = a^{-1} = \dfrac{1}{a}$

25. c. This function models exponential growth and passes through the point (0, 1).

27. d. This function models exponential decay and passes through the point (2, 1).

29. Since $y = 1$ when $x = 0$, $1 = Ca^0 \Rightarrow C = 1$. Since $y = 2$ when $x = 1$, $2 = 1(a)^1 \Rightarrow a = 2$.

31. Since $y = 4$ when $x = 0$, $4 = Ca^0 \Rightarrow C = 4$. Since $y = 1$ when $x = 1$, $1 = 4(a)^1 \Rightarrow a = \dfrac{1}{4}$.

33. See Figure 33. The graph illustrates exponential growth.

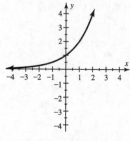

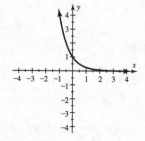

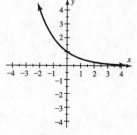

Figure 33   Figure 35   Figure 37

35. See Figure 35. The graph illustrates exponential decay.

37. See Figure 37. The graph illustrates exponential decay.

39. See Figure 39. The graph illustrates exponential growth.

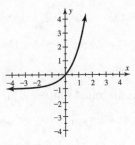

Figure 39

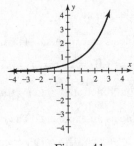

Figure 41

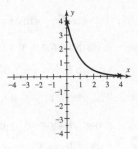

Figure 43

41. See Figure 41. The graph illustrates exponential growth.

43. See Figure 43. The graph illustrates exponential decay.

45. (a) Exponential decay. For each unit increase in x, $f(x)$ decreases by a factor of $\dfrac{1}{4}$.

(b) Since $f(x) = 64$ when $x = 0$, $f(x) = 64\left(\dfrac{1}{4}\right)^x$

47. (a) Linear growth. For each unit increase in x, $f(x)$ increases by 3 units.

(b) Since $f(x) = 8$ when $x = 0$, $f(x) = 3x + 8$

49. (a) Exponential growth. For each unit increase in x, $f(x)$ increases by a factor of 1.25.

(b) Since $f(x) = 4$ when $x = 0$, $f(x) = 4(1.25)^x$

51. (a) $\dfrac{B-A}{A} \times 100 \Rightarrow \dfrac{400-200}{200} \times 100 = \dfrac{200}{200} \times 100 = 100\%$

(b) $\dfrac{A-B}{B} \times 100 \Rightarrow \dfrac{200-400}{400} \times 100 = \dfrac{-200}{400} \times 100 = -50\%$

53. (a) $\dfrac{B-A}{A} \times 100 \Rightarrow \dfrac{30-150}{150} \times 100 = \dfrac{-120}{150} \times 100 = -80\%$

(b) $\dfrac{A-B}{B} \times 100 \Rightarrow \dfrac{150-30}{30} \times 100 = \dfrac{120}{30} \times 100 = 400\%$

55. (a) $rA = 1.2(1000) = 1200$ The account increased by \$1200.

(b) The new value of the account is $A + rA = \$1000 + \$1200 = \$2200$.

(c) The account increased by a factor of $a = 1 + r = 1 + 1.2 = 2.2$.

57. (a) $rA = 0.2(650) = 130$ The account increased by \$130.

(b) The new value of the account is $A + rA = \$650 + \$130 = \$780$.

(c) The account increased by a factor of $a = 1 + r = 1 + 0.2 = 1.2$.

59. (a) $rA = -0.1(800) = -80$ The account decreased by \$80.

(b) The new value of the account is $A + rA = \$800 + \$(-80) = \$720$.

(c) The account decreased by a factor of $a = 1 + r = 1 + (-0.1) = 0.9$.

61. The general form of an exponential function is $f(x) = Ca^x$, where C is the initial value, a is the growth factor, and x is the variable as an exponent. The initial value is $C = 9$, since $a > 1$ we have a growth factor $a = 1.07$. $a = 1 + r \Rightarrow 1.07 = 1 + r \Rightarrow r = 0.7 \Rightarrow$ percent change $R = 7\%$

63. The general form of an exponential function is $f(x) = Ca^x$, where C is the initial value, a is the growth factor, and x is the variable as an exponent. The initial value is $C = 1.5$, since $0 < a < 1$ we have a decay factor $a = 0.45$. $a = 1 + r \Rightarrow 0.45 = 1 + r \Rightarrow r = -0.55 \Rightarrow$ percent change $R = -55\%$

65. $1500(1 + 0.09)^{10} = \$3551.05$

67. $200(1 + 0.20)^{50} = \$1,820,087.63$

69. $560(1 + 0.014)^{25} = \$792.75$

71. Yes. This is equivalent to having two accounts, each containing $1000 initially.

73. $A = P\left(1 + \dfrac{r}{n}\right)^{nt} \Rightarrow A = 700\left(1 + \dfrac{0.04}{4}\right)^{4 \cdot 3} = 700(1.01)^{12} \approx 788.78$. The result is $788.78.

75. $A = P\left(1 + \dfrac{r}{n}\right)^{nt} \Rightarrow A = 1200\left(1 + \dfrac{0.025}{12}\right)^{12 \cdot 7} \approx 1200(1.00208)^{84} \approx 1429.24$. The result is $1429.24.

77. $f(1.2) = e^{1.2} \approx 3.32$

79. $f(-2) = 1 - e^{-2} \approx 0.86$

81. Graph $Y_1 = e \wedge (0.5X)$ in $[-4, 4, 1]$ by $[0, 8, 1]$. See Figure 81.

    The graph illustrates exponential growth.

    [-4, 4, 1] by [0, 8, 1]   [-4, 4, 1] by [0, 8, 1]

    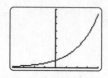   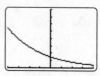

    Figure 81   Figure 83

83. Graph $Y_1 = 1.5e \wedge (-0.32X)$ in $[-4, 4, 1]$ by $[0, 8, 1]$. See Figure 83.

    The graph illustrates exponential decay.

85. The initial value is $C = 5000$, $a = 1 + r \Rightarrow a = 1 + (-0.25) \Rightarrow a = 0.75 \Rightarrow f(x) = 5000(0.75)^x$

    $f(4) = 5000(0.75)^4 \approx 1582$

87. The initial value is $C = 50$, $a = 1 + r \Rightarrow a = 1 + (0.1) \Rightarrow a = 1.1 \Rightarrow f(x) = 50(1.1)^x$

    $f(4) = 50(1.1)^4 \approx 73$

89. A 20% raise is calculated as $A = 50,000(1 + 0.2)^1 = 60,000$ A $20 raise is calculated as

    $A = 50,000 + 20 = 50,020$. A 20% raise per year is much better.

91. (a) $a = 1 + r \Rightarrow a = 1 + (-0.4) \Rightarrow a = 0.6$

    (b)   The initial value is $C = 0.07$ and from part (a) we know $a = 0.6$. $B(x) = 0.07(0.6)^x$

    (c) $B(2) = 0.07(0.6)^2 = 0.0252$. After 2 hours the blood alcohol is 0.0252.

93. (a)   The monthly growth factor is $a = 1.242$.

    (b)   In July 2008 there were about 0.5 million tweets per month.

    (c) $T(24) = 0.5(1.242)^{24} \approx 90.8$. After 24 months there were 90.8 million tweets per month.

95. (a)   Since $f(t) = 500$ when $t = 0$, $500 = Ca^0 \Rightarrow C = 500$. Since $f(t) = 1000$ when $t = 50$,

$$1000 = 500a^{50/50} \Rightarrow a = \frac{1000}{500} = 2$$

    (b) $f(170) = 500(2)^{170/50} \approx 5278$ thousand bacteria per milliliter or 5.278 million bacteria per milliliter

    (c)   The growth in the number of bacteria is exponential and doubles every 50 minutes.

97. Table $Y_1 = (0.905)^\wedge X$ with TblStart = 0 and ΔTbl = 10. See Figure 97.

    (a) $f(0) = (0.905)^0 = 1$. The probability that no vehicle will enter the intersection during a period of zero seconds is 1.

    (b)   Since $f(30) = (0.905)^{30} = 0.05006$ and $f(31) = (0.905)^{31} = 0.04530$, this occurs after about 30 seconds.

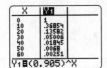

Figure 97

99. (a) $f(x) = 2.7e^{0.014x}$

    (b) $f(10) = 2.7e^{0.014(10)} \approx 3.1$; In 2020. the population of Nevada will be about 3.1 million.

101. (a) $f(x) = 38e^{0.0102x}$

    (b) $f(10) = 38e^{0.0102(10)} \approx 42$; In 2020, the population of California will be about 42 million.

103. (a)   Example: Using a 3% rate of interest. $I = 1200(0.03)(1) = 36$. The result is $36.

    (b)   Example: $I = 1200(0.03)(x)$. In this case we are assuming simple interest.

Checking Basic Concepts Sections 9.1 and 9.2

1. (a) $f(1) = 2(1)^2 + 5(1) - 1 = 6$, and so $(g \circ f)(1) = g(f(1)) = g(6) = 6 + 1 = 7$

(b) $(f \circ g)(x) = f(g(x)) = f(x+1) = 2(x+1)^2 + 5(x+1) - 1 = 2x^2 + 9x + 6$

2. See Figure 2.

(a)   No, this is not a one-to-one function because it does not pass the horizontal line test.

(b)   No, this function does not have an inverse because it is not one-to-one.

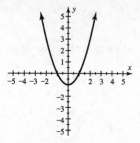   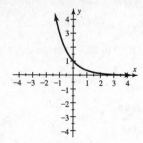

          Figure 2   Figure 5

3. $f(x) = 4x - 3 \Rightarrow y = 4x - 3$, interchange x and y and solve for y.

$$x = 4y - 3 \Rightarrow 4y = x + 3 \Rightarrow y = \frac{x+3}{4} \Rightarrow f^{-1}(x) = \frac{x+3}{4}$$

4. $f(-2) = 3(2^{-2}) = 3 \cdot \frac{1}{2^2} = 3 \cdot \frac{1}{4} = \frac{3}{4}$

5. See Figure 5.

6. Since $y = 2$ when $x = 0$, $2 = Ca^0 \Rightarrow C = 2$. Since $y = 1$ when $x = 1$, $1 = 2(a)^1 \Rightarrow a = \frac{1}{2}$.

Section 9.3 Logarithmic Functions

1. 10

3. D: $\{x \mid x > 0\}$; R: all real numbers

5. k

7. x

9. $\log 5$

11. $\log 1 = 0$ because $10^0 = 1$.

13.

x	10^{-5}	10^0	$10^{0.5}$	$10^{2.2}$
$\log x$	-5	0	0.5	2.2

15.

x	e^{-6}	e^{-1}	$e^{5/7}$	e^{π}
$\ln x$	-6	-1	5/7	π

17. $\log 10^5 = 5$

19. $\log 10^{-4} = -4$

21. $\log 1 = \log 10^0 = 0$

23. $\log \dfrac{1}{100} = \log 10^{-2} = -2$

25. $\log 10^{4.7} = 4.7$

27. $\log 10,000 = \log 10^4 = 4$

29. $\log \sqrt{10} = \log \sqrt{10} = \log (10)^{1/2} = \dfrac{1}{2}$

31. $\log (-23) \Rightarrow$ undefined, the domain of logarithmic functions are all positive real numbers.

33. $\log 0.001 = \log 10^{-3} = -3$

35. $10^{\log 2} = 2$

37. $10^{\log x^2} = x^2$

39. $10^{\log 5} = 5$

41. $\log 10^{(2x-7)} = 2x - 7$

43. $\log 25 \approx 1.398$

45. $\log 1.45 \approx 0.161$

47. See Figure 47. Compared to the graph of $y = \log x$, this graph is shifted 1 unit downward.

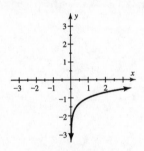

Figure 47

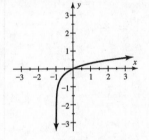

Figure 49

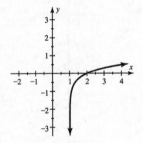

Figure 51

49. See Figure 49. Compared to the graph of $y = \log x$, this graph is shifted 1 unit to the left.

51. See Figure 51. Compared to the graph of $y = \log x$, this graph is shifted 1 unit to the right.

53. See Figure 53. Compared to the graph of $y = \log x$, this graph increases faster.

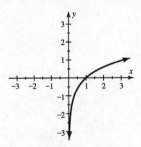

Figure 53

55. $\ln 1 = \ln e^0 = 0$

57. $\ln e^{-5x} = -5x$

59. $e^{\ln x^2} = x^2$, $x > 0$

61. $\ln 7 \approx 1.946$

63. $\ln \dfrac{4}{7} \approx -0.560$

65. Graph $Y_1 = \ln\left(\mathrm{abs}\left(X\right)\right)$ in $[-4, 4, 1]$ by $[-4, 4, 1]$. See Figure 65. The graph is a reflection across

the y-axis together with the graph of $y = \ln x$. The domain is $\left\{x \mid x \neq 0\right\}$.

$[-4, 4, 1]$ by $[-4, 4, 1]$ $[-4, 4, 1]$ by $[-4, 4, 1]$

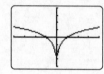   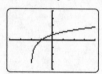

Figure 65   Figure 67

67. Graph $Y_1 = \ln\left(X + 2\right)$ in $[-4, 4, 1]$ by $[-4, 4, 1]$. See Figure 67. The graph is shifted 2 units to the

left. The domain is $\left\{x \mid x > -2\right\}$.

69. $\log_5 5^{6x} = 6x$

71. $\log_2 \sqrt{\dfrac{1}{8}} = \log_2 \sqrt{2^{-3}} = \log_2 2^{-3/2} = -\dfrac{3}{2}$

73. $\log_2 2^8 = 8$

75. $\log_2 \sqrt{8} = \log_2 \left(2^3\right)^{1/2} = \log_2 2^{3/2} = \dfrac{3}{2}$

77. $\log_2 \sqrt[3]{\dfrac{1}{4}} = \log_2 \left(2^{-2}\right)^{1/3} = \log_2 2^{-2/3} = -\dfrac{2}{3}$

79. $\log_2 -8$ is undefined.

81. $\log_2 4 = \log_2 2^2 = 2$

83. $\log_2 \dfrac{1}{16} = \log_2 2^{-4} = -4$

85. $\log_3 \dfrac{1}{9} = \log_3 3^{-2} = -2$

87. $\log_5 \dfrac{1}{25} = \log_5 5^{-2} = -2$

89. $5^{\log_5 17} = 17$

91. $4^{\log_4 (2x)^2} = \left(2x\right)^2$, $x \neq 0$

93. $5^{\log_5 0.6z} = 0.6z$, $z > 0$

95. See Figure 95.

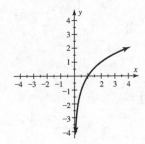

Figure 97

x	$\frac{1}{4}$	$\frac{1}{2}$	1	$\sqrt{2}$	64
$\log_2 x$	-2	-1	0	$\frac{1}{2}$	6

Figure 95

97. See Figure 97.

99. See Figure 99.

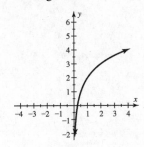

Figure 99

101. d. The graph of this function passes through the point $(1, 0)$ but does not pass through $(3, 1)$.

103. a. The graph of this function is shifted 2 units upward.

105. $f\left(10^{-4}\right) = 160 + 10\log\left(10^{-4}\right) = 160 + 10(-4) = 160 + (-40) = 120$ db.   Yes, this could cause pain.

107. (a)   The runway length increases but it does not double when the weight of the plane doubles.

   (b) $L(50) = 1.3\ln(50) \approx 5.086$; A 50-thousand pound plane needs a runway of at least 5086 feet

109. (a) $M = 6 - 2.5\log\dfrac{10}{1} = 6 - 2.5\log(10) = 6 - 2.5(1) = 3.5$

   (b) $M = 6 - 2.5\log\dfrac{100}{1} = 6 - 2.5\log(100) = 6 - 2.5(2) = 1$

   (c)   The intensity of the star decreases by 2.5.

111. (a) $\log\dfrac{x}{1} = 6.0 \Rightarrow \log x = 6.0 \Rightarrow x = 10^6$ and $\log\dfrac{x}{1} = 8.0 \Rightarrow \log x = 8.0 \Rightarrow x = 10^8$

   (b) $10^8 \div 10^6 = 100$ times

Section 9.4 Properties of Logarithms

1. 4

3. 3

5. $\log m - \log n$

7. No

9. No; $\log(xy) = \log x + \log y$

11. $\log_a x = \dfrac{\log x}{\log a}$ or $\log_a x = \dfrac{\ln x}{\ln a}$

13. $\ln(15) = \ln(3 \cdot 5) = \ln 3 + \ln 5$

15. $\log xy = \log x + \log y$

17. $\log y^2 = \log(y \cdot y) = \log y + \log y$

19. $\log \dfrac{7}{3} = \log 7 - \log 3$

21. $\ln \dfrac{x}{y} = \ln x - \ln y$

23. $\log_2 \dfrac{45}{x} = \log_2 45 - \log_2 x$

25. $\log 45 + \log 5 = \log(45 \cdot 5) = \log 225$

27. $\ln x + \ln y = \ln xy$

29. $\ln 7x^2 + \ln 2x = \ln(7x^2 \cdot 2x) = \ln 14x^3$

31. $\ln x + \ln y^2 - \ln y = \ln xy^2 - \ln y = \ln \dfrac{xy^2}{y} = \ln xy$

33. $\log 20 - \log 4 = \log\left(\dfrac{20}{4}\right) = \log 5$

35. $\ln x^4 - \ln x^2 = \ln\left(\dfrac{x^4}{x^2}\right) = \ln x^2$

37. $\log 300x - \log 3x = \log\left(\dfrac{300x}{3x}\right) = \log 100 = 2$

39. $\log 3^6 = 6 \log 3$

41. $\ln 2^x = x \ln 2$

43. $\log_2 5^{1/4} = \dfrac{1}{4} \log_2 5$

45. $\log_4 \sqrt[3]{z} = \log_4 z^{1/3} = \dfrac{1}{3} \log_4 z$

47. $\log x^{y-1} = (y-1) \log x$

49. $4 \log z - \log z^3 = \log z^4 - \log z^3 = \log \dfrac{z^4}{z^3} = \log z$

51. $\log x + 2\log x + 2\log y = \log x + \log x^2 + \log y^2 = \log\left(x \cdot x^2 \cdot y^2\right) = \log x^3 y^2$

53. $\log x - 2\log \sqrt{x} = \log x - \log\left(\sqrt{x}\right)^2 = \log x - \log x = 0$

55. $\ln 2^{x+1} - \ln 2 = \ln \dfrac{2^{x+1}}{2} = \ln 2^x$

57. $\ln \sqrt[3]{x} + \ln \sqrt{x} = \ln x^{1/3} + \ln x^{1/2} = \ln\left(x^{1/3} \cdot x^{1/2}\right) = \ln x^{5/6}$

59. $2\log_a (x+1) - \log_a\left(x^2 - 1\right) = \log_a (x+1)^2 - \log_a\left(x^2 - 1\right) = \log_a \dfrac{(x+1)(x+1)}{(x+1)(x-1)} = \log_a \dfrac{x+1}{x-1}$

61. $\log xy^2 = \log x + \log y^2 = \log x + 2\log y$

63. $\ln \dfrac{x^4 y}{z} = \ln x^4 y - \ln z = \ln x^4 + \ln y - \ln z = 4\ln x + \ln y - \ln z$

65. $\log \dfrac{\sqrt[3]{z}}{\sqrt{y}} = \log \dfrac{z^{1/3}}{y^{1/2}} = \log z^{1/3} - \log y^{1/2} = \dfrac{1}{3}\log z - \dfrac{1}{2}\log y$

67. $\log\left(x^4 y^3\right) = \log x^4 + \log y^3 = 4\log x + 3\log y$

69. $\ln \dfrac{1}{y} - \ln \dfrac{1}{x} = \ln y^{-1} - \ln x^{-1} = -1\ln y - (-1)\ln x = \ln x - \ln y$

71. $\log_4 \sqrt{\dfrac{x^3 y}{z^2}} = \log_4\left(\dfrac{x^3 y}{z^2}\right)^{1/2} = \dfrac{1}{2}\log_4 \dfrac{x^3 y}{z^2} = \dfrac{1}{2}\left(\log_4 x^3 + \log_4 y - \log_4 z^2\right) =$

$= \dfrac{1}{2}\left(3\log_4 x + \log_4 y - 2\log_4 z\right) = \dfrac{3}{2}\log_4 x + \dfrac{1}{2}\log_4 y - \log_4 z$

73. Graph $Y_1 = \log\left(X^3\right)$ and $Y_2 = 3\log\left(X\right)$ in $[-6, 6, 1]$ by $[-4, 4, 1]$. See Figures 73a & 73b.

By the power rule $\log x^3 = 3\log x$.

[–6, 6, 1] by [–4, 4, 1]   [–6, 6, 1] by [–4, 4, 1]   [–6, 6, 1] by [–4, 4, 1]   [–6, 6, 1] by [–4, 4, 1]

     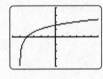

Figure 73a   Figure 73b   Figure 75a   Figure 75b

75. Graph $Y_1 = \ln\left(X + 5\right)$ and $Y_2 = \ln\left(X\right) + \ln\left(5\right)$ in $[-6, 6, 1]$ by $[-4, 4, 1]$. See Figure 75a & 75b.

Not the same.

77. $\log 16 = \log 2^4 = 4\log 2 = 4(0.3) = 1.2$

79. $\log 65 = \log\left(5 \cdot 13\right) = \log 5 + \log 13 = 0.7 + 1.1 = 1.8$

81. $\log 130 = \log\left(2 \cdot 5 \cdot 13\right) = \log 2 + \log 5 + \log 13 = 0.3 + 0.7 + 1.1 = 2.1$

83. $\log \dfrac{5}{2} = \log 5 - \log 2 = 0.7 - 0.3 = 0.4$

85. $\log \dfrac{1}{13} = \log 13^{-1} = -\log 13 = -1.1$

87. $\log_3 5 = \dfrac{\log 5}{\log 3} \approx 1.46$

89. $\log_2 25 = \dfrac{\log 25}{\log 2} \approx 4.64$

91. $\log_9 102 = \dfrac{\log 102}{\log 9} \approx 2.10$

93. $f(x) = 10 \log\left(10^{16}\,x\right) = 10\left(\log 10^{16} + \log x\right) = 10\left(16 + \log x\right) = 160 + 10 \log x$

Checking Basic Concepts Sections 9.3 and 9.4

1. (a) $\log 10^4 = 4$

   (b) $\ln e^x = x$

   (c) $\log_2 \dfrac{1}{8} = \log_2 2^{-3} = -3$

   (d) $\log_5 \sqrt{5} = \log_5 5^{1/2} = \dfrac{1}{2}$

2. See Figure 2.

   (a) $D: \{x \mid x > 0\};\ R:$ all real numbers

   (b) $f(1) = 0$

   (c)   Yes, for example $\log \dfrac{1}{10} = -1$.

   (d)   No, since negative numbers are not in the domain of $f(x) = \log x$.

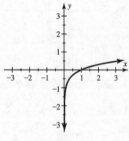

Figure 2

3. (a) $\log xy = \log x + \log y$

   (b) $\ln \dfrac{x}{yz} = \ln x - \ln yz = \ln x - \left(\ln y + \ln z\right) = \ln x - \ln y - \ln z$

(c) $\ln x^2 = 2 \ln x$

(d) $\log \dfrac{x^2 y^3}{\sqrt{z}} = \log x^2 y^3 - \log z^{1/2} = \log x^2 + \log y^3 - \log z^{1/2} = 2 \log x + 3 \log y - \dfrac{1}{2} \log z$

4. (a) $\log x + \log y = \log xy$

(b) $\ln 2x - 3 \ln y = \ln 2x - \ln y^3 = \ln \dfrac{2x}{y^3}$

(c) $2 \log_2 x + 3 \log_2 y - \log_2 z = \log_2 x^2 + \log_2 y^3 - \log_2 z = \log_2 x^2 y^3 - \log_2 z = \log_2 \dfrac{x^2 y^3}{z}$

Section 9.5 Exponential and Logarithmic Equations

1. Add 5 to both sides.

3. Take the common logarithm of both sides.

5. x

7. $2x$

9. No, $\log \dfrac{5}{4} = \log 5 - \log 4$

11. One

13. $10^x = 1000 \Rightarrow 10^x = 10^3 \Rightarrow x = 3$

15. $2^x = 64 \Rightarrow 2^x = 2^6 \Rightarrow x = 6$

17. $2^{x-3} = 8 \Rightarrow 2^{x-3} = 2^3 \Rightarrow x - 3 = 3 \Rightarrow x = 6$

19. $4^x + 3 = 259 \Rightarrow 4^x = 256 \Rightarrow 4^x = 4^4 \Rightarrow x = 4$

21. $10^{0.4x} = 124 \Rightarrow \log 10^{0.4x} = \log 124 \Rightarrow 0.4x = \log 124 \Rightarrow x = \dfrac{\log 124}{0.4} \approx 5.23$

23. $e^{-x} = 1 \Rightarrow e^{-x} = e^0 \Rightarrow -x = 0 \Rightarrow x = 0$

25. $e^x = 25 \Rightarrow \ln e^x = \ln 25 \Rightarrow x = \ln 25 \Rightarrow x \approx 3.22$

27. $0.4^x = 2 \Rightarrow \ln 0.4^x = \ln 2 \Rightarrow x \ln 0.4 = \ln 2 \Rightarrow x = \dfrac{\ln 2}{\ln 0.4} \Rightarrow x \approx -0.76$

29. $e^x - 1 = 6 \Rightarrow e^x = 7 \Rightarrow \ln e^x = \ln 7 \Rightarrow x = \ln 7 \approx 1.95$

31. $2(10)^{x+2} = 35 \Rightarrow 10^{x+2} = \dfrac{35}{2} \Rightarrow \log 10^{x+2} = \log \dfrac{35}{2} \Rightarrow x + 2 = \log \dfrac{35}{2} \Rightarrow x = \log \dfrac{35}{2} - 2 \approx -0.76$

33. $3.1^{2x} - 4 = 16 \Rightarrow 3.1^{2x} = 20 \Rightarrow \log_{3.1} 3.1^{2x} = \log_{3.1} 20 \Rightarrow 2x = \log_{3.1} 20 \Rightarrow x = \dfrac{\log 20}{2 \log 3.1} \approx 1.32$

35. $e^{3x} = e^{2x-1} \Rightarrow 3x = 2x - 1 \Rightarrow x = -1$

37. $5^{4x} = 5^{x^2-5} \Rightarrow 4x = x^2 - 5 \Rightarrow x^2 - 4x - 5 = 0 \Rightarrow (x+1)(x-5) = 0 \Rightarrow x = -1$ or 5

39. $e^{2x} \cdot e^x = 10 \Rightarrow e^{2x+x} = 10 \Rightarrow e^{3x} = 10 \Rightarrow \ln e^{3x} = \ln 10 \Rightarrow 3x = \ln 10 \Rightarrow x = \dfrac{\ln 10}{3} \approx 0.77$

41. $e^x = 2^{x+2} \Rightarrow \ln e^x = \ln 2^{x+2} \Rightarrow x = (x+2) \ln 2 \Rightarrow x = x \ln 2 + 2 \ln 2 \Rightarrow x - x \ln 2 = 2 \ln 2 \Rightarrow$

$x(1 - \ln 2) = 2 \ln 2 \Rightarrow x = \dfrac{2 \ln 2}{1 - \ln 2} \approx 4.52$

43. $4^{0.5x} = 5^{x+2} \Rightarrow \log 4^{0.5x} = \log 5^{x+2} \Rightarrow 0.5x \log 4 = (x+2) \log 5 \Rightarrow 0.5x \log 4 = x \log 5 + 2 \log 5 \Rightarrow$

$0.5x \log 4 - x \log 5 = 2 \log 5 \Rightarrow x(0.5 \log 4 - \log 5) = 2 \log 5 \Rightarrow x = \dfrac{2 \log 5}{0.5 \log 4 - \log 5} \approx -3.51$

45. (a)   The solution is the x-coordinate of the intersection point, $x = 1$.

    (b) $0.2(10^x) = 2 \Rightarrow 10^x = 10 \Rightarrow x = 1$

47. (a)   The solution is the x-coordinate of the intersection point, $x = -2$.

    (b) $2^{-x} = 4 \Rightarrow 2^{-x} = 2^2 \Rightarrow -x = 2 \Rightarrow x = -2$

49. $10^x = 0.1 \Rightarrow 10^x = 10^{-1} \Rightarrow x = -1$ For numerical support, table $Y_1 = 10^{\wedge}X$ and $Y_2 = 0.1$ with

    TblStart $= -3$ and ΔTbl $= 1$. See Figure 49.

$[-1, 1, 1]$ by $[0, 3, 1]$

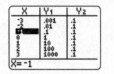   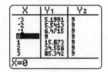   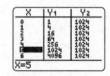   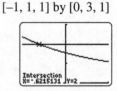

Figure 49   Figure 51   Figure 53   Figure 55

51. $4e^x + 5 = 9 \Rightarrow 4e^x = 4 \Rightarrow e^x = 1 \Rightarrow e^x = e^0 \Rightarrow x = 0$ For numerical support, table

    $Y_1 = 4e^{\wedge}X + 5$ and $Y_2 = 9$ with TblStart $= -3$ and ΔTbl $= 1$. See Figure 51.

53. $4^x = 1024 \Rightarrow 4^x = 4^5 \Rightarrow x = 5$ For numerical support, table $Y_1 = 4^{\wedge}X$ and $Y_2 = 1024$ with

    TblStart $= 0$ and ΔTbl $= 1$. See Figure 53.

55. $(0.55)^x + 0.55 = 2 \Rightarrow 0.55^x = 1.45 \Rightarrow \log_{0.55} 0.55^x = \log_{0.55} 1.45 \Rightarrow x = \dfrac{\log 1.45}{\log 0.55} \approx -0.62$

    For graphical support, graph $Y_1 = 0.55^{\wedge}X + 0.55$ and $Y_2 = 2$ in $[-1, 1, 1]$ by $[0, 3, 1]$.

    See Figure 55.

57. Graph $Y_1 = e^{\wedge}X - X$ and $Y_2 = 2$ in $[-5, 5, 1]$ by $[-5, 5, 1]$. See Figures 57a & 57b.

    The solutions are the x-coordinates of the intersection points, $x \approx -1.84$ and $x \approx 1.15$.

[–5, 5, 1] by [–5, 5, 1]   [–5, 5, 1] by [–5, 5, 1]

Figure 57a   Figure 57b

59. Graph $Y_1 = \ln(X)$ and $Y_2 = e^\wedge(-X)$ in [–5, 5, 1] by [–5, 5, 1]. See Figure 59.

The solution is the x-coordinate of the intersection point, $x \approx 1.31$.

[–5, 5, 1] by [–5, 5, 1]

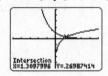

Figure 59

61. $\log x = 2 \Rightarrow 10^{\log x} = 10^2 \Rightarrow x = 100$

63. $\ln x = 5 \Rightarrow e^{\ln x} = e^5 \Rightarrow x = e^5 \approx 148.41$

65. $\log 2x = 7 \Rightarrow 10^{\log 2x} = 10^7 \Rightarrow 2x = 10,000,000 \Rightarrow x = 5,000,000$

67. $\log_2 x = 4 \Rightarrow 2^{\log_2 x} = 2^4 \Rightarrow x = 16$

69. $\log_2 5x = 2.3 \Rightarrow 2^{\log_2 5x} = 2^{2.3} \Rightarrow 5x = 2^{2.3} \Rightarrow x = \dfrac{2^{2.3}}{5} \approx 0.98$

71. $2\log x + 5 = 7.8 \Rightarrow 2\log x = 2.8 \Rightarrow \log x = 1.4 \Rightarrow 10^{\log x} = 10^{1.4} \Rightarrow x = 10^{1.4} \approx 25.12$

73. $5\ln(2x+1) = 55 \Rightarrow \ln(2x+1) = 11 \Rightarrow e^{\ln(2x+1)} = e^{11} \Rightarrow 2x+1 = e^{11} \Rightarrow x = \dfrac{e^{11}-1}{2} \approx 29,936.57$

75. $\log x^2 = \log x \Rightarrow x^2 = x \Rightarrow x^2 - x = 0 \Rightarrow x(x-1) = 0 \Rightarrow x = 0$ or 1

The solution $x = 0$ causes an undefined expression in the original equation. The only solution is 1.

77. $\ln x + \ln(x+1) = \ln 30 \Rightarrow \ln x(x+1) = \ln 30 \Rightarrow x(x+1) = 30 \Rightarrow x^2 + x - 30 = 0 \Rightarrow$

$(x+6)(x-5) = 0 \Rightarrow x = -6$ or 5 The solution $x = -6$ causes an undefined expression in the original

equation. The only solution is 5.

79. $\log_3 3x - \log_3(x+2) = \log_3 2 \Rightarrow \log_3 \dfrac{3x}{x+2} = \log_3 2 \Rightarrow \dfrac{3x}{x+2} = 2 \Rightarrow 3x = 2x+4 \Rightarrow x = 4$

81. $\log_2(x-1) + \log_2(x+1) = 3 \Rightarrow \log_2(x^2-1) = 3 \Rightarrow 2^{\log_2(x^2-1)} = 2^3 \Rightarrow x^2 - 1 = 8 \Rightarrow$

$x^2 - 9 = 0 \Rightarrow (x+3)(x-3) = 0 \Rightarrow x = -3$ or 3 The solution $x = -3$ causes an undefined expression

in the original equation. The only solution is 3.

83. $\log x = 1.6 \Rightarrow 10^{\log x} = 10^{1.6} \Rightarrow x = 10^{1.6} \approx 39.81$ For graphical support, graph

$Y_1 = \log(X)$ and $Y_2 = 1.6$ in [0, 50, 10] by [–2, 2, 1]. See Figure 83.

[0, 50, 10] by [−2, 2, 1]   [−1, 2, 1] by [−2, 2, 1]

Figure 83   Figure 85

85. $\ln(x+1) = 1 \Rightarrow e^{\ln(x+1)} = e^1 \Rightarrow x+1 = e \Rightarrow x = e - 1 \approx 1.72$ For graphical support, graph

$Y_1 = \ln(X+1)$ and $Y_2 = 1$ in [−1, 2, 1] by [−2, 2, 1]. See Figure 85.

87. $17 - 6\log_3 x = 5 \Rightarrow 6\log_3 x = 12 \Rightarrow \log_3 x = 2 \Rightarrow 3^{\log_3 x} = 3^2 \Rightarrow x = 3^2 = 9$ For graphical support,

graph $Y_1 = 17 - 6\left(\ln(X)/\ln(3)\right)$ and $Y_2 = 5$ in [0, 12, 1] by [0, 12, 1]. See Figure 87.

[0, 12, 1] by [0, 12, 1]   [

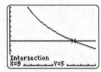

Figure 87

89. (a)   The solution is the x-coordinate of the intersection point, $x = 2$.

(b) $\ln x = 0.7 \Rightarrow e^{\ln x} = e^{0.7} \Rightarrow x = e^{0.7} \approx 2.01$

91. (a)   The solution is the x-coordinate of the intersection point, $x = 2$.

(b) $5\log 2x = 3 \Rightarrow \log 2x = 0.6 \Rightarrow 10^{\log 2x} = 10^{0.6} \Rightarrow 2x = 10^{0.6} \Rightarrow x = \dfrac{10^{0.6}}{2} \approx 1.99$

93. (a) $W(5) = 1.73\left(10^{0.276(5)}\right) = 1.73\left(10^{1.38}\right) \approx 41.5$, In 2010 China generated about 41.5 gigawatts.

(b) $22 = 1.73\left(10^{0.276(x)}\right) \Rightarrow \dfrac{22}{1.73} = \left(10^{0.276x}\right) \Rightarrow \log\left(\dfrac{22}{1.73}\right) = 0.276x \Rightarrow \dfrac{\log\left(\dfrac{22}{1.73}\right)}{0.276} = x \Rightarrow$

$x \approx 4$ Since x is the number of years after 2005 the result is 2009.

95. $20 = 5(1.15)^x \Rightarrow 4 = (1.15)^x \Rightarrow \ln 4 = x\ln(1.15) \Rightarrow \dfrac{\ln 4}{\ln(1.15)} = x \Rightarrow x \approx 10$ hours

97. $2000(1+0.15)^t = 6000 \Rightarrow 1.15^t = 3 \Rightarrow \log_{1.15} 1.15^t = \log_{1.15} 3 \Rightarrow t = \dfrac{\log 3}{\log 1.15} \approx 7.86 \approx 8$ years

99. (a) $f(1994) = 2339(1.24)^{(1994-1988)} \approx 8503$; In 1994 there were about 8503 people waiting for

liver transplants.

(b) $2339(1.24)^{(x-1988)} = 20,000 \Rightarrow (1.24)^{(x-1988)} = \dfrac{20,000}{2339} \Rightarrow$

$\log_{1.24}(1.24)^{(x-1988)} = \log_{1.24}\left(\dfrac{20,000}{2339}\right) \Rightarrow x - 1988 = \log_{1.24}\left(\dfrac{20,000}{2339}\right) \Rightarrow$

$$x = \log_{1.24}\left(\frac{20,000}{2339}\right) + 1988 = \frac{\log\left(\frac{20,000}{2339}\right)}{\log 1.24} + 1988 \approx 1998$$

101. $3 \log x = 3.960 \Rightarrow \log x = 1.320 \Rightarrow 10^{\log x} = 10^{1.320} \Rightarrow x = 10^{1.320} \approx 20.893 = 20,893$ lb

103. (a) $f(1) = 230\left(10^{-0.055 \cdot 1}\right) \approx 203$; In 1975 there were about 203 thousand bluefin tuna.

       (b)   About 1979.

       (c) $230\left(10^{-0.055x}\right) = 115 \Rightarrow 10^{-0.055x} = 0.5 \Rightarrow \log 10^{-0.055x} = \log 0.5 \Rightarrow -0.055x = \log 0.5 \Rightarrow$

               $x = \dfrac{\log 0.5}{-0.055} \approx 5.47$ or about 1979.

105. From the data point (1, 25), $25 = a + b \log 1 \Rightarrow 25 = a + b(0) \Rightarrow a = 25$.

       Using the data point (10, 28) and the fact that $a = 25$, $28 = 25 + b \log 10 \Rightarrow 28 = 25 + b \Rightarrow b = 3$

107. (a)   Graph $Y_1 = 645 \log(X+1) + 1925$ and $Y_2 = 2200$ in [0, 4, 1] by [0, 3000, 1000].

               See Figure 107. A person consuming 2200 calories would typically own about 1.67 acres.

       (b) $645 \log(x+1) + 1925 = 2200 \Rightarrow 645 \log(x+1) = 275 \Rightarrow \log(x+1) = \dfrac{275}{645} \Rightarrow$

               $10^{\log(x+1)} = 10^{275/645} \Rightarrow x+1 = 10^{275/645} \Rightarrow x = 10^{275/645} - 1 \approx 1.67$ acres

[0, 4, 1] by [0, 3000, 1000]

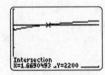

       Figure 107

109. (a)   The data is nonlinear. It does not increase at a constant rate.

       (b)   Each year the amount of fertilizer increases by a factor of 1.06 or 6%.

       (c) $5(1.06)^{(x-1950)} = 15 \Rightarrow 1.06^{(x-1950)} = 3 \Rightarrow \log_{1.06} 1.06^{(x-1950)} = \log_{1.06} 3 \Rightarrow$

               $x - 1950 = \log_{1.06} 3 \Rightarrow x = \log_{1.06} 3 + 1950 = \dfrac{\log 3}{\log 1.06} + 1950 \approx 1968.85$ or in 1968

111. $160 + 10 \log x = 100 \Rightarrow 10 \log x = -60 \Rightarrow \log x = -6 \Rightarrow 10^{\log x} = 10^{-6} \Rightarrow x = 10^{-6}$ w/cm^2

113. $0.48 \ln(x+1) + 27 = 28 \Rightarrow 0.48 \ln(x+1) = 1 \Rightarrow \ln(x+1) = \dfrac{1}{0.48} \Rightarrow e^{\ln(x+1)} = e^{1/0.48} \Rightarrow$

       $x + 1 = e^{1/0.48} \Rightarrow x = e^{1/0.48} - 1 \approx 7.03$ or about 7 miles

Checking Basic Concepts Section 9.5

1. (a) $2\left(10^x\right) = 40 \Rightarrow 10^x = 20 \Rightarrow \log 10^x = \log 20 \Rightarrow x = \log 20 \approx 1.30$

(b) $2^{3x}+3=150 \Rightarrow 2^{3x}=147 \Rightarrow \log_2 2^{3x}=\log_2 147 \Rightarrow 3x=\dfrac{\log 147}{\log 2} \Rightarrow x=\dfrac{\log 147}{3 \log 2} \approx 2.40$

(c) $\ln x=4.1 \Rightarrow e^{\ln x}=e^{4.1} \Rightarrow x=e^{4.1} \approx 60.34$

(d) $4 \log 2x=12 \Rightarrow \log 2x=3 \Rightarrow 10^{\log 2x}=10^3 \Rightarrow 2x=1000 \Rightarrow x=500$

2. $\log(x+4)+\log(x-4)=\log 48 \Rightarrow \log(x^2-16)=\log 48 \Rightarrow x^2-16=48 \Rightarrow$

    $x^2-64=0 \Rightarrow (x+8)(x-8)=0 \Rightarrow x=-8$ or 8 The solution $x=-8$ causes an undefined expression

    in the original equation. The only solution is 8.

3. $500(1.03)^x=900 \Rightarrow 1.03^x=\dfrac{9}{5} \Rightarrow \log_{1.03} 1.03^x=\log_{1.03}\left(\dfrac{9}{5}\right) \Rightarrow x=\dfrac{\log\left(\frac{9}{5}\right)}{\log 1.03} \approx 20$ years

Chapter 9 Review

1. (a) $f(-2)=2(-2)^2-4(-2)=16,$ then $(g \circ f)(-2)=g(f(-2))=g(16)=5(16)+1=81$

   (b) $(f \circ g)(x)=f(g(x))=f(5x+1)=2(5x+1)^2-4(5x+1)=50x^2-2$

2. (a) $f(-2)=\sqrt[3]{-2-6}=-2,$ then $(g \circ f)(-2)=g(f(-2))=g(-2)=4(-2)^3=-32$

   (b) $(f \circ g)(x)=f(g(x))=f(4x^3)=\sqrt[3]{4x^3-6}$

3. (a) $(f \circ g)(2)=f(g(2))=f(3)=0$

   (b) $(g \circ f)(1)=g(f(1))=g(2)=3$

4. (a) $(f \circ g)(-1)=f(g(-1))=f(2)=3$

   (b) $(g \circ f)(2)=g(f(2))=g(3)=-2$

   (c) $(f \circ f)(1)=f(f(1))=f(0)=-1$

5. $f(1)=f(-1)=2$

6. $f(0)=f(2)=1$

7. This graph does not pass the horizontal line test. The function is not one-to-one.

8. This graph passes the horizontal line test. The function is one-to-one.

9. $(f \circ f^{-1})(x)=f(f^{-1}(x))=f\left(\dfrac{x+9}{2}\right)=2\left(\dfrac{x+9}{2}\right)-9=x+9-9=x$

    $(f^{-1} \circ f)(x)=f^{-1}(f(x))=f^{-1}(2x-9)=\dfrac{(2x-9)+9}{2}=\dfrac{2x}{2}=x$

10. $(f \circ f^{-1})(x)=f(f^{-1}(x))=f\left(\sqrt[3]{x-1}\right)=\left(\sqrt[3]{x-1}\right)^3+1=x-1+1=x$

$$\left(f^{-1}\circ f\right)(x)=f^{-1}\left(f(x)\right)=f^{-1}\left(x^3+1\right)=\sqrt[3]{\left(x^3+1\right)-1}=\sqrt[3]{x^3}=x$$

11. $f(x)=5x\Rightarrow y=5x,$ interchange x and y and solve for y. $x=5y\Rightarrow y=\dfrac{x}{5}\Rightarrow f^{-1}(x)=\dfrac{x}{5}$

12. $f(x)=x-11\Rightarrow y=x-11,$ interchange x and y and solve for y.

$$x=y-11\Rightarrow y=x+11\Rightarrow f^{-1}(x)=x+11$$

13. $f(x)=2x+7\Rightarrow y=2x+7,$ interchange x and y and solve for y.

$$x=2y+7\Rightarrow 2y=x-7\Rightarrow y=\dfrac{x-7}{2}\Rightarrow f^{-1}(x)=\dfrac{x-7}{2}$$

14. $f(x)=\dfrac{4}{x}\Rightarrow y=\dfrac{4}{x},$ interchange x and y and solve for y. $x=\dfrac{4}{y}\Rightarrow xy=4\Rightarrow y=\dfrac{4}{x}\Rightarrow f^{-1}(x)=\dfrac{4}{x}$

15. See Figure 15. $D=\{3,7,8,10\};\ R=\{0,1,2,3\}$

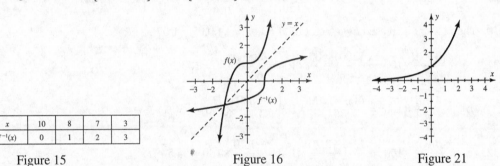

x	10	8	7	3
$f^{-1}(x)$	0	1	2	3

Figure 15   Figure 16   Figure 21

16. The graph of f^{-1} is a reflection of the graph of f across the line $y=x$. See Figure 16.

17. $f(-1)=6^{-1}=\dfrac{1}{6}$ and $f(2)=6^2=36$

18. $f(0)=5\left(2^0\right)=5(1)=5$ and $f(3)=5\left(2^{-3}\right)=5\left(\dfrac{1}{8}\right)=\dfrac{5}{8}$

19. $f(-1)=\left(\dfrac{1}{3}\right)^{-1}=3$ and $f(4)=\left(\dfrac{1}{3}\right)^4=\dfrac{1}{3^4}=\dfrac{1}{81}$

20. $f(0)=3\left(\dfrac{1}{6}\right)^0=3(1)=3$ and $f(1)=3\left(\dfrac{1}{6}\right)^1=3\left(\dfrac{1}{6}\right)=\dfrac{3}{6}=\dfrac{1}{2}$

21. See Figure 21. The graph illustrates exponential growth.

22. See Figure 22. The graph illustrates exponential decay.

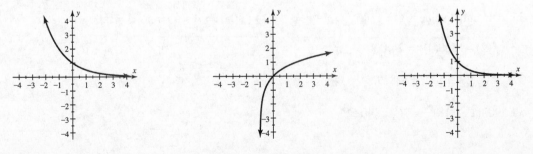

Figure 22   Figure 23   Figure 24

23. See Figure 23. The graph illustrates logarithmic growth.

24. See Figure 24. The graph illustrates exponential decay.

25. (a)   Exponential growth. For each unit increase in x, $f(x)$ increases by a factor of 2.

   (b)   Since $f(x) = 5$ when $x = 0$, $f(x) = 5(2)^x$

26. (a)   Linear growth. For each unit increase in x, $f(x)$ increases by 5 units.

   (b)   Since $f(x) = 5$ when $x = 0$, $f(x) = 5x + 5$

27. Since $y = \dfrac{1}{2}$ when $x = 0$, $\dfrac{1}{2} = Ca^0 \Rightarrow C = \dfrac{1}{2}$. Since $y = 1$ when $x = 1$, $1 = \dfrac{1}{2}(a)^1 \Rightarrow a = 2$.

28. Since $y = 2$ when $x = 2$, $2 = k\log_2 2 \Rightarrow 2 = k(1) \Rightarrow k = 2$.

29. $\dfrac{A-B}{B} \times 100 \Rightarrow \dfrac{120-150}{150} \times 100 = \dfrac{-30}{150} \times 100 = -20\%$

30. The account increased by a factor of $a = 1 + r = 1 + 0.07 = 1.07$.

31. (a) $\dfrac{B-A}{A} \times 100 \Rightarrow \dfrac{1200-600}{600} \times 100 = \dfrac{600}{600} \times 100 = 100\%$

   (b) $\dfrac{A-B}{B} \times 100 \Rightarrow \dfrac{600-1200}{1200} \times 100 = \dfrac{-600}{1200} \times 100 = -50\%$

32. (a) $\dfrac{B-A}{A} \times 100 \Rightarrow \dfrac{1.00-2.20}{2.20} \times 100 = \dfrac{-1.20}{2.20} \times 100 = 54.\overline{54}\%$

   (b) $\dfrac{A-B}{B} \times 100 \Rightarrow \dfrac{2.20-1.00}{1.00} \times 100 = \dfrac{1.20}{1.00} \times 100 = 120\%$

33. (a) $rA = 2.10(500) = 1050$ The account increased by \$1050.

   (b)   The new value of the account is $A + rA = \$500 + \$1050 = \$1550$.

   (c)   The account increased by a factor of $a = 1 + r = 1 + 2.1 = 3.1$.

34. (a) $rA = -0.25(700) = -175$ The account decreased by \$175.

   (b)   The new value of the account is $A + rA = \$700 + \$(-175) = \$525$.

   (c)   The account decreased by a factor of $a = 1 + r = 1 + (-0.25) = 0.75$.

35. The initial value is $C = 20{,}000$, $a = 1 + r \Rightarrow a = 1 + (-0.05) \Rightarrow a = 0.95 \Rightarrow f(x) = 20{,}000(0.95)^x$

   $f(2) = 20{,}000(0.95)^2 = 18{,}050$

36. The initial value is $C = 1500$, $a = 1 + r \Rightarrow a = 1 + (3.00) \Rightarrow a = 4 \Rightarrow f(x) = 1500(4)^x$

   $f(2) = 1500(4)^2 = 24{,}000$

37. $1200(1+0.10)^9 = \$2829.54$

38. $900(1+0.18)^{40} = \$675{,}340.51$

39. $f(5.3) = 2e^{5.3} - 1 \approx 399.67$

40. $f(2.1) = 0.85^{2.1} \approx 0.71$

41. $f(55) = 2 \log 55 \approx 3.48$

42. $f(23) = \ln(2 \cdot 23 + 3) \approx 3.89$

43. $\log 0.001 = \log 10^{-3} = -3$

44. $\log \sqrt{10,000} = \log 100 = 2$

45. $\ln e^{-4} = -4$

46. $\log_4 16 = \log_4 4^2 = 2$

47. $\log 65 \approx 1.813$

48. $\ln 0.85 \approx -0.163$

49. $\ln 120 \approx 4.787$

50. $\log \dfrac{2}{5} \approx -0.398$

51. $10^{\log 7} = 7$

52. $\log_2 2^{5/9} = \dfrac{5}{9}$

53. $\ln e^{6-x} = 6 - x$

54. $e^{2 \ln x} = e^{\ln x^2} = x^2, \, x > 0$

55. $\ln xy = \ln x + \ln y$

56. $\log \dfrac{x}{y} = \log x - \log y$

57. $\ln x^2 y^3 = \ln x^2 + \ln y^3 = 2 \ln x + 3 \ln y$

58. $\log \dfrac{\sqrt{x}}{z^3} = \log \dfrac{x^{1/2}}{z^3} = \log x^{1/2} - \log z^3 = \dfrac{1}{2} \log x - 3 \log z$

59. $\log_2 \dfrac{x^2 y}{z} = \log_2 x^2 y - \log_2 z = \log_2 x^2 + \log_2 y - \log_2 z = 2 \log_2 x + \log_2 y - \log_2 z$

60. $\log_3 \sqrt[3]{\dfrac{x}{y}} = \log_3 \left(\dfrac{x}{y}\right)^{1/3} = \dfrac{1}{3} \log_3 \left(\dfrac{x}{y}\right) = \dfrac{1}{3}\left(\log_3 x - \log_3 y\right) = \dfrac{1}{3} \log_3 x - \dfrac{1}{3} \log_3 y$

61. $\log 45 + \log 5 - \log 3 = \log(45 \cdot 5) - \log 3 = \log 225 - \log 3 = \log \dfrac{225}{3} = \log 75$

62. $\log_4 2x + \log_4 5x = \log_4 (2x \cdot 5x) = \log_4 \left(10x^2\right)$

63. $2 \ln x - 3 \ln y = \ln x^2 - \ln y^3 = \ln \dfrac{x^2}{y^3}$

64. $\log x^4 - \log x^3 + \log y = \log \dfrac{x^4}{x^3} + \log y = \log x + \log y = \log xy$

65. $\log 6^3 = 3 \log 6$

66. $\ln x^2 = 2 \ln x$

67. $\log_2 5^{2x} = (2x) \log_2 5$

68. $\log_4 (0.6)^{x+1} = (x+1) \log_4 0.6$

69. $10^x = 100 \Rightarrow 10^x = 10^2 \Rightarrow x = 2$

70. $2^{2x} = 256 \Rightarrow 2^{2x} = 2^8 \Rightarrow 2x = 8 \Rightarrow x = 4$

71. $3e^x + 1 = 28 \Rightarrow 3e^x = 27 \Rightarrow e^x = 9 \Rightarrow \ln e^x = \ln 9 \Rightarrow x = \ln 9 \approx 2.20$

72. $0.85^x = 0.2 \Rightarrow \log_{0.85} 0.85^x = \log_{0.85} 0.2 \Rightarrow x = \log_{0.85} 0.2 = \dfrac{\log 0.2}{\log 0.85} \approx 9.90$

73. $5 \ln x = 4 \Rightarrow \ln x = 0.8 \Rightarrow e^{\ln x} = e^{0.8} \Rightarrow x = e^{0.8} \approx 2.23$

74. $\ln 2x = 5 \Rightarrow e^{\ln 2x} = e^5 \Rightarrow 2x = e^5 \Rightarrow x = \dfrac{e^5}{2} \approx 74.21$

75. $2 \log x = 80 \Rightarrow \log x = 40 \Rightarrow 10^{\log x} = 10^{40} \Rightarrow x = 10^{40}$

76. $3 \log x - 5 = 1 \Rightarrow 3 \log x = 6 \Rightarrow \log x = 2 \Rightarrow 10^{\log x} = 10^2 \Rightarrow x = 100$

77. $2^{x+4} = 3^x \Rightarrow \log 2^{x+4} = \log 3^x \Rightarrow (x+4) \log 2 = x \log 3 \Rightarrow x \log 2 + 4 \log 2 = x \log 3 \Rightarrow$

 $x \log 3 - x \log 2 = 4 \log 2 \Rightarrow x(\log 3 - \log 2) = 4 \log 2 \Rightarrow x = \dfrac{4 \log 2}{\log 3 - \log 2} \approx 6.84$

78. $\ln (2x+1) + \ln (x-5) = \ln 13 \Rightarrow \ln (2x+1)(x-5) = \ln 13 \Rightarrow (2x+1)(x-5) = 13 \Rightarrow$

 $2x^2 - 9x - 5 = 13 \Rightarrow 2x^2 - 9x - 18 = 0 \Rightarrow (2x+3)(x-6) = 0 \Rightarrow x = -\dfrac{3}{2}$ or 6 The solution $x = -\dfrac{3}{2}$

 causes an undefined expression in the original equation. The only solution is 6.

79. (a)   The solution is the x-coordinate of the intersection point, $x = 3$.

 (b) $\dfrac{1}{2}(2^x) = 4 \Rightarrow 2^x = 8 \Rightarrow 2^x = 2^3 \Rightarrow x = 3$

80. (a)   The solution is the x-coordinate of the intersection point, $x = 4$.

 (b) $\log_2 2x = 3 \Rightarrow 2^{\log_2 2x} = 2^3 \Rightarrow 2x = 8 \Rightarrow x = 4$

81. (a) $(S \circ r)(8) = S(r(8)) = S\left(\sqrt{2 \cdot 8}\right) = S(4) = 4\pi(4)^2 = 64\pi$

 After 8 seconds, the balloon has a surface area of $64\pi \approx 201$ square inches.

(b) $(S \circ r)(t) = S(r(t)) = S(\sqrt{2t}\,) = 4\pi(\sqrt{2t})^2 = 4\pi \cdot 2t = 8\pi t$

82. (a) Yes, this is a one-to-one function because different inputs result in different outputs.

(b) $f(x) = 0.08x \Rightarrow y = 0.08x,$ interchange x and y and solve for y.

$$x = 0.08y \Rightarrow y = \frac{x}{0.08} \Rightarrow f^{-1}(x) = \frac{x}{0.08}$$

This formula calculates the cost of an item whose sales tax is x dollars.

83. $1500(1+0.11)^t = 3000 \Rightarrow 1.11^t = 2 \Rightarrow \log_{1.11} 1.11^t = \log_{1.11} 2 \Rightarrow t = \dfrac{\log 2}{\log 1.11} \approx 6.64 \approx 7$ years

84. $100 = a + b \log 1 \Rightarrow 100 = a + b(0) \Rightarrow a = 100$

$150 = 100 + b \log 10 \Rightarrow 150 = 100 + b(1) \Rightarrow b = 50$

85. $3 = Ca^0 \Rightarrow C = 3$ and $6 = 3a^1 \Rightarrow a = 2$

86. $\log \dfrac{x}{1} = 7 \Rightarrow 10^{\log x} = 10^7 \Rightarrow x = 10^7$

87. (a) Graph $Y_1 = 2e \wedge (0.051X)$ in $[0, 10, 2]$ by $[0, 4, 1]$. The function represents exponential growth.

[0, 10, 2] by [0, 4, 1]

(b) In 2010, $x = 10$ thus $f(10) = 2e^{0.051(10)} \approx 3.3$ million

(c) $2e^{0.051x} = 3 \Rightarrow e^{0.051x} = \dfrac{3}{2} \Rightarrow \ln e^{0.051x} = \ln\left(\dfrac{3}{2}\right) \Rightarrow 0.051x = \ln(1.5) \Rightarrow x = \dfrac{\ln(1.5)}{0.051} \approx 7.95$

That is about 8 years after 2000, which is 2008.

88. (a) $N(0) = 1000e^{0.0014(0)} = 1000;$ There were initially 1000 bacteria.

(b) $1000e^{0.0014x} = 2000 \Rightarrow e^{0.0014x} = 2 \Rightarrow \ln e^{0.0014x} = \ln 2 \Rightarrow 0.0014x = \ln 2 \Rightarrow$

$x = \dfrac{\ln 2}{0.0014} \approx 495.11$ min.

89. (a) $f(5) = 1.2 \ln 5 + 5 \approx 6.93$ m/sec

(b) $1.2 \ln x + 5 = 8 \Rightarrow 1.2 \ln x = 3 \Rightarrow \ln x = 2.5 \Rightarrow e^{\ln x} = e^{2.5} \Rightarrow x = e^{2.5} \approx 12.18$ meters

Chapter 9 Test

1. $f(1) = 4(1)^3 - 5(1) = -1,$ then $(g \circ f)(1) = g(f(1)) = g(-1) = (-1) + 7 = 6$

$$(f \circ g)(x) = f(g(x)) = f(x+7) = 4(x+7)^3 - 5(x+7)$$

2. (a) $(f \circ g)(-1) = f(g(-1)) = f(-1) = 3$

    (b) $(g \circ f)(1) = g(f(1)) = g(1) = 3$

3. Two different inputs result in the same output. For example $-5 \neq 5$, but $f(-5) = f(5) = 0$.

4. $f(x) = 5 - 2x \Rightarrow y = 5 - 2x$, interchange x and y and solve for y.

$$x = 5 - 2y \Rightarrow 2y = 5 - x \Rightarrow y = \frac{5-x}{2} \Rightarrow f^{-1}(x) = \frac{5-x}{2}$$

5. The graph of f^{-1} is a reflection of the graph of f across the line $y = x$. See Figure 5.

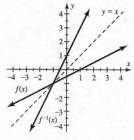

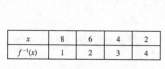

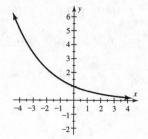

Figure 5   Figure 6   Figure 8

6. See Figure 6. $D = \{2, 4, 6, 8\}$; $R = \{1, 2, 3, 4\}$

7. $f(2) = 3\left(\dfrac{1}{4}\right)^2 = 3\left(\dfrac{1}{16}\right) = \dfrac{3}{16}$

8. See Figure 8. This graph represents exponential decay.

9. (a)   Exponential growth. For each unit increase in x, $f(x)$ increases by a factor of 2.

    (b)   Since $f(x) = 3$ when $x = 0$, $f(x) = 3(2)^x$

10. (a)   Linear growth. For each unit increase in x, $f(x)$ increases by 1.5 units.

    (b)   Since $f(x) = -1$ when $x = 0$, $f(x) = 1.5x - 1$

11. $1 = Ca^0 \Rightarrow C = 1$ and $2 = 1a^{-1} \Rightarrow a = \dfrac{1}{2}$

12. $\dfrac{A-B}{B} \times 100 \Rightarrow \dfrac{900-600}{600} \times 100 = \dfrac{300}{600} \times 100 = 50\%$

13. The account increased by a factor of $a = 1 + r = 1 + (0.05) = 1.05$.

14. $750(1+0.07)^5 = \$1051.91$

15. $f(21) = 1.5 \ln(21-5) = 1.5 \ln 16 \approx 4.16$

16. $\log \sqrt{10} = \log 10^{1/2} = \dfrac{1}{2}$

17. $\log_2 43 = \dfrac{\log 43}{\log 2} \approx 5.426$

18. See Figure 18. The graph is shifted 2 units to the right.

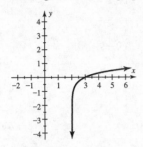

Figure 18

19. $\log \dfrac{x^3 y^2}{\sqrt{x}} = \log \dfrac{x^3 y^2}{x^{1/2}} = \log x^3 + \log y^2 - \log x^{1/2} = 3 \log x + 2 \log y - \dfrac{1}{2} \log z$

20. $4 \ln x - 5 \ln y + \ln z = \ln x^4 - \ln y^5 + \ln z = \ln \dfrac{x^4}{y^5} + \ln z = \ln \dfrac{x^4 z}{y^5}$

21. $\log 7^{2x} = 2x \log 7$

22. $\ln e^{1-3x} = 1 - 3x$

23. $2e^x = 50 \Rightarrow e^x = 25 \Rightarrow \ln e^x = \ln 25 \Rightarrow x = \ln 25 \approx 3.22$

24. $3(10)^x - 7 = 143 \Rightarrow 3(10)^x = 150 \Rightarrow 10^x = 50 \Rightarrow \log 10^x = \log 50 \Rightarrow x = \log 50 \approx 1.70$

25. $5 \log x = 9 \Rightarrow \log x = 1.8 \Rightarrow 10^{\log x} = 10^{1.8} \Rightarrow x = 10^{1.8} \approx 63.10$

26. $3 \ln 5x = 27 \Rightarrow \ln 5x = 9 \Rightarrow e^{\ln 5x} = e^9 \Rightarrow 5x = e^9 \Rightarrow x = \dfrac{e^9}{5} \approx 1620.62$

27. $5 = a + b \log 1 \Rightarrow 5 = a + b(0) \Rightarrow a = 5$ and $8 = 5 + b \log 10 \Rightarrow 8 = 5 + b(1) \Rightarrow b = 3$

28. (a) $f(0) = 4(1.09)^0 = 4(1) = 4$ million

(b) $f(5) = 4(1.09)^{(5)} \approx 6.15$; After 5 hours there were about 6.15 million bacteria.

(c)   This represents exponential growth.

(d) $4(1.09)^x = 8 \Rightarrow (1.09)^x = 2 \Rightarrow \ln (1.09)^x = \ln 2 \Rightarrow \ln(1.09)x = \ln 2 \Rightarrow x = \dfrac{\ln 2}{\ln(1.09)} \approx 8$

There were 6 million bacteria after about 8 hours.

29. (a)   The initial value C is 5000. The account decreased by a factor of $a = 1 + r = 1 + (-0.2) = 0.98$.

The function is $A(x) = 5000(0.98)^x$.

(b) $A(3) = 5000(0.98)^3 = 4705.96$ The result is \$4705.96.

(c) $4500 = 5000(0.98)^x \Rightarrow 0.9 = 0.98^x \Rightarrow \ln(0.9) = \ln(0.98)x \Rightarrow \dfrac{\ln(0.9)}{\ln(0.98)} = x \Rightarrow x \approx 5.2$

Chapter 9 Extended and Discovery Exercises

1. $a^{5730} = 0.5 \Rightarrow \left(a^{5730}\right)^{1/5730} = 0.5^{1/5730} \Rightarrow a \approx 0.9998790392$

2. $P(10,000) = 0.9998790392^{10,000} \approx 0.298$ or 29.8%

3. $0.9998790392^{x} = 0.9 \Rightarrow \log_{0.9998790392} 0.9998790392^{x} = \log_{0.9998790392} 0.9 \Rightarrow$

$x = \dfrac{\log 0.9}{\log 0.9998790392} \Rightarrow x \approx 871$ years

4. $0.9998790392^{x} = 0.01 \Rightarrow \log_{0.9998790392} 0.999879^{x} = \log_{0.9998790392} 0.01 \Rightarrow$

$x = \dfrac{\log 0.01}{\log 0.9998790392} \Rightarrow x \approx 38,069$ years   (38,100 rounded to the nearest 100 years)

5. Plot the data and graph $Y_1 = 0.133\left(0.878\left(0.73 \wedge X\right) + 0.122\left(0.92 \wedge X\right)\right)$ in [0, 25, 5] by

    [0, 0.11, 0.01]. See Figure 5. The fit is quite good.

    [0, 25, 5] by [0, 0.11, 0.01]   [0, 25, 5] by [0, 0.11, 0.01]

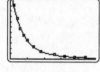

      Figure 5   Figure 8

6. $f(0) = 0.133\left(0.878\left(0.73^{0}\right) + 0.122\left(0.92^{0}\right)\right) = 0.133\left(0.878 + 0.122\right) = 0.133(1) = 0.133$

    The initial concentration is 0.133 mg/mL.

7. The concentration decreases to 0 as the body eliminates the dye from the blood stream.

8. Note that 40% of 0.133 is 0.0532. Graph $Y_1 = 0.133\left(0.878\left(0.73 \wedge X\right) + 0.122\left(0.92 \wedge X\right)\right)$

    and $Y_2 = 0.0532$ in [0, 25, 5] by [0, 0.11, 0.01]. See Figure 8. This happens after about 3.33

    minutes. Solving this problem symbolically would be very difficult.

9. $f\left(10^{-4.7}\right) = -\log 10^{-4.7} = -(-4.7) = 4.7;$ This rain could cause the pH to drop below 5.6.

10. The ion concentration in seawater is $-\log x = 8.2 \Rightarrow \log x = -8.2 \Rightarrow 10^{\log x} = 10^{-8.2} \Rightarrow x = 10^{-8.2}$.

    This is $\dfrac{10^{-4.7}}{10^{-8.2}} \approx 3162$ times greater.

11. $A = 100\left[\dfrac{\left(1 + \frac{0.09}{26}\right)^{260} - 1}{\frac{0.09}{26}}\right] \approx \$42,055.97$

12. A 19-year-old student would have 46 years to deposit money before age 65. *Answers may vary.*

$x\left[\dfrac{\left(1 + \frac{0.12}{26}\right)^{1196} - 1}{\frac{0.12}{26}}\right] = 1,000,000 \Rightarrow x = 1,000,000\left[\dfrac{\frac{0.12}{26}}{\left(1 + \frac{0.12}{26}\right)^{1196} - 1}\right] = \$18.80$

Chapters 1-9 Cumulative Review Exercises

1. Move the decimal point 4 places to the right: $0.000429 = 4.29 \times 10^{-4}$.

2. Natural: none; Whole: 0; Integer: $-3, 0$; Rational: $-\dfrac{11}{7}, -3, 0, 5.\overline{18}$; Irrational: $\sqrt{6}, \pi$

3. By substitution of values, the formula that best fits the data is (c).

4. This equation illustrates the commutative property for addition.

5. $\left(\dfrac{1}{d^2}\right)^{-2} = \left(d^2\right)^2 = d^4$

6. $\left(\dfrac{8a^2}{2b^3}\right)^{-3} = \left(\dfrac{4a^2}{b^3}\right)^{-3} = \left(\dfrac{b^3}{4a^2}\right)^3 = \dfrac{\left(b^3\right)^3}{\left(4a^2\right)^3} = \dfrac{b^9}{4^3\left(a^2\right)^3} = \dfrac{b^9}{64a^6}$

7. $\dfrac{\left(2x^{-2}y^3\right)^2}{xy^{-2}} = \dfrac{2^2\left(x^{-2}\right)^2\left(y^3\right)^2}{xy^{-2}} = \dfrac{4x^{-4}y^6}{xy^{-2}} = 4x^{-4-1}y^{6-(-2)} = 4x^{-5}y^8 = \dfrac{4y^8}{x^5}$

8. $\dfrac{x^{-3}y}{4x^2y^{-3}} = \dfrac{1}{4}x^{-3-2}y^{1-(-3)} = \dfrac{1}{4}x^{-5}y^4 = \dfrac{y^4}{4x^5}$

9. The graph rises 5 units for each 4 units of run. The slope is $\dfrac{5}{4}$. The y-intercept is 1. So $y = \dfrac{5}{4}x + 1$.

10. The function is defined for all values of the variable except -3. The domain is $\left\{x \mid x \neq -3\right\}$.

11. The slope is $m = \dfrac{5-1}{2-1} = \dfrac{4}{1} = 4$. Since $f(x) = -3$ when $x = 0$ the y-intercept is -3.

    Here $f(x) = 4x - 3$.

12. Vertical lines have equations of the form $x = k$. The equation of the vertical line passing through $(4, 7)$ is $x = 4$.

13. $m = \dfrac{-3-(-1)}{2-4} = \dfrac{-2}{-2} = 1$

14. See Figure 14.

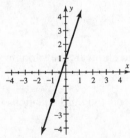

Figure 14

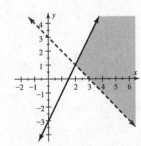

Figure 25

15. Since the line is perpendicular to $y = -\dfrac{1}{7}x - 8$, the slope is $m = 7$. Using the point-slope form gives

    $$y = 7(x-1) + 1 \Rightarrow y = 7x - 7 + 1 \Rightarrow y = 7x - 6$$

16. Since the line is parallel to $y = 3x - 1$, the slope is $m = 3$. The y-intercept is given, so $b = 5$.

    The equation is $y = 3x + 5$.

17. The graphs intersect at the point $(-1, 1)$, so $y_1 = y_2$ when $x = -1$.

18. Here $y < -4$ when $x < 0$.

19. $\frac{2}{3}(x-3) + 8 = -6 \Rightarrow \frac{2}{3}(x-3) = -14 \Rightarrow x - 3 = -21 \Rightarrow x = -18$

20. $\frac{1}{3}z + 6 < \frac{1}{4}z - (5z - 6) \Rightarrow 4z + 72 < 3z - 12(5z - 6) \Rightarrow 4z + 72 < 3z - 60z + 72 \Rightarrow$

    $4z + 72 < -57z + 72 \Rightarrow 61z < 0 \Rightarrow z < 0$. The interval is $(-\infty, 0)$.

21. $\left(\dfrac{t+2}{3}\right) - 10 = \frac{1}{3}t - (5t + 8) \Rightarrow t + 2 - 30 = t - 3(5t + 8) \Rightarrow t - 28 = t - 15t - 24 \Rightarrow 15t = 4 \Rightarrow t = \dfrac{4}{15}$

22. $-10 \le -\frac{3}{5}x - 4 < -1 \Rightarrow -6 \le -\frac{3}{5}x < 3 \Rightarrow 10 \ge x > -5 \Rightarrow -5 < x \le 10$. The interval is $(-5, 10]$.

23. First divide each side of $-2|t - 4| \ge -12$ by -2 to obtain $|t - 4| \le 6$. The solutions to $|t - 4| \le 6$

    satisfy $c \le t \le d$ where c and d are the solutions to $|t - 4| = 6$. $|t - 4| = 6$ is equivalent to $t - 4 = -6 \Rightarrow$

    $t = -2$ and $t - 4 = 6 \Rightarrow t = 10$ The interval is $[-2, 10]$.

24. $\left|\dfrac{1}{2}x - 5\right| = 3 \Rightarrow \frac{1}{2}x - 5 = -3 \Rightarrow \frac{1}{2}x = 2 \Rightarrow x = 4$ or $\frac{1}{2}x - 5 = 3 \Rightarrow \frac{1}{2}x = 8 \Rightarrow x = 16$

25. See Figure 25.

26. $\det A = -1(4) - 3(-2) = -4 + 6 = 2$

27. Multiply the first equation by 2 and the second equation by 3. Add the equations to eliminate the

    $$\begin{array}{r} 8x - 6y = 2 \\ 15x + 6y = 21 \\ \hline 23x = 23 \end{array}$$

    variable y. Thus, $x = 1$. And so $4(1) - 3y = 1 \Rightarrow y = 1$. The solution is $(1, 1)$.

28. Multiplying the first equation by 3 and adding the equations will eliminate both variables.

    $$\begin{array}{r} 6x - 9y = -6 \\ -6x + 9y = 5 \\ \hline 0 = -1 \end{array}$$

    Thus, the system has no solutions.

29. Multiply the first equation by 5 and add the first and second equations to eliminate the variable y.

    $$\begin{array}{r} 10x - 5y + 15z = -10 \\ x + 5y - 2z = -8 \\ \hline 11x + 13z = -18 \end{array}$$

    Multiply the third equation by 5 and add the second and third equations to

    $$\begin{array}{r} x + 5y - 2z = -8 \\ -15x - 5y - 15z = 30 \\ \hline -14x - 17z = 22 \end{array}$$

    eliminate the variable y. Multiply the first new equation by 14 and the second

$$154x + 182z = -252$$

new equation by 11. Add the equations to eliminate x. $\dfrac{-154x - 187z = \ \ 242}{-5z = -10}$ And so $z = 2$.

Substitute $z = 2$ into the first *new* equation: $11x + 13(2) = -18 \Rightarrow x = -4$ Substitute

$x = -4$ and $z = 2$ into the *original* first equation: $2(-4) - y + 3(2) = -2 \Rightarrow y = 0$

The solution is $(-4, 0, 2)$.

30. $\begin{bmatrix} 1 & 1 & -1 & | & -1 \\ -1 & -1 & -1 & | & -1 \\ 1 & -2 & 1 & | & 1 \end{bmatrix} \begin{matrix} \\ R_2 + R_1 \rightarrow \\ R_3 - R_1 \rightarrow \end{matrix} \begin{bmatrix} 1 & 1 & -1 & | & -1 \\ 0 & 0 & -2 & | & -2 \\ 0 & -3 & 2 & | & 2 \end{bmatrix} \begin{matrix} \\ Exchange \\ R_2 \leftrightarrow R_3 \end{matrix} \begin{bmatrix} 1 & 1 & -1 & | & -1 \\ 0 & -3 & 2 & | & 2 \\ 0 & 0 & -2 & | & -2 \end{bmatrix}$

$\begin{matrix} \\ \\ (-1/2)R_3 \rightarrow \end{matrix} \begin{bmatrix} 1 & 1 & -1 & | & -1 \\ 0 & -3 & 2 & | & 2 \\ 0 & 0 & 1 & | & 1 \end{bmatrix} \begin{matrix} R_1 + R_3 \rightarrow \\ R_2 - 2R_3 \rightarrow \\ \ \end{matrix} \begin{bmatrix} 1 & 1 & 0 & | & 0 \\ 0 & -3 & 0 & | & 0 \\ 0 & 0 & 1 & | & 1 \end{bmatrix} \begin{matrix} R_1 + (1/3)R_2 \rightarrow \\ (-1/3)R_2 \rightarrow \\ \ \end{matrix} \begin{bmatrix} 1 & 0 & 0 & | & 0 \\ 0 & 1 & 0 & | & 0 \\ 0 & 0 & 1 & | & 1 \end{bmatrix}$

The solution is $(0, 0, 1)$.

31. From the graph of the region of feasible solutions (not shown), the vertices are $(0, 0)$, $(0, 4)$, $(3, 3)$, and $(4, 0)$. The maximum value of R occurs at one of the vertices. For $(0, 0)$, $R = 2(0) + 5(0) = 0$.

For $(0, 4)$, $R = 2(0) + 5(4) = 20$. For $(3, 3)$, $R = 2(3) + 5(3) = 21$. For $(4, 0)$, $R = 2(4) + 5(0) = 8$.

The maximum value is $R = 21$.

32. The triangle has vertices $(-2, -3)$, $(-1, 2)$ and $(2, 1)$. The matrix needed is $A = \begin{bmatrix} -2 & -1 & 2 \\ -3 & 2 & 1 \\ 1 & 1 & 1 \end{bmatrix}$.

The area is $D = \left| \dfrac{1}{2} \det([A]) \right| = 8 \text{ in}^2$.

33. $2x^3 - 4x^2 + 2x = 2x(x^2 - 2x + 1) = 2x(x - 1)^2$

34. $4a^2 - 25b^2 = (2a)^2 - (5b)^2 = (2a - 5b)(2a + 5b)$

35. $8t^3 - 27 = (2t)^3 - 3^3 = (2t - 3)(4t^2 + 6t + 9)$

36. $4a^3 - 2a^2 + 10a - 5 = 2a^2(2a - 1) + 5(2a - 1) = (2a^2 + 5)(2a - 1)$

37. $6x^2 - 7x - 10 = 0 \Rightarrow (6x + 5)(x - 2) = 0 \Rightarrow x = -\dfrac{5}{6}$ or $x = 2$

38. $9x^2 = 4 \Rightarrow 9x^2 - 4 = 0 \Rightarrow (3x + 2)(3x - 2) = 0 \Rightarrow x = -\dfrac{2}{3}$ or $x = \dfrac{2}{3}$

39. $x^4 - 2x^3 = 15x^2 \Rightarrow x^4 - 2x^3 - 15x^2 = 0 \Rightarrow x^2(x + 3)(x - 5) = 0 \Rightarrow x = -3, 0, \text{ or } 5$

40. $5x - 10x^2 = 0 \Rightarrow 5x(1 - 2x) = 0 \Rightarrow x = 0$ or $x = \dfrac{1}{2}$

41. $\dfrac{x^2+5x+6}{x^2-9} \cdot \dfrac{x-3}{x+2} = \dfrac{(x+2)(x+3)(x-3)}{(x-3)(x+3)(x+2)} = 1$

42. $\dfrac{x^2-2x-8}{x^2+x-12} \div \dfrac{(x-4)^2}{x^2-16} = \dfrac{x^2-2x-8}{x^2+x-12} \cdot \dfrac{x^2-16}{(x-4)^2} = \dfrac{(x+2)(x-4)(x-4)(x+4)}{(x+4)(x-3)(x-4)(x-4)} = \dfrac{x+2}{x-3}$

43. $\dfrac{2}{x+2} - \dfrac{1}{x-2} = \dfrac{-3}{x^2-4} \Rightarrow 2(x-2)-1(x+2) = -3 \Rightarrow x-6 = -3 \Rightarrow x = 3$

44. $\dfrac{3y}{y^2+y-2} = \dfrac{1}{y-1} - 2 \Rightarrow 3y = 1(y+2) - 2(y+2)(y-1) \Rightarrow 3y = y+2-2y^2-2y+4 \Rightarrow$

   $2y^2+4y-6 = 0 \Rightarrow 2(y+3)(y-1) = 0 \Rightarrow y = -3$ or $y = 1$. The value $y = 1$ is not valid, thus $y = -3$.

45. $P = \dfrac{J+2z}{J} \Rightarrow JP = J+2z \Rightarrow JP - J = 2z \Rightarrow J(P-1) = 2z \Rightarrow J = \dfrac{2z}{P-1}$

46. $\dfrac{\dfrac{3}{x^2}+x}{x-\dfrac{3}{x^2}} = \dfrac{\dfrac{3}{x^2}+x}{x-\dfrac{3}{x^2}} \cdot \dfrac{x^2}{x^2} = \dfrac{3+x^3}{x^3-3} = \dfrac{x^3+3}{x^3-3}$

47. $y = kx \Rightarrow 15 = 3k \Rightarrow k = 5 \Rightarrow y = 5x$, so when $x = 8$, $y = 5(8) = 40$

48.
$$
\begin{array}{r}
3x^2+6x+10 \\
x-2\overline{)3x^3+0x^2-2x-15} \\
\underline{3x^3-6x^2} \\
6x^2-2x \\
\underline{6x^2-12x} \\
10x-15 \\
\underline{10x-20} \\
5
\end{array}
$$

   The solution is: $3x^2+6x+10+\dfrac{5}{x-2}$

49. $\left(\dfrac{x^6}{y^9}\right)^{2/3} = \dfrac{\left(x^6\right)^{2/3}}{\left(y^9\right)^{2/3}} = \dfrac{x^{6\cdot(2/3)}}{y^{9\cdot(2/3)}} = \dfrac{x^4}{y^6}$

50. $\sqrt[3]{-x^4} \cdot \sqrt[3]{-x^5} = \sqrt[3]{(-x^4)(-x^5)} = \sqrt[3]{x^9} = x^{9/3} = x^3$

51. $\sqrt{5ab} \cdot \sqrt{20ab} = \sqrt{5 \cdot 20 \cdot ab \cdot ab} = \sqrt{100(ab)^2} = \sqrt{(10ab)^2} = 10ab$

52. $2\sqrt{24} - \sqrt{54} = 2\sqrt{4\cdot6} - \sqrt{9\cdot6} = 2\sqrt{4}\cdot\sqrt{6} - \sqrt{9}\cdot\sqrt{6} = 4\sqrt{6} - 3\sqrt{6} = \sqrt{6}$

53. $\sqrt[3]{a^5b^4} + 3\sqrt[3]{a^5b} = \sqrt[3]{(ab)^3 \cdot a^2b} + 3\sqrt[3]{a^3 \cdot a^2b} = ab\sqrt[3]{a^2b} + 3a\sqrt[3]{a^2b} = (b+3)a\sqrt[3]{a^2b}$

54. $(5+\sqrt{5})(5-\sqrt{5}) = 5^2 - (\sqrt{5})^2 = 25-5 = 20$

55. $\dfrac{2}{5-\sqrt{3}} = \dfrac{2}{5-\sqrt{3}} \cdot \dfrac{5+\sqrt{3}}{5+\sqrt{3}} = \dfrac{10+2\sqrt{3}}{25-3} = \dfrac{10+2\sqrt{3}}{22} = \dfrac{2\left(5+\sqrt{3}\right)}{2\cdot 11} = \dfrac{5+\sqrt{3}}{11}$

56. For the function to be defined, $x-4>0 \Rightarrow x>4$. The interval is $(4,\infty)$.

57. $2(x+1)^2 = 50 \Rightarrow (x+1)^2 = 25 \Rightarrow x+1 = \pm\sqrt{25} \Rightarrow x+1 = \pm 5 \Rightarrow x = -6$ or 4

58. $\sqrt{x+6} = x \Rightarrow x+6 = x^2 \Rightarrow x^2 - x - 6 = 0 \Rightarrow (x+2)(x-3) = 0 \Rightarrow x = -2$ or 3

   The value $x = -2$ does not check. The only solution is 3.

59. $(-2+3i)-(-5-2i) = -2-(-5)+3i+2i = 3+5i$

60. $\dfrac{3-i}{1+3i} = \dfrac{3-i}{1+3i} \cdot \dfrac{1-3i}{1-3i} = \dfrac{3-9i-i+3i^2}{1-9i^2} = \dfrac{3-10i-3}{1+9} = \dfrac{-10i}{10} = -i$

61. $-\dfrac{b}{2a} = -\dfrac{-12}{2(3)} = \dfrac{12}{6} = 2;\ f(2) = 3(2)^2 - 12(2) + 13 = 1.$ The vertex is $(2,1)$.

62. $-\dfrac{b}{2a} = -\dfrac{6}{2(-2)} = \dfrac{6}{4} = \dfrac{3}{2};\ f\left(\dfrac{3}{2}\right) = -2\left(\dfrac{3}{2}\right)^2 + 6\left(\dfrac{3}{2}\right) - 1 = \dfrac{7}{2}.$ The vertex is $\left(\dfrac{3}{2},\dfrac{7}{2}\right)$. The maximum

   value is $\dfrac{7}{2}$.

63. Compared to $y = x^2$, the graph of $f(x)$ is shifted right 3 units and up 2 units.

64. $y = x^2 + 6x - 2 \Rightarrow y = \left(x^2 + 6x + 9\right) - 2 - 9 \Rightarrow y = (x+3)^2 - 11.$ The vertex is $(-3,-11)$.

65. $x^2 - 13x + 40 = 0 \Rightarrow (x-5)(x-8) = 0 \Rightarrow x = 5$ or $x = 8$

66. $2d^2 - 5 = d \Rightarrow 2d^2 - d - 5 = 0.$ Let $a = 2, b = -1$ and $c = -5$ in the quadratic formula.

   $$d = \dfrac{-(-1) \pm \sqrt{(-1)^2 - 4(2)(-5)}}{2(2)} = \dfrac{1 \pm \sqrt{41}}{4}$$

67. $z^2 - 4z = -2 \Rightarrow z^2 - 4z + 4 = -2 + 4 \Rightarrow (z-2)^2 = 2 \Rightarrow z - 2 = \pm\sqrt{2} \Rightarrow z = 2 \pm \sqrt{2}$

68. $x^4 - 10x^2 + 24 = 0 \Rightarrow \left(x^2 - 6\right)\left(x^2 - 4\right) = 0 \Rightarrow x^2 - 6 = 0$ or $x^2 - 4 = 0 \Rightarrow x = \pm\sqrt{6}$ or $x = \pm 2$

69. (a)   The graph intersects the x-axis at -1 and 3.

   (b)   Because the parabola opens downward, $a < 0$.

   (c)   Because there are two real solutions, the discriminant is positive.

70. $x^2 + 5x - 14 = 0 \Rightarrow (x+7)(x-2) = 0 \Rightarrow x = -7$ or $x = 2$

   Since the parabola opens upward, the solution is $(-\infty, -7] \cup [2, \infty)$.

71. (a) $g(1) = 2(1) + 1 = 3$, then $(f \circ g)(1) = f(g(1)) = f(3) = (3)^2 - 2 = 7$

   (b) $(g \circ f)(x) = g(f(x)) = g\left(x^2 - 2\right) = 2\left(x^2 - 2\right) + 1 = 2x^2 - 4 + 1 = 2x^2 - 3$

72. $f(-4) = f(3) = 6$

73. $f(x) = \dfrac{3}{x} \Rightarrow y = \dfrac{3}{x}$, interchange x and y and solve for y. $\quad x = \dfrac{3}{y} \Rightarrow xy = 3 \Rightarrow y = \dfrac{3}{x} \Rightarrow f^{-1}(x) = \dfrac{3}{x}$

74. $A = 800(1+0.075)^{15} \approx \$2367.10$

75. $\log_3 81 = \log_3 3^4 = 4$

76. $e^{\ln(2x)} = 2x,\ x > 0$

77. $\log \dfrac{\sqrt{x}}{y^2} = \log \dfrac{x^{1/2}}{y^2} = \log x^{1/2} - \log y^2 = \dfrac{1}{2}\log x - 2\log y$

78. $2\ln x + \ln 5x = \ln x^2 + \ln 5x = \ln\left(x^2 \cdot 5x\right) = \ln\left(5x^3\right)$

79. $6\log x - 2 = 9 \Rightarrow 6\log x = 11 \Rightarrow \log x = \dfrac{11}{6} \Rightarrow 10^{\log x} = 10^{11/6} \Rightarrow x \approx 68.13$

80. $2^{3x} = 17 \Rightarrow \log_2 2^{3x} = \log_2 17 \Rightarrow 3x = \dfrac{\log 17}{\log 2} \Rightarrow x = \dfrac{\log 17}{3\log 2} \approx 1.36$

81. $12{,}000(1+r)^5 = 14{,}600 \Rightarrow (1+r)^5 = \dfrac{73}{60} \Rightarrow 1+r = \sqrt[5]{\dfrac{73}{60}} \Rightarrow r = \sqrt[5]{\dfrac{73}{60}} - 1 \approx 0.04$ or 4%

82. $27.4\sqrt[3]{W} = 36 \Rightarrow \sqrt[3]{W} = \dfrac{36}{27.4} \Rightarrow W = \left(\dfrac{36}{27.4}\right)^3 \approx 2.27$ pounds

83. (a) $\quad -\dfrac{b}{2a} = -\dfrac{-975}{2(0.25)} = 1950$

    (b) $\quad f(1950) = 0.25(1950)^2 - 975(1950) + 950{,}845 = 220$ million Btu

84. $\dfrac{x^2}{12} = 350 \Rightarrow x^2 = 4200 \Rightarrow x = \sqrt{4200} \approx 64.8$ miles per hour

85. $8000(1+r)^{45} = 1{,}000{,}000 \Rightarrow (1+r)^{45} = 125 \Rightarrow 1+r = 125^{1/45} \Rightarrow r = 125^{1/45} - 1 \approx 0.113$ or 11.3%

86. (a) $\quad f(8) = 1.4\ln 8 + 7 \approx 9.91$ meters per second

    (b) $\quad 1.4\ln x + 7 = 10 \Rightarrow 1.4\ln x = 3 \Rightarrow \ln x = \dfrac{3}{1.4} \Rightarrow e^{\ln x} = e^{3/1.4} \Rightarrow x \approx 8.52$ meters

Chapter 10: Conic Sections

10.1: Parabolas and Circles

1. Parabola, ellipse and hyperbola

3. No, it does not pass the vertical line test for functions.

5. No, it does not pass the vertical line test for functions.

7. left

9. circle; (h, k)

11. d; The equation fits the form $x = a(y-k)^2 + h$ where (h, k) is the vertex.

13. c; The equation fits the form $(x-h)^2 + (y-k)^2 = r^2$ where (h, k) is the center and r is the radius.

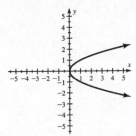
Figure 15

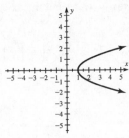
Figure 17

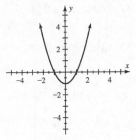
Figure 19

15. Since $x = (y-0)^2 + 0$, the vertex is (0, 0) and the axis of symmetry is $y = 0$. See Figure 15.

17. Since $x = (y-0)^2 + 1$, the vertex is (1, 0) and the axis of symmetry is $y = 0$. See Figure 17.

19. Since $y = x^2 - 1 = (x-0)^2 - 1$, the vertex is $(0, -1)$ and the axis of symmetry is $x = 0$.

See Figure 19.

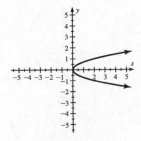
Figure 21

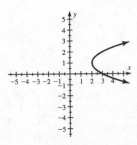
Figure 23

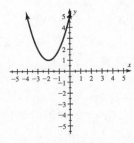
Figure 25

21. Since $x = 2(y-0)^2 + 0$, the vertex is (0, 0) and the axis of symmetry is $y = 0$. See Figure 21.

23. Since $x = (y-1)^2 + 2$, the vertex is (2, 1) and the axis of symmetry is $y = 1$. See Figure 23.

25. Since $y = (x+2)^2 + 1$, the vertex is (–2, 1) and the axis of symmetry is $x = -2$. See Figure 25.

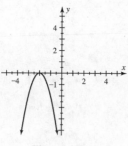

Figure 27

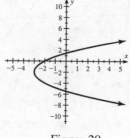

Figure 29

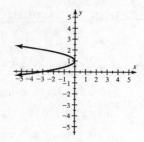

Figure 31

27. Since $y = -2(x+2)^2$, the vertex is (–2, 0) and the axis of symmetry is $x = -2$. See Figure 27.

29. Since $x = \dfrac{1}{2}(y+1)^2 - 3$, the vertex is (–3, –1) and the axis of symmetry is $y = -1$. See Figure 29.

31. Since $x = -3(y-1)^2 + 0$, the vertex is (0, 1) and the axis of symmetry is $y = 1$. See Figure 31.

33. See Figure 33. $x = -\dfrac{b}{2a} = -\dfrac{-1}{2(2)} = \dfrac{1}{4}$ and $y = 2\left(\dfrac{1}{4}\right)^2 - \left(\dfrac{1}{4}\right) + 1 = \dfrac{7}{8}$. Vertex: $\left(\dfrac{1}{4}, \dfrac{7}{8}\right)$.

Axis of symmetry: $x = \dfrac{1}{4}$.

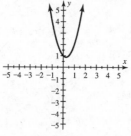

Figure 33

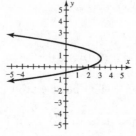

Figure 35

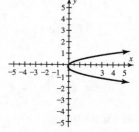

Figure 37

35. See Figure 35. $y = -\dfrac{b}{2a} = -\dfrac{3}{2(-2)} = \dfrac{3}{4}$ and $x = -2\left(\dfrac{3}{4}\right)^2 + 3\left(\dfrac{3}{4}\right) + 2 = \dfrac{25}{8}$. Vertex: $\left(\dfrac{25}{8}, \dfrac{3}{4}\right)$.

Axis of symmetry: $y = \dfrac{3}{4}$.

37. See Figure 37. $y = -\dfrac{b}{2a} = -\dfrac{1}{2(3)} = -\dfrac{1}{6}$ and $x = 3\left(-\dfrac{1}{6}\right)^2 + \left(-\dfrac{1}{6}\right) = -\dfrac{1}{12}$. Vertex: $\left(-\dfrac{1}{12}, -\dfrac{1}{6}\right)$.

Axis of symmetry: $y = -\dfrac{1}{6}$.

39. See Figure 39. $y = -\dfrac{b}{2a} = -\dfrac{2}{2(1)} = -\dfrac{2}{2} = -1$ and $x = (-1)^2 + 2(-1) + 1 = 0$.

Vertex: $(0, -1)$. Axis of symmetry: $y = -1$

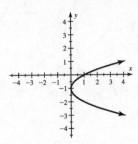

Figure 39

41. Since the parabola opens upward and the vertex is (0, 0), the equation has the form $y = a(x-0)^2 + 0$.

Since the parabola passes through (1, 1), $1 = a(1-0)^2 + 0 \Rightarrow a = 1$. The equation is $y = x^2$.

43. Since the parabola opens to the right and the vertex is (–2, –1), the equation has the form

$x = a(y+1)^2 - 2$. Since the parabola passes through (–1, 0), $-1 = a(0+1)^2 - 2 \Rightarrow a = 1$.

The equation is $x = (y+1)^2 - 2$.

45. By plotting the points by hand, we see that the parabola must open upward.

47. By plotting the points and axis by hand, we see that the parabola must open downward.

49. Since the parabola opens to the right and the vertex is (0, 0), the possible x-values are $x \geq 0$.

51. Since the parabola opens to the right and the vertex is to the left of the y-axis, the parabola has two
y-intercepts.

53. $x = 3(0)^2 - (0) + 1 \Rightarrow x = 1$

55. $(x-0)^2 + (y-0)^2 = 1^2 \Rightarrow x^2 + y^2 = 1$

57. $(x-(-1))^2 + (y-5)^2 = 3^2 \Rightarrow (x+1)^2 + (y-5)^2 = 9$

59. $(x-(-4))^2 + (y-(-6))^2 = (\sqrt{2})^2 \Rightarrow (x+4)^2 + (y+6)^2 = 2$

61. Since the center is (0, 0) and the radius is 4, the equation is $x^2 + y^2 = 16$.

63. Since the center is (–3, 2) and the radius is 1, the equation is $(x+3)^2 + (y-2)^2 = 1$.

65. The radius is 2 and the center is (0, 0). Solving the equation for y results in $y = \pm\sqrt{4 - x^2}$.

See Figure 65.

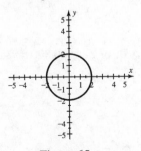

Figure 65

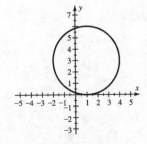

Figure 67

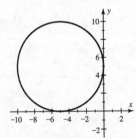

Figure 69

67. The radius is 3 and the center is (1, 3). Solving the equation for y results in $y = 3 \pm \sqrt{9 - (x-1)^2}$.

See Figure 67.

69. The radius is 5 and the center is (–5, 5). Solving the equation for y results in $y = 5 \pm \sqrt{25 - (x+5)^2}$.

See Figure 69.

71. $x^2 + 6x + y^2 - 2y = -1 \Rightarrow x^2 + 6x + 9 + y^2 - 2y + 1 = -1 + 9 + 1 \Rightarrow (x+3)^2 + (y-1)^2 = 9$

The radius is 3 and the center is (–3, 1). Solving the equation for y results in $y = 1 \pm \sqrt{9 - (x+3)^2}$.

See Figure 71.

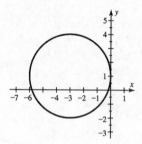
Figure 71

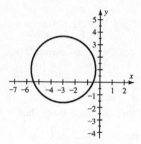
Figure 73

73. $x^2 + 6x + y^2 - 2y + 3 = 0 \Rightarrow x^2 + 6x + 9 + y^2 - 2y + 1 = -3 + 9 + 1 \Rightarrow (x+3)^2 + (y-1)^2 = 7$

The radius is $\sqrt{7}$ and the center is (–3, 1). Solving the equation for y results in $y = 1 \pm \sqrt{7 - (x+3)^2}$.

See Figure 73.

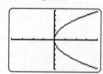
Figure 75

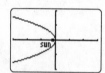
Figure 77

75. (a)   Graph $Y_1 = (32/11025)X^2$ using the DrawInv feature in [–40, 40, 10] by [–120, 120, 20].

   See Figure 75.

   (b)   When $y = 105$, $x = \dfrac{32}{11,025}(105)^2 = 32$ feet.

77. (a)   Plot (–0.1, 0) and graph $Y_1 = -2.5X^2$ using the DrawInv feature in [–1.5, 1.5, 0.5] by

   [–1, 1, 0.5]. See Figure 77.

   (b) $d = \sqrt{(-2.5 - (-0.1))^2 + (1-0)^2} = \sqrt{(-2.4)^2 + 1^2} = \sqrt{6.76} = 2.6$ A.U. or 241,800,000 miles.

Section 10.2 Ellipses and Hyperbolas

1. See Figure 1.

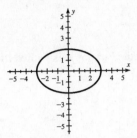

Figure 1

3. horizontal

5. 2

7. left and right

9. They are the diagonals extended.

11. a; Since the given equation is in the form $\dfrac{x^2}{b^2} + \dfrac{y^2}{a^2} = 1$ where the vertices are $(0, a)$ and $(0, -a)$.

13. The ellipse has a vertical major axis with vertices $(0, \pm 5)$ and minor axis endpoints $(\pm 3, 0)$.

   See Figure 13.

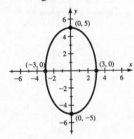

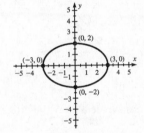
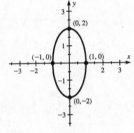

Figure 13   Figure 15   Figure 17

15. The ellipse has a horizontal major axis with vertices $(\pm 3, 0)$ and minor axis endpoints $(0, \pm 2)$.

   See Figure 15.

17. The ellipse has a vertical major axis with vertices $(0, \pm 2)$ and minor axis endpoints $(\pm 1, 0)$.

   See Figure 17.

19. The ellipse has a horizontal major axis with vertices $\left(\pm \sqrt{7}, 0\right)$ and minor axis endpoints $\left(0, \pm \sqrt{5}\right)$.

   See Figure 19.

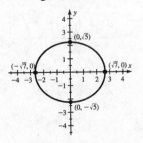
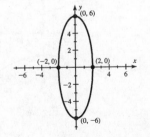
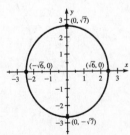

Figure 19   Figure 21   Figure 23

21. $36x^2 + 4y^2 = 144 \Rightarrow \dfrac{36x^2}{144} + \dfrac{4y^2}{144} = 1 \Rightarrow \dfrac{x^2}{4} + \dfrac{y^2}{36} = 1$ The ellipse has a vertical major axis with

    vertices $(0, \pm 6)$ and minor axis endpoints $(\pm 2, 0)$. See Figure 21.

23. $6y^2 + 7x^2 = 42 \Rightarrow \dfrac{6y^2}{42} + \dfrac{7x^2}{42} = 1 \Rightarrow \dfrac{y^2}{7} + \dfrac{x^2}{6} = 1$ The ellipse has a vertical major axis with vertices

    $\left(0, \pm\sqrt{7}\right)$ and minor axis endpoints $\left(\pm\sqrt{6}, 0\right)$. See Figure 23.

25. Horizontal major axis with vertices $(\pm 3, 0)$ and minor axis endpoints $(0, \pm 2)$ $\Rightarrow \dfrac{x^2}{9} + \dfrac{y^2}{4} = 1$

27. Vertical major axis with vertices $(0, \pm 5)$ and minor axis endpoints $(\pm 4, 0)$ $\Rightarrow \dfrac{y^2}{25} + \dfrac{x^2}{16} = 1$

29. The hyperbola has a horizontal transverse axis with vertices $(\pm 2, 0)$ and asymptotes $y = \pm\dfrac{3}{2}x$.

    See Figure 29.

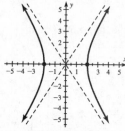

Figure 29

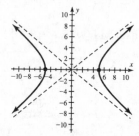

Figure 31

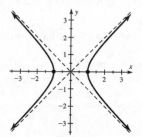

Figure 33

31. The hyperbola has a horizontal transverse axis with vertices $(\pm 5, 0)$ and asymptotes $y = \pm\dfrac{4}{5}x$.

    See Figure 31.

33. The hyperbola has a horizontal transverse axis with vertices $(\pm 1, 0)$ and asymptotes $y = \pm x$.

    See Figure 33.

35. The hyperbola has a horizontal transverse axis with vertices $\left(\pm\sqrt{3}, 0\right)$ and asymptotes $y = \pm\dfrac{2}{\sqrt{3}}x$.

    See Figure 35.

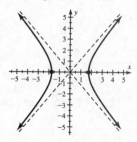

Figure 35

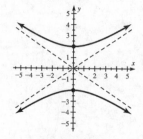

Figure 37

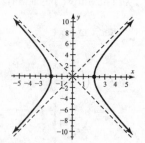

Figure 39

37. $9y^2 - 4x^2 = 36 \Rightarrow \dfrac{9y^2}{36} - \dfrac{4x^2}{36} = 1 \Rightarrow \dfrac{y^2}{4} - \dfrac{x^2}{9} = 1$ The hyperbola has a vertical transverse axis with

vertices $(0, \pm 2)$ and asymptotes $y = \pm\dfrac{2}{3}x$. See Figure 37.

39. $16x^2 - 4y^2 = 64 \Rightarrow \dfrac{16x^2}{64} - \dfrac{4y^2}{64} = 1 \Rightarrow \dfrac{x^2}{4} - \dfrac{y^2}{16} = 1$ The hyperbola has a horizontal transverse axis

with vertices $(\pm 2, 0)$ and asymptotes $y = \pm 2x$. See Figure 39.

41. Horizontal transverse axis with vertices $(\pm 1, 0)$ and asymptotes $y = \pm\dfrac{1}{1}x \Rightarrow x^2 - y^2 = 1$

43. Vertical transverse axis with vertices $(0, \pm 2)$ and asymptotes $y = \pm\dfrac{2}{3}x \Rightarrow \dfrac{y^2}{4} - \dfrac{x^2}{9} = 1$

45. (a) $A = \pi(5)(4) \approx 62.83;\ P = 2\pi\sqrt{\dfrac{5^2 + 4^2}{2}} \approx 28.45$

(b) $A = \pi\left(\sqrt{7}\right)\left(\sqrt{2}\right) \approx 11.75;\ P = 2\pi\sqrt{\dfrac{\left(\sqrt{7}\right)^2 + \left(\sqrt{2}\right)^2}{2}} \approx 13.33$

47. (a) $\dfrac{x^2}{39.44^2} + \dfrac{y^2}{38.20^2} = 1 \Rightarrow \dfrac{y^2}{38.20^2} = 1 - \dfrac{x^2}{39.44^2} \Rightarrow y = \pm\sqrt{38.20^2\left(1 - \dfrac{x^2}{39.44^2}\right)}$

Graph $Y_1 = \sqrt{\left(38.20^2\left(1 - \left(X^2/39.44^2\right)\right)\right)}$ and $Y_2 = -\sqrt{\left(38.20^2\left(1 - \left(X^2/39.44^2\right)\right)\right)}$ and plot the

point $(9.81, 0)$ in $[-60, 60, 10]$ by $[-40, 40, 10]$. See Figure 47.

(b) $P = 2\pi\sqrt{\dfrac{39.44^2 + 38.20^2}{2}} \approx 243.9$ A.U. or about 2.27×10^{10} miles

$A = \pi(39.44)(38.20) \approx 4733$ A.U. or about 4.09×10^{19} square miles

$[-60, 60, 10]$ by $[-40, 40, 10]$ \qquad $[-21, 21, 5]$ by $[-14, 14, 5]$

Figure 47

49. The maximum and minimum heights occur when $y = 0$. These values are calculated below.

Earth: $\dfrac{(x - 164)^2}{3960^2} + 0 = 1 \Rightarrow (x - 164)^2 = 3960^2 \Rightarrow x - 164 = \pm 3960 \Rightarrow x = 4124$ or $x = -3796$.

Explorer VII: $\dfrac{x^2}{4464^2} + 0 = 1 \Rightarrow x^2 = 4464^2 \Rightarrow x = \pm 4464 \Rightarrow x = 4464$ or $x = -4464$.

The maximum height is $-3796 - (-4464) = 668$ miles. The minimum height is

$4464 - 4124 = 340$ miles.

51. The height is half the length of the minor axis and the width is the full length of the major axis.

$$400x^2 + 10,000y^2 = 4,000,000 \Rightarrow \frac{400x^2}{4,000,000} + \frac{10,000y^2}{4,000,000} = 1 \Rightarrow \frac{x^2}{10,000} + \frac{y^2}{400} = 1$$

The height is $\sqrt{400} = 20$ feet and the width is $2\sqrt{10,000} = 2(100) = 200$ feet.

53. Example: Satellite and Planet orbits, *Answers may vary.*

Checking Basic Concepts Sections 10.1 and 10.2

1. See Figure 1. Vertex: $(1, 2)$. Axis of symmetry: $y = 2$.

2. The equation is $(x-1)^2 + (y+2)^2 = 4$.

$$(x-1)^2 + (y+2)^2 = 4 \Rightarrow (y+2)^2 = 4 - (x-1)^2 \Rightarrow y = -2 \pm \sqrt{4 - (x-1)^2}.\quad \text{See Figure 2}$$

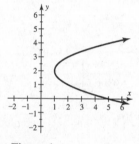

Figure 1   Figure 2

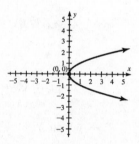

Figure 4a

3. x-intercepts: $\dfrac{x^2}{4} + \dfrac{0^2}{9} = 1 \Rightarrow \dfrac{x^2}{4} = 1 \Rightarrow x^2 = 4 \Rightarrow x = \pm 2$

   y-intercepts: $\dfrac{0^2}{4} + \dfrac{y^2}{9} = 1 \Rightarrow \dfrac{y^2}{9} = 1 \Rightarrow y^2 = 9 \Rightarrow y = \pm 3$

4. (a) This parabola has vertex $(0, 0)$ and axis of symmetry $y = 0$. See Figure 4a.

   (b) This ellipse has a vertical major axis with vertices $(0, \pm 5)$ and minor axis endpoints $(\pm 4, 0)$.

        See Figure 4b.

   (c) This hyperbola has a horizontal transverse axis with vertices $(\pm 2, 0)$ and asymptotes

        $y = \pm \dfrac{3}{2}x$. See Figure 4c.

   (d) This circle is centered at $(1, -2)$ and has radius 3. See Figure 4d.

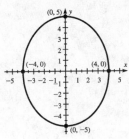

Figure 4b

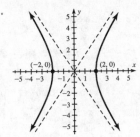

Figure 4c

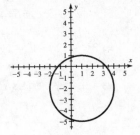

Figure 4d

Section 10.3 Nonlinear Systems of Equations and Inequalities

1. Any number.

3. Two, the line intersects the circle twice.

5. No. $5(-2)^2 - 2(-1)^2 = 5(4) - 2(1) = 18 \not> 18$

7. See Figure 7. *Answers may vary.*

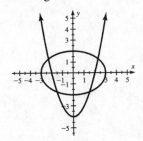

Figure 7

9. The solutions are the intersection points (1, 3) and (–1, –3). Both solutions check.

11. The solutions are the intersection points (0, –1) and (0, 1). Both solutions check.

13. Substitute $y = 2x$ into the second equation.

$$x^2 + (2x)^2 = 45 \Rightarrow x^2 + 4x^2 = 45 \Rightarrow 5x^2 = 45 \Rightarrow x^2 = 9 \Rightarrow x = \pm 3$$

When $x = -3$, $y = 2(-3) = -6$. When $x = 3$, $y = 2(3) = 6$. The solutions are (–3, –6) and (3, 6).

15. From the first equation, $y = 1 - x$. Substitute $y = 1 - x$ into the second equation.

$$x^2 - (1-x)^2 = 3 \Rightarrow x^2 - 1 + 2x - x^2 = 3 \Rightarrow 2x - 1 = 3 \Rightarrow 2x = 4 \Rightarrow x = 2$$

When $x = 2$, $y = 1 - (2) = -1$ The solution is (2, –1).

17. From the first equation, $x^2 = y$. Substitute $x^2 = y$ into the second equation.

$$y + y^2 = 6 \Rightarrow y^2 + y - 6 = 0 \Rightarrow (y+3)(y-2) = 0 \Rightarrow \text{Either } y = -3 \text{ or } y = 2$$

When $y = -3$, $x = \pm\sqrt{-3}$, (not real numbers). When $y = 2$, $x = \pm\sqrt{2}$.

The solutions are $\left(-\sqrt{2}, 2\right)$ and $\left(\sqrt{2}, 2\right)$.

19. Solve the second equation for $x : x = y - 2$. Substitute $x = y - 2$ into the first equation:

$$3(y-2)^2 + 2y^2 = 5 \Rightarrow 3(y^2 - 4y + 4) + 2y^2 = 5$$

$$\Rightarrow 3y^2 - 12y + 12 + 2y^2 = 5 \Rightarrow 5y^2 - 12y + 7 = 0 \Rightarrow (5y - 7)(y - 1) = 0$$

$$\Rightarrow y = \frac{7}{5} \text{ or } y = 1 \Rightarrow x = -\frac{3}{5} \text{ or } x = -1. \text{ The solutions are } \left(-\frac{3}{5}, \frac{7}{5}\right) \text{ and } (-1, 1).$$

21. Multiply the second equation by -1 and add to the first equation. $\begin{array}{r} x^2 + y^2 = 4 \\ -x^2 + 9y^2 = -9 \\ \hline 10y^2 = -5 \end{array}$

$y^2 = -\frac{1}{2} \Rightarrow y = \pm\sqrt{-\frac{1}{2}}$, which is not a real number. There are no real solutions.

23. Add the equations. $\begin{array}{r} x^2 + y^2 = 10 \\ 2x^2 - y^2 = 17 \\ \hline 3x^2 \quad\quad = 27 \end{array} \Rightarrow x^2 = 9 \Rightarrow x = \pm 3$

$(3)^2 + y^2 = 10 \Rightarrow y^2 = 1 \Rightarrow y = \pm 1$

$(-3)^2 + y^2 = 10 \Rightarrow y^2 = 1 \Rightarrow y = \pm 1$ The solutions are $(\pm 3, \pm 1)$

25. Solve the second equation for y. $2x^2 - y = 1 - 3x \Rightarrow y = 2x^2 + 3x - 1$

Graph $Y_1 = X^2 - 3$ and $Y_2 = 2X^2 + 3X - 1$ in $[-5, 5, 1]$ by $[-5, 5, 1]$. See Figures 25a & 25b.

The solutions are the intersection points $(-2, 1)$ and $(-1, -2)$.

$[-5, 5, 1]$ by $[-5, 5, 1]$ $[-5, 5, 1]$ by $[-5, 5, 1]$

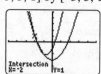   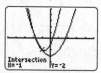

       Figure 25a   Figure 25b

27. Solve both equations for y. $y - x = -4 \Rightarrow y = x - 4$ and $x - y^2 = -2 \Rightarrow y^2 = x + 2 \Rightarrow y = \pm\sqrt{x + 2}$

Graph $Y_1 = X - 4$, $Y_2 = \sqrt{(X + 2)}$ and $Y_3 = -\sqrt{(X + 2)}$ in $[-9.4, 9.4, 1]$ by $[-6.2, 6.2, 1]$.

See Figures 27a & 27b. The solutions are the intersection points $(7, 3)$ and $(2, -2)$.

$[-9.4, 9.4, 1]$ by $[-6.2, 6.2, 1]$ $[-9.4, 9.4, 1]$ by $[-6.2, 6.2, 1]$

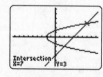   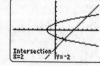

      Figure 27a   Figure 27b

29. (a)   Substitute $y = -2x$ into the second equation.

      $x^2 + (-2x) = 3 \Rightarrow x^2 - 2x - 3 = 0 \Rightarrow (x + 1)(x - 3) = 0 \Rightarrow$ Either $x = -1$ or $x = 3$

When $x = -1$, $y = -2(-1) = 2$. When $x = 3$, $y = -2(3) = -6$.

The solutions are $(-1, 2)$ and $(3, -6)$.

(b)   Solve the second equation for y. $x^2 + y = 3 \Rightarrow y = 3 - x^2$ Graph $Y_1 = -2X$ and $Y_2 = 3 - X^2$ in

      $[-10, 10, 1]$ by $[-10, 10, 1]$. See Figures 29a & 29b.

(c)   Table $Y_1 = -2X$ and $Y_2 = 3 - X^2$ with TblStart $= -3$ and ΔTbl $= 1$. See Figure 29c.

$[-10, 10, 1]$ by $[-10, 10, 1]$ $[-10, 10, 1]$ by $[-10, 10, 1]$

      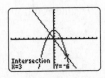   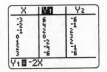

     Figure 29a   Figure 29b   Figure 29c

31. (a)   From the second equation, $y = x$. Substitute $y = x$ into the first equation.

      $x \cdot x = 1 \Rightarrow x^2 = 1 \Rightarrow x = \pm 1$ When $x = -1$, $y = -1$. When $x = 1$, $y = 1$.

      The solutions are $(-1, -1)$ and $(1, 1)$.

(b)   Solve both equations for y. $xy = 1 \Rightarrow y = \dfrac{1}{x}$ and $x - y = 0 \Rightarrow y = x$

      Graph $Y_1 = 1/X$ and $Y_2 = X$ in $[-4.7, 4.7, 1]$ by $[-3.1, 3.1, 1]$. See Figures 31a & 31b.

(c)   Table $Y_1 = 1/X$ and $Y_2 = X$ with TblStart $= -3$ and ΔTbl $= 1$. See Figure 31c.

$[-4.7, 4.7, 1]$ by $[-3.1, 3.1, 1]$ $[-4.7, 4.7, 1]$ by $[-3.1, 3.1, 1]$

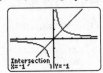   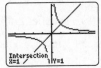   

     Figure 31a   Figure 31b   Figure 31c

33. Sketch the parabola given by $y = x^2$ using a solid line. Try test point $(0, 2)$ to determine shading.

     Since $2 \geq 0^2$, we shade the portion of the xy-plane containing the point $(0, 2)$. See Figure 33.

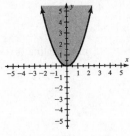   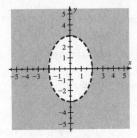   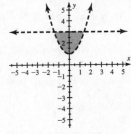

     Figure 33   Figure 35   Figure 37

35. Sketch the ellipse given by $\dfrac{x^2}{4} + \dfrac{y^2}{9} = 1$ using a dashed line. Try test point $(4, 0)$ to determine

     shading. Since $\dfrac{4^2}{4} + \dfrac{0^2}{9} = 4 > 1$, we shade the portion of the xy-plane containing the point $(4, 0)$.

     See Figure 35.

37. Sketch the parabola given by $y = x^2 + 1$ using a dashed line and the line given by $y = 3$ using a dashed line. Since the point $(0, 2)$ satisfies both inequalities, we shade the portion of the xy-plane containing the point $(0, 2)$. See Figure 37. One solution is $(0, 2)$. *Answers may vary.*

39. Sketch the circle given by $x^2 + y^2 = 1$ using a solid line and the line given by $y = x$ using a dashed line. Since the point $\left(\dfrac{1}{2}, -\dfrac{1}{2}\right)$ satisfies both inequalities, we shade the portion of the xy-plane containing the point $\left(\dfrac{1}{2}, -\dfrac{1}{2}\right)$. See Figure 39. One solution is $(0.5, -0.5)$. *Answers may vary.*

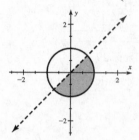

Figure 39

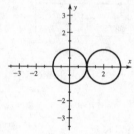

Figure 41

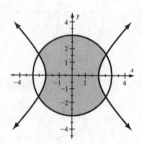

Figure 43

41. Sketch the circle given by $x^2 + y^2 = 1$ using a solid line and the circle given by $(x-2)^2 + y^2 = 1$ using a solid line. Since the point $(1, 0)$ is the only point that satisfies both inequalities, we mark only the point $(1, 0)$ on the graph. See Figure 41. The only solution is $(1, 0)$.

43. Sketch the hyperbola given by $x^2 - y^2 = 4$ using a solid line and the circle given by $x^2 + y^2 = 9$ using a solid line. Since the point $(0, 0)$ satisfies both inequalities, we shade the portion of the xy plane containing the point $(0, 0)$. See Figure 43. One solution is $(0, 0)$. *Answers may vary.*

45. a

47. A parabola with vertex $(0, 0)$ has an equation of the form $y = a(x-0)^2 + 0$ or $y = ax^2$. Since the parabola also passes through the point $(1, 1)$, $1 = a \cdot 1^2 \Rightarrow a = 1$. The equation of the parabola is $y = x^2$. Since the line passes through $(0, 4)$ and $(4, 0)$, its slope is $m = \dfrac{0-4}{4-0} = \dfrac{-4}{4} = -1$.

The y-intercept is $(0, 4)$. The equation of the line is $y = -x + 4$ or $y = 4 - x$. Since the line is dashed and the parabola is solid, the system of inequalities is $y \geq x^2$ and $y < 4 - x$.

49. See Figure 49.

[–10, 10, 1] by [–10, 10, 1]

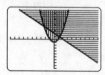

Figure 49

[0, 5, 1] by [0, 10, 1]

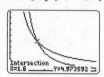

Figure 51

51. The necessary equations are $\pi r^2 h = 40$ and $2\pi r h = 50$. Start by solving each of these equations

    for h. $\pi r^2 h = 40 \Rightarrow h = \dfrac{40}{\pi r^2}$ and $2\pi r h = 50 \Rightarrow h = \dfrac{50}{2\pi r}$

    (a) Graph $Y_1 = 40/(\pi X^2)$ and $Y_2 = 50/(2\pi X)$ in $[0, 5, 1]$ by $[0, 10, 1]$. See Figure 51.

    The graphs intersect near the point $(1.6, 4.97)$. The dimensions are

    $r = 1.6$ inches and $h \approx 4.97$ inches.

    (b) $\dfrac{40}{\pi r^2} = \dfrac{50}{2\pi r} \Rightarrow 80\pi r = 50\pi r^2 \Rightarrow 8r = 5r^2 \Rightarrow 5r^2 - 8r = 0 \Rightarrow r(5r - 8) = 0 \Rightarrow r = 0 \text{ or } r = \dfrac{8}{5}$

    Since $r = 0$ has no physical meaning, $r = \dfrac{8}{5} = 1.6$. And so $h = \dfrac{50}{2\pi\left(\frac{8}{5}\right)} = \dfrac{125}{8\pi} \approx 4.97$.

    The dimensions are $r = 1.6$ inches and $h \approx 4.97$ inches.

53. (a) $xy = 143$ and $2x + 2y = 48$

    (b) Graph $Y_1 = 143/X$ and $Y_2 = 24 - X$ in $[0, 20, 4]$ by $[0, 20, 4]$. See Figure 53.

    The dimensions of the room are $x = 11$ ft and $y = 13$ ft.

$[0, 20, 4]$ by $[0, 20, 4]$

Figure 53

Checking Basic Concepts Section 10.3

1. Symbolically: From the second equation $y = 2x - 3$. Substitute $y = 2x - 3$ into the first equation.

    $x^2 - (2x - 3) = 2x \Rightarrow x^2 - 4x + 3 = 0 \Rightarrow (x - 1)(x - 3) = 0 \Rightarrow x = 1 \text{ or } x = 3$

    When $x = 1$, $y = 2(1) - 3 = -1$. When $x = 3$, $y = 2(3) - 3 = 3$. The solutions are $(1, -1)$ and $(3, 3)$.

    Graphically: Solve each equation for y. $x^2 - y = 2x \Rightarrow y = x^2 - 2x$ and $2x - y = 3 \Rightarrow y = 2x - 3$

    Graph $Y_1 = X^2 - 2X$ and $Y_2 = 2X - 3$ in $[-5, 5, 1]$ by $[-5, 5, 1]$. See Figures 1a & 1b.

$[-5, 5, 1]$ by $[-5, 5, 1]$ $[-5, 5, 1]$ by $[-5, 5, 1]$

          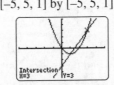

Figure 1a   Figure 1b

2. Since graphing these equations would result in a parabola intersected twice by a line, there are two

    solutions.

3. (a)   The point (0, 3) is in the shaded region and is a solution. The point (4, 4) is not in the shaded
         region and is not a solution. *Answers may vary.*

   (b)   A parabola with vertex (0, 4) has an equation of the form

         $y = a(x-0)^2 + 4$ or $y = ax^2 + 4$. Since the parabola also passes through the point (2, 0),

         $0 = a \cdot 2^2 + 4 \Rightarrow 4a = -4 \Rightarrow a = -1$. The equation of the parabola is

         $y = -x^2 + 4$ or $y = 4 - x^2$.

         Since the line passes through (0, 2) and (2, 0), its slope is $m = \dfrac{0-2}{2-0} = \dfrac{-2}{2} = -1$. The y-

         intercept is (0, 2). The equation of the line is $y = -x + 2$ or $y = 2 - x$. Since both the line and

         the parabola are solid, the system of inequalities is $y \le 4 - x^2$ and $y \ge 2 - x$.

4. Sketch the circle given by $x^2 + y^2 = 4$ using a solid line and the line given by $y = 1$ using a dashed
   line. The point (0, 0) satisfies both inequalities. Shade the portion of the *xy*-plane containing
   (0, 0). See Figure 4.

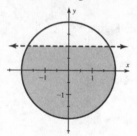

   Figure 4

Chapter 10 Review

1. Since $x = 2(y+0)^2 + 0$, the vertex is (0, 0) and the axis of symmetry is $y = 0$. See Figure 1.

2. Since $x = -(y+1)^2 + 0$, the vertex is (0, −1) and the axis of symmetry is $y = -1$. See Figure 2.

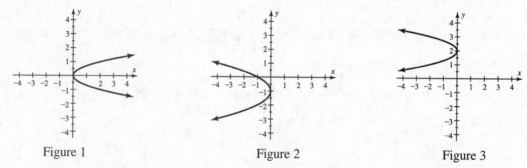

Figure 1   Figure 2   Figure 3

3. Since $x = -2(y-2)^2 + 0$, the vertex is (0, 2) and the axis of symmetry is $y = 2$. See Figure 3.

4. Since $x = (y+2)^2 - 1$, the vertex is (−1, −2) and the axis of symmetry is $y = -2$. See Figure 4.

5. Since $x = -3(y-0)^2 + 1$, the vertex is (1, 0) and the axis of symmetry is $y = 0$. See Figure 5.

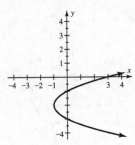
Figure 4

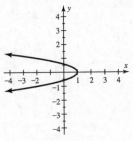
Figure 5

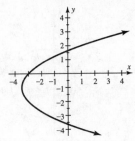
Figure 6

6. Since $x = \frac{1}{2}(y+1)^2 - \frac{7}{2}$, the vertex is $\left(-\frac{7}{2}, -1\right)$ and the axis of symmetry is $y = -1$. See Figure 6.

7. Since the vertex is (0, 0), the parabola has an equation of the form $x = a(x-0)^2 + 0$ or $x = ay^2$.

   Since the parabola passes through the point (1, 1), $1 = a(1) \Rightarrow a = 1$. The equation is $x = y^2$.

8. This is a circle of radius 4 centered at (–2, 2). The equation is $(x+2)^2 + (y-2)^2 = 16$.

9. $(x-0)^2 + (y-0)^2 = 1^2 \Rightarrow x^2 + y^2 = 1$

10. $(x-2)^2 + (y-(-3))^2 = 4^2 \Rightarrow (x-2)^2 + (y+3)^2 = 16$

11. The radius is 5 and the center is (0, 0). Solving the equation for y results in $y = \pm\sqrt{25 - x^2}$.

    See Figure 11.

12. The radius is 3 and the center is (2, 0). Solving the equation for y results in $y = \pm\sqrt{9 - (x-2)^2}$.

    See Figure 12.

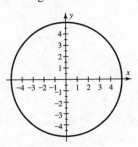
Figure 11

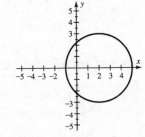
Figure 12

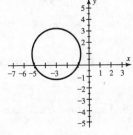
Figure 13

13. The radius is $\sqrt{5}$ and the center is (–3, 1). Solving the equation for y results in $y = 1 \pm \sqrt{5 - (x+3)^2}$.

    See Figure 13.

14. $x^2 - 2x + y^2 + 2y = 7 \Rightarrow x^2 - 2x + 1 + y^2 + 2y + 1 = 7 + 1 + 1 \Rightarrow (x-1)^2 + (y+1)^2 = 9$ The radius is 3

    and the center is (1, –1). Solving the equation for y results in $y = -1 \pm \sqrt{9 - (x-1)^2}$. See Figure 14.

15. The ellipse has a vertical major axis with vertices $(0, \pm 5)$ and minor axis endpoints $(\pm 2, 0)$.

    See Figure 15.

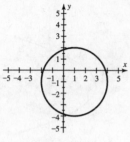

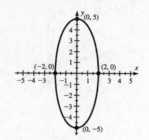

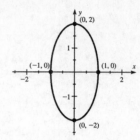

Figure 14   Figure 15   Figure 16

16. The ellipse has a vertical major axis with vertices $(0, \pm 2)$ and minor axis endpoints $(\pm 1, 0)$.

   See Figure 16.

17. $25x^2 + 20y^2 = 500 \Rightarrow \dfrac{25x^2}{500} + \dfrac{20y^2}{500} = 1 \Rightarrow \dfrac{x^2}{20} + \dfrac{y^2}{25} = 1$ The ellipse has a vertical major axis with

   vertices $(0, \pm 5)$ and minor axis endpoints $\left(\pm \sqrt{20}, 0 \right)$. See Figure 17.

18. $4x^2 + 9y^2 = 36 \Rightarrow \dfrac{4x^2}{36} + \dfrac{9y^2}{36} = 1 \Rightarrow \dfrac{x^2}{9} + \dfrac{y^2}{4} = 1$ The ellipse has a horizontal major axis with

   vertices $(\pm 3, 0)$ and minor axis endpoints $(0, \pm 2)$. See Figure 18.

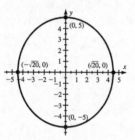

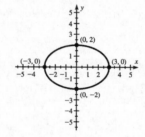

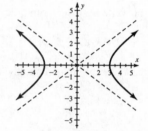

Figure 17   Figure 18   Figure 21

19. Vertical major axis with vertices $(0, \pm 4)$ and minor axis endpoints $(\pm 2, 0)$ $\Rightarrow \dfrac{y^2}{16} + \dfrac{x^2}{4} = 1$

20. Horizontal transverse axis with vertices $(\pm 1, 0)$ and asymptotes $y = \pm 2x$ $\Rightarrow x^2 - \dfrac{y^2}{4} = 1$

21. The hyperbola has a horizontal transverse axis with vertices $(\pm 3, 0)$ and asymptotes $y = \pm \dfrac{2}{3} x.$

   See Figure 21.

22. The hyperbola has a vertical transverse axis with vertices $(0, \pm 5)$ and asymptotes $y = \pm \dfrac{5}{4} x.$

   See Figure 22.

23. The hyperbola has a vertical transverse axis with vertices $(0, \pm 1)$ and asymptotes $y = \pm x.$

   See Figure 23.

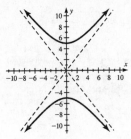
Figure 22

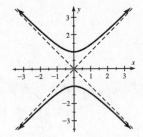
Figure 23

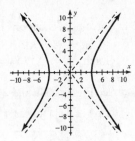
Figure 24

24. $25x^2 - 16y^2 = 400 \Rightarrow \dfrac{25x^2}{400} - \dfrac{16y^2}{400} = 1 \Rightarrow \dfrac{x^2}{16} - \dfrac{y^2}{25} = 1$ The hyperbola has a horizontal transverse

axis with vertices $(\pm 4, 0)$ and asymptotes $y = \pm\dfrac{5}{4}x$. See Figure 24.

25. The solutions are the intersection points $(0, 3)$ and $(3, 0)$. Both solutions check.

26. The solutions are the intersection points $(-1, -2)$ and $(1, 2)$. Both solutions check.

27. The solutions are the intersection points $(0, 0)$ and $(2, 2)$. Both solutions check.

28. The solutions are the intersection points $(-2, -1)$, $(-2, 1)$, $(2, -1)$ and $(2, 1)$. All four solutions check.

29. Substitute $y = x$ into the second equation. $x^2 + (x)^2 = 32 \Rightarrow 2x^2 = 32 \Rightarrow x^2 = 16 \Rightarrow x = \pm 4$

When $x = -4$, $y = -4$. When $x = 4$, $y = 4$. The solutions are $(-4, -4)$ and $(4, 4)$.

30. From the first equation, $y = x - 4$. Substitute $y = x - 4$ into the second equation.

$x^2 + (x - 4)^2 = 16 \Rightarrow x^2 + x^2 - 8x + 16 = 16 \Rightarrow 2x^2 - 8x = 0 \Rightarrow 2x(x - 4) = 0 \Rightarrow x = 0 \text{ or } x = 4$

When $x = 0$, $y = (0) - 4 = -4$. When $x = 4$, $y = (4) - 4 = 0$. The solutions are $(0, -4)$ and $(4, 0)$.

31. Substitute $y = x^2$ into the second equation. $2x^2 + (x^2) = 3 \Rightarrow 3x^2 = 3 \Rightarrow x^2 = 1 \Rightarrow x = \pm 1$

When $x = -1$, $y = (-1)^2 = 1$. When $x = 1$, $y = (1)^2 = 1$. The solutions are $(-1, 1)$ and $(1, 1)$.

32. Substitute $y = x^2 + 1$ into the second equation.

$2x^2 - (x^2 + 1) = 3x - 3 \Rightarrow x^2 - 3x + 2 = 0 \Rightarrow (x - 1)(x - 2) = 0 \Rightarrow x = 1 \text{ or } x = 2$

When $x = 1$, $y = (1)^2 + 1 = 2$. When $x = 2$, $y = (2)^2 + 1 = 5$. The solutions are $(1, 2)$ and $(2, 5)$.

33. Solve both equations for y. $2x - y = 4 \Rightarrow y = 2x - 4$ and $x^2 + y = 4 \Rightarrow y = 4 - x^2$

Graph $Y_1 = 2X - 4$ and $Y_2 = 4 - X^2$ in $[-8, 8, 1]$ by $[-25, 10, 5]$. See Figures 33a & 33b.

The solutions are the intersection points $(-4, -12)$ and $(2, 0)$.

$[-8, 8, 1]$ by $[-25, 10, 5]$ $[-8, 8, 1]$ by $[-25, 10, 5]$

Figure 33a

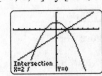
Figure 33b

34. Solve both equations for y. $x^2 + y = 0 \Rightarrow y = -x^2$ and $x^2 + y^2 = 2 \Rightarrow y = \pm\sqrt{2-x^2}$

    Graph $Y_1 = -X^2$, $Y_2 = \sqrt{(2-X^2)}$ and $Y_3 = -\sqrt{(2-X^2)}$ in $[-4.7, 4.7, 1]$ by $[-4.1, 2.1, 1]$.

    See Figures 34a and 34b. The solutions are the intersection points $(-1, -1)$ and $(1, -1)$.

    $[-4.7, 4.7, 1]$ by $[-4.1, 2.1, 1]$ $[-4.7, 4.7, 1]$ by $[-4.1, 2.1, 1]$

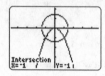

    Figure 34a   Figure 34b

35. (a)   Substitute $y = x$ into the second equation.

    $$x^2 + 2(x) = 8 \Rightarrow x^2 + 2x - 8 = 0 \Rightarrow (x+4)(x-2) = 0 \Rightarrow x = -4 \text{ or } x = 2$$

    When $x = -4$, $y = -4$. When $x = 2$, $y = 2$. The solutions are $(-4, -4)$ and $(2, 2)$.

    (b)   Solve the second equation for y. $x^2 + 2y = 8 \Rightarrow y = 4 - \dfrac{x^2}{2}$

    Graph $Y_1 = X$ and $Y_2 = 4 - (X^2/2)$ in $[-10, 10, 1]$ by $[-10, 10, 1]$. See Figures 35a & 35b.

    (c)   Table $Y_1 = X$ and $Y_2 = 4 - (X^2/2)$ with TblStart $= -2$ and ΔTbl $= 1$. See Figure 35c.

    $[-10, 10, 1]$ by $[-10, 10, 1]$ $[-10, 10, 1]$ by $[-10, 10, 1]$

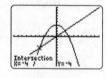  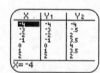

    Figure 35a   Figure 35b   Figure 35c

36. (a)   Substitute $y = x^3$ into the second equation. $x^2 - x^3 = 0 \Rightarrow x^2(1-x) = 0 \Rightarrow x = 0$ or $x = 1$

    When $x = 0$, $y = 0^3 = 0$. When $x = 1$, $y = 1^3 = 1$. The solutions are $(0, 0)$ and $(1, 1)$.

    (b)   Solve the second equation for y. $x^2 - y = 0 \Rightarrow y = x^2$

    Graph $Y_1 = X \wedge 3$ and $Y_2 = X^2$ in $[-1.5, 1.5, 1]$ by $[-1.5, 1.5, 1]$. See Figures 36a & 36b.

    (c)   Table $Y_1 = X \wedge 3$ and $Y_2 = X^2$ with TblStart $= -1$ and ΔTbl $= 0.5$. See Figure 36c.

    $[-1.5, 1.5, 1]$ by $[-1.5, 1.5, 1]$ $[-1.5, 1.5, 1]$ by $[-1.5, 1.5, 1]$

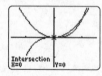 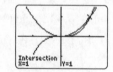 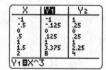

    Figure 36a   Figure 36b   Figure 36c

37. Sketch the parabola given by $y = 2x^2$ using a solid line. Try test point $(0, 2)$ to determine shading.

Since $2 \geq 2\left(0^2\right)$, we shade the portion of the xy-plane containing the point (0, 2). See Figure 37.

38. Sketch the line given by $y = 2x - 3$ using a dashed line. Try test point (3, 0) to determine shading.

Since $0 < 2(3) - 3$, we shade the portion of the xy-plane containing the point (3, 0). See Figure 38.

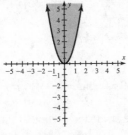
Figure 37

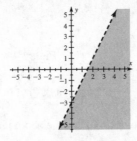
Figure 38

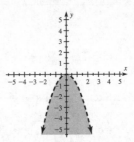
Figure 39

39. Sketch the parabola given by $y = -x^2$ using a dashed line. Try test point (0, –1) to determine

shading. Since $-1 < -(0)^2$, we shade the portion of the xy-plane containing the point (0, –1).

See Figure 39.

40. Sketch the ellipse given by $\dfrac{x^2}{9} + \dfrac{y^2}{16} = 1$ using a solid line. Try test point (0, 0) to determine shading.

Since $0 + 0 \leq 1$, we shade the portion of the xy-plane containing the point (0, 0). See Figure 40.

41. Sketch the parabola given by $y = x^2 + 1$ using a solid line and the line given by $y = 2$ using a solid

line. Since the point (0, 1.5) satisfies both inequalities, we shade the portion of the xy-plane

containing the point (0, 1.5). See Figure 41.

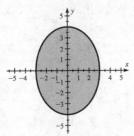
Figure 40

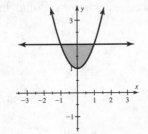
Figure 41

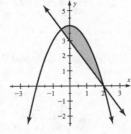
Figure 42

42. Sketch the parabola given by $y = 4 - x^2$ using a solid line and the line given by $3x + 2y = 6$ using a

solid line. Since the point (1, 2) satisfies both inequalities, we shade the portion of the xy-plane

containing the point (1, 2). See Figure 42.

43. Sketch the parabola given by $y = x^2$ using a dashed line and the parabola given by $y = 4 - x^2$ using

a dashed line. Since the point (0, 2) satisfies both inequalities, we shade the portion of the xy-plane

containing the point (0, 2). See Figure 43.

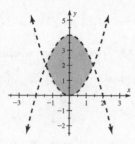

Figure 43

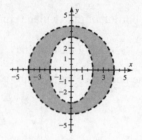

Figure 44

44. Sketch the ellipse given by $\dfrac{x^2}{4} + \dfrac{y^2}{9} = 1$ using a dashed line and the circle given by $x^2 + y^2 = 16$

    using a dashed line. Since the point $(3, 0)$ satisfies both inequalities, we shade the portion of the xy

    plane containing the point $(3, 0)$. See Figure 44.

45. A parabola with vertex $(0, -2)$ has an equation of the form $y = a(x-0)^2 - 2$ or $y = ax^2 - 2$. Since

    the parabola also passes through the point $(1, -1)$, $-1 = a \cdot 1^2 - 2 \Rightarrow a = 1$. The equation of the

    parabola is $y = x^2 - 2$. Since the line passes through $(0, 2)$ and $(2, 0)$, its slope is

    $m = \dfrac{0-2}{2-0} = \dfrac{-2}{2} = -1$. The y-intercept is $(0, 2)$. The equation of the line is $y = -x + 2$ or $y = 2 - x$.

    Since both the line and the parabola are solid, the system of inequalities is $y \ge x^2 - 2$ and $y \le 2 - x$.

46. A circle of radius 2 centered at $(0, 0)$ has the equation $x^2 + y^2 = 4$. Since the line passes through the

    points $(0, 0)$ and $(1, 1)$, its slope is $m = \dfrac{1-0}{1-0} = \dfrac{1}{1} = 1$. The y-intercept is $(0, 0)$. The equation of the

    line is $y = x$. Since both the circle and the line are solid, the system of inequalities is

    $y \ge x$ and $x^2 + y^2 \le 4$.

47. (a) $xy = 1000$ and $2x + 2y = 130$

    (b)   Solve each equation for y. $xy = 1000 \Rightarrow y = \dfrac{1000}{x}$ and $2x + 2y = 130 \Rightarrow y = 65 - x$

          Graph $Y_1 = 1000/X$ and $Y_2 = 65 - X$ in $[0, 50, 10]$ by $[0, 50, 10]$. See Figure 47.

          The table has dimensions $x = 25$ inches and $y = 40$ inches.

    (c)   From the second equation, $y = 65 - x$. Substitute $y = 65 - x$ into the first equation.

          $x(65 - x) = 1000 \Rightarrow x^2 - 65x + 1000 = 0 \Rightarrow (x - 25)(x - 40) = 0 \Rightarrow x = 25$ or $x = 40$

          When $x = 25$, $y = 65 - 25 = 40$. When $x = 40$, $y = 65 - 40 = 25$.

          These answers are equivalent.

[0, 50, 10] by [0, 50, 10]   [0, 16, 4] by [0, 16, 4]   [0, 5, 1] by [0, 50, 10]

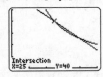   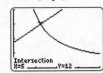

Figure 47   Figure 48   Figure 49

48. (a) $xy = 60$ and $y - x = 7$

(b) Solve each equation for y. $xy = 60 \Rightarrow y = \dfrac{60}{x}$ and $y - x = 7 \Rightarrow y = x + 7$

Graph $Y_1 = 60/X$ and $Y_2 = X + 7$ in [0, 16, 4] by [0, 16, 4]. See Figure 48.

The numbers are $x = 5$ and $y = 12$.

(c) From the second equation, $y = x + 7$. Substitute $y = x + 7$ into the first equation.

$x(x+7) = 60 \Rightarrow x^2 + 7x - 60 = 0 \Rightarrow (x+12)(x-5) = 0 \Rightarrow x = -12$ or $x = 5$

The only positive value is . $x = 5$. When $x = 5$, $y = 5 + 7 = 12$.

49. Solve each equation for h. $V = \pi r^2 h \Rightarrow h = \dfrac{V}{\pi r^2}$ and $A = 2\pi rh \Rightarrow h = \dfrac{A}{2\pi r}$

Graph $Y_1 = 50/\!\left(\pi X^2\right)$ and $Y_2 = 100/(2\pi X)$ in [0, 5, 1] by [0, 50, 10]. See Figure 49.

The unique answer is $r = 1$ foot and $h \approx 15.92$ feet.

50. Solve each equation for h. $V = \pi r^2 h \Rightarrow h = \dfrac{V}{\pi r^2}$ and $A = 2\pi rh + 2\pi r^2 \Rightarrow h = \dfrac{A - 2\pi r^2}{2\pi r}$

Graph $Y_1 = 35/\!\left(\pi X^2\right)$ and $Y_2 = \left(80 - 2\pi X^2\right)/(2\pi X)$ in [0, 4, 1] by [−5, 25, 5].

See Figures 50a & 50b. Two solutions are possible. The answer is not unique.

Either $r \approx 0.94$ inches and $h \approx 12.60$ inches or $r \approx 3.00$ inches and $h \approx 1.23$ inches.

[0, 4, 1] by [−5, 25, 5]   [0, 4, 1] by [−5, 25, 5]   [−7.5, 7.5, 1] by [−5, 5, 1]   [−3, 3, 1] by [−2, 2, 1]

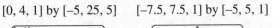   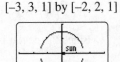

Figure 50a   Figure 50b   Figure 51   Figure 52

51. (a) $\dfrac{x^2}{5} + \dfrac{y^2}{12} = 1 \Rightarrow \dfrac{y^2}{12} = 1 - \dfrac{x^2}{5} \Rightarrow y^2 = 12\left(1 - \dfrac{x^2}{5}\right) \Rightarrow y = \pm\sqrt{12\left(1 - \dfrac{x^2}{5}\right)}$

Graph $Y_1 = \sqrt{\left(12\left(1 - X^2/5\right)\right)}$ and $Y_2 = -\sqrt{\left(12\left(1 - X^2/5\right)\right)}$ in [−7.5, 7.5, 1] by [−5, 5, 1].

See Figure 51.

(b) $A = \pi\left(\sqrt{12}\right)\left(\sqrt{5}\right) \approx 24.33$ square units and $P = 2\pi\sqrt{\dfrac{5+12}{2}} \approx 18.32$ units

52. (a) $\dfrac{x^2}{1.524^2}+\dfrac{y^2}{1.517^2}=1 \Rightarrow \dfrac{y^2}{1.517^2}=1-\dfrac{x^2}{1.524^2} \Rightarrow y=\pm\sqrt{1.517^2\left(1-\dfrac{x^2}{1.524^2}\right)}$

Plot the point $(0.15, 0)$ and Graph

$Y_1=\sqrt{\left(1.517^2\left(1-X^2/1.524^2\right)\right)}$ and $Y_2=-\sqrt{\left(1.517^2\left(1-X^2/1.524^2\right)\right)}$ in $[-3, 3, 1]$ by

$[-2, 2, 1]$. See Figure 52.

(b) $P=2\pi\sqrt{\dfrac{1.524^2+1.517^2}{2}}\approx 9.55$ A.U. or about 8.9×10^8 miles

$A=\pi(1.524)(1.517)\approx 7.26$ square A.U. or about 6.3×10^{16} square miles

Chapter 10 Test

1. Since $y=-(x-1)^2+2$, the vertex is $(1, 2)$ and the axis of symmetry is $x=1$. See Figure 1.

2. Since $x=(y-4)^2-2$, the vertex is $(-2, 4)$ and the axis of symmetry is $y=4$. See Figure 2.

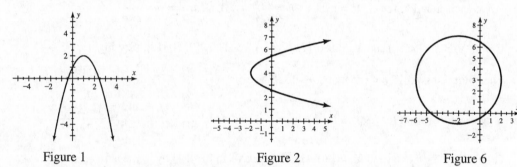

Figure 1   Figure 2   Figure 6

3. Since the parabola opens to the left and the vertex is $(1, 0)$, the equation has the form

$x=a(y-0)^2+1.$ Since the parabola passes through $(0, 1)$, $0=a(1-0)^2+1 \Rightarrow a=-1.$

The equation is $x=-y^2+1.$

4. Since the center is $(2, -4)$ and the radius is 2, the equation is $(x-2)^2+(y+4)^2=4.$

5. $\left(x-(-5)\right)^2+(y-2)^2=10^2 \Rightarrow (x+5)^2+(y-2)^2=100$

6. $x^2+4x+y^2-6y=3 \Rightarrow x^2+4x+4+y^2-6y+9=3+4+9 \Rightarrow (x+2)^2+(y-3)^2=16$

The radius is 4 and the center is $(-2, 3)$. Solving the equation for y results in $y=3\pm\sqrt{16-(x+2)^2}.$

See Figure 6.

7. The ellipse has a vertical major axis with vertices $(0, \pm7)$ and minor axis endpoints $(\pm4, 0)$.

See Figure 7.

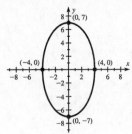

Figure 7

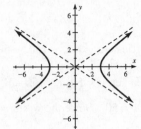

Figure 9

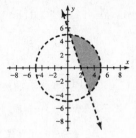

Figure 13

8. Horizontal major axis with vertices $(\pm 10, 0)$ and minor axis endpoints $(0, \pm 8)$ $\Rightarrow \dfrac{x^2}{100} + \dfrac{y^2}{64} = 1$

9. $4x^2 - 9y^2 = 36 \Rightarrow \dfrac{4x^2}{36} - \dfrac{9y^2}{36} = 1 \Rightarrow \dfrac{x^2}{9} - \dfrac{y^2}{4} = 1$ The hyperbola has a horizontal transverse axis with

   vertices $(\pm 3, 0)$ and asymptotes $y = \pm \dfrac{2}{3} x.$ See Figure 9

10. The solutions are the intersection points $(0, -4)$ and $(4, 0)$. Both solutions check.

11. From the first equation, $y = x - 3$. Substitute $y = x - 3$ into the second equation.

   $x^2 + (x - 3)^2 = 17 \Rightarrow x^2 + x^2 - 6x + 9 = 17 \Rightarrow 2(x - 4)(x + 1) = 0 \Rightarrow x = -1 \text{ or } x = 4$

   When $x = -1$, $y = (-1) - 3 = -4$. When $x = 4$, $y = (4) - 3 = 1$. The solutions are $(-1, -4)$ and $(4, 1)$.

12. Solve both equations for y. $2x^2 - y = 4 \Rightarrow y = 2x^2 - 4$ and $x^2 + y = 8 \Rightarrow y = 8 - x^2$

   Graph $Y_1 = 2X^2 - 4$, $Y_2 = 8 - X^2$ in $[-10, 10, 1]$ by $[-10, 10, 1]$. See Figures 12a & 12b.

   The solutions are the intersection points $(-2, 4)$ and $(2, 4)$.

   $[-10, 10, 1]$ by $[-10, 10, 1]$ $[-10, 10, 1]$ by $[-10, 10, 1]$

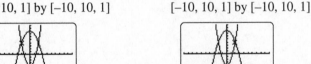

   Figure 12a   Figure 12b

13. Sketch the line given by $3x + y = 6$ using a dashed line and the circle given by $x^2 + y^2 = 25$ using a

   dashed line. Since the point $(4, 0)$ satisfies both inequalities, we shade the portion of the xy-plane

   containing the point $(4, 0)$. See Figure 13.

14. A parabola with vertex $(0, -4)$ has an equation of the form $y = a(x - 0)^2 - 4$ or $y = ax^2 - 4$. Since

   the parabola also passes through the point $(2, 0)$, $0 = a \cdot 2^2 - 4 \Rightarrow a = 1$. The equation of the

   parabola is $y = x^2 - 4$. The other parabola is a reflection of this parabola across the x-axis. Its

   equation is $y = 4 - x^2$. Since both the parabolas are solid, the system of inequalities is

   $y \le 4 - x^2$ and $y \ge x^2 - 4$.

15. (a) $xy = 5000$ and $2x + 2y = 300$

(b) $\dfrac{5000}{x} = 150 - x \Rightarrow 5000 = 150x - x^2 \Rightarrow x^2 - 150x + 5000 = 0 \Rightarrow (x - 50)(x - 100) = 0 \Rightarrow$

Either $x = 50$ or $x = 100$.   When $x = 50$, $y = 150 - 50 = 100$.

When $x = 100$, $y = 150 - 100 = 50$.   Since the width is shorter than the length,

$x = 50$ and $y = 100$.   The solution is 50 by 100 feet.

16. From the hint, the two equations to graph are $y = \dfrac{1183}{x^2}$ and $y = \dfrac{702 - x^2}{4x}$.   Figures not shown.   There

are two possible solutions: Either $x \approx 22.08$ and $y \approx 2.43$ or $x \approx 7.29$ and $y \approx 22.24$.   The answer is

not unique.

17. (a) $\dfrac{x^2}{19.18^2} + \dfrac{y^2}{19.16^2} = 1 \Rightarrow \dfrac{y^2}{19.16^2} = 1 - \dfrac{x^2}{19.18^2} \Rightarrow y = \pm\sqrt{19.16^2\left(1 - \dfrac{x^2}{19.18^2}\right)}$

Plot the point (0.9, 0) and graph

$Y_1 = \sqrt{\left(19.16^2\left(1 - X^2/19.18^2\right)\right)}$ and $Y_2 = -\sqrt{\left(19.16^2\left(1 - X^2/19.18^2\right)\right)}$ in [–30, 30, 10] by

[–20, 20, 10].   See Figure 17.

(b)   The minimum distance occurs when $x = 19.18$ A.U.   The distance is $19.18 - 0.9 = 18.28$ A.U.

This is about 1,700,040,000 miles.

Figure 17

Chapter 10 Extended and Discovery Exercises

1. (a) $x^2 = 4y \Rightarrow x^2 = 4(1)y$.   Since $p = 1$, the focus is (0, 1).   See Figure 1a.

(b) $y^2 = -8x \Rightarrow y^2 = 4(-2)x$.   Since $p = -2$, the focus is (–2, 0).   See Figure 1b.

(c) $x = 2y^2 \Rightarrow y^2 = \dfrac{1}{2}x \Rightarrow y^2 = 4\left(\dfrac{1}{8}\right)x$.   Since $p = \dfrac{1}{8}$, the focus is $\left(\dfrac{1}{8}, 0\right)$.   See Figure 1c.

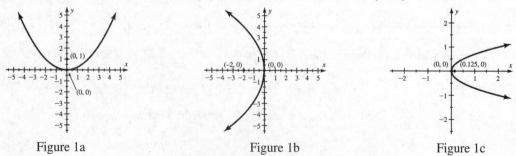

Figure 1a   Figure 1b   Figure 1c

2. (a)   Since the vertex of the parabola is (0, 0), the cross section of the dish has an equation of the

form $y = ax^2$. Since the parabola passes through $(150, 44)$,

$$44 = a(150)^2 \Rightarrow 44 = 22{,}500a \Rightarrow a = \frac{44}{22{,}500} = \frac{11}{5625}. \quad \text{The equation is } y = \frac{11}{5625}x^2.$$

(b)   Writing this equation in the form $x^2 = 4py$ yields $x^2 = 4\left(\dfrac{5625}{44}\right)y$. The focus is located at

$\left(0, \dfrac{5625}{44}\right)$. That is, the focus is $\dfrac{5625}{44} \approx 127.8$ feet from the vertex.

3. (a)   This is an ellipse with horizontal major axis, centered at $(3, 1)$. See Figure 3a.

(b)   This is an ellipse with vertical major axis, centered at $(-1, -2)$. See Figure 3b.

(c)   This is a hyperbola centered at $(-1, 3)$. The asymptotes are $y = \pm\dfrac{3}{2}(x+1)+3$. See Figure 3c.

(d)   This is a hyperbola centered at $(-1, 4)$. The asymptotes are $y = \pm 2(x+1)+4$. See Figure 3d.

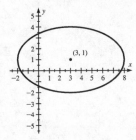

Figure 3a

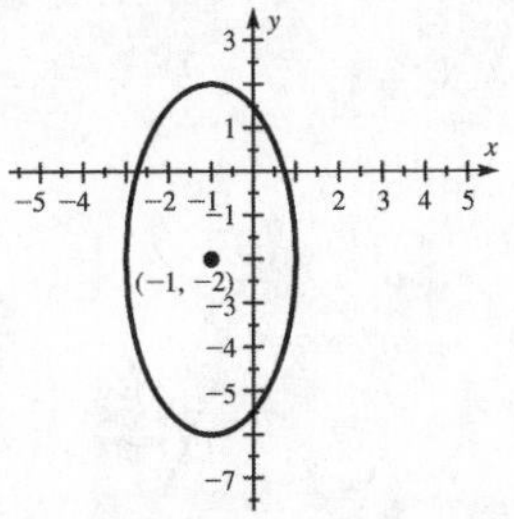

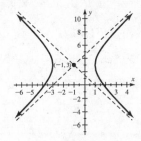

Figure 3b

Figure 3c

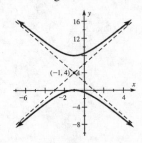

Figure 3d

4. (a) $\dfrac{(x+3)^2}{16} + \dfrac{(y-5)^2}{4} = 1$

(b) $\dfrac{(y+3)^2}{25} + \dfrac{(x-2)^2}{9} = 1$

5. (a) $9x^2 - 18x + 4y^2 + 24y + 9 = 0 \Rightarrow 9(x^2 - 2x + 1) + 4(y^2 + 6y + 9) = -9 + 9 + 36 \Rightarrow$

$9(x-1)^2 + 4(y+3)^2 = 36 \Rightarrow \dfrac{(x-1)^2}{4} + \dfrac{(y+3)^2}{9} = 1 \Rightarrow \text{Center: } (1, -3)$

(b) $25x^2 + 150x - 16y^2 + 32y - 191 = 0 \Rightarrow 25(x^2 + 6x + 9) - 16(y^2 - 2y + 1) = 191 + 225 - 16 \Rightarrow$

$$25(x+3)^2 - 16(y-1)^2 = 400 \Rightarrow \frac{(x+3)^2}{16} - \frac{(y-1)^2}{25} = 1 \Rightarrow \text{Center: } (-3, 1)$$

Chapters 1-10 Cumulative Review Exercises

1. $K = (4)^2 + (-3)^2 = 16 + 9 = 25$

2. $\dfrac{\left(a^{-2}b\right)^2}{a^{-1}\left(b^3\right)^{-2}} = \dfrac{a^{-4}b^2}{a^{-1}b^{-6}} = a^{-4-(-1)}b^{2-(-6)} = a^{-3}b^8 = \dfrac{b^8}{a^3}$

3. $7.345 \times 10^{-3} = 0.007345$

4. $f(-4) = \dfrac{-4}{-4-4} = \dfrac{-4}{-8} = \dfrac{1}{2}$; the denominator cannot equal zero, so $x - 4 \neq 0 \Rightarrow x \neq 4$.

5. $f(3) = 4$

6. See Figure 6.

7. See Figure 7.

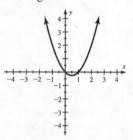 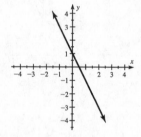

Figure 6   Figure 7

8. A line perpendicular to $y = -\dfrac{2}{3}x + 1$ has slope $m = \dfrac{3}{2}$.

$$y - (-2) = \frac{3}{2}(x-2) \Rightarrow y = \frac{3}{2}x - 3 - 2 \Rightarrow y = \frac{3}{2}x - 5$$

9. $2(1-x) - 4x = x \Rightarrow 2 - 2x - 4x - x = 0 \Rightarrow -7x = -2 \Rightarrow x = \dfrac{2}{7}$

10. $-5 \le 1 - 2x < 3 \Rightarrow -6 \le -2x < 2 \Rightarrow 3 \ge x > -1 \Rightarrow -1 < x \le 3; (-1, 3]$

11. $x^2 - 4 \le 0$; replace the inequality symbol with an equals sign and solve the resulting equation.

    $x^2 - 4 = 0 \Rightarrow (x-2)(x+2) = 0 \Rightarrow x = 2$ or $x = -2$. The graph of $y = x^2 - 4$ is a parabola that lies on

    or below the x-axis when $x \ge -2$ and $x \le 2$. The solution is $[-2, 2]$.

12. $|1 - x| \ge 2 \Rightarrow 1 - x \ge 2$ or $1 - x \le -2 \Rightarrow -x \ge 1$ or $-x \le -3 \Rightarrow x \le -1$ or $x \ge 3$;

    The solution is $(-\infty, -1] \cup [3, \infty)$.

13. Add the equations.
$$-2x+y=1$$
$$\underline{5x-y=2}$$
$$3x\quad=3$$
$\Rightarrow x=1.$ Substitute $x=1$ into the first equation.

$-2(1)+y=1 \Rightarrow -2+y=1 \Rightarrow y=3.$ The solution is $(1, 3)$.

14. Note that $x+y\le 4 \Rightarrow y\le -x+4$ and $x-y\ge 2 \Rightarrow y\le x-2.$ See Figure 14.

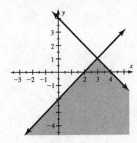

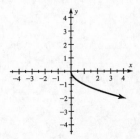

Figure 14   Figure 27

15. $(2x-1)(x+5)=2x^2+10x-x-5=2x^2+9x-5$

16. $xy(2x-3y^2+1)=2x^2y-3xy^3+xy$

17. $6x^2-13x-5=(3x+1)(2x-5)$

18. $x^3-4x=x(x^2-4)=x(x-2)(x+2)$

19. $x^2+3x+2=0 \Rightarrow (x+2)(x+1)=0 \Rightarrow x=-2$ or $x=-1$

20. $x^2+1=-3x \Rightarrow x^2+3x+1=0;\ a=1,b=3,c=1$ $\quad x=\dfrac{-b\pm\sqrt{b^2-4ac}}{2a}=\dfrac{-3\pm\sqrt{3^2-4(1)(1)}}{2(1)}=\dfrac{-3\pm\sqrt5}{2}$

21. $\dfrac{x-2}{x+2}\div\dfrac{2x-4}{3x+6}=\dfrac{x-2}{x+2}\cdot\dfrac{3(x+2)}{2(x-2)}=\dfrac{3}{2}$

22. $\dfrac{1}{x+1}+\dfrac{1}{x-1}=\dfrac{1}{x+1}\cdot\dfrac{x-1}{x-1}+\dfrac{1}{x-1}\cdot\dfrac{x+1}{x+1}=\dfrac{x-1}{x^2-1}+\dfrac{x+1}{x^2-1}=\dfrac{x-1+x+1}{x^2-1}=\dfrac{2x}{x^2-1}$

23. $\sqrt{8x^2}=\sqrt{4x^2\cdot 2}=2x\sqrt2$

24. $8^{2/3}=\left(\sqrt[3]{8}\right)^2=2^2=4$

25. $\sqrt[3]{2x}\cdot\sqrt[3]{32x^2}=\sqrt[3]{2x\cdot 32x^2}=\sqrt[3]{64x^3}=4x$

26. $3\sqrt{3x}+\sqrt{12x}=3\sqrt{3x}+2\sqrt{3x}=5\sqrt{3x}$

27. See Figure 27.

28. $d=\sqrt{(x_2-x_1)^2+(y_2-y_1)^2}=\sqrt{(-2-2)^2+(0-(-3))^2}=\sqrt{(-4)^2+(3)^2}=\sqrt{16+9}=\sqrt{25}=5$

29. $(2+3i)(2-3i)=(2)^2-(3i)^2=4-9i^2=4-9(-1)=4+9=13$

30. $\sqrt{x+2}=x \Rightarrow x+2=x^2 \Rightarrow x^2-x-2=0 \Rightarrow (x-2)(x+1)=0 \Rightarrow x=2\ (x=-1\text{ does not check.})$

31. $x = -\dfrac{b}{2a} = -\dfrac{(-6)}{2(1)} = \dfrac{-6}{-2} = 3;\ y = (3)^2 - 6(3) + 3\ \Rightarrow y = 9 - 18 + 3 \Rightarrow y = -6.$ The vertex is $(3, -6)$.

32. $f(x) = x^2 - 2x + 3 \Rightarrow f(x) = (x^2 - 2x + 1) + 3 - 1\ \Rightarrow f(x) = (x-1)^2 + 2$

33. The graph of $f(x)$ is shifted 4 units right.

34. $x(3-x) = 2 \Rightarrow 3x - x^2 - 2 = 0 \Rightarrow x^2 - 3x + 2 = 0\ \Rightarrow (x-2)(x-1) = 0 \Rightarrow x = 2$ or $x = 1$

35. $x^3 + x = 0 \Rightarrow x(x^2 + 1) = 0 \Rightarrow x = 0$ or $x^2 = -1\ \Rightarrow x = \pm\sqrt{-1} \Rightarrow x = \pm i$

36. (a) $\log 10{,}000 = 4$ because $10^4 = 10{,}000$

    (b) $\log_2 8 = 3$

    (c) $\log_3 3^x = x$

    (d) $e^{\ln 6} = 6$

    (e) $\log 2 + \log 50 = \log(2 \cdot 50) = \log(100) = 2$ because $10^2 = 100$

    (f) $\log_2 24 - \log_2(3) = \log_2\left(\dfrac{24}{3}\right) = \log_2(8) = 3$

37. (a) $(f \circ g)(2) = f(g(2)) = f(2 \cdot 2) = f(4) = 4^2 + 1\ = 16 + 1 = 17$

    (b) $(g \circ f)(x) = g(f(x)) = g(x^2 + 1) = 2(x^2 + 1) = 2x^2 + 2$

38. Let $x = 2 - 3y$. Then $x - 2 = -3y \Rightarrow y = \dfrac{x-2}{-3}$ or $f^{-1}(x) = -\dfrac{1}{3}x + \dfrac{2}{3}$

39. $A = 1000(1 + 0.05)^6 = 1000(1.05)^6 \approx \$1340.10$

40. $\log\dfrac{x^2\sqrt{y}}{z^3} = \log\left(x^2\sqrt{y}\right) - \log\left(z^3\right) = \log\left(x^2\right) + \log\left(y^{1/2}\right) - \log\left(z^3\right) = 2\log x + \dfrac{1}{2}\log y - 3\log z$

41. $2e^x - 1 = 17 \Rightarrow 2e^x = 18 \Rightarrow e^x = 9 \Rightarrow \ln e^x = \ln 9\ \Rightarrow x = \ln 9$

42. $3 + \log 4x = 5 \Rightarrow \log 4x = 2 \Rightarrow 10^{\log 4x} = 10^2\ \Rightarrow 4x = 100 \Rightarrow x = 25$

43. See Figures 43a, 43b, 43c, 43d

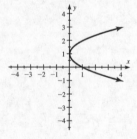

Figure 43a

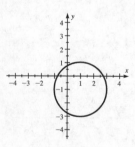

Figure 43b

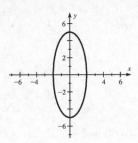

Figure 43c

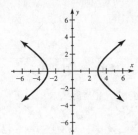

Figure 43d

44. Multiply the first equation by –1 and add to the second equation.

$$-x^2 - y^2 = -1$$
$$x^2 + 9y^2 = 9 \Rightarrow y^2 = 1 \Rightarrow y = \pm 1$$
$$8y^2 = 8$$

$x^2 + (\pm 1)^2 = 1 \Rightarrow x^2 + 1 = 1 \Rightarrow x^2 = 0 \Rightarrow x = 0$ The solutions are (0, 1) and (0, –1).

45. See Figure 45.

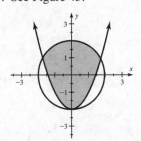

Figure 45

46. (a) $D(0) = 400 - 50(0) = 400$; initially, the driver is 400 miles from home.

(b) $400 - 50x = 0 \Rightarrow 400 = 50x \Rightarrow x = 8$; after 8 hours the driver arrives at home.

(c) -50; the driver is traveling 50 miles per hour toward home.

47. Let x, y, z represent the amounts invested at 5%, 6%, and 7%, respectively. The system needed is

$$x + y + z = 2000 \qquad\qquad x + y + z = 2000$$
$$y = x + 500 \qquad\qquad \Rightarrow -x + y \qquad = 500$$
$$0.05x + 0.06y + 0.07z = 120 \qquad 5x + 6y + 7z = 12,000$$

Multiply the first equation by –7 and add to the third equation.

$$-7x - 7y - 7z = -14,000$$
$$\underline{5x + 6y + 7z = 12,000}\quad \text{Add this new equation to the second equation.}$$
$$-2x - y = -2000$$

$$-2x - y = -2000$$
$$\underline{-x + y = 500}\quad \Rightarrow x = 500 \Rightarrow y = 500 + 500 = 1000$$
$$-3x = -1500$$

Substitute $x = 500$ and $y = 1000$ into the first equation. $500 + 1000 + z = 2000 \Rightarrow z = 500$,

$500 is invested at 5%, $1000 is invested at 6%, and $500 is invested at 7%.

48. Let x represent the length of the garden. Then the area is $x(600-x)=600x-x^2$. Find the x-value of

the vertex of $y=-x^2+600x$. $x=-\dfrac{b}{2a}=-\dfrac{600}{2(-1)}=\dfrac{-600}{-2}=300$

The dimensions should be 300 feet by 300 feet.

49. $2e^{0.02x}=4\Rightarrow e^{0.02x}=2\Rightarrow \ln e^{0.02x}=\ln 2 \Rightarrow 0.02x=\ln 2\Rightarrow x=\dfrac{\ln 2}{0.02}\Rightarrow 50\ln 2\approx 34.7$ years

50. $V=\pi r^2 h=60$ and $S=2\pi rh=50$, solve for $h\Rightarrow h=\dfrac{50}{2\pi r}$, substitute into $\pi r^2 h=60$

$\Rightarrow \pi r^2\left(\dfrac{50}{2\pi r}\right)=60\Rightarrow 25r=60\Rightarrow r=\dfrac{60}{25}=2.4$ inches

Chapter 11: Sequences and Series

Section 11.1: Sequences

1. 1, 2, 3, 4; *Answers may vary.*

3. function; natural numbers

5. 6

7. $f(2)$

9. $f(1) = 1^2 = 1, f(2) = 2^2 = 4, f(3) = 3^2 = 9, f(4) = 4^2 = 16 \Rightarrow 1, 4, 9, 16$

11. $f(1) = \dfrac{1}{1+5} = \dfrac{1}{6}, f(2) = \dfrac{1}{2+5} = \dfrac{1}{7}, f(3) = \dfrac{1}{3+5} = \dfrac{1}{8}, f(4) = \dfrac{1}{4+5} = \dfrac{1}{9} \Rightarrow \dfrac{1}{6}, \dfrac{1}{7}, \dfrac{1}{8}, \dfrac{1}{9}$

13. $f(1) = 5\left(\dfrac{1}{2}\right)^1 = \dfrac{5}{2}, f(2) = 5\left(\dfrac{1}{2}\right)^2 = \dfrac{5}{4}, f(3) = 5\left(\dfrac{1}{2}\right)^3 = \dfrac{5}{8}, f(4) = 5\left(\dfrac{1}{2}\right)^4 = \dfrac{5}{16} \Rightarrow \dfrac{5}{2}, \dfrac{5}{4}, \dfrac{5}{8}, \dfrac{5}{16}$

15. $f(1) = 9, f(2) = 9, f(3) = 9, f(4) = 9 \Rightarrow 9, 9, 9, 9$

17. $a_1 = 1^3 = 1, a_2 = 2^3 = 8, a_3 = 3^3 = 27 \Rightarrow 1, 8, 27$

19. $a_1 = \dfrac{4(1)}{3+1} = 1, a_2 = \dfrac{4(2)}{3+2} = \dfrac{8}{5}, a_3 = \dfrac{4(3)}{3+3} = 2 \Rightarrow 1, \dfrac{8}{5}, 2$

21. $a_1 = 2(1)^2 + 1 - 1 = 2, a_2 = 2(2)^2 + 2 - 1 = 9, a_3 = 2(3)^2 + 3 - 1 = 20 \Rightarrow 2, 9, 20$

23. $a_1 = -2, a_2 = -2, a_3 = -2 \Rightarrow -2, -2, -2$

25. $a_1 = b(1) + c = b + c; a_2 = b(2) + c = 2b + c$

27. $\dfrac{1}{2}(a_1 + a_4) = \dfrac{1}{2}(10 + 4) = \dfrac{1}{2}(14) = 7$

29. The points shown are (1, 3), (2, 4), (3, 5), (4, 3) and (5, 1). The sequence is 3, 4, 5, 3, 1.

31. The points shown are (1, 6), (2, 5), (3, 4), (4, 3), (5, 2) and (6, 1). The sequence is 6, 5, 4, 3, 2, 1.

33. Numerical: See Figure 33a. Graphical: See Figure 33b.

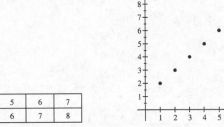

n	1	2	3	4	5	6	7
a_n	2	3	4	5	6	7	8

Figure 33a   Figure 33b

35. Numerical: See Figure 35a. Graphical: See Figure 35b.

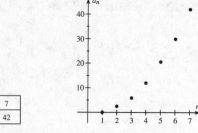

n	1	2	3	4	5	6	7
a_n	0	2	6	12	20	30	42

Figure 35a   Figure 35b

37. Numerical: See Figure 37a. Graphical: See Figure 37b.

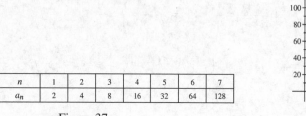

n	1	2	3	4	5	6	7
a_n	2	4	8	16	32	64	128

Figure 37a   Figure 37b

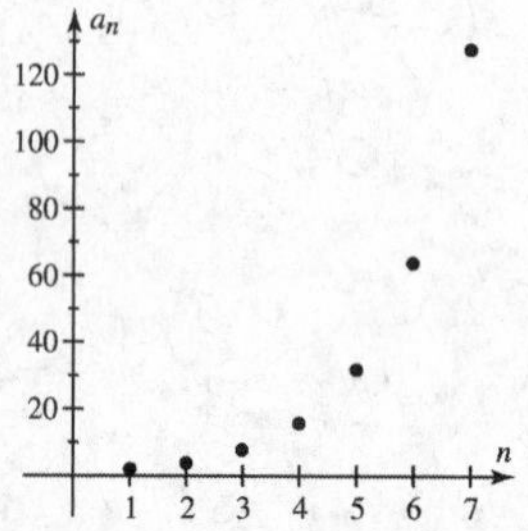

39. Symbolic: $a_n = 30n$ for $n = 1, 2, 3, \ldots, 7$ Numerical: See Figure 39a. Graphical: See Figure 39b.

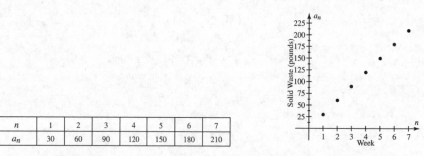

n	1	2	3	4	5	6	7
a_n	30	60	90	120	150	180	210

Figure 39a   Figure 39b

41. (a) $a_1 = 1^2 = 1, a_2 = 2^2 = 4, a_3 = 3^2 = 9, a_4 = 4^2 = 16 \Rightarrow 1, 4, 9, 16$

　　(b) $a_1 = 4(1) = 4, a_2 = 4(2) = 8, a_3 = 4(3) = 12, a_4 = 4(4) = 16 \Rightarrow 4, 8, 12, 16$

43. (a)   After 1 year it is worth $25,000(0.80) = \$20,000$. After 2 years it is worth

　　　　$20,000(0.80) = \$16,000$.

　　(b) $a_n = 25,000(0.8)^n$

　　(c)   See Figure 43.

n	1	2	3	4	5	6	7
a_n	20,000	16,000	12,800	10,240	8192	6553.6	5242.9

Figure 43

45. (a) $a_n = 2048(0.5)^{n-1}$ for $n = 1, 2, 3, \ldots, 7$

　　(b)   See Figure 45b.

　　(c)   See Figure 45.

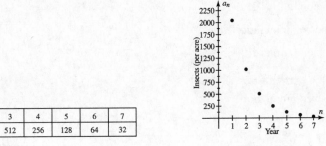

n	1	2	3	4	5	6	7
a_n	2048	1024	512	256	128	64	32

Figure 45b   Figure 45c

Section 11.2 Arithmetic and Geometric Sequences

1. linear

3. $a_n = 3n + 1$; the common difference is 3. *Answers may vary.*

5. add; previous

7. 19; 4

9. $a_n = a_1 (r)^{n-1}$

11. Yes, the common difference is 10.

13. Yes, the common difference is -1.

15. No, there is no common difference.

17. Yes, the common difference is 3.

19. Yes, the common difference is -3.

21. No, there is no common difference.

23. Yes, the common difference is 1.

25. No, there is no common difference.

27. Yes, the common difference is 2.

29. $a_n = 7 + (n-1)(-2) \Rightarrow a_n = 7 - 2n + 2 \Rightarrow a_n = -2n + 9$

31. Note: $d = \dfrac{6 - (-2)}{2} = 4$, thus $a_n = -2 + (n-1)(4) \Rightarrow a_n = -2 + 4n - 4 \Rightarrow a_n = 4n - 6$

33. Note: $d = \dfrac{8 - 16}{4} = -2$ and $a_1 = 16 - 7(-2) = 30$, thus

$a_n = 30 + (n-1)(-2) \Rightarrow a_n = 30 - 2n + 2 \Rightarrow a_n = -2n + 32$

35. $a_{32} = -3 + (32-1)(2) \Rightarrow -3 + (31)(2) = -3 + 62 = 59$

37. Note: $d = 0 - (-3) = 3$, thus $a_9 = -3 + (9-1)(3) \Rightarrow -3 + (8)(3) = -3 + 24 = 21$

39. Yes, the common ratio is 3.

41. Yes, the common ratio is 0.8.

43. No, there is no common ratio.

45. Yes, the common ratio is 2.

47. No, there is no common ratio.

49. Yes, the common ratio is 4.

51. Yes, the common ratio is 2.

53. No, there is no common ratio.

55. $a_n = 1.5(4)^{n-1}$

57. Note: $r = \dfrac{6}{-3} = -2$, thus $a_n = -3(-2)^{n-1}$

59. Note: $16 = 1 \cdot r^2 \Rightarrow r^2 = 16 \Rightarrow r = 4 \,(\text{since } r > 0)$, thus $a_n = 1(4)^{n-1}$

61. $a_8 = 2(3)^{8-1} = 2(3)^7 = 4374$

63. Note: $r = \dfrac{3}{-1} = -3$, thus $a_6 = -1(-3)^{6-1} = -1(-3)^5 = 243$

65. (a)   3000, 6000, 9000, 12,000, 15,000; This sequence is arithmetic.

    (b) $a_n = 3000n$

    (c) $a_{20} = 3000(20) = 60,000;$ When there are 20 people, the ventilation should be 60,000 cubic

          feet per hour.

    (d)   See Figure 65. The points are collinear.

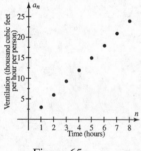

Figure 65

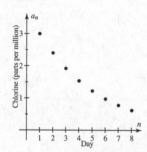

Figure 67

67. (a)   Since 20% of the chlorine dissipates, 80% will remain. The function is $a_n = 3(0.8)^{n-1}$.

    (b)   See Figure 67. The points are not collinear. The sequence is geometric.

69. (a) $a_n = 5(0.85)^{n-1}$

    (b)   The sequence is geometric and the common ratio is 0.85.

    (c) $a_8 = 5(0.85)^{8-1} = 5(0.85)^7 \approx 1.6;$ On the 8th bounce, the ball reaches a maximum height of

          about 1.6 feet.

71. (a)   The number of seats can be modeled by an arithmetic sequence whose common difference is 2.

    (b)   Since $a_1 = 40$ and $d = 2$, $a_n = 40 + 2(n-1)$ or $a_n = 2n + 38$

    (c) $a_{20} = 40 + 2(20-1) = 40 + 2(19) = 40 + 38 = 78$ seats

Checking Basic Concepts Sections 11.1 and 11.2

1. $a_1 = \dfrac{1}{1+4} = \dfrac{1}{5}, a_2 = \dfrac{2}{2+4} = \dfrac{1}{3}, a_3 = \dfrac{3}{3+4} = \dfrac{3}{7}, a_4 = \dfrac{4}{4+4} = \dfrac{1}{2} \Rightarrow \dfrac{1}{5}, \dfrac{1}{3}, \dfrac{3}{7}, \dfrac{1}{2}$

2. Graphical: See Figure 2a. Numerical: See Figure 2b.

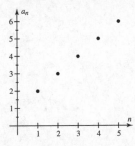

n	1	2	3	4	5
a_n	2	3	4	5	6

   Figure 2a   Figure 2b

3. (a)   Arithmetic. Here $d = 1 - (-2) = 3$ and $a_1 = -2$, thus $a_n = -2 + (n-1)(3) \Rightarrow a_n = 3n - 5$

   (b)   Geometric. Here $r = \dfrac{-6}{3} = -2$ and $a_1 = 3$, thus $a_n = 3(-2)^{n-1}$

4. $a_n = 5 + (n-1)(2) \Rightarrow a_n = 5 + 2n - 2 \Rightarrow a_n = 2n + 3$

5. $a_n = 5(2)^{n-1}$

Section 11.3 Series

1. series

3. arithmetic (the common difference is 2)

5. $n\left(\dfrac{a_1 + a_n}{2}\right)$ or $\dfrac{n}{2}\left(2a_1 + (n-1)d\right)$

7. sum

9. arithmetic

11. $6\left(\dfrac{3+13}{2}\right) = 6(8) = 48$

13. $40\left(\dfrac{1+40}{2}\right) = 40(20.5) = 820$

15. $5\left(\dfrac{-7+5}{2}\right) = 5(-1) = -5$

17. Here $r = 3$ and $n = 7$ thus $S_7 = 3\left(\dfrac{1-3^7}{1-3}\right) = 3\left(\dfrac{-2186}{-2}\right) = 3(1093) = 3279$

19. Here $r = -2$ and $n = 8$ thus $S_8 = 1\left(\dfrac{1-(-2)^8}{1-(-2)}\right) = 1\left(\dfrac{-255}{3}\right) = -85$

21. Here $r = 3$ and $n = 6$ thus $S_6 = 0.5\left(\dfrac{1-3^6}{1-3}\right) = 0.5\left(\dfrac{-728}{-2}\right) = 0.5(364) = 182$

23. $S_{20} = 2000\left(\dfrac{(1+0.08)^{20}-1}{0.08}\right) \approx \$91,523.93$

25. $S_5 = 10,000\left(\dfrac{(1+0.11)^5-1}{0.11}\right) \approx \$62,278.01$

27. $2(1)+2(2)+2(3)+2(4) \Rightarrow 2+4+6+8 = 20$

29. $4+4+4+4+4+4+4+4 = 32$

31. $1^2 + 2^2 + 3^2 + 4^2 + 5^2 + 6^2 + 7^2 \Rightarrow 1+4+9+16+25+36+49 = 140$

33. $(4^2 - 4)+(5^2 - 5) \Rightarrow 12 + 20 = 32$

35. $\displaystyle\sum_{k=1}^{6} k^4$

37. $\displaystyle\sum_{k=1}^{5} \dfrac{1}{k^2}$

39. $\displaystyle\sum_{k=1}^{n} k = n\left(\dfrac{a_1 + a_n}{2}\right) = n\left(\dfrac{1+n}{2}\right) = \dfrac{n(n+1)}{2}$

41. (a) $8518 + 9921 + 10,706 + 14,035 + 14,307 + 12,249$

    (b) $8518 + 9921 + 10,706 + 14,035 + 14,307 + 12,249 = 69,736$

43 (a)   Each air filter removes 80% or 0.8 of the impurities, so 20% or 0.2 passes through it

            100% or 1 represent the amount of impurities entering the first air filter, the amount removed

            by n filters equals $(0.8)(1)+(0.8)(0.2)+(0.8)(0.04)+(0.8)(0.008) +\cdots+(0.8)(0.2)^{n-1}$. In

            summation notation, $\displaystyle\sum_{k=1}^{n} 0.8(0.2)^{k-1}$

    (b)   To remove 96% or 0.96 of the impurities requires 2 filters because

            $\displaystyle\sum_{k=1}^{2} 0.8(0.2)^{k-1} = (0.8)(1)+(0.8)(0.2) = 0.8 + 0.16 = 0.96$

45. (a)   Each successive square has half the area of the square before it. $1, \dfrac{1}{2}, \dfrac{1}{4}, \dfrac{1}{8}, \dfrac{1}{16}$

    (b)   Here $r = \dfrac{1}{2}$ and $n = 10$ thus $S_5 = 1\left(\dfrac{1-\left(\frac{1}{2}\right)^{10}}{1-\left(\frac{1}{2}\right)}\right) = \left(\dfrac{\frac{1023}{1024}}{\frac{1}{2}}\right) = \dfrac{1023}{512}.$

47. The sum is $14+13+12+11+10+9+8+7+6$. This is an arithmetic series with $a_1 = 14$ and $a_9 = 6$.

   The sum is $S_9 = 9\left(\dfrac{14+6}{2}\right) = 9(10) = 90$ logs.

49. This is an arithmetic series with $a_1 = 35,000$, $n = 20$ and $d = 2000$

   The sum is $S_{20} = \dfrac{20}{2}\left(2(35,000)+(20-1)(2000)\right) = 10(70,000+38,000) =$

   $10(108,000) = \$1,080,000$

51. This is a geometric sequence given by $a_n = 10(0.75)^n$. The distance it falls is

   $a_4 = 10(0.75)^4 \approx 3.16$ feet.

53. *Answers may vary.*

Section 11.4 The Binomial Theorem

1. 5

3. Row 4

5. $4! = 1 \cdot 2 \cdot 3 \cdot 4 = 24$

7. $\dfrac{n!}{(n-r)!r!}$

9. $\dfrac{n!}{(n-1)!} = \dfrac{n \cdot (n-1)!}{(n-1)!} = n$

11. Row 4 of Pascal's triangle is 1, 3, 3, 1. $(x+y)^3 = x^3 + 3x^2 y + 3xy^2 + y^3$

13. Row 5 of Pascal's triangle is 1, 4, 6, 4, 1.

   $(2x+1)^4 = (2x)^4 + 4(2x)^3(1) + 6(2x)^2(1)^2 + 4(2x)(1)^3 + (1)^4 \Rightarrow$

   $(2x+1)^4 = 16x^4 + 32x^3 + 24x^2 + 8x + 1$

15. Row 6 of Pascal's triangle is 1, 5, 10, 10, 5, 1. $(a-b)^5 = a^5 - 5a^4 b + 10a^3 b^2 - 10a^2 b^3 + 5ab^4 - b^5$

17. Row 4 of Pascal's triangle is 1, 3, 3, 1. $\left(x^2+1\right)^3 = \left(x^2\right)^3 + 3\left(x^2\right)^2(1) + 3\left(x^2\right)(1)^2 + (1)^3 \Rightarrow$

   $\left(x^2+1\right)^3 = x^6 + 3x^4 + 3x^2 + 1$

19. $3! = 1 \cdot 2 \cdot 3 = 6$

21. $\dfrac{4!}{3!} = \dfrac{1 \cdot 2 \cdot 3 \cdot 4}{1 \cdot 2 \cdot 3} = 4$

23. $\dfrac{2!}{0!} = \dfrac{1 \cdot 2}{1} = 2$

25. $\dfrac{5!}{2!3!} = \dfrac{1 \cdot 2 \cdot 3 \cdot 4 \cdot 5}{(1 \cdot 2)(1 \cdot 2 \cdot 3)} = 2 \cdot 5 = 10$

27. $_5C_4 = \dfrac{5!}{4!1!} = \dfrac{1 \cdot 2 \cdot 3 \cdot 4 \cdot 5}{(1 \cdot 2 \cdot 3 \cdot 4)(1)} = 5$

29. $_6C_5 = \dfrac{6!}{5!1!} = \dfrac{1 \cdot 2 \cdot 3 \cdot 4 \cdot 5 \cdot 6}{(1 \cdot 2 \cdot 3 \cdot 4 \cdot 5)(1)} = 6$

31. $_4C_0 = \dfrac{4!}{0!4!} = \dfrac{1 \cdot 2 \cdot 3 \cdot 4}{(1)(1 \cdot 2 \cdot 3 \cdot 4)} = 1$

33. $_{12}C_7 = 792$

34. $_{13}C_8 = 1287$

35. $_9C_5 = 126$

37. $_{19}C_{11} = 75,582$

39. $(m+n)^3 = (_3C_0)m^3 + (_3C_1)m^2n + (_3C_2)mn^2 + (_3C_3)n^3 \Rightarrow (m+n)^3 = m^3 + 3m^2n + 3mn^2 + n^3$

41. $(x-y)^4 = (_4C_0)x^4 - (_4C_1)x^3y + (_4C_2)x^2y^2 - (_4C_3)xy^3 + (_4C_4)y^4 \Rightarrow$

   $(x-y)^4 = x^4 - 4x^3y + 6x^2y^2 - 4xy^3 + y^4$

43. $(2a+1)^3 = (_3C_0)(2a)^3 + (_3C_1)(2a)^2(1) + (_3C_2)(2a)(1)^2 + (_3C_3)(1)^3 \Rightarrow$

   $(2a+1)^3 = 8a^3 + 12a^2 + 6a + 1$

45. $(x+2)^5 = (_5C_0)x^5 + (_5C_1)x^4(2) + (_5C_2)x^3(2)^2 + (_5C_3)x^2(2)^3 + (_5C_4)x(2)^4 + (_5C_5)(2)^5 \Rightarrow$

   $(x+2)^5 = x^5 + 10x^4 + 40x^3 + 80x^2 + 80x + 32$

47. $(3+2m)^4 = (_4C_0)(3)^4 + (_4C_1)(3)^3(2m) + (_4C_2)(3)^2(2m)^2 + (_4C_3)(3)(2m)^3 + (_4C_4)(2m)^4 \Rightarrow$

   $(3+2m)^4 = 81 + 216m + 216m^2 + 96m^3 + 16m^4$

49. $(2x-y)^3 = (_3C_0)(2x)^3 - (_3C_1)(2x)^2y + (_3C_2)(2x)y^2 - (_3C_3)y^3 \Rightarrow$

   $(2x-y)^3 = 8x^3 - 12x^2y + 6xy^2 - y^3$

51. Here $r = 0$ and $n = 8$. The first term is $(_8C_0)a^{8-0}b^0 = a^8$.

53. Here $r = 3$ and $n = 7$. The fourth term is $(_7C_3)x^{7-3}y^3 = 35x^4y^3$.

55. Here $r = 0$ and $n = 9$. The first term is $(_9C_0)(2m)^{9-0}n^0 = 512m^9$.

Checking Basic Concepts Sections 11.3 and 11.4

1. (a)　Geometric. The common ratio is $\dfrac{1}{2}$.

(b)   Arithmetic. The common difference is 2.

2. $12\left(\dfrac{4+48}{2}\right)=12\left(26\right)=312$

3. Here $r=-2$ and $n=10$ thus $S_{10}=1\left(\dfrac{1-\left(-2\right)^{10}}{1-\left(-2\right)}\right)=1\left(\dfrac{-1023}{3}\right)=-341$

4. Row 5 of Pascal's triangle is 1, 4, 6, 4, 1. $\left(x-y\right)^{4}=x^{4}-4x^{3}y+6x^{2}y^{2}-4xy^{3}+y^{4}$

5. $\left(x+2\right)^{3}=\left(_{3}C_{0}\right)x^{3}+\left(_{3}C_{1}\right)x^{2}\left(2\right)+\left(_{3}C_{2}\right)x\left(2\right)^{2}+\left(_{3}C_{3}\right)\left(2\right)^{3}\Rightarrow\left(x+2\right)^{3}=x^{3}+6x^{2}+12x+8$

Chapter 11 Review

1. $f\left(1\right)=1^{3}=1, f\left(2\right)=2^{3}=8, f\left(3\right)=3^{3}=27, f\left(4\right)=4^{3}=64\Rightarrow 1, 8, 27, 64$

2. $f\left(1\right)=5-2\left(1\right)=3, f\left(2\right)=5-2\left(2\right)=1, f\left(3\right)=5-2\left(3\right)=-1, f\left(4\right)=5-2\left(4\right)=-3\Rightarrow 3, 1, -1, -3$

3. $f\left(1\right)=\dfrac{2\left(1\right)}{1^{2}+1}=1, f\left(2\right)=\dfrac{2\left(2\right)}{2^{2}+1}=\dfrac{4}{5}, f\left(3\right)=\dfrac{2\left(3\right)}{3^{2}+1}=\dfrac{3}{5}, f\left(4\right)=\dfrac{2\left(4\right)}{4^{2}+1}=\dfrac{8}{17}\Rightarrow 1, \dfrac{4}{5}, \dfrac{3}{5}, \dfrac{8}{17}$

4. $f\left(1\right)=\left(-2\right)^{1}=-2, f\left(2\right)=\left(-2\right)^{2}=4, f\left(3\right)=\left(-2\right)^{3}=-8, f\left(4\right)=\left(-2\right)^{4}=16\Rightarrow -2, 4, -8, 16$

5. The points shown are $(1, -2), (2, 0), (3, 4),$ and $(4, 2)$. The sequence is $-2, 0, 4, 2$.

6. The points shown are $(1, 5), (2, 3), (3, 2),$ and $(4, 1)$. The sequence is $5, 3, 2, 1$.

7. Numerical: See Figure 7a. Graphical: See Figure 7b.

n	1	2	3	4	5	6	7
a_n	2	4	6	8	10	12	14

Figure 7a

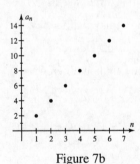

Figure 7b

8. Numerical: See Figure 8a. Graphical: See Figure 8b.

n	1	2	3	4	5	6	7
a_n	-3	0	5	12	21	32	45

Figure 8a

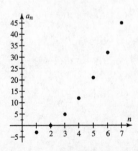

Figure 8b

9. Numerical: See Figure 9a. Graphical: See Figure 9b.

n	1	2	3	4	5	6	7
a_n	2	1	0.5	0.25	0.125	0.0625	0.0313

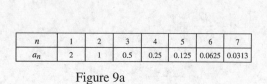

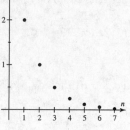

Figure 9a

Figure 9b

10. Numerical: See Figure 10a. Graphical: See Figure 10b.

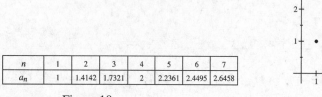

n	1	2	3	4	5	6	7
a_n	1	1.4142	1.7321	2	2.2361	2.4495	2.6458

Figure 10a

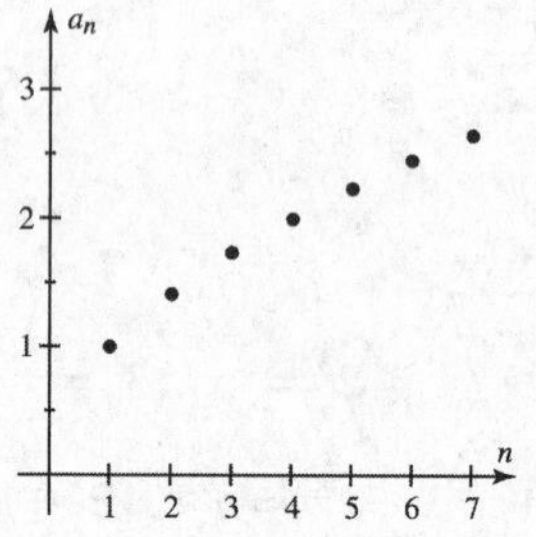

Figure 10b

11. Yes, the common difference is 5.
12. No, there is no common difference.
13. No, there is no common difference.
14. Yes, the common difference is $-\frac{1}{3}$.
15. Yes, the common difference is –3.
16. No, there is no common difference.
17. Yes, the common difference is –1.
18. No, there is no common difference.
19. $a_n = -3 + (n-1)(4) \Rightarrow a_n = -3 + 4n - 4 \Rightarrow a_n = 4n - 7$
20. Note: $d = -3 - 2 = -5$, thus $a_n = 2 + (n-1)(-5) \Rightarrow a_n = 2 - 5n + 5 \Rightarrow a_n = -5n + 7$
21. Yes, the common ratio is 4.
22. No, there is no common ratio.
23. No, there is no common ratio.
24. Yes, the common ratio is 0.7.
25. No, there is no common ratio.
26. Yes, the common ratio is $-\frac{1}{3}$.
27. No, there is no common ratio.
28. Yes, the common ratio is 2.
29. $a_n = 5(0.9)^{n-1}$

30. Note: $r = \dfrac{8}{2} = 4$, thus $a_n = 2(4)^{n-1}$

31. $9\left(\dfrac{4+44}{2}\right) = 9(24) = 216$

32. $5\left(\dfrac{4.5+(-1.5)}{2}\right) = 5(1.5) = 7.5$

33. Here $r = -4$ and $n = 7$ thus $S_7 = 1\left(\dfrac{1-(-4)^7}{1-(-4)}\right) = \left(\dfrac{16,385}{5}\right) = 3277$

34. Here $r = \dfrac{1}{2}$ and $n = 9$ thus $S_9 = 1\left(\dfrac{1-\left(\frac{1}{2}\right)^9}{1-\left(\frac{1}{2}\right)}\right) = \left(\dfrac{\frac{511}{512}}{\frac{1}{2}}\right) = \dfrac{511}{256}$

35. $(2(1)+1)+(2(2)+1)+(2(3)+1)+(2(4)+1)+(2(5)+1) \Rightarrow 3+5+7+9+11$

36. $\dfrac{1}{1+1}+\dfrac{1}{2+1}+\dfrac{1}{3+1}+\dfrac{1}{4+1} \Rightarrow \dfrac{1}{2}+\dfrac{1}{3}+\dfrac{1}{4}+\dfrac{1}{5}$

37. $1^3 + 2^3 + 3^3 + 4^3 \Rightarrow 1+8+27+64$

38. $(1-2)+(1-3)+(1-4)+(1-5)+(1-6)+(1-7) \Rightarrow -1+(-2)+(-3)+(-4)+(-5)+(-6)$

39. $\displaystyle\sum_{k=1}^{20} k$

40. $\displaystyle\sum_{k=1}^{20} \dfrac{1}{k}$

41. $\displaystyle\sum_{k=1}^{9} \dfrac{k}{k+1}$

42. $\displaystyle\sum_{k=1}^{7} k^2$

43. Row 4 of Pascal's triangle is 1, 3, 3, 1. $(x+4)^3 = x^3 + 3x^2(4) + 3x(4)^2 + (4)^3 \Rightarrow$

 $(x+4)^3 = x^3 + 12x^2 + 48x + 64$

44. Row 5 of Pascal's triangle is 1, 4, 6, 4, 1.

 $(2x+1)^4 = (2x)^4 + 4(2x)^3(1) + 6(2x)^2(1)^2 + 4(2x)(1)^3 + (1)^4 \Rightarrow$

 $(2x+1)^4 = 16x^4 + 32x^3 + 24x^2 + 8x + 1$

45. Row 6 of Pascal's triangle is 1, 5, 10, 10, 5, 1. $(x-y)^5 = x^5 - 5x^4 y + 10x^3 y^2 - 10x^2 y^3 + 5xy^4 - y^5$

46. Row 7 of Pascal's triangle is 1, 6, 15, 20, 15, 6, 1.

 $(a-1)^6 = a^6 - 6a^5(1) + 15a^4(1)^2 - 20a^3(1)^3 + 15a^2(1)^4 - 6a(1)^5 + (1)^6 \Rightarrow$

$$(a-1)^6 = a^6 - 6a^5 + 15a^4 - 20a^3 + 15a^2 - 6a + 1$$

47. $3! = 1 \cdot 2 \cdot 3 = 6$

48. $\dfrac{5!}{3!2!} = \dfrac{1 \cdot 2 \cdot 3 \cdot 4 \cdot 5}{(1 \cdot 2 \cdot 3)(1 \cdot 2)} = 2 \cdot 5 = 10$

49. $_6C_3 = \dfrac{6!}{3!3!} = \dfrac{1 \cdot 2 \cdot 3 \cdot 4 \cdot 5 \cdot 6}{(1 \cdot 2 \cdot 3)(1 \cdot 2 \cdot 3)} = 2 \cdot 5 \cdot 2 = 20$

50. $_4C_3 = \dfrac{4!}{3!1!} = \dfrac{1 \cdot 2 \cdot 3 \cdot 4}{(1 \cdot 2 \cdot 3)(1)} = 4$

51. $(m+2)^4 = \left(_4C_0\right)m^4 + \left(_4C_1\right)m^3(2) + \left(_4C_2\right)m^2(2)^2 + \left(_4C_3\right)m(2)^3 + \left(_4C_4\right)(2)^4 \Rightarrow$

$(m+2)^4 = m^4 + 8m^3 + 24m^2 + 32m + 16$

52. $(a+b)^5 = \left(_5C_0\right)a^5 + \left(_5C_1\right)a^4b + \left(_5C_2\right)a^3b^2 + \left(_5C_3\right)a^2b^3 + \left(_5C_4\right)ab^4 + \left(_5C_5\right)b^5 \Rightarrow$

$(a+b)^5 = a^5 + 5a^4b + 10a^3b^2 + 10a^2b^3 + 5ab^4 + b^5$

53. $(x-3y)^4 = \left(_4C_0\right)x^4 - \left(_4C_1\right)x^3(3y) + \left(_4C_2\right)x^2(3y)^2 - \left(_4C_3\right)x(3y)^3 + \left(_4C_4\right)(3y)^4 \Rightarrow$

$(x-3y)^4 = x^4 - 12x^3y + 54x^2y^2 - 108xy^3 + 81y^4$

54 $(3x-2)^3 = \left(_3C_0\right)(3x)^3 - \left(_3C_1\right)(3x)^2(2) + \left(_3C_2\right)(3x)(2)^2 - \left(_3C_3\right)(2)^3 \Rightarrow$

$(3x-2)^3 = 27x^3 - 54x^2 + 36x - 8$

55. Symbolic: $a_n = 45,000(1.10)^{\wedge}(n-1)$ for $n = 1, 2, 3, \ldots, 7$. This is a geometric sequence.

Numerical: See Figure 55a. Graphical: See Figure 55b.

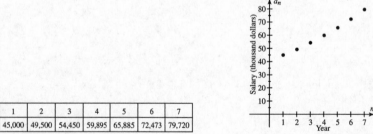

n	1	2	3	4	5	6	7
a_n	45,000	49,500	54,450	59,895	65,885	72,473	79,720

Figure 55a   Figure 55b

56. Symbolic: $a_n = 45,000 + 5000(n-1)$ for $n = 1, 2, 3, \ldots, 7$. This is an arithmetic sequence.

Numerical: See Figure 56a. Graphical: See Figure 56b.

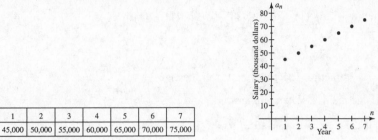

n	1	2	3	4	5	6	7
a_n	45,000	50,000	55,000	60,000	65,000	70,000	75,000

Figure 56a   Figure 56b

57. Symbolic: $a_n = 49n$ for $n = 1, 2, 3, \ldots, 7$

Graphical: See Figure 57a. Numerical: See Figure 57b.

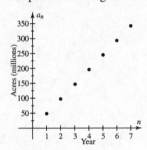

Figure 57a

n	1	2	3	4	5	6	7
a_n	49	98	147	196	245	294	343

Figure 57b

58. (a) $a_n = 1087(1.025)^{n-1}$

(b) The sequence is geometric. The common ratio is 1.025.

(c) $a_5 = 1087(1.025)^{5-1} \approx 1200$; The average mortgage payment in 2000 was about $1200.

(d) See Figure 58.

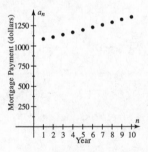

Figure 58

n	1	2	3	4	5	6	7
a_n	0	2	6	12	20	30	42

Figure 3

Chapter 11 Test

1. $f(1) = \dfrac{1^2}{1+1} = \dfrac{1}{2}, f(2) = \dfrac{2^2}{2+1} = \dfrac{4}{3}, f(3) = \dfrac{3^2}{3+1} = \dfrac{9}{4}, f(4) = \dfrac{4^2}{4+1} = \dfrac{16}{5} \Rightarrow \dfrac{1}{2}, \dfrac{4}{3}, \dfrac{9}{4}, \dfrac{16}{5}$

2. The points shown are (1, –3), (2, 2), (3, 1), (4, –2) and (5, 3). The sequence is –3, 2, 1, –2, 3.

3. See Figure 3.

4. Row 5 of Pascal's triangle is 1, 4, 6, 4,1.

$$(2x-1)^4 = (2x)^4 - 4(2x)^3(1) + 6(2x)^2(1)^2 - 4(2x)(1)^3 + (1)^4 \Rightarrow$$

$$(2x-1)^4 = 16x^4 - 32x^3 + 24x^2 - 8x + 1$$

5. The sequence is arithmetic. The common difference is -3.

6. The sequence is geometric. The common ratio is -2.

7. $a_n = 2 + (n-1)(-3) \Rightarrow a_n = 2 - 3n + 3 \Rightarrow a_n = -3n + 5$

8. Note: $2 \cdot r^2 = 4.5 \Rightarrow r^2 = \dfrac{4.5}{2} \Rightarrow r = 1.5$, thus $a_n = 2(1.5)^{n-1}$

9. Yes, the common ratio is 2.5.

10. No, there is no common ratio.

11. $9\left(\dfrac{-1+23}{2}\right) = 9(11) = 99$

12. Here $r = -\dfrac{2}{3}$ and $n = 7$ thus $S_7 = 1\left(\dfrac{1-\left(-\frac{2}{3}\right)^7}{1-\left(-\frac{2}{3}\right)}\right) = \left(\dfrac{\frac{2315}{2187}}{\frac{5}{3}}\right) = \dfrac{463}{729}$

13. $3(2) + 3(3) + 3(4) + 3(5) + 3(6) + 3(7) \Rightarrow 6 + 9 + 12 + 15 + 18 + 21$

14. $\displaystyle\sum_{k=1}^{60} k^3$

15. $\dfrac{7!}{4!3!} = \dfrac{1 \cdot 2 \cdot 3 \cdot 4 \cdot 5 \cdot 6 \cdot 7}{(1 \cdot 2 \cdot 3 \cdot 4)(1 \cdot 2 \cdot 3)} = 5 \cdot 7 = 35$

16. ${}_5C_3 = \dfrac{5!}{3!2!} = \dfrac{1 \cdot 2 \cdot 3 \cdot 4 \cdot 5}{(1 \cdot 2 \cdot 3)(1 \cdot 2)} = 2 \cdot 5 = 10$

17. This is an arithmetic series with $a_1 = 50$ and $d = 7$. The total number of seats is

$$S_{45} = \dfrac{45}{2}\left(2(50) + (45-1)(7)\right) = 22.5(408) = 9180 \text{ seats.}$$

18. Symbolic: $a_n = 180,000(0.96)\wedge(n-1)$ for $n = 1, 2, 3, 4, 5$. This is a geometric sequence.

Numerical: See Figure 18a. Graphical: See Figure 18b.

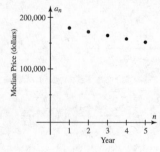

n	1	2	3	4	5
a_n	180,000	172,800	165,888	159,252	152,882

Figure 18a   Figure 18b

19. (a) $a_n = 2000(2)^{n-1}$

(b)   The sequence is geometric. The common ratio is 2.

(c) $a_6 = 2000(2)^{6-1} \approx 64{,}000$; After 30 days, there are 64,000 caterpillars.

(d)   See Figure 19.

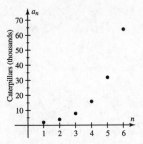

Figure 19

Chapter 11 Extended and Discovery Exercises

1. Each term of the sequence is obtained by adding the two previous terms. 1, 1, 2, 3, 5, 8, 13, 21, 34, 55, 89, 144

2. (a)   Table $u(n) = 2.85u(n-1) - 0.19(u(n-1))^2$ with TblStart = 1 and ΔTbl = 1. See Figure 2a.

   Note: be sure to set nMin = 1 and u(nMin) = {1}.

   (b)   Graph $u(n) = 2.85u(n-1) - 0.19(u(n-1))^2$ in [0, 22, 2] by [0, 11, 1]. See Figure 2b.

   Note: be sure to set nMin = 1 and nMax = 20 in the WINDOW settings.

   The moth population increases and then oscillates until it settles to a constant number (about 9.737 thousand).

[0, 22, 2] by [0, 11, 1]

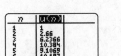   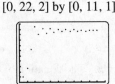

Figure 2a   Figure 2b

3. (a) $\pi \approx \left[90 \left(\dfrac{1}{1^4} + \dfrac{1}{2^4} + \dfrac{1}{3^4} + \dfrac{1}{4^4} \right) \right]^{1/4} \approx 3.138997889$

   (b) $\pi \approx 3.141590776$; This is correct to 5 decimal places.

4. (a) $S = \dfrac{2}{1-\left(-\frac{1}{2}\right)} = \dfrac{2}{\frac{3}{2}} = \dfrac{4}{3}$

   (b) $S = \dfrac{1}{1-\left(\frac{1}{3}\right)} = \dfrac{1}{\frac{2}{3}} = \dfrac{3}{2}$

(c) $S = \dfrac{0.1}{1-(0.1)} = \dfrac{0.1}{0.9} = \dfrac{1}{9} = 0.\overline{1}$

(d) $S = \dfrac{0.12}{1-(0.01)} = \dfrac{0.12}{0.99} = \dfrac{4}{33} = 0.\overline{12}$

Chapters 1-11 Cumulative Review Exercises

1. This equation illustrates a distributive property.

2. The domain corresponds to the x-coordinates and the range corresponds to the y-coordinates.

   $D = \{-6, -2, 0, 2\};\ R = \{0, 1, 3, 5\}$

3. $\dfrac{x^{-2}y^3}{\left(3xy^{-2}\right)^3} = \dfrac{x^{-2}y^3}{3^3 x^3 \left(y^{-2}\right)^3} = \dfrac{x^{-2}y^3}{27x^3 y^{-6}} = \dfrac{1}{27}x^{-2-3}y^{3-(-6)} = \dfrac{1}{27}x^{-5}y^9 = \dfrac{y^9}{27x^5}$

4. $\left(\dfrac{3b}{6a^2}\right)^{-4} = \left(\dfrac{b}{2a^2}\right)^{-4} = \left(\dfrac{2a^2}{b}\right)^4 = \dfrac{\left(2a^2\right)^4}{b^4} = \dfrac{2^4 \left(a^2\right)^4}{b^4} = \dfrac{16a^8}{b^4}$

5. $\left(\dfrac{1}{z^2}\right)^{-5} = \left(z^2\right)^5 = z^{10}$

6. $\dfrac{8x^{-3}y^2}{4x^3 y^{-1}} = 2x^{-3-3}y^{2-(-1)} = 2x^{-6}y^3 = \dfrac{2y^3}{x^6}$

7. The function is defined for all values of the variable except 8. The domain is $\{x \mid x \neq 8\}$.

8. The slope is $m = \dfrac{1-5}{0-(-2)} = \dfrac{-4}{2} = -2$. Since $f(x) = 1$ when $x = 0$ the y-intercept is 1.

   Here $f(x) = -2x + 1$.

9. Horizontal lines have equations of the form $y = k$. The equation of the horizontal line passing through $(2, 3)$ is $y = 3$.

10. The equation is in the form $f(x) = mx + b$. The slope is -3 and the y-intercept is 5.

11. Since the line is perpendicular to $y = -\dfrac{2}{3}x - 4$, the slope is $m = \dfrac{3}{2}$. Using the point-slope form

    gives $y = \dfrac{3}{2}(x-1) + 4 \Rightarrow y = \dfrac{3}{2}x - \dfrac{3}{2} + 4 \Rightarrow y = \dfrac{3}{2}x + \dfrac{5}{2}$

12. Since the line is parallel to $y = 2x - 7$, the slope is $m = 2$. Using the point-slope form gives

    $y = 2(x-5) + 2 \Rightarrow y = 2x - 10 + 2 \Rightarrow y = 2x - 8$

13. $\dfrac{2}{5}(x-4) = -12 \Rightarrow x - 4 = -30 \Rightarrow x = -26$

14. $\dfrac{2}{5}z + \dfrac{1}{4}z > 2 - (z-1) \Rightarrow 8z + 5z > 40 - 20(z-1) \Rightarrow 13z > 40 - 20z + 20 \Rightarrow$

$13z > 60 - 20z \Rightarrow 33z > 60 \Rightarrow z > \dfrac{60}{33} \Rightarrow z > \dfrac{20}{11}.$ The interval is $\left(\dfrac{20}{11}, \infty\right).$

15. First divide each side of $-3|t-5| \leq -18$ by -3 to obtain $|t-5| \geq 6$.

The solutions to $|t-5| \geq 6$ satisfy $t \leq c$ or $t \geq d$ where c and d are the solutions to $|t-5| = 6$.

$|t-5| = 6$ is equivalent to $t - 5 = -6 \Rightarrow t = -1$ and $t - 5 = 6 \Rightarrow t = 11$.

The interval is $(-\infty, -1] \cup [11, \infty)$.

16. $\left|4 + \dfrac{2}{3}x\right| = 6 \Rightarrow 4 + \dfrac{2}{3}x = -6 \Rightarrow \dfrac{2}{3}x = -10 \Rightarrow x = -15$ or $4 + \dfrac{2}{3}x = 6 \Rightarrow \dfrac{2}{3}x = 2 \Rightarrow x = 3$

17. $\dfrac{1}{4}t - (2t+5) + 6 = \dfrac{t+3}{4} \Rightarrow \dfrac{1}{4}t - 2t + 1 = \dfrac{t+3}{4} \Rightarrow t - 8t + 4 = t + 3 \Rightarrow -8t = -1 \Rightarrow t = \dfrac{1}{8}$

18. $-3 \leq \dfrac{2}{3}x + 5 < 11 \Rightarrow -8 \leq \dfrac{2}{3}x < 6 \Rightarrow -12 \leq x < 9.$ The interval is $[-12, 9)$.

19. By substitution, $(3, -2)$ is a solution to the given system of equations.

20. See Figure 20.

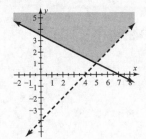

Figure 20

21. Multiply the first equation by 2 and add the equations to eliminate the variable x.

$2x - 4y = 2$
$\underline{-2x + 7y = 4}$ Thus, $y = 2$. And so $x - 2(2) = 1 \Rightarrow x = 5$. The solution is $(5, 2)$.
$3y = 6$

22. $\begin{bmatrix} 1 & 1 & 1 & | & 5 \\ -2 & -1 & 1 & | & -10 \\ 1 & 2 & 8 & | & 1 \end{bmatrix} \begin{matrix} \\ R_2 + 2R_1 \rightarrow \\ R_3 - R_1 \rightarrow \end{matrix} \begin{bmatrix} 1 & 1 & 1 & | & 5 \\ 0 & 1 & 3 & | & 0 \\ 0 & 1 & 7 & | & -4 \end{bmatrix} \begin{matrix} R_1 - R_2 \rightarrow \\ \\ R_3 - R_2 \rightarrow \end{matrix} \begin{bmatrix} 1 & 0 & -2 & | & 5 \\ 0 & 1 & 3 & | & 0 \\ 0 & 0 & 4 & | & -4 \end{bmatrix}$

$\begin{matrix} \\ \\ (1/4)R_3 \rightarrow \end{matrix} \begin{bmatrix} 1 & 0 & -2 & | & 5 \\ 0 & 1 & 3 & | & 0 \\ 0 & 0 & 1 & | & -1 \end{bmatrix} \begin{matrix} R_1 + 2R_3 \rightarrow \\ R_2 - 3R_3 \rightarrow \\ \end{matrix} \begin{bmatrix} 1 & 0 & 0 & | & 3 \\ 0 & 1 & 0 & | & 3 \\ 0 & 0 & 1 & | & -1 \end{bmatrix}$ The solution is $(3, 3, -1)$.

23. From the graph of the region of feasible solutions (not shown), the vertices are $(0, 0)$, $(0, 2.5)$, $(2, 2)$, and $(2.5, 0)$. The maximum value of R occurs at one of the vertices. For $(0, 0)$, $R = 3(0) + 8(0) = 0$.

For $(0, 2.5)$, $R = 3(0) + 8(2.5) = 20$. For $(2, 2)$, $R = 3(2) + 8(2) = 22$.

For $(2.5, 0)$, $R = 3(2.5) + 8(0) = 7.5$. The maximum value is $R = 22$.

24. $\det A = 4(2) - 3(-3) = 8 + 9 = 17$

25. $2x^3 \left(4x^4 - 3x^3 + 5\right) = 8x^7 - 6x^6 + 10x^3$

26. $(2z - 7)(3z + 4) = 6z^2 + 8z - 21z - 28 = 6z^2 - 13z - 28$

27. $4x^2 - 9y^2 = (2x)^2 - (3y)^2 = (2x - 3y)(2x + 3y)$

28. $2a^3 - a^2 + 8a - 4 = a^2(2a - 1) + 4(2a - 1) = \left(a^2 + 4\right)(2a - 1)$

29. $4x^2 - x - 3 = 0 \Rightarrow (4x + 3)(x - 1) = 0 \Rightarrow x = -\dfrac{3}{4}$ or $x = 1$

30. $x^4 - 10x^3 = -24x^2 \Rightarrow x^4 - 10x^3 + 24x^2 = 0 \Rightarrow x^2(x - 4)(x - 6) = 0 \Rightarrow x = 0, 4,$ or 6

31. $\dfrac{x^2 - 7x + 10}{x^2 - 25} \cdot \dfrac{x + 5}{x + 1} = \dfrac{(x - 2)(x - 5)(x + 5)}{(x - 5)(x + 5)(x + 1)} = \dfrac{x - 2}{x + 1}$

32. $\dfrac{x^2 + 7x + 12}{x^2 - 9} \div \dfrac{x^2 - 5x + 6}{(x - 3)^2} = \dfrac{x^2 + 7x + 12}{x^2 - 9} \cdot \dfrac{(x - 3)^2}{x^2 - 5x + 6} = \dfrac{(x + 4)(x + 3)(x - 3)(x - 3)}{(x - 3)(x + 3)(x - 3)(x - 2)} = \dfrac{x + 4}{x - 2}$

33. $\dfrac{2}{x + 5} = \dfrac{-3}{x^2 - 25} + \dfrac{1}{x - 5} \Rightarrow 2(x - 5) = -3 + 1(x + 5) \Rightarrow 2x - 10 = x + 2 \Rightarrow x = 12$

34. $\dfrac{2y}{y^2 - 3y + 2} = \dfrac{1}{y - 2} + 2 \Rightarrow 2y = 1(y - 1) + 2(y - 2)(y - 1) \Rightarrow 2y = y - 1 + 2y^2 - 6y + 4 \Rightarrow$

$2y^2 - 7y + 3 = 0 \Rightarrow (2y - 1)(y - 3) = 0 \Rightarrow y = \dfrac{1}{2}$ or $y = 3$.

35. $R = \dfrac{3C - 2W}{5} \Rightarrow 5R = 3C - 2W \Rightarrow 5R - 3C = -2W \Rightarrow \dfrac{5R - 3C}{-2} = W \Rightarrow W = \dfrac{3C - 5R}{2}$

36. $\dfrac{\dfrac{1}{x^2} + \dfrac{2}{x}}{\dfrac{1}{x^2} - \dfrac{4}{x}} = \dfrac{\dfrac{1}{x^2} + \dfrac{2}{x}}{\dfrac{1}{x^2} - \dfrac{4}{x}} \cdot \dfrac{x^2}{x^2} = \dfrac{1 + 2x}{1 - 4x}$

37. $\sqrt[3]{x^4 y^4} - 2\sqrt[3]{xy} = \sqrt[3]{(xy)^3 \cdot xy} - 2\sqrt[3]{xy} = xy\sqrt[3]{xy} - 2\sqrt[3]{xy} = (xy - 2)\sqrt[3]{xy}$

38. $\left(4 + \sqrt{2}\right)\left(4 - \sqrt{2}\right) = 4^2 - \left(\sqrt{2}\right)^2 = 16 - 2 = 14$

39. $8(x - 3)^2 = 200 \Rightarrow (x - 3)^2 = 25 \Rightarrow x - 3 = \pm\sqrt{25} \Rightarrow x - 3 = \pm 5 \Rightarrow x = -2$ or 8

40. $3\sqrt{2x + 6} = 6x \Rightarrow \sqrt{2x + 6} = 2x \Rightarrow 2x + 6 = 4x^2 \Rightarrow 4x^2 - 2x - 6 = 0 \Rightarrow 2(2x - 3)(x + 1) \Rightarrow$

$x = \dfrac{3}{2}$ or $x = -1$. The value $x = -1$ does not check. The only solution is $\dfrac{3}{2}$.

41. $(-3+i)(-4-2i)=12+6i-4i-2i^2=12+2i+2=14+2i$

42. $\dfrac{2-6i}{1+2i}=\dfrac{2-6i}{1+2i}\cdot\dfrac{1-2i}{1-2i}=\dfrac{2-4i-6i+12i^2}{1-4i^2}=\dfrac{2-10i-12}{1+4}=\dfrac{-10-10i}{5}=-2-2i$

43. $-\dfrac{b}{2a}=-\dfrac{8}{2(3)}=-\dfrac{8}{6}=-\dfrac{4}{3}$; $f\left(-\dfrac{4}{3}\right)=3\left(-\dfrac{4}{3}\right)^2+8\left(-\dfrac{4}{3}\right)+5=-\dfrac{1}{3}$. The vertex is $\left(-\dfrac{4}{3},-\dfrac{1}{3}\right)$.

The minimum value is $-\dfrac{1}{3}$.

44. $y=2x^2+8x+17\Rightarrow y=2\left(x^2+4x+4\right)+17-8\Rightarrow y=2(x+2)^2+9$. The vertex is $(-2,9)$.

45. $x^2-4x+13=0\Rightarrow x^2-4x+4=-13+4\Rightarrow (x-2)^2=-9\Rightarrow x-2=\pm\sqrt{-9}\Rightarrow x=2\pm3i$

46. $z^2-4z=32\Rightarrow z^2-4z-32=0\Rightarrow (z+4)(z-8)=0\Rightarrow z=-4$ or $z=8$

47. (a) The graph intersects the x-axis at -3 and 1.

　　(b) Because the parabola opens upward, $a>0$.

　　(c) Because there are two real solutions, the discriminant is positive.

48. $x^2+2x-3=0\Rightarrow (x+3)(x-1)=0\Rightarrow x=-3,1$. Because the parabola opens up the

　　interval is $(-3,1)$.

49. (a) $g(-2)=3(-2)-2=-8$, then $(f\circ g)(-2)=f\left(g(-2)\right)=f(-8)=(-8)^2+1=65$

　　(b) $(g\circ f)(x)=g\left(f(x)\right)=g\left(x^2+1\right)=3\left(x^2+1\right)-2=3x^2+3-2=3x^2+1$

50. $f(x)=\dfrac{3x+1}{2}\Rightarrow y=\dfrac{3x+1}{2}$, interchange x and y and solve for y.

　　$x=\dfrac{3y+1}{2}\Rightarrow 2x=3y+1\Rightarrow 3y=2x-1\Rightarrow y=\dfrac{2x-1}{3}\Rightarrow f^{-1}(x)=\dfrac{2x-1}{3}$

51. $\ln\left(x^3\sqrt{y}\right)=\ln\left(x^3y^{1/2}\right)=\ln x^3+\ln y^{1/2}=3\ln x+\dfrac{1}{2}\ln y$

52. $2\log x-\log 4xy=\log x^2-\log 4xy=\log\dfrac{x^2}{4xy}=\log\dfrac{x}{4y}$

53. $8\log x+3=17\Rightarrow 8\log x=14\Rightarrow \log x=\dfrac{7}{4}\Rightarrow 10^{\log x}=10^{7/4}\Rightarrow x\approx56.23$

54. $4^{2x}=5\Rightarrow \log_4 4^{2x}=\log_4 5\Rightarrow 2x=\log_4 5\Rightarrow 2x=\dfrac{\log 5}{\log 4}\Rightarrow x=\dfrac{\log 5}{2\log 4}\approx0.58$

55. See Figure 55. The vertex is $(1,3)$, and the axis of symmetry is $y=3$.

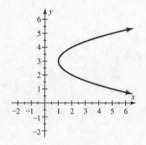

Figure 55

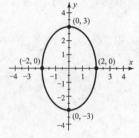

Figure 57

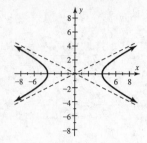

Figure 58

56. $x^2 - 6x + y^2 + 2y = -6 \Rightarrow \left(x^2 - 6x + 9\right) + \left(y^2 + 2y + 1\right) = -6 + 9 + 1 \Rightarrow \left(x - 3\right)^2 + \left(y + 1\right)^2 = 4$

The center is $(3, -1)$, and the radius is 2.

57. See Figure 57.

58. See Figure 58.

59. Vertical transverse axis with vertices $(0, \pm 2)$ and asymptotes $y = \pm \dfrac{2}{4} x \Rightarrow \dfrac{y^2}{4} - \dfrac{x^2}{16} = 1$

60. Horizontal major axis with vertices $(\pm 4, 0)$ and minor axis endpoints $(0, \pm 2) \Rightarrow \dfrac{x^2}{16} + \dfrac{y^2}{4} = 1$

61. Substitute $y = x^2 + 1$ in the second equation and solve for x.

$x^2 + 2\left(x^2 + 1\right) = 5 \Rightarrow 3x^2 - 3 = 0 \Rightarrow 3\left(x^2 - 1\right) = 0 \Rightarrow 3(x + 1)(x - 1) = 0 \Rightarrow x = -1 \text{ or } x = 1$

When $x = -1$, $y = (-1)^2 + 1 = 2$. When $x = 1$, $y = (1)^2 + 1 = 2$. The solutions are $(-1, 2)$ and $(1, 2)$.

62. See Figure 62.

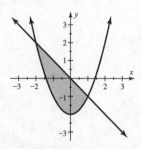

Figure 62

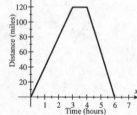

Figure 74

63. This sequence is arithmetic, the common difference is -2.

64. This sequence is geometric, the common ratio is 0.2.

65. This sequence is geometric, the common ratio is 4.

66. This sequence is arithmetic, the common difference is 6.

67. Note: $d = 5 - 2 = 3$, thus $a_n = 2 + (n - 1)(3) \Rightarrow a_n = 2 + 3n - 3 \Rightarrow a_n = 3n - 1$

68. Note: $r = \dfrac{12}{4} = 3$, thus $a_n = 4(3)^{n-1}$

69. $9\left(\dfrac{3+35}{2}\right)=9(19)=171$

70. Here $r=-2$ and $n=11$ thus $S_{11}=1\left(\dfrac{1-(-2)^{11}}{1-(-2)}\right)=1\left(\dfrac{2049}{3}\right)=683$

71. $(2x+3)^4=(_4C_0)(2x)^4+(_4C_1)(2x)^3(3)+(_4C_2)(2x)^2(3^2)+(_4C_3)(2x)(3^3)+(_4C_4)(3^4)\Rightarrow$

$(2x+3)^4=16x^4+96x^3+216x^2+216x+81$

72. $(2a-5b)^3=(_3C_0)(2a)^3-(_3C_1)(2a)^2(5b)+(_3C_2)(2a)(5b)^2-(_3C_3)(5b)^3\Rightarrow$

$(2a-5b)^3=8a^3-60a^2b+150ab^2-125b^3$

73. $r=\sqrt{\dfrac{14}{\pi}}\approx2.11$ inches

74. See Figure 74.

75. (a) $f(x)=\dfrac{170}{2}+0.4x\Rightarrow f(x)=0.4x+85$

(b) $0.4(90)+\dfrac{W}{2}=130\Rightarrow36+\dfrac{W}{2}=130\Rightarrow\dfrac{W}{2}=94\Rightarrow W=188$ pounds

76. Let x and y represent the speed of the airplane and the speed of the wind respectively. Then the system needed is $x-y=360$ and $x+y=400$. Adding the two equations will eliminate y.

$x-y=360$
$\underline{x+y=400}$ Thus, $x=380$. And so $(380)+y=400\Rightarrow y=20$.
$2x=760$

The speed of the airplane is 380 mph and the speed of the wind is 20 mph.

77. Let x represent the width of the tent floor. Then $2x-6$ represents the length.

$x(2x-6)=108\Rightarrow2x^2-6x-108=0\Rightarrow x^2-3x-54=0\Rightarrow(x+6)(x-9)=0\Rightarrow x=-6$ or 9

Since the value $x=-6$ has no physical meaning, the solution is 9. The dimensions are 9 feet by 12 feet.

78. Let x represent time required to weed the garden if they worked together. Then $\dfrac{x}{60}+\dfrac{x}{90}=1$.

$180\cdot\left(\dfrac{x}{60}+\dfrac{x}{90}\right)=1\cdot180\Rightarrow3x+2x=180\Rightarrow5x=180\Rightarrow x=\dfrac{180}{5}=36$ minutes

79. (a) $xy = 96$ and $3x - y = 12$

(b)   Solve the second equation for y and substitute the result in the first

equation. $3x - y = 12 \Rightarrow y = 3x - 12$

$x(3x - 12) = 96 \Rightarrow 3x^2 - 12x - 96 = 0 \Rightarrow x^2 - 4x - 32 = 0 \Rightarrow (x + 4)(x - 8) = 0 \Rightarrow$

$x = -4$ or $x = 8$. Since the numbers must be positive, the solution is $x = 8$. The numbers are 8

and 12.

80. Find the sum of the series $1 + 3 + 5 + \cdots + 23$. The sum is $12\left(\dfrac{1 + 23}{2}\right) = 12(12) = 144$ musicians.